Modelling and Optimization of Distributed Parameter Systems

IFIP – The International Federation for Information Processing

IFIP was founded in 1960 under the auspices of UNESCO, following the First World Computer Congress held in Paris the previous year. An umbrella organization for societies working in information processing, IFIP's aim is two-fold: to support information processing within its member countries and to encourage technology transfer to developing nations. As its mission statement clearly states,

> IFIP's mission is to be the leading, truly international, apolitical organization which encourages and assists in the development, exploitation and application of information technology for the benefit of all people.

IFIP is a non-profitmaking organization, run almost solely by 2500 volunteers. It operates through a number of technical committees, which organize events and publications. IFIP's events range from an international congress to local seminars, but the most important are:

- the IFIP World Computer Congress, held every second year;
- open conferences;
- working conferences.

The flagship event is the IFIP World Computer Congress, at which both invited and contributed papers are presented. Contributed papers are rigorously refereed and the rejection rate is high.

As with the Congress, participation in the open conferences is open to all and papers may be invited or submitted. Again, submitted papers are stringently refereed.

The working conferences are structured differently. They are usually run by a working group and attendance is small and by invitation only. Their purpose is to create an atmosphere conducive to innovation and development. Refereeing is less rigorous and papers are subjected to extensive group discussion.

Publications arising from IFIP events vary. The papers presented at the IFIP World Computer Congress and at open conferences are published as conference proceedings, while the results of the working conferences are often published as collections of selected and edited papers.

Any national society whose primary activity is in information may apply to become a full member of IFIP, although full membership is restricted to one society per country. Full members are entitled to vote at the annual General Assembly, National societies preferring a less committed involvement may apply for associate or corresponding membership. Associate members enjoy the same benefits as full members, but without voting rights. Corresponding members are not represented in IFIP bodies. Affiliated membership is open to non-national societies, and individual and honorary membership schemes are also offered.

Modelling and Optimization of Distributed Parameter Systems Applications to engineering

Selected Proceedings of the IFIP WG7.2 on Modelling and Optimization of Distributed Parameter Systems with Applications to Engineering, June 1995

Edited by
Kazimierz Malanowski, Zbigniew Nahorski and Małgorzata Peszyńska
Systems Research Institute
Polish Academy of Sciences
Warsaw
Poland

Published by Chapman and Hall on behalf of the
International Federation for Information Processing (IFIP)

CHAPMAN & HALL
London · Glasgow · Weinheim · New York · Tokyo · Melbourne · Madras

Published by Chapman & Hall, 2–6 Boundary Row, London SE1 8HN, UK

Chapman & Hall, 2–6 Boundary Row, London SE1 8HN, UK

Blackie Academic & Professional, Wester Cleddens Road, Bishopbriggs, Glasgow G64 2NZ, UK

Chapman & Hall GmbH, Pappelallee 3, 69469 Weinheim, Germany

Chapman & Hall USA, 115 Fifth Avenue, New York, NY 10003, USA

Chapman & Hall Japan, ITP-Japan, Kyowa Building, 3F, 2-2-1 Hirakawacho, Chiyoda-ku, Tokyo 102, Japan

Chapman & Hall Australia, 102 Dodds Street, South Melbourne, Victoria 3205, Australia

Chapman & Hall India, R. Seshadri, 32 Second Main Road, CIT East, Madras 600 035, India

First edition 1996

Printed in Great Britain by TJ Press Ltd, Padstow, Cornwall

ISBN 0 412 72700 5

A catalogue record for this book is available from the British Library

Printed on permanent acid-free text paper, manufactured in accordance with ANSI/NISO Z39.48-1992 and ANSI/NISO Z39.48-1984 (Permanence of Paper).

CONTENTS

Preface

This volume contains a selection of papers presented at the conference on

Modelling and Optimization of Distributed Parameter Systems with Applications to Engineering,

held in Warsaw on July 17-21, 1995.

This conference was a consecutive one in the series of conferences sponsored by the IFIP Working Group WG 7.2 *"Computational Techniques in Distributed Systems"*, chaired by Irena Lasiecka. It was organized by the *Systems Research Institute of the Polish Academy of Sciences* and supported financially by the following institutions:

- *European Community on Computational Methods in Applied Sciences,*
- *Fundacja Stefana Batorego,*
- *International Mathematical Union,*
- *Telekomunikacja Polska S.A.*

The following scientists took an active part in preparation of the scientific program of the conference, organizing or helping to organize special sessions:

- E.Casas and I.Lasiecka (*Optimization and Optimal Control*),
- Z.Mróz (*Mechanical Applications*),
- M.Niezgódka (*Properties of Solutions to P.D.E.s*),
- L.Pandolfi (*Hamilton and Riccati Equation Approaches to Optimization*),
- K.Sobczyk and J.Zabczyk (*Stochastic Systems*),
- J.Sokołowski and J.-P.Zolésio (*Shape Optimization*),
- J.Waśniewski (*Scientific Computation*).

In the conference participated 133 scientists from 22 countries. Ten invited plenary lectures and 103 contributed papers have been presented.

This volume contains a part of the presented material. The core of it is constituted by papers devoted to control and optimization of distributed parameter systems. Other selection will be included in a special issue of the quarterly *Control & Cybernetics* to be published in 1996.

We would like to express our gratitude to all referees of the papers. Without their kind help the selection of the material and the preparation of the volume would not be possible in such a short time.

Special thanks are to Ms.Krystyna Warzywoda and Ms.Agnieszka Jóźwiak for their excelent secretarial work as well as to Ms.Bożena Łopuch for her very efficient and skilful technical assistance in preparation of the camera-ready manuscript.

K.Malanowski
Z.Nahorski
M.Peszyńska

International Program Committee

I.Lasiecka USA *chairperson*

A.Bermudez	Spain	**K.Malanowski**	Poland
A.Bogobowicz	Canada	**Z.Mróz**	Poland
G.Da Prato	Italy	**M.Niezgódka**	Poland
M.Delfour	Canada	**J.Puel**	France
L.Demkowicz	USA	**J.Simon**	France
M.Dryja	Poland	**J.Sokołowski**	France/Poland
R.Glowinski	USA	**F.Tröltzsch**	Germany
K.-H.Hoffmann	Germany	**J.Waśniewski**	Denmark
W.Krabs	Germany	**A.Wierzbicki**	Poland
A.Kurzhanskii	Russia	**J.-P.Yvon**	France
G.Leugering	Germany	**J.P.Zolésio**	France

Local Organizing Committee

K.Malanowski *chairman*

P.Holnicki		**R.Ostrowski**	
A.Myśliński		**M.Peszyńska**	*secretary*
Z.Nahorski	*vice-chairman*	**A.Żochowski**	

PART ONE

Invited Lectures

1

Optimal control of periodic systems

V. Barbu
Department of Mathematics, University of Iaşi
6600 Iaşi, Romania.
Phone: (40) 32-213093. Fax: 32-146330. E–mail: barbu@uaic.ro

Abstract
This work is concerned with existence and maximum principle for optimal control problems with linear periodic state systems in Hilbert spaces.

Keywords
Linear periodic operators, p–stabilizable, p–detectable, parabolic control systems, periodic wave equation

1 INTRODUCTION

We shall study here infinite dimensional optimal control problems of the form

Minimize

$$\int_0^T (g(Cy(t)) + h(u(t)))dt \tag{1}$$

subject to $u \in L^2(0,T;U)$ and $y \in C([0,T];H)$ *solution to state system*

$$\frac{dy}{dt} + Ay = Bu + f, \; t \in (0,T). \tag{2}$$

Here $-A$ is the generator of a C_0–semigroup e^{-At} on a real Hilbert space H, $B \in L(U,H)$, $C \in L(H,Z)$ and $g : Z \to \bar{R} = (-\infty, +\infty]$, $h : U \to \bar{R}$ are lower semicontinous convex functions; U and Z are real Hilbert spaces with the norms $|\cdot|_U$, $|\cdot|_Z$ and scalar products denoted $(\cdot,\cdot)_U$ and $(\cdot,\cdot)_Z$, respectively. The norm of H is denoted by $|\cdot|$ and the scalar product by $(\cdot,\cdot)$.

This is the general form of periodic optimal control problems governed by parabolic and hyperbolic linear equations with distributed controllers. Problems of this type arise quite often in applications since the solutions of most dissipative systems with periodic forcing asymptotically converge to a periodic orbit. On the other hand, the performance of a steady state control system can be improved by forcing the control input to vary periodically. As a matter of fact (1) is a singular control system and the most common way to avoid some delicate problems due to resonance phenomenon is to assume that the pairs (A, B) and (A, C) are stabilizable and respectively, detectable (see e.g. Da Prato (1987), Bensoussan et al. (1993), Barbu and Pavel (1995)). Though these assumptions are quite natural for infinite horizon control problems they are too restrictive for linear periodic control problems of the form (1). Here we shall weaken them to a closed range property assumption (Definition 1) which seems to be the most convenient framework to treat existence and maximum principle in problem (1) in resonant case. We shall use standard notation for function spaces and assume familiarity with basic results of convex analysis.

2 LINEAR PERIODIC OPERATORS WITH CLOSED RANGE

Let $\mathcal{A} : D(\mathcal{A}) \subset L^2(0,T;H) \to L^2(0,T;H)$ be the linear operator defined by

$$\begin{array}{l} \mathcal{A}y = f \text{ iff } \displaystyle\int_0^T ((y(t), \varphi'(t) - A^*\varphi(t)) + (f(t), \varphi(t)))dt = 0 \\ \forall \varphi \in X = \{\varphi \in W^{1,2}([0,T]; H);\ A^*\varphi \in L^2(0,T;H), \varphi(0) = \varphi(T)\} \end{array} \tag{3}$$

Here $W^{1,2}([0,T];H)$ is the space of absolutely continous functions $\varphi : [0,T] \to H$ such that $\varphi' = \dfrac{d\varphi}{dt} \in L^2(0,T;H)$. It is readily seen that $\mathcal{A}$ is closed and densely defined. A function $y \in L^2(0,T;H)$ which satisfies (3) is called weak solution to periodic problem

$$\frac{dy}{dt} + Ay = f,\ t \in (0,T);\ y(0) = y(T). \tag{4}$$

We shall call $\mathcal{A}$ the linear periodic operator associated with A. It is readily seen that the adjoint $\mathcal{A}^*$ of $\mathcal{A}$ is given by

$$\mathcal{A}^*z = g \ \ iff \ \int_0^T ((z(t), \varphi'(t) + A\varphi(t)) - (g(t), \varphi(t)))dt = 0 \tag{5}$$

for all $\varphi \in Y = \{\varphi \in W^{1,2}([0,T];H);\ A\varphi \in L^2(0,T;H)\}$. If the range $R(\mathcal{A})$ of $\mathcal{A}$ is closed in $L^2(0,T;H)$ then we have

$$L^2(0,T;H) = R(\mathcal{A}) \oplus N(\mathcal{A}^*),\ \ R(\mathcal{A}) = N(\mathcal{A}^*)^\perp \tag{6}$$

and the operator $\mathcal{A}^{-1}$ is continous from $R(\mathcal{A})$ to $R(\mathcal{A}^*) = N(\mathcal{A})^\perp$. (Here $N(\mathcal{A})$ and $N(\mathcal{A}^*)$ are the null spaces of $\mathcal{A}$ and $\mathcal{A}^*$.) A simple criterion for $R(\mathcal{A})$ to be closed is given in Proposition 1 below which is related to some earlier result of Prüss (1985).

Proposition 1 *Assume that for each $m \in Z$, the range Y_m of $\mu_m iI + A$ is closed in H and*

$$\sup\{\|(\mu_m iI + A)^{-1}\|_{L(Y_m,H)};\ m \in Z\} < \infty \tag{7}$$

where $\mu_m = 2m\pi T^{-1}$. Then $R(\mathcal{A})$ is closed.

Proof. We have denoted again A the realization of A in the complexified space H. If $f \in R(\mathcal{A})$ and $\mathcal{A}y = f$ then we have

$$y(t) = \sum_m y_m \exp(\mu_m it), t \in (0,T) \tag{8}$$

where $y_m = (\mu_m i + A)^{-1} f_m$ and $f_m = T^{-\frac{1}{2}} \int_0^T \exp(-\mu_m it) f(t)dt$. Then by (7) we see that

$$\|y\|_{L^2(0,T;H)} \leq C\|f\|_{L^2(0,T;H)} \tag{9}$$

which implies that $R(\mathcal{A})$ is closed. □

Let $\mathcal{A}_0 : D(\mathcal{A}_0) \subset L^2(0,T;H) \to L^2(0,T;H)$ be the linear operator

$$\mathcal{A}_0 f = f \;\; iff \;\; y(t) = e^{-At} y(T) + \int_0^t e^{-A(t-s)} f(s)ds, \forall t \in [0,T] \tag{10}$$

In other words, $\mathcal{A}_0 y = f$ if y is a periodic mild solution to equation (4). Clearly $\mathcal{A}_0$ is closed and densely defined.

Proposition 2 *We have* $\mathcal{A}_0 = \mathcal{A}$.

Proof. It is readily seen that $\mathcal{A}_0 y = \mathcal{A}y, \forall y \in D(\mathcal{A}_0) \subset D(\mathcal{A})$. We have

$$y(t) = \sum_{m \in Z} y_m \exp(\mu_m it) \;\; in \;\; L^2(0,T;H); \;\; (\mu_m i + A)y_m = f_m.$$

Then the sequence

$$y_n(t) = \sum_{|m| \leq n} y_m \exp(i\mu_m t)$$

is convergent to y in $L^2(0,T;H)$ and for each n, y_n is a mild solution to (4) where $f = \sum_{|m| \leq n} f_m \exp(i\mu_m t)$. Letting $n \to +\infty$ we get that y satisfies (10) as claimed. □

By Proposition 2 and (9) we have

$$R(\mathcal{A}) = \{f \in L^2(0,T;H); \int_0^T e^{-A(T-t)} f(t)dt \in R(I - e^{-AT})\} \tag{11}$$

$$N(\mathcal{A}) = \{y \in L^2(0,T;H); y(t) = e^{-At} y_0, (I - e^{-AT})y_0 = 0\} \tag{12}$$

and so we have

Corollary 1 *If* $R(I - e^{-AT})$ *is closed in* H *then* $R(\mathcal{A})$ *is closed in* $L^2(0,T;H)$.

In particular, by Riesz–Fredholm theorem we have

Corollary 2 *If e^{-AT} is compact then $R(\mathcal{A})$ is closed and $N(\mathcal{A})$ is finite dimensional.*

Note also that by Proposition 2 if $R(\mathcal{A})$ is closed and $y \in L^2(0,T;H)$ is a weak solution to (4) then $y \in C([0,T];H)$ and

$$\|\mathcal{A}^{-1}y\|_{C([0,T];H)} \le C\|f\|_{L^2(0,T;H)}, \forall f \in R(\mathcal{A}). \tag{13}$$

Note also that if $R(I - e^{-AT})$ is closed in H then

$$\|\mathcal{A}^{-1}y\|_{C([0,T];H)} \le C\|f\|_{L^1(0,T;H)}, \forall f \in R(\mathcal{A}) \tag{14}$$

Moreover, the adjoint operator $\mathcal{A}^*$ (see (5),(6)) can be equivalently defined as

$$\mathcal{A}^*z = g \text{ iff } z(t) = e^{-A^*(T-t)}z(0) + \int_t^T e^{-A^*(s-t)}g(s)ds, \forall t \in [0,T]. \tag{15}$$

Throughout in the sequel by solution y to (2) we mean weak solution i.e., a solution to $\mathcal{A}y = f$ (equivalently $\mathcal{A}_0 y = f$).

Definition 1 *The pair (A,B) is said to be p-stabilizable if either $R(\mathcal{A})$ is closed and $N(\mathcal{A})$, $N(\mathcal{A}^*)$ are finite dimensional or there is $F \in L(U,H)$ such that $R(I - e^{-(A+BF)T})$ is closed in H and $N(\mathcal{A}_F)$, $N(\mathcal{A}_F^*)$ are finite dimensional.*

The pair (A,C) is said to be p-detectable if either $R(\mathcal{A})$ is closed and $\dim N(\mathcal{A}) < \infty$ or $\exists K \in L(Z,H)$ such that $R(I - e^{-(A+KC)T})$ is closed in H and $\dim N(\mathcal{A}_K) < \infty$.

Here $\mathcal{A}_F = \mathcal{A} + BF$ and $\mathcal{A}_K = \mathcal{A} + KC$. We note that $N(\mathcal{A})$ is finite dimensional if and only if so is $N(I - e^{-AT})$.

In particular, (A,B) is p–stabilizable in each of the following situations:

1. *(A,B) is stabilizable.*
2. *$R(I - e^{-AT})$ is closed and $\dim N(I - e^{-A^*T}) < \infty$. (In particular if e^{-AT} is compact.)*

3 EXISTENCE OF OPTIMAL CONTROLLERS

We shall study here existence in problem (1) under the following hypotheses:

(i) The pair (A,C) is p–detectable and

$$N(\mathcal{A}) \cap \{y \in L^2(0,T;H); Cy(t) = 0 \text{ a.e. } t \in (0,T)\} = \{0\} \tag{16}$$

(ii) $g : Z \to \bar{R}, h : U \to \bar{R}$ are convex, lower semicontinous and

$$g(z) \ge \alpha|z|_Z + \beta, \quad \forall z \in Z \tag{17}$$

$$h(u) \ge \omega|u|_U^2 + \gamma, \quad \forall u \in U \tag{18}$$

where $\alpha, \omega > 0$ and $\beta, \gamma \in R$.

Theorem 1 *Under Hypotheses* (i), (ii) *problem* (1) *has at least one optimal pair* $(y^*, u^*) \in C([0,T]; H) \times L^2(0,T; U)$.

Proof. Let $\{(y_n, u_n) \in C([0,T]; H) \times L^2(0,T; U)$ be such that $\mathcal{A}y_n = Bu_n + f$ and

$$\inf(1.1) = d < \int_0^T (g(y_n(t)) + h(u_n(t)))dt \leq d + n^{-1}. \tag{19}$$

By (17), (18) we have

$$\|Cy_n\|_{L^1(0.T;H)} + \|u_n\|_{L^2(0,T;U)} \leq C_1. \tag{20}$$

By Hypothesis (i) there is $K \in L(Z, H)$ such that $R(\mathcal{A}_K)$ is closed ($\mathcal{A}_K = \mathcal{A} + KC$), estimate (14) holds and $\dim N(\mathcal{A} + KC) < \infty$. We have

$$\mathcal{A}_K y_n = Bu_n + KCy_n + f.$$

We set $y_n = y_n^1 + y_n^2$ where $y_n^1 = \mathcal{A}_K^{-1}(Bu_n + KCy_n + f) \in R(\mathcal{A}_K^*)$ and $y_n^2 \in N(\mathcal{A}_K)$. Then by (16) and (20) we have

$$\|y_n^2\|_{C([0,T];H)} \leq C_2, \ \forall n \in N. \tag{21}$$

If $R(\mathcal{A})$ is closed we get the same estimate. Then, on a subsequence, again denoted n, we have

$$\begin{array}{lcll} u_n & \longrightarrow & u^* & \text{weakly in } L^2(0,T;U) \\ y_n(t) & \longrightarrow & y^*(t) & \text{weakly in } L^2(0,T;H) \text{ and} \\ & & & \text{weakly in } H \text{ for } t \in [0,T]. \end{array}$$

clearly $\mathcal{A}y^* = Bu^* + f$ and since the convex integrand is weakly lower semicontinous we get

$$d = \int_0^T (g(y^*(t)) + h(u^*(t)))dt$$

as claimed. □

4 THE MAXIMUM PRINCIPLE

Here we shall assume that

(j) The pair (A, B) is p–stabilizable and

$$N(\mathcal{A}^*) \cap \{p \in L^2(0,T;H); B^*p(t) = 0 \text{ a.e. } t \in (0,T)\} = \{0\}. \tag{22}$$

(jj) The function $g : Z \to R$ is convex and continuous, $h : U \to \bar{R}$ is convex and lower semicontinuous, $\text{int}\, D(h) \neq \emptyset$.

(jjj) $f(t) = Bf_0(t)$ where $f_0 \in C([0,T];U)$ and $f_0(t) \in \text{int}\, D(h)$, $\forall t \in [0,T]$. Here $D(h) = \{u \in U;\ h(u) > +\infty\}$.

Theorem 2 *Assume that Hypotheses* (j), (jj), (jjj) *hold. Then the pair* $(y^*, u^*) \in C([0,T];H) \times L^2(0,T;U)$ *is optimal in problem* (1) *if and only if there are* $p \in C([0,T];H)$ *and* $\eta \in L^\infty(0,T;Z)$ *such that*

$$\frac{dp}{dt} - A^*p = C^*\eta, t \in (0,T); p(0) = p(T). \tag{23}$$

$$\eta(t) \in \partial g(Cy^*(t)), \text{ a.e. } t \in (0,T). \tag{24}$$

$$u^*(t) \in \partial h^*(B^*p(t)), \text{ a.e. } t \in (0,T). \tag{25}$$

Here $\partial g : Z \to Z$ and $\partial h^* : U \to U$ stand for subdifferentials of g and h^* (the conjugate function of h). The system (23) is considered of course in the weak sense (5) or (15), i.e.,

$$\mathcal{A}^*p = -C^*\eta. \tag{26}$$

Proof. It is readily seen that equations (23)~(25) are sufficient for optimality. To prove necessity we fix an optimal pair (y^*, u^*) and consider the approximating control problem

$$\begin{aligned}\text{Min}\{\int_0^T &(g_\varepsilon(Cy(t)) + |y(t) - y^*(t)|^2 + |u(t) - u^*(t)|_U^2 + h(u(t)) + \varepsilon^{-1}|v(t)|^2)dt;\\ &\mathcal{A}y = Bu + v + f, u \in L^2(0,T;U),\ v \in L^2(0,T;H), y \in C([0,T];H)\},\end{aligned} \tag{27}$$

where

$$g_\varepsilon(z) = \inf\{(2\varepsilon)^{-1}|z - \bar{z}|_Z^2 + g(\bar{z}); \bar{z} \in Z\}, \forall z \in Z.$$

(Recall that $g_\varepsilon \in C^1(Z)$ is a smooth approximation of g.) Since $\mathcal{A}$ is closed it is easily seen that problem (27) has a unique solution $(y_\varepsilon, u_\varepsilon, v_\varepsilon)$. Moreover, by standard device (see e.g. Barbu (1993)) we get

$$\begin{aligned} u_\varepsilon &\longrightarrow u^* \text{ strongly in } L^2(0,T;U)\\ y_\varepsilon &\longrightarrow y^* \text{ strongly in } L^2(0,T;H)\\ v_\varepsilon &\longrightarrow v^* \text{ strongly in } L^2(0,T;H)\end{aligned} \tag{28}$$

Next by (27) we have

$$\int_0^T ((C^*\nabla g_\varepsilon(Cy_\varepsilon), z) + 2(y_\varepsilon - y^*, z) + 2(u_\varepsilon - u^*, w)_U + h'(u_\varepsilon, w) + 2\varepsilon^{-1}(v_\varepsilon(t), v(t)))dt \geq 0 \tag{29}$$

for all $(z, w, v) \in C([0,T]; H) \times L^2(0,T;U) \times L^2(0,T;H)$ such that $\mathcal{A}z = Bw + v$. We set $p_\varepsilon = \varepsilon^{-1}v_\varepsilon$. Then (29) yields

$$\int_0^T ((C^*\nabla g_\varepsilon(Cy_\varepsilon) + 2(y_\varepsilon - y^*), z) + 2(u_\varepsilon - u^*, w)_U + h'(u_\varepsilon, w) + (p_\varepsilon, \mathcal{A}z - Bw))dt \geq 0, \quad (30)$$

$\forall z \in D(\mathcal{A})$, $\forall w \in L^2(0,T;U)$. (Here h' is the directional derivative of h.) This yields

$$\mathcal{A}^*p_\varepsilon = -C^*\nabla g_\varepsilon(Cy_\varepsilon) - 2(y_\varepsilon - y^*). \quad (31)$$

$$B^*p_\varepsilon(t) + 2(u^*(t) - u_\varepsilon(t)) \in \partial h(u_\varepsilon(t)), \text{ a.e. } t \in (0,T). \quad (32)$$

We may rewrite the equation

$$\mathcal{A}y_\varepsilon = Bu_\varepsilon + f$$

as

$$\mathcal{A}_F y_\varepsilon = Bu_\varepsilon + BFy_\varepsilon + f \quad (33)$$

where $F \in L(H,U)$ is as in Definition 1. Then by (28) and Proposition 2 we see that

$$y_\varepsilon \longrightarrow y^* \text{ strongly in } C([0,T]; H). \quad (34)$$

Since ∂g is locally bounded on H and $\nabla g_\varepsilon \in \partial g(1 + \varepsilon\partial g)^{-1}$ we infer that

$$|\nabla g_\varepsilon(Cy_\varepsilon(t))|_Z \leq C_3, \ \forall t \in [0,T]. \quad (35)$$

On the other hand, by assumption (jjj) we have

$$h^*(B^*p_\varepsilon + 2(u^* - u_\varepsilon)) + (f, p_\varepsilon) \geq$$
$$\geq \rho|B^*p_\varepsilon|_U + C_4|u^* - u_\varepsilon|_U - h(-f_0 + \rho|B^*p_\varepsilon|_U|B^*p_\varepsilon|_U^{-1}) + C_4, \quad a.e. \ t \in (0,T).$$

If ρ is sufficiently small, this yields

$$\rho\int_0^T |B^*p_\varepsilon(t)|_U dt \leq \int_0^T (h^*(B^*p_\varepsilon(t) + 2(u^*(t) - u_\varepsilon(t))) + (f(t), p_\varepsilon(t)))dt + C_5 \leq$$
$$\leq -\int_0^T ((\nabla g_\varepsilon(Cy_\varepsilon(t)), Cy_\varepsilon(t))_Z + h(u_\varepsilon(t)) + 2(y_\varepsilon(t) - y^*(t), y_\varepsilon(t)))dt \leq C_6$$

because ∇g_ε is monotone and h is bounded from below. Here we have also used the equation

$$h^*(w) + h(u) = (w,u)_U, \ \forall w \in \partial h(u).$$

Hence

$$\int_0^T |B^*p_\varepsilon(t)|_U dt \leq C_7, \ \forall \varepsilon > 0. \quad (36)$$

Now we rewrite (31) as

$$\mathcal{A}_F^* p_\varepsilon = -C^*\nabla g_\varepsilon(Cy_\varepsilon) - 2(y_\varepsilon - y^*) + F^*B^*p_\varepsilon$$

and set $p_\varepsilon = p_\varepsilon^1 + p_\varepsilon^2$ where $p_\varepsilon^1 \in R(\mathcal{A}_F) = N(\mathcal{A}_F^*)^\perp$, $p_\varepsilon^2 \in N(\mathcal{A}_F^*)$. By (14, (28), (35), (36) we have

$$\|p_\varepsilon^1\|_{C[0,T];H)} \leq C_8, \ \forall \varepsilon > 0. \tag{37}$$

whilst (22) and (36) imply that $\{p_\varepsilon^2\}$ is bounded in $L^2(0,T;H)$. Hence selecting further subsequence we have,

$$\begin{array}{lcll} p_\varepsilon & \longrightarrow & p & \text{weakly star in } L^2(0,T;H) \\ \nabla g_\varepsilon(Cy_\varepsilon) & \longrightarrow & \eta & \text{weak star in } L^\infty(0,T;Z). \end{array} \tag{38}$$

Since ∂g and ∂h are maximal monotone (and therefore weakly–strongly closed) we may pass to limit in (31), (32) to get the optimality system (23)–(25). □

The previous result remains true by an easy adaptation of the proof if B is unbounded but is in $L(U,(D(\mathcal{A}^*))')$ and satisfies one of the following two conditions (Lasiecka and Triggiani (1981))

$$\begin{array}{ll} \|B^*e^{-A^*t}\|_{L(H,U)} \leq C_9 t^{\gamma-1}, & 0 < \gamma < 1 \\ \int_0^t |B^*e^{-A^*s}|^2 ds \leq C_t |x|^2, & \forall x \in H, t > 0. \end{array}$$

In particular, it follows by Theorem 2 that the pair (y^*, u^*) is optimal in problem (1) if and only if the dual problem

$$\text{Min}\{\int_0^T (\tilde{g}(-v(t)) + h^*(B^*p(t)) + (f(t), p(t)))dt; \mathcal{A}^*p = v\} \tag{39}$$

has a solution (p^*, v^*) and Min(1)+Min(39) = 0. (Here $\tilde{g}$ is the concept of $y \to g(Cy)$.) It should be noted that since the pair (A^*, I) is stabilizable the dual problem (39) is simpler than the primal problem (1).

Examples. Throughout in the following Ω is an open bounded subset of R^n with C^2 boundary $\partial\Omega$.

1°. **Parabolic control problems.** Consider the system

$$\begin{array}{ll} \dfrac{\partial y}{\partial t} - \Delta y + b(x)\cdot\nabla y + c(x)y = Bu + f(x,t), & (x,t) \in \Omega \times R \\ y(x,t) = 0, & \forall (x,t) \in \partial\Omega \times R \\ y(x,t+T) = y(x,t), & \forall (x,t) \in \Omega \times R, \end{array}$$

where $b \in W^{1,\infty}(\Omega; R^n)$, $c \in L^\infty(\Omega)$, $f \in L^2_{\text{loc}}(R; L^2(\Omega))$ is T-periodic in t and $B \in L(U, L^2(\Omega))$ where U is a real Hilbert space (the controller space).

The operator $Ay = -\Delta y + b\cdot\nabla y + cy$, $D(A) = H_0^1(\Omega)\cap H^2(\Omega)$, generates a C_0–compact semigroup on $H = L^2(\Omega)$.

Then as seen earlier $R(\mathcal{A})$ is closed and $N(\mathcal{A}^*)$ is finite dimensional.

To be more specific we take $U = L^2(\Omega)$ and $Bu = mu$, $\forall u \in L^2(\Omega)$ where $m \in C(\overline{\Omega})$. Then condition (22) holds. Indeed if $p \in N(\mathcal{A}^*)$ then p is the solution to boundary value problem

$$\begin{aligned}&\frac{\partial p}{\partial t} + \Delta p + \operatorname{div}(bp) - cp = 0, && \text{in } \Omega \times R\\ &p = 0, && \text{in } \partial\Omega \times R\end{aligned}$$

and $B^*p = mp$.

Let ω_0 be an open subset of ω such that $m(x) \neq 0$, $\forall x \in \overline{\omega}_0$. We have therefore

$$p(x,t) = 0,\ \forall x \in \omega_0,\ t \in R$$

and by the unique continuation property of solutions to parabolic equations (see e.g. Saut and Scheurer (1981)) we infer that $p \equiv 0$ as claimed. Hence Theorem 2 is applicable in the present case.

2°. **Linear delay control systems.** Consider the delay system

$$\begin{aligned}&y'(t) + A_0 y(t) + A_1 y(t-h) = B_0 u(t) + f(t)\\ &y(t) = y(t+T) \qquad\qquad \forall t \in R,\end{aligned}$$

where A_0, A_1 are $n \times n$ matrices, B_0 is a $n \times m$ matrix, $f \in L^2_{\text{loc}}(R; R^n)$, $f(t+T) = f(t)$, $u \in L^2_{\text{loc}}(R; R^n)$, $u(t) = u(t+T)$. It is well known that this system can be written in the form (2) where $H = M_2 = R^n \times L^2(-h, 0; R^n)$ and

$$\begin{aligned}&A(y_0, y^0) = \{A_0 y_0 + A_1 y^0(-h), \frac{dy^0}{ds}\}\\ &D(A) = \{(y_0, y^0) \in R^n \times W^{1,2}([-h,0]; R^n), y_0 = y^0(0)\}\\ &Bu = (B_0 u, 0),\ u \in R^m = U.\end{aligned}$$

In virtue of Proposition 1, the corresponding operator $\mathcal{A}$ has closed range in $L^2(-h, 0; R^n)$. Moreover, $N(\mathcal{A}^*)$ is finite dimensional and if

$$N(miI - A_0^* - e^{-mih}A_1^*) \cap N(B_0^*) = \{0\},\ \forall m \in Z \tag{40}$$

then assumption (j) holds and so theorem 2 is applicable. It should be mentioned that condition (40) does not imply the stabilizability of the pair (A, B) as the following example shows

$$x_1'(t) = x_1(t) + x_3(t-h),\ x_2'(t) = x_2(t) + x_3(t),\ x_3'(t) = u(t).$$

This system is not stabilizable (Manitius and Triggiani (1978)). However it is p-stabilizable and condition (40) holds.

5 THE OPTIMAL CONTROL OF WAVE EQUATION

The optimal control problem

Minimize

$$\int_0^T (g(Cy(t)) + h(u(t)))dt \tag{41}$$

subject to $u \in L^2(0,T;H), y \in L^2(0,T;U)$

$$\begin{aligned} &y'' + Ay = Bu + f,\ t \in (0,T) \\ &y(0) = y(T),\ y'(0) = y'(T) \end{aligned} \tag{42}$$

where A is a self–adjoint linear and positively defined operator on the Hilbert space H and $B \in L(U,H)$ can be treated as a special case of problem (1) in the product space $D(A^{1/2}) \times H$. However, a direct approach to this problem allows to weaken hypotheses (i), (j) in this specific case. For simplicity we shall discuss only the maximum principle for problem (41).

By weak solution to equation (42) we mean a function $y \in L^2(0,T;H)$ such that

$$\int_0^T (y(t), \varphi''(t) + A\varphi(t))dt = \int_0^T (f(t) + Bu(t), \varphi(t))dt, \tag{43}$$

for all $\varphi \in Y = \{\varphi \in C^2([0,T];H) \cap C([0,T];D(A));\ \varphi(0) = \varphi(T),\ \varphi'(0) = \varphi'(T)\}$. Equivalently,

$$\mathcal{W}y = Bu + f \tag{44}$$

where $\mathcal{W} : D(\mathcal{W}) \subset L^2(0,T;H) \to L^2(0,T;H)$ is the linear operator defined by

$$\mathcal{W}y = f_0 \text{ iff } \int_0^T (y(t), \varphi''(t) + A\varphi(t))dt = \int_0^T (f_0(t), \varphi(t))dt,\ \forall \varphi \in Y.$$

We shall present now two variants of maximum principle for problem (41) involving two different basic assumptions on system (42) and function $h : U \to \bar{R}$.

Theorem 3 *Assume that*

(k) $R(\mathcal{W})$ *is closed,* $\dim N(\mathcal{W}) < \infty$ *and*

$$N(\mathcal{W}) \cap \{p \in L^2(0,T;H); B^*p(t) = 0 \quad a.e. \quad t \in (0,T)\} = \{0\}. \tag{45}$$

If h and g satisfy Hypotheses (jj), (jjj), *and ∂g has sublinear growth, then the pair* $(y^*, u^*) \in L^2(0,T;H) \times L^2(0,T;U)$ *is optimal in problem* (41) *if and only if there are* $p, \eta \in L^2(0,T;H)$ *such that*

$$\mathcal{W}p = -C^*\eta, \eta(t) \in \partial g(y^*(t)) \ a.e.\ t \in (0,T) \tag{46}$$

$$u^*(t) \in \partial h^*(B^*p(t)) \text{ a.e. } t \in (0,T) \tag{47}$$

Theorem 4 *Assume that*

(kk) *$R(\mathcal{W})$ is closed and*

$$\|p\|_{L^2(0,T;H)} \leq M\|B^*p\|_{L^2(0,T;U)}, \ \forall p \in N(\mathcal{W}) \tag{48}$$

Then if g, h satisfy assumptions (jj), *∂g has sublinear growth, $f \in C([0,T];H)$ and*

$$h(u) \leq \alpha_1|u|_U^2 + \beta_1, \ \forall u \in U \tag{49}$$

the pair $(y^*, u^*) \in L^2(0,T;H) \times L^2(0,T;U)$ is optimal in problem (41) *if and only if it satisfies system* (46),(47).

The proof of Theorem 3 is identical with that of Theorem 2. As regards Theorem 4, if denote by $(y_\varepsilon, u_\varepsilon, p_\varepsilon)$ the sequence defined in the proof of Theorem 2 where $\mathcal{A}$ is replaced by $\mathcal{W}$ we note that by (49) we have

$$\|B^*p_\varepsilon\|_{L^2(0,T;U)} \leq C_{10}$$

and so by (48) it follows that $\{p_\varepsilon^2\}$ is bounded in $L^2(0,T;H)$. From now on the proof is identical with that of Theorem 2.

3°. **The one dimensional wave equation.** Consider the control system

$$\begin{array}{ll} y_{tt}(x,t) - v^{-1}(x)(v(x)y_x(x,t))_x = m(x)u(x,t) + f(x,t), & \\ y(0,t) = y(\pi,t) = 0, & x \in (0,\pi) \times R \\ y(x,t+T) = y(x,t), y_t(x,t+T) = y_t(x,t), & (x,t) \in (0,\pi) \times R \end{array} \tag{50}$$

where $v \in H^2(0,T)$, $v(x) > 0$, $\forall x \in [0,\pi]$, $m \in C([0,T])$ $m \not\equiv 0$ and

$$\text{ess inf}\{(v'(x))^2 - 2v''(x)v(x);\ x \in (0,\pi)\} < 0. \tag{51}$$

In this case $H = L^2(0,\pi)$, $Ay = -v^{-1}(vy_x)_x$, $D(A) = H_0^1(0,\pi)$, $U = L^2(0,\pi)$, $Bu = mu$, $\forall u \in U$. Equation (50) models the forced vibrations of a nonhomogeneous string as well as the propagation of waves in nonisotropic media. If T is a rational multiple of π then $R(\mathcal{W})$ is closed and $N(\mathcal{W})$ is finite dimensional (Barbu and Pavel (to appear)). If $y \in N(\mathcal{W})$ and $m(x)y(x,t) = 0$ *a.e.* $(x,t) \in (0,\pi) \times (0,T)$ then $y(x,t) = 0$ *a.e.* $(x,t) \in (\alpha,\beta) \times (0,T)$. Indeed,

$$y(x,t) = \sum_{k=1}^{n_0} \alpha_k e^{i\mu_k t}\varphi_k(x), \quad \forall (x,t) \in Q = (0,\pi) \times (0,T)$$

where $\mu_k = 2k\pi/T$ and φ_k are the eigen functions of the Sturm–Liouville problem associated to operator $y \longrightarrow v^{-1}(x)(v(x)y_x)_x$ we infer that $\alpha_k = 0$, $\forall k$. Hence assumption (k) holds in this case.

4°. **The n–dimensional wave equation.** Consider the control system

$$y_{tt} - \Delta y = m(x)u + f,\ x \in \Omega, t \in R$$
$$y = 0,\ \text{in } \partial\Omega \in R; y(x,t+T) = y(x,t), y_t(x,t+T) = y_t(x,t) \tag{52}$$

where $\Omega = (0,\pi)^n$, $m \in C(\overline{\Omega})$, $m \not\equiv 0$ and $f \in C([0,T]; L^2(\Omega))$, $f(x,t+T) \equiv f(x,t)$. We may write (52) in the form (42) where $H = L^2(\Omega)$, $A = -\Delta$, $D(A) = H_0^1(\Omega) \cap H^2(\Omega)$ and $Bu = mu$, $\forall u \in U = L^2(\Omega)$.

If T is a rational multiple of π then the corresponding operator $\mathcal{W} : L^2(Q) \longrightarrow L^2(Q)$, $Q = \Omega \times (0,T)$ has closed range. Here is the argument (see also N.Pavel (to appear)). If

$$\mathcal{W}y = f,\ (y,f) \in L^2(Q) \times L^2(Q)$$

then

$$y = \sum_{m\in Z, k\in N^n} f_{mk}(\mu_m^2 - \lambda_k^2)^{-1} e^{i\mu_m t}\varphi_k \tag{53}$$

where $\mu_m = 2m\pi T^{-1}$, $\lambda_k^2 = k_1^2 + k_2^2 + \cdots + k_n^2$, $k_i \in N$ are the eigenvalues of A and φ_k are the corresponding eigen functions; f_{mk} are the Fourier coefficients of f. If T is a rational multiple of π we see by (53) that

$$\|y\|_{L^2(Q)} \le C_{11}\|f\|_{L^2(Q)},\ \forall f \in R(\mathcal{W})$$

as claimed.

Let us check now condition (48). If $\{p_\varepsilon\} \subset L^2(0,T; L^2(\Omega))$ is in $N(\mathcal{W})$, i.e., $\mathcal{W}p_\varepsilon = 0$ and $\{mp_\varepsilon\}$ is bounded in $L^2(0,T; L^2(\Omega))$ we have for some interval $(a,b) \subset (0,\pi)$

$$\int_a^b dx \int_0^T p_\varepsilon^2(x,t)dx\,dt \le C_{12},\quad \forall\varepsilon > 0$$
$$p_\varepsilon(x,t) = \sum_k a_{\varepsilon k} e^{i\mu_k t}\varphi_k(x),\qquad \forall (x,t) \in Q \tag{54}$$

where $|\mu_k| = \lambda_k$. By (54) we get

$$\sum_k a_{\varepsilon k}^2 \int_a^b \varphi_k^2(x)dx \le C_{13},\ \forall\varepsilon > 0.$$

Hence

$$\sum_{|\mu_k|=\lambda_k} a_{\varepsilon k}^2 < C_{14},\ \forall\varepsilon > 0$$

and this implies that $\{p_\varepsilon\}$ is bounded in $L^2((0,T) \times \Omega)$ as claimed.

Remark. In particular, the cost functional (1) in $L^2(Q)$ is of the form

$$\int_Q g_0(x, y(x,t))dx\,dt + \int_0^T h(u(t))dt$$

where $g_0 : \Omega \times R \to R$ is continuous, convex and has quadratic growth in y. In the special case $m = 1$ the maximum principle for this problem follows for a general domain Ω under assumption (jjj) as a consequence of Theorem 2 because the pair (A, B) is stabilizable. In the case $n = 1$ assumption (jjj) can be weakened to an interiority condition in $L^\infty(\Omega)$ (Barbu (to appear)). The general case $\Omega \subset R^n$ and $m \equiv 1$ is open.

REFERENCES

Barbu, V. (1993) *Analysis and control of nonlinear infinite dimensional systems.* Academic Press, Boston.

Barbu, V. (to appear) Optimal control of the one dimensional periodic wave equation. *Applied Mathematics and Optimization.*

Barbu, V. and Pavel, N. (1995) Optimal control of periodic systems. *Applied Mathematics and Optimization.*

Barbu, V and Pavel, N. (submitted) Periodic solutions to nonlinear one dimensional wave equation with x-dependent coefficients.

Bensoussan, A., Da Prato, G, Delfour, M., Mitter, S.K., (1993) *Representation and control of infinite dimensional control systems.* Birkhäuser, Boston, Basel, Berlin.

Da Prato, G., (1987) Synthesis of optimal control for an infinte dimensional periodic problem. SIAM *Journal on Control and Optimization,* 25, 706-14.

Lasiecka, I. and Triggiani, R. (1991) *Algebraic Riccati equations with applications to boundary joint control problems. Continuous theory and approximation theory.* LNCIS. Springer Verlag, Berlin, New York.

Prüss, J. (1984) *On the spectrum of* C_0*–semigroup.* Transactions American Mathematical Society, 284, 847-57.

Manitius, A. and Triggiani, R. (1978) Function space controllability of linear retarded systems: a derivation from abstract operator condition. SIAM *Journal on Control* 16, 599-645.

Pavel, N., (to appear) Periodic solutions to nonlinear $2 - D$ wave equations.

Saut, C. and Scheurer, B. (1987) Unique continuation for some evolution equations. *Journal Differential Equations*, 16, 118–39.

2
Intrinsic modelling of shells

Michel C. Delfour
Centre de recherches mathématiques, Université de Montréal
CP 6128, Succ. Centre-ville, Montréal, Canada H3C 3J7.
and
Jean-Paul Zolésio
CNRS Institut Non Linéaire de Nice
1361 route des Lucioles, 06904 Sophia-Antipolis, France.

Abstract
We present recent results on the use of the oriented boundary (signed, algebraic) distance function and the tangential differential calculus in the intrinsic modelling of thin/shallow shells. We provide the link with covariant operators and show how to express them without Christoffel symbols. Such models are mathematically more tractable than classical ones.

Keywords
Shell, distance function, intrinsic differential operators, tangential calculus

1 INTRODUCTION

In this paper we review and announce some recent results on intrinsic models of linear shells by Delfour and Zolésio (1994-4, 1995-1 to 3) and show how to reformulate the classical models of Naghdi, Koiter, and the asymptotic membrane model of Ciarlet and Sanchez-Palencia (1993) in terms of intrinsic differential operators. This link was so far missing making comparisons difficult between our intrinsic model and the classical ones. It is now our belief that models expressed in terms of intrinsic differential operators are mathematically more tractable and natural to use in associated control, optimal design, and shape sensitivity problems (cf. for instance Lions (1968), Lagnese and Lions (1988), Sokolowski and Zolésio (1992)). It can also open the way to different parametrizations of the mean surface in the numerical analysis of partial differential equations on submanifolds of $\mathbb{R}^N$. An illustration of this viewpoint is given in the companion paper of Delfour and Zolésio (1995-5).

The new linear model of Delfour and Zolésio (1995-2) only uses two assumptions: (i) the displacement vector is equal to the displacement of the mean surface plus a tangential vector times the normal coordinate z to the mean surface, and (ii) a truncation of the infinite expansion with respect to z of the corresponding strain (deformation) tensor after its second power in z. No other approximations is involved. The simplest rheological law has been used, but our development readily extends to more complex laws. This approach was developed in a sequence of papers (Delfour and Zolésio (1994-4, 1995-1)) starting with a truncation of the strain tensor after its first power in z . However for this model it was difficult to exactly recover the rigid displacements as the kernel of the deformation tensor. The difficulty completely disappeared by going to the second power of z. It was possible to give a completely self-contained treatment for both static and dynamical models extending to thin/shallow shells the "Natural Theory" and the Love-Kirchhoff theory of plates (cf. for instance Germain (1986)) in the general spirit of completely intrinsic methods of Valid (1981). Finally it is interesting to emphasize that the Love-Kirchhoff theory comes out of the analysis as a special case of the natural theory by looking at the same variational equation over a closed linear subspace of the Hilbert space $\mathcal{V}$ associated with the natural theory.

2 DEFINITIONS AND NOTATION

2.1 Oriented boundary distance function

Let $\mathbb{R}^N$ be the N-dimensional Euclidean space for some integer $N > 1$ (in practice $N = 3$). Let Ω be a subset of $\mathbb{R}^N$ with a boundary $\partial\Omega$ which is a C^2 $(N-1)$-dimensional submanifold of $\mathbb{R}^N$. Associate with Ω the *oriented boundary (resp. algebraic or signed) distance function*

$$b_\Omega(x) \stackrel{\text{def}}{=} d_\Omega(x) - d_{\Omega^c}(x), \quad \Omega^c \stackrel{\text{def}}{=} \mathbb{R}^N \backslash \Omega = \{x \in \mathbb{R}^N \, : \, x \notin \Omega\}, \tag{1}$$

where d_A is the usual distance function to a subset A of $\mathbb{R}^N$. This function captures the geometrical properties of the boundary $\partial\Omega$. Moreover for any integer $k \geq 2$, a domain Ω has a C^k boundary $\partial\Omega$ if and only if in each point $X \in \partial\Omega$ there exists a bounded open neighbourhood $N(X)$ of X such that $b_\Omega \in C^k(N(X))$ (cf. Gilbarg and Trudinger (1983), Delfour and Zolésio (1994-1, 1994-2)). At each point X of $\partial\Omega$, its gradient $\nabla b_\Omega(X)$ coincides with the unitary exterior normal n to $\partial\Omega$ and the eigenvalues of the symmetrical matrix of second order partial derivatives $D^2 b_\Omega$ are 0 and the *principal curvatures*, κ_i, $1 \leq i \leq N-1$, of the submanifold $\partial\Omega$. The trace of $D^2 b_\Omega(X)$ is the *mean curvature*

$$H(X) \stackrel{\text{def}}{=} \operatorname{tr}(D^2 b_\Omega(X)) = \Delta b_\Omega(X), \tag{2}$$

up to the multiplying factor $(N-1)$ which is used as a normalization factor to make the mean curvature of the unit sphere equal to one in all dimensions. *We choose to modify the classical definition since it is the term Δb_Ω which will naturally occur and not $\Delta b_\Omega/(N-1)$*. If we really want to make a distinction, our definition of $H(X)$ would be the *additive curvature.* The trace of the matrix of cofactors $M(D^2 b_\Omega)$ is the *total* or *Gaussian curvature*

$$K(X) \stackrel{\text{def}}{=} \operatorname{tr} M(D^2 b_\Omega(X)). \tag{3}$$

Since the domain Ω is fixed throughout this paper, we shall now drop the subscript Ω. For each $X \in \partial\Omega$, the *projection mapping* $p : N(X) \to \partial\Omega$ and its Jacobian matrix are obtained directly from the oriented distance function b as

$$p(x) = x - b(x)\nabla b(x), \qquad Dp(x) = I - b(x)D^2 b(x) - \nabla b(x)\,{}^*\nabla b(x), \tag{4}$$

where ${}^*\nabla b(x)$ is the transposed of the vector $\nabla b(x)$ and I is the identity matrix. For $x \in N(X)$, the *linear projector* onto the *tangent plane* $T_{p(x)}\partial\Omega$ at $p(x)$ of $\partial\Omega$ is given by

$$P(x) = I - \nabla b(x)\,{}^*\nabla b(x). \tag{5}$$

2.2 Definition of the "shell"

In practice *the mean surface* Γ is the mean surface of a thin piece of material called the shell, but in order to make sense of all the differential operators defined on Γ it is a posteriori assumed that it is a sufficiently smooth submanifold of $\mathbb{R}^N$. So we might as well start from a mathematical description of the mean surface and *build the shell around it.* Therefore a *(mathematical) shell* is characterized by its *mean surface* Γ and its *thickness (function)* $\tilde{h} : \Gamma \to \mathbb{R}^+$. Assume that Γ is a bounded open domain in the $(N-1)$-submanifold $\partial\Omega$ of $\mathbb{R}^N$. When $\Gamma = \partial\Omega$ (hence $\partial\Omega$ is compact), the shell has no boundary. When $\Gamma \neq \partial\Omega$, the (relative) boundary $\partial_{\partial\Omega}\Gamma$ is assumed to be uniformly Lipschitzian in $\partial\Omega$. Since Γ is bounded and $\partial\Omega$ is C^2, there exist $h > 0$ and a bounded neighbourhood

$$S_h \stackrel{\text{def}}{=} \{x \in \mathbb{R}^N : p(x) \in \Gamma,\ |b(x)| < h\}, \tag{6}$$

where b is C^2. The set S_h is a bounded open domain in $\mathbb{R}^N$ with a Lipschitzian boundary. When $\Gamma \neq \partial\Omega$, S_h has a *lateral boundary*

$$\Sigma_h = \{x \in \mathbb{R}^N : p(x) \in \partial_{\partial\Omega}\Gamma,\ |b(x)| < h\} \tag{7}$$

which is an $(N-1)$-dimensional surface normal to the mean surface Γ. It is important to keep in mind that we use the distance function $b = b_\Omega$ and not the distance function to Γ.

2.3 Flow of the gradient of b and local coordinates

Since $\nabla b \in C^1(S_h)$, consider the *flow mapping*

$$T_z(X) = x(z), \quad \frac{dx}{dz}(z) = \nabla b(x(z)), \quad |z| < h, \quad x(0) = X. \tag{8}$$

It is a homeomorphism from Γ onto $\Gamma_z = \{x \in \mathbb{R}^N : b(x) = z,\ p(x) \in \Gamma\}$. In particular

$$T_z(X) = X + z\nabla b(X), \qquad DT_z = I + zD^2 b \tag{9}$$

for $|z| < h$. This induces a *curvilinear coordinate system* $(X, z) \in \Gamma \times]-h, h[$ in S_h. The points on the level set Γ_z are given by $\{X + z\nabla b(X) : X \in \Gamma\}$ and for each

$(X,z) \in \Gamma \times]-h,h[$, $\nabla b(T_z(X)) = \nabla b(X)$. Therefore for all $x \in S_h$, $\nabla b(x) = n(p(x))$, where n is the normal to Γ at $p(x)$. *So we shall often simply write n instead of ∇b.* Moreover $\det DT_z(X)$ is a polynomial of degree at most $N-1$

$$j(z) \stackrel{\text{def}}{=} \det DT_z(X) = \sum_{i=0}^{N-1} K_i(X) z^i, \tag{10}$$

where $K_0 = 1$, $K_1 = H$ for $N \geq 2$, and $K_{N-1} = K$ for $N \geq 3$. For $N = 3$, $j(z) = 1 + zH(X) + z^2 K(X)$.

3 INTRINSIC TANGENTIAL CALCULUS

Given a scalar function $w : \Gamma \to \mathbb{R}$, denote by $\nabla_\Gamma w$ the *tangential gradient*

$$\nabla_\Gamma w = \nabla W|_\Gamma - \frac{\partial W}{\partial n} n \tag{11}$$

defined in terms of an extension W of w to S_h. This definition is independent of the choice of the extension W. Moreover it is easy to check that

$$\nabla(w \circ p) = Dp\,(\nabla_\Gamma w) \circ p = [I - bD^2 b]\, \nabla_\Gamma w \circ p \quad \text{and} \quad \nabla(w \circ p)|_\Gamma = \nabla_\Gamma w, \tag{12}$$

where $\circ$ denotes the composition of two functions. The *tangential Jacobian matrix* of a vector $v : \Gamma \to \mathbb{R}^N$ is defined through an extension V of v or through the transposed ${}^*D_\Gamma v = (\nabla_\Gamma v_1, \ldots, \nabla_\Gamma v_N)$ in terms of the column tangential gradients of its components. In particular

$$D(v \circ p) = (D_\Gamma v) \circ p\, Dp = (D_\Gamma v) \circ p\, [I - bD^2 b] \quad \text{and} \quad D(v \circ p)|_\Gamma = D_\Gamma v. \tag{13}$$

In the same way define the *tangential divergence* in term of an extension V of v to a neighbourhood of Γ

$$\operatorname{div}_\Gamma v \stackrel{\text{def}}{=} \operatorname{div} V|_\Gamma - DVn \cdot n \quad \text{and} \quad \operatorname{div}(v \circ p)|_\Gamma = \operatorname{div}_\Gamma v = \operatorname{tr} D_\Gamma(v), \tag{14}$$

where $\cdot$ is the inner product in $\mathbb{R}^N$. Similarly the *tangential strain tensor* is defined as

$$\varepsilon_\Gamma(v) \stackrel{\text{def}}{=} \frac{1}{2}(D_\Gamma v + {}^*D_\Gamma v) \qquad \text{and} \qquad \varepsilon_\Gamma(v) = \varepsilon(v \circ p)|_\Gamma. \tag{15}$$

In view of identities (12) and (14) the composition of $\operatorname{div}_\Gamma$ and ∇_Γ yields the *Laplace-Beltrami* operator

$$\Delta_\Gamma w \stackrel{\text{def}}{=} \operatorname{div}_\Gamma(\nabla_\Gamma w) \quad \text{and} \quad \Delta(w \circ p)|_\Gamma = \Delta_\Gamma w, \tag{16}$$

but the *matrix of tangential second order derivatives* is generally not symmetrical

$$\begin{aligned} &D_\Gamma(\nabla_\Gamma w) - D^2 b\, \nabla_\Gamma w\, {}^*n = D^2(w \circ p)|_\Gamma = D^*_\Gamma(\nabla_\Gamma w) - n\, {}^*(D^2 b\, \nabla_\Gamma w) \\ &\varepsilon(\nabla(w \circ p))|_\Gamma = \varepsilon_\Gamma(\nabla_\Gamma(w)) - \tfrac{1}{2}[D^2 b\, \nabla_\Gamma w\, {}^*n + n\, {}^*(D^2 b\, \nabla_\Gamma w)]. \end{aligned} \tag{17}$$

Another tangential operator which will naturally occur is

$$\varepsilon^P_\Gamma(v) \stackrel{\text{def}}{=} P\varepsilon_\Gamma(v)P = \varepsilon^P_\Gamma(v) = \varepsilon_\Gamma(v) - [\varepsilon_\Gamma(v)n\, {}^*n + n^*(\varepsilon_\Gamma(v)n)]. \tag{18}$$

If we denote by w and u the normal and tangential components of v, $w \stackrel{\text{def}}{=} v \cdot n$ and $u \stackrel{\text{def}}{=} v - w\,n$, then

$$\varepsilon^P_\Gamma(v) = \varepsilon^P_\Gamma(u) + wD^2 b \qquad \varepsilon^P_\Gamma(u) = \varepsilon_\Gamma(u) + \frac{1}{2}[D^2 b\, u\, {}^*n + n\, {}^*(D^2 b\, u)]. \tag{19}$$

In the sequel the following identity will be useful

$${}^*D_\Gamma v\, n = \nabla_\Gamma(v \cdot n) - D^2 b\, v = \nabla_\Gamma(w) - D^2 b\, u \quad \text{on } \Gamma. \tag{20}$$

4 INTRINSIC LINEAR MODEL FOR SHELLS

4.1 Preliminary results

For simplicity we work with a shell of constant thickness and make the following standard mechanical assumption on the displacement vector.

Assumption 1 *At each point x of the shell the displacement vector $U(x)$ is of the form*

$$U(x) = u(p(x)) + b(x)\,\ell(p(x)) \qquad (U = u \circ p + b\,\ell \circ p) \text{ in } S_h, \tag{21}$$

for vector-valued mappings u and ℓ from Γ to $\mathbb{R}^N$ such that ℓ is a tangent field, that is $\ell(X) \cdot n(X) = 0$, for all $X \in \Gamma$.

With the help of the tangential calculus the Jacobian matrix DU in S_h is given by

$$\begin{aligned} &DU = [D_\Gamma(u) \circ p + b\, D_\Gamma(\ell) \circ p + \ell \circ p\, {}^*\nabla b]\,[I - b\, D^2 b] \\ &DU\, \nabla b = \ell \circ p \end{aligned} \tag{22}$$

and the *strain tensor* $\varepsilon(U)$ by

$$\begin{aligned} &2\varepsilon(U) = D(U) + {}^*D(U) \\ &= [D_\Gamma(u) \circ p + b\, D_\Gamma(\ell) \circ p + \ell \circ p\, {}^*\nabla b]\,[I - b\, D^2 b] \\ &+[I - b\, D^2 b]\, {}^*[D_\Gamma(u) \circ p + b\, D_\Gamma(\ell) \circ p + \ell \circ p\, {}^*\nabla b] \end{aligned} \tag{23}$$

$$2\varepsilon(U)\, n = [I - b\, D^2 b]\,[2\varepsilon_\Gamma(u) \circ p\, n + \ell \circ p], \tag{24}$$

where $\varepsilon_\Gamma(e)$ is the *tangential strain tensor* defined in (15). In the (X, z) coordinate system the above expression becomes

$$DU \circ T_z = [D_\Gamma(u) + z\, D_\Gamma(\ell) + \ell\, {}^*\nabla b]\,[I + z\, D^2 b]^{-1} \quad (25)$$
$$DU \circ T_z\, \nabla b = \ell$$

$$2\varepsilon(U) \circ T_z = [D_\Gamma(u) + z\, D_\Gamma(\ell) + \ell\, {}^*n]\,[I + z\, D^2 b]^{-1}$$
$$+ [I + z\, D^2 b]^{-1}\, {}^*[D_\Gamma(u) + z\, D_\Gamma(\ell) + \ell\, {}^*n]$$
$$2\varepsilon(U) \circ T_z\, n = [I + z D^2 b]^{-1} [2\, \varepsilon_\Gamma(u)\, n + \ell]. \quad (26)$$

Under Assumption 1 we always have the identity

$$\varepsilon(U) \circ T_z\, n \cdot n = 0.$$

This identity is often introduced as an assumption in the literature. It is a direct consequence of the choice of U and the fact that the vector ℓ is tangential. The identity

$$2\varepsilon_\Gamma(u)\, n + \ell = 0 \text{ on } \Gamma$$

characterizes the *Love-Kirchhoff* models. When $u \cdot \nabla b = u \cdot n \in H^1(\Gamma)$, this identity can be written as

$$\nabla_\Gamma(u \cdot n) - D^2 b\, u + \ell = 0 \text{ on } \Gamma$$

extending to shells the identity for plates in Germain (1986). If ℓ and u belong to $H^1(\Gamma)^N$ and Γ is C^3, then $b \in C^3$ and $u \cdot n \in H^2(\Gamma)$.

The nonlinear part of $\varepsilon(U) \circ T_z$ with respect to the variable z is contained in the matrix $[I + z\, D^2 b]^{-1}$. So for $h\, \|D^2 b\| < 1$, that is

$$h \max_{1 \le i \le N-1} |\kappa_i(X)| < 1, \quad \forall X \in \Gamma,$$

the inverse is given by

$$[I + z D^2 b]^{-1} = \sum_{i=0}^{\infty} (-D^2 b)^i\, z^i, \quad (27)$$

and we get

$$\varepsilon(U) \circ T_z = \varepsilon(u \circ p + b\, \ell \circ p) \circ T_z = \sum_{i=0}^{\infty} \varepsilon^i(u, \ell)\, z^i, \quad (28)$$

where

$$\begin{aligned}
2\, \varepsilon^0(u, \ell) &= 2\, \varepsilon_\Gamma(u) + \ell\, {}^*n + n\, {}^*\ell \\
2\, \varepsilon^1(u, \ell) &= 2\, \varepsilon_\Gamma(\ell) - D_\Gamma(u)\, D^2 b - D^2 b\, {}^*D_\Gamma(u) \\
2\, \varepsilon^2(u, \ell) &= [D_\Gamma(\ell) - D_\Gamma(u)\, D^2 b](-D^2 b) + (-D^2 b)\, [{}^*D_\Gamma(\ell) - D^2 b\, {}^*D_\Gamma(u)] \\
2\, \varepsilon^i(u, \ell) &= [D_\Gamma(\ell) - D_\Gamma(u)\, D^2 b](-D^2 b)^{i-1} + (-D^2 b)^{i-1}\, [{}^*D_\Gamma(\ell) - D^2 b\, {}^*D_\Gamma(u)]
\end{aligned} \quad (29)$$

for $i \ge 3$. The next theorem characterizes the *rigid displacements*.

Theorem 1 *Given u and ℓ in $H^1(\Gamma)^N$, the following statements are equivalent:*

(i) $\varepsilon(u \circ p + b\,\ell \circ p) = 0 \quad \text{in } S_h$
(ii) $\varepsilon^0(u,\ell) = \varepsilon^1(u,\ell) = \frac{1}{2}[{}^*D_\Gamma(\ell)\, D^2 b + D^2 b\, D_\Gamma(\ell)] = 0 \quad \text{on } \Gamma$
(iii) $\varepsilon^0(u,\ell) = \varepsilon^1(u,\ell) = \varepsilon^2(u,\ell) = 0 \quad \text{on } \Gamma$
(iv) $\exists\, a \in \mathbb{R}^N$ *and an* $N \times N$ *matrix* B, $B + {}^*B = 0$, *such that*

$$\ell(X) = B\,n(X), \quad u(X) = a + BX, \quad \forall X \in \Gamma. \tag{30}$$

Therefore ℓ is tangential and $n \cdot (B\,n) = 0$ on Γ.

4.2 The second order model in the thickness variable

In order to preserve the *rigid displacements* the series is truncated as

$$\tilde{\varepsilon}(U) \circ T_z \stackrel{\text{def}}{=} \varepsilon^0(u,\ell) + \varepsilon^1(u,\ell)\, z + \varepsilon^2(u,\ell)\, z^2 \tag{31}$$

It is natural to associate with $\tilde{\varepsilon}$ the following Hilbert spaces

$$\begin{aligned} \mathcal{H} &= \{(u,\ell) \in L^2(\Gamma)^N \times L^2(\Gamma)^N : \ell \cdot n = 0 \text{ on } \Gamma\} \\ \mathcal{V} &= \left\{(u,\ell) \in \mathcal{H} : \varepsilon^i(u,\ell) \in L^2(\Gamma)^{N\times N},\ 0 \le i \le 2\right\} \\ \mathcal{N} &= \{(u,\ell) \in \mathcal{V} : \varepsilon^i(u,\ell) = 0 \text{ on } \Gamma,\ 0 \le i \le 2\} = \textstyle\bigcap_{i=0}^{i=2} \operatorname{Ker} \varepsilon^i \end{aligned} \tag{32}$$

with norms

$$|(e,\ell)|^2_{\mathcal{H}} = |e|^2_{L^2(\Gamma)} + |\ell|^2_{L^2(\Gamma)} \quad \text{and} \quad \|(e,\ell)\|^2_{\mathcal{V}} = |(e,\ell)|^2_{\mathcal{H}} + \sum_{i=0}^{2} \|\varepsilon^i(e,\ell)\|^2_{L^2(\Gamma)}. \tag{33}$$

From Theorem 1

$$\begin{aligned} \mathcal{N} = \{(u,\ell) \in \mathcal{V} :\ & u(X) = a + BX,\ \ell(X) = B\,n(X), \quad \forall a \in \mathbb{R}^N, \\ & \forall B \text{ an } N \times N \text{ matrix such that } B + {}^*B = 0\}. \end{aligned} \tag{34}$$

In order to complete the characterization of $\mathcal{V}$ we use the following *Korn's inequality.*

Theorem 2 *Assume that Γ is a bounded open domain in the C^2 $(N-1)$-dimensional submanifold $\partial\Omega$ of $\mathbb{R}^N$ with a Lipschitzian boundary $\partial_{\partial\Omega}\Gamma$ in $\partial\Omega$. As h goes to zero,* there exists a constant $c(h) > 0$ such that for all $(u,\ell) \in \mathcal{V}$

$$\begin{aligned} &\int_\Gamma 2\,h\,[|\ell|^2 + \|D_\Gamma(u)\|^2] + 2\,\frac{h^3}{3}\,\|D_\Gamma(\ell)\|^2\, d\Gamma \\ &\le c(h)^2 \int_\Gamma 2\,h\,|u|^2 + 2\,\frac{h^3}{3}\,|\ell|^2 + 2\,h\,\|\varepsilon^0(u,\ell)\|^2 + 2\,\frac{h^3}{3}\,\|\varepsilon^1(u,\ell)\|^2 + 2\,\frac{h^5}{5}\,\|\varepsilon^2(u,\ell)\|^2\, d\Gamma, \end{aligned} \tag{35}$$

where

$$\|A\|^2 = \sum_{i,j=1}^{N} A_{ij} A_{ji}, \quad |a|^2 = \sum_{i=1}^{N} a_i^2.$$

In particular

$$\mathcal{V} = \left\{(u,\ell) \in H^1(\Gamma)^N \times H^1(\Gamma)^N \,:\, \ell \cdot n = 0\right\}. \tag{36}$$

Remark 1 *(Spherical shells) For a spherical shell $\{x \in \mathbb{R}^N \,:\, |x| = R\}$ of radius R in $\mathbb{R}^N$*

$$b(x) = |x| - R, \quad \nabla b(x) = \frac{x}{|x|}, \quad p(x) = R\,\nabla b(x).$$

As a result

$$Dp(x) = R\,D^2b(x), \quad P(x) = |x|\,D^2b(x)$$

and since $P^2(x) = P(x)$

$$D^2b(x) = |x|\,(D^2b(x))^2.$$

Moreover

$$\begin{array}{ll} \frac{1}{2}[D_\Gamma v\, D^2b + D^2b^* D_\Gamma v] = \frac{1}{R}\varepsilon_\Gamma(v), & \frac{1}{2}[D_\Gamma v\,(D^2b)^2 + (D^2b)^{2\,*} D_\Gamma v] = \frac{1}{R^2}\varepsilon_\Gamma(v) \\ \varepsilon^0(u,\ell) = \varepsilon_\Gamma(u) + \frac{1}{2}[\ell^*\nabla b + \nabla b^*\ell], & \varepsilon^1(u,\ell) = \varepsilon_\Gamma(\ell) - \frac{1}{R}\varepsilon_\Gamma(u) \\ \varepsilon^2(u,\ell) = -\frac{1}{R}\varepsilon^1(u,\ell) & \\ \tilde{\varepsilon}(U)\circ T_z = [1 - \frac{z}{R} + \left(\frac{z}{R}\right)^2]\,\varepsilon_\Gamma(u) + \frac{1}{2}[\ell^*\nabla b + \nabla b^*\ell] + [1 - \frac{z}{R}]\,z\,\varepsilon_\Gamma(\ell). & \end{array} \tag{37}$$

By introducing the simple rheological law

$$\sigma = 2\,\mu\,\tilde{\varepsilon} + \lambda\,\mathrm{tr}\,\tilde{\varepsilon}, \quad \mu > 0,\ \lambda \geq 0,$$

we obtain the bilinear form associated with the strain energy in terms of polynomials $\alpha_n(h)$ in h and bilinear forms a_n. The polynomials $\alpha_n(h)$ of odd powers of h are functions of X on Γ defined as

$$\alpha_n(h) \stackrel{\text{def}}{=} h^{n+1} \sum_{i=0}^{N-1} [1 - (-1)^{n+i+1}] \frac{h^i}{n+i+1} K_i, \quad 0 \leq n \leq 4. \tag{38}$$

For $N = 3$

$$\begin{array}{l} \alpha_0 = 2\,h + 2\,\frac{h^3}{3}\,K \quad \alpha_1 = 2\,\frac{h^3}{3}\,H \quad \alpha_2 = 2\,\frac{h^3}{3} + 2\,\frac{h^5}{5}\,K \quad \alpha_3 = 2\,\frac{h^5}{5}\,H \\ \alpha_4 = 2\,\frac{h^5}{5} + 2\,\frac{h^7}{7}\,K. \end{array} \tag{39}$$

If as in Naghdi's model the assumption of zero normal constraint, $\sigma \circ T_z\, n \cdot n = 0$, is used then since $\tilde{\varepsilon}\, n \cdot n = 0$

$$\mathrm{tr}\,\tilde{\varepsilon} = 0 \text{ and } \mathrm{tr}\,\varepsilon^i = 0, \quad 0 \leq i \leq 2.$$

This is equivalent to

$$\operatorname{div}_\Gamma(u) = 0, \quad \operatorname{div}_\Gamma(\ell) - \operatorname{tr}(D_\Gamma(u)D^2 b) = 0 \quad \operatorname{tr}(D_\Gamma(\ell)D^2 b) - \operatorname{tr}(D_\Gamma(u)(D^2 b))^2 = 0.$$

The first condition is some kind of *inextensibility* of the mean surface. The spaces $\mathcal{H}$, $\mathcal{V}$, and $\mathcal{N}$, and their associated norms and seminorm have been defined in (33). Now define the bilinear operator $A : \mathcal{V} \to \mathcal{V}'$ and the linear operator $\mathcal{B} : L^2(\Gamma)^N \times L^2(\Gamma)^N \to \mathcal{H}'$: for all (u, ℓ) and $(\overline{u}, \overline{\ell})$ in $\mathcal{V}$

$$\langle A(u,\ell), (\overline{u}, \overline{\ell})\rangle_{\mathcal{V}} = \sum_{n=0}^{4} \int_\Gamma \alpha_n(h)\, a_n((u,\ell),(\overline{u},\overline{\ell}))\, d\Gamma, \tag{40}$$

where for all $\varepsilon^i = \varepsilon^i(u,\ell)$ and $\overline{\varepsilon}^i = \varepsilon^i(\overline{u},\overline{\ell})$

$$\begin{aligned}
a_0((u,\ell),(\overline{u},\overline{\ell})) &= 2\mu\, \varepsilon^0 \cdot\cdot\, \overline{\varepsilon}^0 + \lambda \operatorname{tr}\varepsilon^0 \operatorname{tr}\overline{\varepsilon}^0 \\
a_1((u,\ell),(\overline{u},\overline{\ell})) &= 2\mu\, [\varepsilon^0 \cdot\cdot\, \overline{\varepsilon}^1 + \overline{\varepsilon}^0 \cdot\cdot\, \varepsilon^1] + \lambda\, [\operatorname{tr}\varepsilon^0 \operatorname{tr}\overline{\varepsilon}^1 + \operatorname{tr}\overline{\varepsilon}^0 \operatorname{tr}\varepsilon^1] \\
a_2((u,\ell),(\overline{e},\overline{\ell})) &= 2\mu\, [\varepsilon^1 \cdot\cdot\, \overline{\varepsilon}^1 + \varepsilon^0 \cdot\cdot\, \overline{\varepsilon}^2 + \overline{\varepsilon}^0 \cdot\cdot\, \varepsilon^2] + \lambda\, [\operatorname{tr}\varepsilon^1 \operatorname{tr}\overline{\varepsilon}^1 + \operatorname{tr}\varepsilon^0 \operatorname{tr}\overline{\varepsilon}^2 + \operatorname{tr}\overline{\varepsilon}^0 \operatorname{tr}\varepsilon^2] \\
a_3((u,\ell),(\overline{u},\overline{\ell})) &= 2\mu\, [\varepsilon^1 \cdot\cdot\, \overline{\varepsilon}^2 + \overline{\varepsilon}^1 \cdot\cdot\, \varepsilon^2] + \lambda\, [\operatorname{tr}\varepsilon^1 \operatorname{tr}\overline{\varepsilon}^2 + \operatorname{tr}\overline{\varepsilon}^1 \operatorname{tr}\varepsilon^2] \\
a_4((u,\ell),(\overline{u},\overline{\ell})) &= 2\mu\, \varepsilon^2 \cdot\cdot\, \overline{\varepsilon}^2 + \lambda \operatorname{tr}\varepsilon^2 \operatorname{tr}\overline{\varepsilon}^2
\end{aligned}$$

and

$$\langle \mathcal{B}(f,m), (u,\ell)\rangle_{\mathcal{H}} = \int_\Gamma \alpha_0(h)\, [f \cdot u + m \cdot \ell] + \alpha_1(h)\, f \cdot \ell\, d\Gamma. \tag{41}$$

By construction A is symmetrical and positive and

Lemma 1 *There exists $\overline{h} > 0$ and $\alpha > 0$ such that for all $0 < h < \overline{h}$*

$$\forall (u,\ell) \in \mathcal{V}, \quad \langle A(u,\ell), (u,\ell)\rangle_{\mathcal{V}} \geq 2\mu\, h\, \alpha \sum_{n=0}^{2} h^{2n}\, \|\varepsilon^n(u,\ell)\|^2. \tag{42}$$

If the elements of the dual $\mathcal{H}'$ of $\mathcal{H}$ are identified with those of $\mathcal{H}$, then the lemma says that A is a $\mathcal{V}$-$\mathcal{H}$ coercive operator.

Theorem 3 *Given $\overline{h} > 0$ as specified in Lemma 1 and assuming that the following condition is verified*

$$\forall (u,\ell) \in \mathcal{N}, \quad \int_\Gamma \alpha_0(h)\, [f \cdot u + m \cdot \ell] + \alpha_1(h)\, f \cdot \ell\, d\Gamma = 0, \tag{43}$$

then for all h, $0 < h \leq \overline{h}$, there exists a unique solution $(\hat{u}, \hat{\ell}) \in \mathcal{V}/\mathcal{N}$ to the variational equation:

$$\forall (u,\ell) \in \mathcal{V}, \quad \langle A(\hat{u},\hat{\ell}), (u,\ell)\rangle_{\mathcal{V}} + \langle \mathcal{B}(f,m), (u,\ell)\rangle_{\mathcal{H}} = 0. \tag{44}$$

For a shell with boundary and homogeneous Dirichlet boundary conditions the results are analogous to the ones of Theorem 3 without condition (43) in the space

$$\mathcal{V}_0 \stackrel{\text{def}}{=} \left\{(u,\ell) \in H_0^1(\Gamma)^N \times H_0^1(\Gamma)^N \,:\, \ell \cdot n = 0\right\}. \tag{45}$$

Theorem 4 *Given $\overline{h} > 0$ as specified in Lemma 1 and h, $0 < h \leq \overline{h}$,* there exists a unique solution $(\hat{e}, \hat{\ell}) \in \mathcal{V}_0$ to the variational equation:

$$\forall (u,\ell) \in \mathcal{V}_0, \quad \langle A(\hat{u},\hat{\ell}),(u,\ell)\rangle_{\mathcal{V}} + \langle \mathcal{B}(f,m),(u,\ell)\rangle_{\mathcal{H}} = 0. \tag{46}$$

5 INTRINSIC TANGENTIAL AND COVARIANT DERIVATIVES

Associate with the space $\mathbb{R}^N$ an orthonormal basis $\{e_1, \ldots, e_N\}$ at the origin. Let Ω be a subset of $\mathbb{R}^N$ with a boundary $\partial\Omega$ of class C^2 and let Γ be a bounded subset of $\partial\Omega$ with relative boundary $\partial\Gamma \stackrel{\text{def}}{=} \partial_{\partial\Omega}\Gamma$ in the $(N-1)$-submanifold $\partial\Omega$.

5.1 Local coordinates

Assume the existence of a C^2-map

$$\xi' \stackrel{\text{def}}{=} (\xi^1, \ldots, \xi^{N-1}) \mapsto \Phi(\xi') \,:\, \bar{A} \subset \mathbb{R}^{N-1} \to \overline{\Gamma} \subset \mathbb{R}^N, \tag{47}$$

where A is a bounded open connected domain in $\mathbb{R}^{N-1}$ with Lipschitzian boundary ∂A (located on the same side of ∂A). Further assume that in each point of $\bar{A}$ the vectors

$$a_\alpha = \frac{\partial \Phi}{\partial \xi^\alpha}, \qquad 1 \leq \alpha \leq N-1, \tag{48}$$

are linearly independent. In addition assume that Γ is oriented and select a unit normal a_N to Γ. For instance for $N = 3$ the choice is usually

$$a_3 = \frac{a_1 \times a_2}{|a_1 \times a_2|}. \tag{49}$$

But if we want the unit sphere to have a positive curvature, it is necessary to choose for a_N the inward unit normal to the unit ball. We shall follow the usual convention that a Greek index ranges from 1 to $N-1$ and that a Roman index ranges from 1 to N. The contravariant basis is defined as

$$a^i \cdot a_j = \delta_{ij}, \quad \Rightarrow a_N = a^N, \tag{50}$$

where δ_{ij} is the Kronecker index function. *Finally we choose a domain Ω such that*

$$a_N = a^N = -\nabla b_\Omega \circ \Phi. \tag{51}$$

5.2 Partial derivatives and fundamental forms

Consider a C^1-function $w : \Gamma \to \mathbb{R}$ and its extension $W = w \circ p$ in a neighbourhood of Γ. By definition the *partial derivative* of w is given by

$$w_{,\alpha} \stackrel{\text{def}}{=} \frac{\partial}{\partial \xi^\alpha}(w \circ \Phi) \quad \text{and} \quad w_{,N} \stackrel{\text{def}}{=} 0, \tag{52}$$

since $w \circ \Phi$ is independent of the normal displacement to the submanifold $\partial\Omega$. Note that

$$w \circ \Phi = (w \circ p) \circ \Phi = W \circ \Phi$$

and consequently

$$\frac{\partial}{\partial \xi^\alpha}(w \circ \Phi) = (\nabla W \circ \Phi) \cdot \frac{\partial \Phi}{\partial \xi^\alpha} = [\nabla(w \circ p) \circ \Phi] \cdot a_\alpha.$$

But $\nabla(w \circ p)|_\Gamma = \nabla_\Gamma w$ and

$$w_{,\alpha} = \frac{\partial}{\partial \xi^\alpha}(w \circ \Phi) = [\nabla_\Gamma w \circ \Phi] \cdot a_\alpha.$$

In the sequel it will be convenient to use the notation a_i for both a_i and $a_i \circ \Phi^{-1}$ and $\nabla_\Gamma w$ for both $\nabla_\Gamma w$ and $\nabla_\Gamma w \circ \Phi$ whenever no confusion arises. Hence

$$w_{,\alpha} = \nabla_\Gamma w \cdot a_\alpha \quad \text{and} \quad w_{,N} = 0 = \nabla_\Gamma w \cdot a_N. \tag{53}$$

This extends to C^1 vector functions $v : \Gamma \to \mathbb{R}^N$ the extension $V = v \circ p$ to a neighbourhood of Γ

$$v_{,\alpha} \stackrel{\text{def}}{=} \frac{\partial}{\partial \xi^\alpha}(v \circ \Phi), \quad v_{,N} \stackrel{\text{def}}{=} 0 \quad \Rightarrow \quad v_{,i} = [D_\Gamma v \circ \Phi]\, a_i. \tag{54}$$

As an application of the above identity, we get the *second and third fundamental forms*

$$b_{\alpha\beta} \stackrel{\text{def}}{=} -a_\beta \cdot a_{N,\alpha} = a_\beta \cdot (D^2 b\, a_\alpha) \quad \text{and} \quad c_{\alpha\beta} \stackrel{\text{def}}{=} b_\alpha^\lambda\, b_{\lambda\beta} = a_\beta \cdot ((D^2 b)^2\, a_\alpha). \tag{55}$$

5.3 Christoffel symbols and covariant derivatives

All the results below readily extend to tensors of higher order. By definition

$$\Gamma^\alpha_{\beta\gamma} \stackrel{\text{def}}{=} a^\alpha \cdot a_{\beta,\gamma} = a^\alpha \cdot D_\Gamma(a_\beta \circ \Phi^{-1}) \circ \Phi\, a_\gamma \tag{56}$$

and for simplicity we use the notation $D_\Gamma a_j$ for $D_\Gamma(a_j \circ \Phi^{-1}) \circ \Phi$. For a vector $v : \Gamma \to \mathbb{R}^N$

$$\begin{aligned} v_\alpha|_\gamma &\stackrel{\text{def}}{=} v_{\alpha,\gamma} - \Gamma^\lambda_{\alpha\gamma} v_\lambda = D_\Gamma v\, a_\gamma \cdot a_\alpha + D^2 b\, a_\alpha \cdot a_\gamma v_N = [{}^*D_\Gamma v + D^2 b\, v_N] a_\alpha \cdot a_\gamma \\ v^\alpha|_\gamma &\stackrel{\text{def}}{=} v^\alpha_{,\gamma} + \Gamma^\alpha_{\lambda\gamma} v^\lambda = D_\Gamma v\, a_\gamma \cdot a^\alpha + D^2 b\, a^\alpha \cdot a_\gamma v_N = [{}^*D_\Gamma v + D^2 b\, v_N] a^\alpha \cdot a_\gamma \\ \Rightarrow &\quad {}^*D_\Gamma v\, a_\alpha \cdot a_\gamma = v_\alpha|_\gamma - b_{\alpha\gamma} v_N \quad \text{and} \quad {}^*D_\Gamma v\, a^\alpha \cdot a_\gamma = v^\alpha|_\gamma - b^\alpha_\gamma v_N \end{aligned} \tag{57}$$

For tangential vector fields ($v \cdot n = 0$), the covariant derivatives coincide with the bilinear form generated by ${}^*D_\Gamma v$. For non tangent field we have an additional term which arises from the fact that in the definition of $v_\alpha|_\gamma$ and $v^\alpha|_\gamma$ the summation over λ ranges from 1 to $N-1$ missing the normal component v_N.

5.4 A few useful formulas

In the theory of shells some identities will often occur. We summarize some of them below.

Theorem 5 *For all u and v in $H^1(\Gamma)^N$*

$$\begin{aligned}
\varepsilon_\Gamma(v)\, a_\alpha \cdot a_\beta &= \tfrac{1}{2}\,(v_\alpha|_\beta + v_\beta|_\alpha) - b_{\alpha\beta}\, v_N,\\
\varepsilon_\Gamma(v)\, a_\alpha \cdot a^\beta &= \tfrac{1}{2}\,(v^\alpha|_\beta + v^\beta|_\alpha) - b^\beta_\alpha\, v_N,\\
(D^2 b\, D_\Gamma v\, a_\alpha) \cdot a_\beta &= b^\gamma_\beta\,[v_\gamma|_\alpha - b_{\alpha\gamma}\, v_N] = b^\gamma_\beta\, v_\gamma|_\alpha - c_{\alpha\beta}\, v_N,\\
(D_\Gamma v\, D^2 b\, a_\alpha) \cdot a_\beta &= b^\gamma_\alpha\,[v_\beta|_\gamma - b_{\beta\gamma}\, v_N] = b^\gamma_\alpha\, v_\beta|_\gamma - c_{\alpha\beta}\, v_N,
\end{aligned} \tag{58}$$

where $c^\beta_\alpha = c_{\alpha\gamma}\, a^\gamma_\beta$, $\quad c^{\alpha\beta} = c_{\alpha\gamma}\, a^{\gamma\beta}$. *Moreover*

$$\begin{aligned}
a_\beta \cdot \varepsilon_\Gamma(u)\, a^\beta &= \operatorname{tr} \varepsilon_\Gamma(u) = \operatorname{div}_\Gamma u,\\
a^\alpha \cdot \varepsilon_\Gamma(u) a_\beta\, \varepsilon_\Gamma(v)\, a^\beta \cdot a_\alpha &= \varepsilon^0(u, -2\,\varepsilon_\Gamma(u)\, n) \cdot\cdot\, \varepsilon^0(v, -2\,\varepsilon_\Gamma(v)\, n)\\
&= \varepsilon_\Gamma(u) \cdot\cdot\, \varepsilon_\Gamma(v) - 2\,\varepsilon_\Gamma(u)\, n \cdot \varepsilon_\Gamma(v)\, n
\end{aligned} \tag{59}$$

where $\varepsilon^0(u,\ell)$ is given by (29)

6 SOME CLASSICAL LINEAR MODELS

In this section we use the material from §5 to rewrite the linear models of Naghdi and Koiter and the asymptotic model with tangential operators.

6.1 Naghdi's and Koiter's linear models

We use the variational forms and associated definitions from the recent book of Bernadou (1994) (Chapter I, §3). The displacement vector is

$$U = u + \xi^3 \beta_\alpha a^\alpha \tag{60}$$

where $-e < \xi^3 < e$, $2e$ is the thickness of the shell, u and β are maps from Γ to $\mathbb{R}^3$ and

$$\beta = \beta_\alpha\, a^\alpha, \quad \beta \cdot n = 0, \quad \beta_\alpha = \beta \cdot a_\alpha. \tag{61}$$

We use the same notation for a vector $v : \Gamma \to \mathbb{R}^3$ and its 2-dimensional representation $v \circ \Phi$ in the $1-2$ coordinates. We quote the main definitions and give their tangential equivalent

$$
\begin{aligned}
&\varphi_\alpha(u) \stackrel{\text{def}}{=} u_{3,\alpha} + b_\alpha^\lambda u_\lambda = a_\alpha \cdot D^2 b\, u = -D_\Gamma u\, a_\alpha \cdot n,\\
&\gamma_{\alpha\beta}(u) \stackrel{\text{def}}{=} \tfrac{1}{2}\,(u_\alpha|_\beta + u_\beta|_\alpha) - b_{\alpha\beta}\, u_3 \gamma_{\alpha\beta}(u) = \varepsilon_\Gamma(u)\, a_\alpha \cdot a_\beta,\\
&d_{\lambda\alpha}(u) \stackrel{\text{def}}{=} u_\lambda|_\alpha - b_{\lambda\alpha}\, u_3 = D_\Gamma u\, a_\alpha \cdot a_\lambda,\\
&\chi_{\alpha\mu}(u,\beta) \stackrel{\text{def}}{=} \tfrac{1}{2}\,[\beta_\alpha|_\mu + \beta_\mu|_\alpha - b_\alpha^\lambda\, d_{\lambda\mu}(u) - b_\mu^\lambda\, d_{\lambda\alpha}(u)]\\
&\qquad = a_\alpha \cdot \left[\varepsilon_\Gamma(\beta) - \tfrac{1}{2}\,(D^2 b\, D_\Gamma u + {}^*D_\Gamma u\, D^2 b)\right] a_\mu,\\
&\varepsilon_{\alpha\beta}(u,\beta) \stackrel{\text{def}}{=} \gamma_{\alpha\beta}(u) + \xi^3\, \chi_{\alpha\beta}(u,\beta)\\
&\qquad = a_\alpha \cdot \left\{\varepsilon_\Gamma(u) + \xi^3 \left[\varepsilon_\Gamma(\beta) - \tfrac{1}{2}\,(D^2 b\, D_\Gamma u + {}^*D_\Gamma u\, D^2 b)\right]\right\} a_\beta.
\end{aligned}
\tag{62}
$$

The variational problem for Naghdi's model consists in finding (u, β) such that

$$
\forall\,(v,\delta), \quad a^S((u,\beta),(v,\delta)) + b^S((u,\beta),(v,\delta)) = f^S(v,\delta), \tag{63}
$$

where the bilinear forms a^S and b^S and the linear form f^S are given by

$$
\begin{aligned}
&a^S((u,\beta),(v,\delta)) \stackrel{\text{def}}{=} \int_\Gamma e E^{\alpha\beta\lambda\mu} \left\{\gamma_{\alpha\beta}(u)\, \gamma_{\lambda\mu}(v) + \frac{e^2}{12} \chi_{\alpha\beta}(u)\, \chi_{\lambda\mu}(v)\right\} d\Gamma\\
&b^S((u,\beta),(v,\delta)) \stackrel{\text{def}}{=} \int_\Gamma \frac{e E a^{\alpha\beta}}{2(1+\nu)} (\varphi_\alpha(u) + \beta_\alpha)(\varphi_\beta(v) + \delta_\beta)\, d\Gamma\\
&f^S(v,\delta) \stackrel{\text{def}}{=} \int_\Gamma p \cdot v\, d\Gamma + \int_{\partial\Gamma} N \cdot v - M \cdot \delta\, d\gamma
\end{aligned}
\tag{64}
$$

and

$$
E^{\alpha\beta\lambda\mu} = \frac{E}{2\,(1+\nu)} \left[a^{\alpha\lambda}\, a^{\beta\mu} + a^{\alpha\mu}\, a^{\beta\lambda} + \frac{2\,\nu}{1-\nu}\, a^{\alpha\beta}\, a^{\lambda\mu}\right]. \tag{65}
$$

By using the following identifications

$$
h = e, \quad \ell = -\beta, \quad u = u, \quad \bar\ell = -\delta, \quad v = v, \tag{66}
$$

we get an intrinsic reformulation of the variational problem

$$
\begin{aligned}
&\frac{h\,E}{1+\nu} \int_\Gamma \varepsilon^0(u, \ell - \nabla_\Gamma(u\cdot n)) \cdot\cdot\, \varepsilon^0(v, \bar\ell - \nabla_\Gamma(v \cdot n)) + \frac{h^2}{12}\, \varepsilon_t^1(u,\ell) \cdot\cdot\, \varepsilon_t^1(v,\bar\ell)\\
&\qquad + \frac{1}{1-\nu} \left\{\mathrm{div}_\Gamma\, u\, \mathrm{div}_\Gamma\, v + \frac{h^2}{12}\, \mathrm{div}_\Gamma\, \ell\, \mathrm{div}_\Gamma\, \bar\ell\right\} d\Gamma\\
&= \int_\Gamma p \cdot v\, d\Gamma + \int_{\partial\Gamma} N \cdot v - M \cdot \bar\ell\, d\gamma,
\end{aligned}
\tag{67}
$$

where

$$
\begin{aligned}
&\varepsilon^0(u,\ell) \stackrel{\text{def}}{=} \varepsilon_\Gamma(u) + \frac{1}{2}\,[\ell\, {}^*n + n\, {}^*\ell],\\
&\varepsilon_t^1(u,\ell) \stackrel{\text{def}}{=} \varepsilon_\Gamma(\ell) + \frac{1}{2}\,[D^2 b\, D_\Gamma u + {}^*D_\Gamma u\, D^2 b] - \frac{1}{2}\,[D^2 b\, \ell\, {}^*n + n\, {}^*(D^2 b\, \ell)].
\end{aligned}
$$

This model is different from our intrinsic model since it uses the assumption $\sigma \circ T_z\, n \cdot n = 0$. Koiter's linear model is the same model as Naghdi's model with

$$\ell + 2\,\varepsilon_\Gamma(u)\,n = 0 \quad \Rightarrow \quad 0 = \ell + \nabla_\Gamma(u \cdot n) - D^2 b\, u. \tag{68}$$

6.2 Asymptotic membrane equation model

We use the definitions and notation from Ciarlet and Sanchez-Palencia (1993). The model is characterized by the following bilinear and linear forms

$$\begin{aligned} B(\zeta,\eta) &\stackrel{\text{def}}{=} \int_\Gamma A^{\alpha\beta\rho\delta}\,\gamma_{\rho\sigma}(\zeta)\,\gamma_{\alpha\beta}(\eta)\,d\Gamma \\ A^{\alpha\beta\rho\delta} &\stackrel{\text{def}}{=} \frac{4\lambda\mu}{\lambda + 2\,\mu}\,a^{\alpha\beta}\,a^{\rho\delta} + 2\,\mu\,[a^{\alpha\rho}\,a^{\beta\sigma} + a^{\alpha\sigma}\,a^{\beta\rho}] \\ \gamma_{\alpha\beta}(\eta) &\stackrel{\text{def}}{=} \frac{1}{2}\,(\eta_{\alpha,\beta} + \eta_{\beta,\alpha}) - \Gamma^\rho_{\alpha\beta}\,\eta_\beta - b_{\alpha\beta}\,\eta_3. \end{aligned} \tag{69}$$

Clearly $\gamma_{\alpha\beta}(\eta)$ is the same as the one in Naghdi's model and

$$\begin{aligned} B(\zeta,\eta) &= \int_\Gamma 4\,\mu\,\varepsilon^0(\zeta, -2\,\varepsilon_\Gamma(\zeta)\,n) \cdot\cdot\, \varepsilon^0(\eta, -2\,\varepsilon_\Gamma(\eta)\,n) \\ &\quad + \frac{4\lambda\mu}{\lambda + 2\,\mu}\,\mathrm{tr}\,\varepsilon^0(\zeta, -2\,\varepsilon_\Gamma(\zeta)\,n) + \mathrm{tr}\,\varepsilon^0(\eta, -2\,\varepsilon_\Gamma(\eta)\,n)\,d\Gamma \\ B(\zeta,\eta) &= \int_\Gamma 4\,\mu\,\varepsilon^0(\zeta, -2\,\varepsilon_\Gamma(\zeta)\,n) \cdot\cdot\, \varepsilon^0(\eta, -2\,\varepsilon_\Gamma(\eta)\,n) \\ &\quad + \frac{4\lambda\mu}{\lambda + 2\,\mu}\,\mathrm{div}_\Gamma\zeta\,\mathrm{div}_\Gamma\eta\,d\Gamma. \end{aligned} \tag{70}$$

By making the identification $u = \eta, \quad v = \zeta$, we finally get

$$\begin{aligned} &\int_\Gamma 4\,\mu\,\varepsilon^0(u, -2\,\varepsilon_\Gamma(u)\,n) \cdot\cdot\, \varepsilon^0(v, -2\,\varepsilon_\Gamma(v)\,n) \\ &\quad + \frac{4\,\mu\,\lambda}{\lambda + 2\,\mu}\,\mathrm{tr}\,\varepsilon^0(u, -2\,\varepsilon_\Gamma(u)\,n)\,\mathrm{tr}\,\varepsilon^0(v, -2\,\varepsilon_\Gamma(v)\,n)\,d\Gamma = \int_\Gamma f \cdot v\,d\Gamma. \end{aligned} \tag{71}$$

By using the tangential operator $\varepsilon^P_\Gamma(u) = \varepsilon^0(u, -2\,\varepsilon_\Gamma(u)\,n)$, we obtain

$$\int_\Gamma 4\mu\,\varepsilon^P_\Gamma(u) \cdot\cdot\, \varepsilon^P_\Gamma(v) + \frac{4\,\mu\ \lambda}{\lambda + 2\,\mu}\,\mathrm{tr}\,\varepsilon^P_\Gamma(u)\,\mathrm{tr}\,\varepsilon^P_\Gamma(v)\,d\Gamma = \int_\Gamma f \cdot v\,d\Gamma. \tag{72}$$

REFERENCES

Bernadou, M. (1994) *Méthodes d'éléments finis pour les problèmes de coques minces*, Masson, Paris, Milan, Barcelone.

Bernadou, M., Ciarlet, Ph.G., Miara, B. (1994) Existence theorems for two-dimensional linear shell theories. *J. of Elasticity*, 111–138.

Bielski, W. and Telega, J.J. (1988) On existence of solutions for geometrically nonlinear shells and plates. *ZAMM Z. angew. Math. Mech.* **68**, 4, T155-157.

Ciarlet, Ph. G., Sanchez-Palencia, E. (1993) Un théorème d'existence et d'unicité pour les équations des coques membranaires. *C.R. Acad. Sci. Paris, Sér. I*, **317**, 801–805.

Delfour, M.C. and Zolésio, J.P. (1994-1) Shape analysis via oriented distance functions. *J. Functional Analysis*, **123**, 129–201.

Delfour, M.C. and Zolésio, J.P. (1994-2) Functional analytic methods in shape analysis. *In "Boundary Control and Variation"*, (J.-P. Zolésio, ed.), Marcel Dekker, New-York, pp. 105–139.

Delfour, M.C. and Zolésio, J.P. (1994-3) Oriented distance function in shape analysis and optimization. *In "Control and Optimal Design of Distributed Parameter Systems"*, (J. Lagnese, D.L. Russell, L. White, eds.), Springer-Verlag, Berlin, Heidelberg, New York, Tokyo, pp. 39–72.

Delfour, M.C. and Zolésio, J.P. (1994-4) On a variational equation for thin shells. *In "Control and Optimal Design of Distributed Parameter Systems"*, (J. Lagnese, D.L. Russell, L. White, eds.), Springer-Verlag, Berlin, Heidelberg, New York, Tokyo, 1994, pp. 25–37.

Delfour, M.C. and Zolésio, J.P. (1995-1) A boundary differential equation for thin shells. *J. Differential Equations*, in press.

Delfour, M.C. and Zolésio, J.P. (1995-2) Tangential differential equations for dynamical thin/shallow shells. *J. Differential Equations*, to appear.

Delfour, M.C. and Zolésio, J.P. (1995-3) Differential equations for linear shells: comparison between intrinsic and classical models. *CRM-Report, Université de Montréal*, June.

Delfour, M.C. and Zolésio, J.P. (1995-4) Shape analysis via distance functions. II. *CRM-Report, Université de Montréal*, March.

Delfour, M.C. and Zolésio, J.P. (1995-5) On the design and control of systems governed by differential equations on submanifolds. *Control and Cybernetics*, to be published.

Destuynder, Ph. (1990) *Modélisation des coques minces élastiques.* Masson, Paris, Milan, Barcelone.

Destuynder, Ph. (1986) *Une théorie asymptotique des plaques minces en elasticité linéaire.* Masson, Paris, Milan, Barcelone.

Gałka, A. and Telega, J.J. (1992) The complementary energy principle as a dual problem for a specific model of geometrically non-linear elastic shells with an independent rotation vector: general results. *Eur. J. Mech., A/Solids*, **11**, n 2, 245-270.

Germain, P. (1986) *Mécanique, tomes* I *and* II. Ellipses éditeur, Paris.

Gilbarg, D. and Trudinger, N.S. (1983) *Elliptic partial differential equations of second order.* Springer-Verlag, Berlin, Heidelberg, New York, Tokyo.

Lagnese, J.E. and Lions, J.-L. (1988) Modelling, analysis and control of thin plates. *RMA*, Vol. 6, Masson, Paris, Milan, Barcelone, Mexico.

Lions, J.-L. (1968) *Contrôle optimal de systèmes gouvernés par des équations aux dérivées partielles.* Dunod, Gauthier-Villars, Paris.

Sokolowski, J. and Zolésio, J.-P. (1992) *Introduction to shape optimization: Shape sensitivity analysis.* Springer-Verlag, New York, Berlin, Heidelberg.

Valid, R. (1981) *Mechanics of continuous media and analysis of structures.* North-Holland Series in Applied Mathematics and Mechanics, Amsterdam, New York, Oxford.

Zolésio, J.-P. (1981-1) The Material Derivative (or Speed) Method for Shape Optimization. *Optimization of Distributed Parameter Structures, vol. II*, (E.J. Haug and J. Céa, eds.), Sijthoff and Noordhoff, Alphen aan den Rijn, The Netherlands, pp. 1089–1151.

Zolésio, J.-P. (1981-2) Domain variational formulation for free boundary problems. *Optimization of Distributed Parameter Structures, vol. II* (E.J. Haug and J. Céa, eds.) Sijthoff and Noordhoff, Alphen aan den Rijn, The Netherlands, pp. 1152–1194

3

Additive Schwarz Methods for Elliptic Mortar Finite Element Problems

Maksymilian Dryja
Warsaw University, Department of Mathematics,
Banacha 2, 02-097 Warsaw, Poland. E–mail: dryja@mimuw.edu.pl

Abstract
Domain decomposition methods are designed and analyzed as additive Schwarz methods for the linear systems arising from the discretization of elliptic problems. The discretization is obtained by a mortar element method with a finite element approximation on a nonmatching triangulation. The additive Schwarz methods use inexact solvers and they can also be applied to elliptic problems with discontinuous coefficients.

Keywords
Domain decomposition, elliptic mortar finite element problems, preconditioned conjugate gradients, additive Schwarz methods, iterative substructuring.

1 INTRODUCTION

In this paper, we discuss a parallel algorithm for solving systems arising from the discretization of elliptic problems by the mortar element method; see Bernardi, Maday, and Patera (1994), Bernardi and Maday (1995). Our algorithm is a domain decomposition method and it is described as an additive Schwarz method (ASM); see Dryja and Widlund (1987) and Dryja, Smith, and Widlund (1994).

The mortar element method is applied to elliptic self-adjoint second order equations with Dirichlet boundary conditions in 2-D. To simplify the presentation only certain model problems are discussed. The mortar element method is considered in the geometrically conforming case as well as the nonconforming case. The first case corresponds to a partition of the region into triangular or rectangular subregions that form a coarse finite element triangulation, while in the second case a vertex of a subregion can fall in the interior of an edge of one of its neighbors. In the first case a model problem with discontinuous coefficients is discussed while in the second only problems with regular coefficients are considered. A finite element method based on piecewise linear continuous functions is introduced for each subregion without insisting on meshes matching across subregion interfaces.

The resulting linear systems are solved by an iterative substructuring method described as an additive Schwarz method using the general framework given in Dryja, Smith, and Widlund (1994). First, the unknowns corresponding to interior nodal points in each substructure are eliminated. The reduced problem is defined on the interfaces of the substructures and it is solved by an ASM with a special coarse space and inexact solvers. It is shown that the method is almost optimal from the point of view of parallel computations.

There are several papers devoted to solving systems arising from the mortar element method by domain decomposition methods; see Achdou, Maday and Widlund (1995), Widlund (1995) and the literature therein. Some of the results of this paper, which have been obtained independently, are close those of Achdou, Maday, and Widlund (1995). The main results of our paper is the development of simple inexact solvers for both the geometrically conforming and nonconforming case, and the analysis of the method for problems with discontinuous coefficients in the geometrically conforming case. These results are, to our knowledge, new.

The outline of the paper is as follows. In Section 2, the discrete problem is formulated and its properties are discussed including the condition number of stiffness matrix arising from the discrete problem. This is done for the geometrically conforming case. In Section 3, an ASM for the discrete problem is designed. The analysis of it is carried out in Section 5. To do so, several technical tools are needed. They are formulated and proved in Section 4.

The geometrically nonconforming case is discussed in Sections 6 and 7 where a description and analysis of the method can be found. For the analysis, the stability of a special projection is proved for functions which do not vanish at the vertices of the substructures. This very important and new result is crucial for the analysis.

Some of the results of this paper have been obtained in joint work with Olof Widlund.

2 DIFFERENTIAL AND DISCRETE PROBLEMS

To simplify our presentation, we consider only a model problem in 2-D, a special second-order problem with discontinuous, piecewise constant coefficients:

Find $u \in H_0^1(\Omega)$ such that

$$a(u,v) = f(v), \quad v \in H_0^1(\Omega), \tag{1}$$

where

$$a(u,v) = \int_\Omega \rho(x)\nabla u \cdot \nabla v dx, \quad f(v) = \int_\Omega f\, v dx.$$

The region Ω is polygonal and a union of nonoverlapping subregions Ω_i $i = 1,\dots,N$, also called substructures.

We first consider the case when Ω is a union of triangles or rectangles $\Omega_i, i = 1,\dots,N$, which form a coarse triangulation. We assume that jumps of $\rho(x) > 0$ occur only at substructure interfaces and, for simplicity, that the coefficient takes on a constant value $\rho(x) = \rho_i$, $x \in \Omega_i$, in each substructure. We can generalize our results to the case when the relative variation of $\rho(x)$ over each subregion is modest.

To define a discrete problem by the mortar element method, we introduce two partitions of Ω, a coarse and a fine. As mentioned above, the coarse partition is formed by the substructures Ω_i. To simplify our presentation, we assume that it is quasi-uniform with a parameter H, see Ciarlet (1978). To obtain the fine partition of Ω, we introduce a triangulation of each Ω_i with triangular elements $e_i^{(k)}$. The resulting triangulation is non-matching; the triangulation of the different Ω_i generally do not match on the interfaces between subregions. We assume that the triangulation of each Ω_i is quasi-uniform with a parameter h_i. Let $X_i^{h_i}(\Omega_i)$ be the finite element space of piecewise linear continuous functions defined on the triangulation of Ω_i vanishing on $\partial\Omega_i \cap \partial\Omega$. Let

$$X^h(\Omega) = X_1^{h_1}(\Omega_1) \times \ldots \times X_N^{h_N}(\Omega_N)$$

be a Hilbert space with the inner product

$$a(u_h, v_h) = \sum_{i=1}^{N} \rho_i \int_{\Omega_i} \nabla u_{i,h} \nabla v_{i,h} dx,$$

where $u_h = \{u_{i,h}\}_{i=1}^N$, $u_{i,h} \in X_i^{h_i}(\Omega_i)$.

To describe the discrete problem fully, we introduce some auxiliary notations and finite element spaces. Let $\bar{\Gamma}_{ij} = \bar{\Omega}_i \cap \bar{\Omega}_j$ and let $W_{ij}^{h_i}(\Gamma_{ij})$ and $W_{ji}^{h_j}(\Gamma_{ij})$ be the restrictions of $X_i^{h_i}(\Omega_i)$ and $X_j^{h_j}(\Omega_j)$ to Γ_{ij}. Note that the functions of these spaces are discontinuous across Γ_{ij}, a side common to Ω_i and Ω_j. To define weak continuity across the interface, we introduce mortar (master) and nonmortar (slave) sides Γ_{ij}. Let the master side be Γ_{ij} a side of the substructure Ω_i where the coefficient $\rho(x)$ is larger i.e. $\rho_i \geq \rho_j$. Thus Γ_{ij} also forms a slave side of Ω_j. We now introduce the master, slave, and test spaces. The spaces $W_{ij}^{h_i}(\Gamma_{ij})$ and $W_{ij}^{h_j}(\Gamma_{ij})$ are the master and slave ones. The test space, denoted by $M_{ij}^{h_j}(\Gamma_{ij})$, is a subspace of $W_{ij}^{h_j}(\Gamma_{ij})$ and its functions are constant on the elements $e_j^{(k)}$ at the ends of Γ_{ij}; the dimension of $M_{ij}^{h_j}(\Gamma_{ij})$ is two less than that of $W_{ij}^{h_j}(\Gamma_{ij})$. We are now in position to define a space $V^h(\Omega)$ used for the discretization of (1) by the mortar element method. V^h is the space of $u_h \in X^h$ such that for $i = 1, \ldots, N$ and each slave side Γ_{ij} the following mortar condition

$$\int_{\Gamma_{ij}} (u_{i,h} - u_{j,h}) \Psi ds = 0, \quad \Psi \in M_{ij}^{h_j}(\Gamma_{ij}) \tag{2}$$

is satisfied. Here $u_{i,h}$ and $u_{j,h}$ in the integral are the restrictions to Γ_{ij}, the common side of Ω_i and Ω_j, of the the finite element functions. This imposes a weak continuity constraint of $u_{i,h}$ and $u_{j,h}$ on Γ_{ij}. It also means that the L_2-projections of $u_{i,h}$ and $u_{j,h}$, restricted to Γ_{ij}, onto $M_{ij}^{h_j}(\Gamma_{ij})$ are equal to each other.

The discrete problem is of the form:

Find $u_h \in V^h$ such that

$$a(u_h, v_h) = f(v_h), \quad v_h \in V^h. \tag{3}$$

This problem has a unique solution and

$$\sum_{i=1}^{N} ||u - u_{i,h}||^2_{H^1(\Omega_i)} \leq C \sum_{i=1}^{N} h_i^2 |u|^2_{H^2(\Omega_i)}, \tag{4}$$

provided that $u \in H_0^1(\Omega) \cap H^2(\Omega_i)$; see Bernardi, Maday and Patera (1994), Belgacem (1995), Bernardi and Maday (1995).

Our goal is to design and analyze a parallel algorithm for solving the discrete problem. It is a domain decomposition method, more exactly, an additive Schwarz method which is almost optimal from the point of view of parallel computations; see Dryja, Smith, and Widlund (1994). The first question concerns the condition number of the stiffness matrix A of the system resulting from (3). It has been established, independently by Bernardi and Maday (1995) and by Dryja and Widlund (1995) that it is of the same order as in for conforming finite element approximations, i.e.

$$\mathrm{cond}(A) \leq Ch^{-2}. \tag{5}$$

Here h is the smallest of the h_i.

3 ADDITIVE SCHWARZ METHODS

In this section, we describe an iterative substructuring method in terms of an additive Schwarz method (ASM). For that the general framework of ASM is used, see Dryja, Smith, and Widlund (1995).

In a first step of the method considered, all unknowns associated with interior nodal points of Ω_i are eliminated. For $i = 1, \ldots, N$ $u_{i,h} \in X_i^{h_i}(\Omega_i)$ is represented as

$$u_{i,h} = Pu_{i,h} + Hu_{i,h}, \tag{6}$$

where $Pu_{i,h}$ is the $H_0^1(\Omega_i)$ - projection onto $\overset{0}{X}{}_i^{h_i}(\Omega_i)$ while $Hu_{i,h}$ is the discrete harmonic function defined by

$$\left(\nabla Hu_{i,h}, \nabla v_{i,h},\right)_{L_2(\Omega_i)} = 0, \quad v \in \overset{0}{X}{}_i^{h_i}(\Omega_i) \tag{7}$$

with $Hu_{i,h} = u_{i,h}$ on $\partial\Omega_i$. Here $\overset{0}{X}{}_i^{h_i}(\Omega_i)$ is the subspace of functions of $X_i^{h_i}(\Omega_i)$ which vanish on $\partial\Omega_i$. Using this approach, we reduce the problem (3) to the following: (Let $V_1^h(\Omega)$, a subspace of $V^h(\Omega)$, be a space of discrete harmonic functions in each Ω_i.)

Find $u_h \in V_1^h(\Omega)$ such that

$$s(u_h, v_h) = \tilde{f}(v_h), \quad v_h \in V_1^h(\Omega), \tag{8}$$

where

$$s(v_h, w_h) = a(v_h, w_h), \quad u_h, v_h \in V_1^h(\Omega) \qquad \text{and} \qquad \tilde{f}(v_h) = f(v_h) - a(Pu_h, v_h).$$

The problem (8) has a unique solution and it can be proved that the condition number of the matrix S, resulting from (8), is of the order of h^{-1} provided that the h_i are of the order of h.

For solving (8), we design and analyze an ASM. For that we use the general framework, see Dryja, Smith, and Widlund (1994), which describes the method in terms of a decomposition of V_1^h into subspaces, certain bilinear forms given on these subspaces, and the projections onto these subspaces in the sense of the bilinear forms. The decomposition of V_1^h is taken as

$$V_1^h = V_{10} + V_{11}, \tag{9}$$

where V_{11} is a space of functions belonging to V_1^h vanishing at the vertices of the substructures Ω_i, i.e. the nodal points of the coarse triangulation. To define the space V_{10}, let us first discuss the case when functions of V_1^h are continuous at the vertices of Ω_i, which are assumed to be triangles. In this case $V_{10} = V^H$, is a conforming finite element space defined on the coarse triangulation. We now define V_{10} without assuming continuity at the vertices. Let x_{ijl} be a vertex of Ω_i and the end of Γ_{ij} and Γ_{il}, sides of Ω_j and Ω_l common with Ω_i. We should denote that point by $x_{ij\ldots k}$ since it is common to the substructures $\Omega_i, \Omega_j, \ldots, \Omega_k$. For simplicity let us denote only by x_{ijl}. We associate with x_{ijl} a basis function φ_{ijl} defined as follows. Let Γ_{ij}, a side of Ω_i with one end as x_{ijl}, be the master one. We introduce $\varphi_{ij}^{(m)}$ as a linear function defined on Γ_{ij} with the values one at x_{ijl} and zero at the other end of Γ_{ij}. The function $\varphi_{ij}^{(m)}$ is extended to the slave side of Ω_j (common with Ω_i) by using the mortar condition. That extension is denoted by $\varphi_{ij}^{(ms)}$ and defined by

$$\int_{\Gamma_{ij}} \varphi_{ij}^{(ms)} \Psi ds = \int_{\Gamma_{ij}} \varphi_{ij}^{(m)} \Psi ds, \quad \Psi \in M_{ij}^{h_j}(\Gamma_{ij}), \tag{10}$$

with the additional condition that $\varphi_{ij}^{(ms)}(x)$ vanishes at the ends of Γ_{ij}. If Γ_{il} is a slave (nonmortar) side of Ω_i, we associate with x_{ijl} a function $\varphi_{il}^{(s)}$ defined by

$$\int_{\Gamma_{il}} \varphi_{il}^{(s)} \Psi ds = 0, \quad \Psi \in M_{il}^{h_i}(\Gamma_{il}), \tag{11}$$

with the condition that $\varphi_{ij}(x_{ijl}) = 1$ and it vanishes at the other end of Γ_{ij}. This means that $\varphi_{ij}^{(s)}$ is the image of zero defined on the master side Γ_{il} belonging to Ω_l. The function $\varphi_{ijl}(x)$ associated with the vertex x_{ijl} is the discrete harmonic function in each substructure with the following boundary data: $\varphi_{ijl}(x) = 0$ on all sides of the substructures except those for which x_{ijl} is a common vertex. On the sides with the common vertex φ_{ijl} is: for a master side Γ_{ij} it is equal to $\varphi_{ij}^{(m)}$ and for a slave side of Ω_i common with Ω_l (which is the master for Ω_l) it is equal to $\varphi_{il}^{(s)} + \varphi_{il}^{(ms)}$. Of course, $\varphi_{ijl} \in V_1^h$ and

$$V_{10} = \text{span}\{\varphi_{ijl}\}. \tag{12}$$

We now introduce bilinear forms b_{1k}, $k = 0, 1$, related to the inexact solvers over $V_{1k} \times V_{1k}$, which approximate $a(u, v)$. One possibility to define $b_{11}(u, v)$ is

$$b_{11}(u_h, v_h) = \sum_{\Omega_i} \sum_{\Gamma_{ij} \subset \partial\Omega_i} \rho_i \left(K_{ij}^{1/2} \underline{u}_{i,h}^{(m)}, \underline{v}_{i,h}^{(m)} \right)_{R^{N_{ij}}}. \tag{13}$$

Here the sum is taken over the master sides of Ω_i, $u_{i,h}^{(m)} = u_{i,h}$ when Γ_{ij} is the master side of Ω_i. $\underline{u}_{i,h}^{(m)}$ is the vector representation of $u_{i,h}$ on Γ_{ij} using the standard nodal basis functions associated with the interior nodal points of Γ_{ij} corresponding to the triangulation on the master side; the number of nodes is denoted by N_{ij}. The matrix K_{ij} is the matrix representation of $(\nabla u, \nabla v)_{L^2(0,l)}$, approximated by the conforming finite element space of piecewise linear continuous functions which vanish at the ends of $(0, l)$, defined on triangulation on $(0, l)$ with uniform step and N_{ij} the number of interior nodal points. Note that the system with $K_{ij}^{1/2}$ can be solved using FFT, for details see, e.g. Dryja (1982).

The other possibility to define $b_{11}(u_h, v_h)$ is given by

$$b_{11}(u_h, v_h) = \sum_{\Omega_i} \sum_{\Gamma_{ij} \subset \partial\Omega_i} b_{\Gamma_{ij}}(u_{i,h}^{(m)}, v_{i,h}^{(m)}), \tag{14}$$

where the sum, as in (13), is taken over the master sides Γ_{ij} of $\partial\Omega_i$ and

$$b_{\Gamma_{ij}}\left(u_{i,h}^{(m)}, v_{i,h}^{(m)}\right) = \rho_i\left(\nabla u_{i,h}^{(m)}, \nabla v_{i,h}^{(m)}\right)_{L_2(\Omega_i)}. \tag{15}$$

Here Γ_{ij} is the master side of Ω_i and $u_{i,h}^{(m)}$ is equal to $u_{i,h}$ on Γ_{ij} at the interior points and zero on the remaining sides of $\partial\Omega_i$ and others. Note that the solving of a problem with the form b_{11} reduces to solving of local conforming finite element problems defined only on individual substructures and that these problems are independent.

The form $b_{10}(u_h, v_h)$, $u_h, v_h \in V_{10}$, is given by

$$\begin{aligned} b_{10}(u_h, u_h) &= \sum_{\Omega_i}\Big\{\sum_{x\in\nu_i}\Big(1+\log\frac{H}{h_i}\Big)\rho_i\Big(u_{i,h}(x) - \overline{u}_{i,h}\Big)^2 \\ &+ \sum_{\Gamma_{ij}\subset\partial\Omega_i}\sum_{x\in\nu_{ij}}\Big(1+\log\frac{H}{h_j}\Big)\rho_j\Big(u_{i,h}^{(m)}(x) - \overline{u}_{j,h}^{(s)}\Big)^2\Big\}, \end{aligned} \tag{16}$$

where the sum is taken over all substructures Ω_i while in the second term it is taken over all the master sides Γ_{ij} belonging to Ω_i. Here ν_i is a set of vertices of Ω_i and ν_{ij} is the ends of Γ_{ij}; $\overline{u}_{i,h}$ is the average of $u_{i,h}$ at vertices of Ω_i, i.e.

$$\overline{u}_{i,h} = \frac{1}{n_i}\cdot\sum_{x\in\nu_i} u_{i,h}(x), \tag{17}$$

where n_i is the number of vertices of Ω_i; $u_{i,h}^{(m)} = u_{i,h}$ when Γ_{ij} is the master side of Ω_i and $u_{j,h}^{(s)} = u_{j,h}$ when Γ_{ij} is a slave side of Ω_j, common with Ω_i.

We now are in position to define approximate projections $T_{1k} : V_1^h \to V_{1k}$, $k = 0, 1$, as

$$b_{1,k}(T_k u, v) = a(u, v), \quad v \in V_{1k} \tag{18}$$

and we denote

$$T_1 = T_{10} + T_{11}, \tag{19}$$

where $T_1 : V^h \to V^h$.

The original problem (8) can be replaced by

$$T_1 u_h = g_1, \tag{20}$$

where $g_1 = g_{10} + g_{11}$ which are computed as the solutions of

$$b_{1k}(g_{1k}, v) = f(v), \quad v \in V_{1k}. \tag{21}$$

The problems (20) and (8) have the same unique solution since T_1 is invertible; see Theorem 1 below.

To find u_h, the solution of (8), we solve the problem (20) iteratively since the operator

T_1 is symmetric, positive definite and well conditioned, which follows from Theorem 1. We now formulate the main result of this section.

Theorem 1 *For any* $v_h \in V_1^h$

$$c\delta^{-2}a(v_h, v_h) \leq (T_1 v_h, v_h) \leq Ca(v_h, v_h), \tag{22}$$

where c *and* C *are positive constants independent of* H, h_i *and the jumps of* $\rho(x)$, $\delta = (1 + \log \frac{H}{h})$ *and* $h = \inf_i h_i$.

A proof of Theorem 1 is given in Section 5. We now discuss an implementation of the method. For solving (20), we can use the conjugate gradient method since T_1 is symmetric, positive definite and its condition number is almost constant. To simplify our presentation, we use the Richardson method for solving (20):

$$u_h^{n+1} = u_h^n - \tau_{opt}\Big(T_1 u_h^n - g_1\Big),$$

where τ_{opt} is the relaxation parameter. To find u_h^{n+1}, we need to compute

$$r_1 = T_1 u_h^n - g_1 = T_1(u_h^n - u_h) = r_{10} + r_{11},$$

where $r_{1k} = T_{1k}(u_h^n - u_h)$, $k = 0, 1$, are the solutions of

$$b_{1k}(r_{1k}, v_h) = a(u_h^n, v_h) - \tilde{f}(v_h) \equiv F(v_h), \quad v_h \in V_{1k}. \tag{23}$$

These two problems are independent. To find r_{10}, we solve a global system of dimension equal to the number of values of u_h at the vertices of the substructures.

The solution of (23), for $k = 1$, reduces to solving a set of local subproblems defined in one or two substructures. Let us give some details in the case when b_{11} is given by (13). For a fixed Γ_{ij}, we solve the equation

$$\Big(K_{ij}^{1/2}\underline{r}_i^{(m)}, \underline{v}_i^{(m)}\Big) = g(v_h), \quad v_h \in V_{11}, \tag{24}$$

where $r_i^{(m)} = r_{11}$ on Γ_{ij}, using the scheme:

1. For the nodal points $x_K \in \Gamma_{ij}$, the master side of Ω_i, take a basis function $\Phi_K \in V_1$ which is equal to the discrete harmonic function on $\overline{\Omega}_i$ with values one at x_K and zero at the remaining nodal points of $\partial\Omega_i$. On the slave side Γ_{ij} of Ω_j, Φ_K is an extension using the mortar condition (2) with $\Phi_K(x) = 0$ at the ends of Γ_{ij}. Then $\Phi_K(x)$ on Ω_j is the discrete harmonic extension with zero on $\partial\Omega_j \backslash \Gamma_{ij}$. On the remaining substructures $\Phi_K(x) = 0$. The system so obtained is solved by FFT and it gives the values of $r_i^{(m)}(x) = r_{11}(x)$ at the nodal points of Γ_{ij}, the master side.
2. Find $r_j(x)$ at the nodal points of Γ_{ij}, the slave side of Ω_j, solving the problem

$$\int_{\Gamma_{ij}} r_j(x)\Psi ds = \int_{\Gamma_{ij}} r_i^{(m)}(x)\Psi ds, \quad \Psi \in M_{ij}^{h_j}(\Gamma_{ij}),$$

with $r_j(x) = 0$ at the ends of Γ_{ij}. For that, we use the standard nodal basis functions of $W_{ij}^{h_j}(\Gamma_{ij})$ and $M_{ij}^{h_j}(\Gamma_{ij})$. That gives the values $r_j(x) = r_{11}$ on Γ_{ij}, the slave side.

Using the above scheme, we find r_{11} on the boundary of each substructure Ω_i. To get $r_{11} \in V_{11}$, we extend r_{11}, given on $\partial\Omega_i$, to Ω_i as a discrete harmonic function. For that, we solve N independent problems of the form (7).

Summarizing, we see that the algorithm presented is very well suited for parallel computations. The number of iterations to get the solution with an accuracy ε is independent of the jumps of $\rho(x)$ and depends only logarithmically on the ratio of H and h_i.

4 TECHNICAL TOOLS

In this section a number of auxiliary results needed to prove Theorem 1 are formulated and proved. Let us first introduce certain operators similar to L_2 – projections. Let p_j denote a projection from $L_2(\Gamma_{ij})$ onto $M_{ij}^{h_j}(\Gamma_{ij})$ given by

$$\int_{\Gamma_{ij}} p_j v \,\Psi ds = \int_{\Gamma_{ij}} v\Psi ds, \quad \Psi \in M_{ij}^{h_j}(\Gamma_{ij}). \tag{25}$$

It is easy to see that for $v \in H^1(\Gamma_{ij})$

$$||v - p_j v||^2_{L_2(\Gamma_{ij})} \leq C h_j^2 |v|^2_{H^1(\Gamma_{ij})} \tag{26}$$

and

$$|p_j v|^2_{H^1(\Gamma_{ij})} \leq C |v|^2_{H^1(\Gamma_{ij})}. \tag{27}$$

To prove (26), we choose $v_h \in M_{ij}^{h_j}$ equal to v at the interior nodal points of the h_j-triangulation of Γ_{ij} and use the fact that $||v - p_j v||_{L_2} \leq ||v - v_h||_{L_2}$. Using a standard interpolation theorem, element by element, we get (26). The estimate (27) follows from an inverse inequality and (26).

Let Π_j be the projection from $L_2(\Gamma_{ij})$ onto $W_{ij}^{h_j}(\Gamma_{ij}) \cap H_0^1(\Gamma_{ij})$ defined by

$$\int_{\Gamma_{ij}} \Pi_j v \Psi ds = \int_{\Gamma_{ij}} v\Psi ds, \quad \Psi \in M_{ij}^{h_j}(\Gamma_{ij}). \tag{28}$$

Let Γ_{ij}, a side common to Ω_i and Ω_j, be the master and slave side for $u_{i,h}$ and $u_{j,h}$ defined on Ω_i and Ω_j, respectively. Let these functions satisfy the mortar condition:

$$\int_{\Gamma_{ij}} u_{i,h}\Psi ds = \int_{\Gamma_{ij}} u_{j,h}\Psi ds, \quad \Psi \in M_{ij}^{h_j}(\Gamma_{ij}). \tag{29}$$

Let Q_j be the L_2 – projection from $H_0^1(\Gamma_{ij})$ onto $\overset{0}{W}{}_{ij}^{h_i}(\Gamma_{ij})$, i.e.

$$\int_{\Gamma_{ij}} Q_j v \, \varphi ds = \int_{\Gamma_{ij}} v \, \varphi ds, \quad \varphi \in \overset{0}{W}{}_{ij}^{h_i}(\Gamma_{ij}),$$

where $\overset{0}{W}{}_{ij}^{h_i}(\Gamma_{ij})$ is a space of functions from $W_{ij}^{h_i}(\Gamma_{ij})$ which vanish at the ends of Γ_{ij}.
In the lemmas below, we assume that h_i and h_j are of the same order; the constant C is independent of h_i and h_j.

Lemma 1 *Let $u_{i,h}$ and $u_{j,h}$ vanish at the ends of Γ_{ij} and satisfy (29), and let Γ_{ij} be the master and slave sides of Ω_i and Ω_j, respectively. Then*

$$||u_{j,h}||^2_{H_{00}^{1/2}(\Gamma_{ij})} \leq C||u_{i,h}||^2_{H_{00}^{1/2}(\Gamma_{ij})}. \tag{30}$$

Remark 1 *Lemma 1 follows from Lemma 1 of Belgacem (1995). Here we give an alternative proof, which is simpler.*

Proof. We need to show only that

$$||\Pi_j u_{i,h}||^2_{H_{00}^{1/2}(\Gamma_{ij})} \leq C||u_{i,h}||^2_{H_{00}^{1/2}(\Gamma_{ij})} \tag{31}$$

since $\Pi_j u_{i,h} = u_{j,h}$ in view of (28) and (29). We first note that

$$||\Pi_j u_{i,h}||^2_{L_2(\Gamma_{ij})} \leq 3||u_{i,h}||^2_{L_2(\Gamma_{ij})}. \tag{32}$$

This follows from (28) taking Ψ, here denoted by v, which is equal to $\Pi_j u_{i,h}$ at the nodal points of Γ_{ij} and the fact that

$$||\Pi_j u_{i,h}||^2_{L_2(\Gamma_{ij})} \leq \left(\Pi_j u_{i,h}, v\right)_{L_2(\Gamma_{ij})} = \left(u_{i,h}, v\right)_{L_2(\Gamma_{ij})} \quad \text{and} \tag{33}$$

$$||v||^2_{L_2(\Gamma_{ij})} \leq 3||\Pi_j u_{i,h}||^2_{L^2(\Gamma_{ij})}, \tag{34}$$

which are shown straightforwardly. For that, we use the fact that

$$\begin{aligned}(\Pi_j u_{i,h}, v)_{L^2(\Gamma_{ij})} \quad = \quad & \int_{x_0+h_j}^{x_{N_{ij}}-h_j} (\Pi_j u_{i,h})^2 ds + \\ & \frac{h_j}{2}\left\{\left(\Pi_j u_{i,h}\left(x_0+h_j\right)\right)^2 + \left(\Pi_j u_{i,h}\left(x_{N_{ij}}-h_j\right)\right)^2\right\},\end{aligned} \tag{35}$$

where x_0 and $x_{N_{ij}}$ are the ends of Γ_{ij}.
Note that (32) is valid also for a $u_{i,h}$ which does not vanish at the ends of Γ_{ij}. That fact will be used below.

Using the inverse inequality, we get

$$||\Pi_j u_{i,h}||^2_{H^{1/2}_{00}(\Gamma_{ij})} \le C\Big\{\frac{1}{h_j}||\Pi_j u_{i,h} - Q_j u_{i,h}||^2_{L_2} + ||Q_j u_{i,h}||^2_{H^{1/2}_{00}}\Big\}. \tag{36}$$

Note that $\Pi_j Q_j u_{i,h} = Q_j u_{i,h}$. Using now the properties of Q_j, see e.g. Dryja and Widlund (1994), and (32), we get (31). □

Let $\varphi^{(s)}_{im}$ and $\varphi^{(ms)}_{im}$ be the functions defined on Γ_{im}, the slave side, and let φ_{ilm} be defined on $\partial\Omega_i$; they have been introduced in Section 3.

Lemma 2

$$||\varphi^{(s)}_{im}||^2_{L_2(\Gamma_{im})} \le C h_i. \tag{37}$$

Remark 2 *This and the two next lemmas are proved in Achdou, Maday, and Widlund (1995). Here we give alternative proofs, which are simpler.*

Proof. Let $\varphi^{(s)}_{im} = z_0 + z_1$ where z_0 is equal to $\varphi^{(s)}_{im}$ at the ends of Γ_{im} and zero at the remaining nodal points. By the definition of $\varphi^{(s)}_{im}$, see (11),

$$\int_{\Gamma_{im}} z_0 \Psi ds + \int_{\Gamma_{im}} z_1 \Psi ds = 0, \quad \Psi \in M^{h_i}_{im}. \tag{38}$$

Let $\Psi = \tilde{z}_1$ be defined by the nodal values of z_1. ¿From (38) and using the ε-inequality, we get

$$\int_{\Gamma_{im}} z_1 \tilde{z}_1 dx = -\int_{\Gamma_{im}} z_0 \tilde{z}_1 dx \le \frac{1}{2\varepsilon}\Big|\Big|z_0\Big|\Big|^2_{L_2} + \frac{\varepsilon}{2}\Big|\Big|\tilde{z}_1\Big|\Big|^2_{L_2}. \tag{39}$$

Straightforwardly, using (35) for z_1 and $\tilde{z}_1$, we show that, cf. (34) and (33),

$$||\tilde{z}_1||^2_{L^2(\Gamma_{ij})} \le 2\int_{\Gamma_{ij}} z_1 \tilde{z}_1 dx.$$

Hence,

$$\int_{\Gamma_{im}} z_1 \tilde{z}_1 dx \le C||z_0||^2_{L_2}.$$

We also show straightforwardly, using again (35) for z_1 and $\tilde{z}_1$, that, cf. (33),

$$||z_1||^2_{L_2(\Gamma_{im})} \le (z_1, \tilde{z}_1)_{L_2(\Gamma_{im})}.$$

Note that $||z_0||^2_{L^2} \le C h_i$. Using these estimates, we get $||\varphi^{(s)}_{im}||^2_{L_2} \le ||z_0||^2_{L_2} + ||z_1||^2_{L_2} \le C h_i$, which proves (37). □

Corollary 1

$$||\varphi_{im}^{(s)}||^2_{L^\infty(\Gamma_{im})} \le \frac{C}{h_i}||\varphi_{im}^{(s)}||^2_{L_2(\Gamma_{im})} \le C. \tag{40}$$

Lemma 3

$$||\varphi_{im}^{(ms)}||^2_{H_{00}^{1/2}(\Gamma_{im})} \le C\Big(1+\log\frac{H}{h_i}\Big). \tag{41}$$

Proof. Note that $\varphi_{im}^{(ms)} = \Pi_m \varphi_{im}^{(m)}$, where Π_m is defined by (28).

Let $\varphi_A(x)$ be the nodal basis function defined on the triangulation of the master side, associated with the vertex $A = x_{iml}$. We have,

$$||\Pi_m \varphi_{im}^{(m)}||_{H_{00}^{1/2}(\Gamma_{ij})} \le ||\Pi_j(\varphi_{im}^{(m)} - \varphi_A)||_{H_{00}^{1/2}(\Gamma_{ij})} + ||\Pi_j \varphi_A||_{H_{00}^{1/2}(\Gamma_{ij})}. \tag{42}$$

Using the inverse inequality and (32), valid also for functions which do not vanish at the ends of Γ_{im}, we get

$$||\Pi_m \varphi_A||^2_{H_{00}^{1/2}(\Gamma_{ij})} \le \frac{C}{h_m}||\varphi_A||^2_{L_2(\Gamma_{ij})} \le C, \tag{43}$$

provided that h_i and h_m are of the same order. Using now (31), we get

$$||\Pi_m(\varphi_{im}^{(m)} - \varphi_A)||^2_{H_{00}^{1/2}(\Gamma_{ij})} \le C||\varphi_{im}^{(m)} - \varphi_A||^2_{H_{00}^{1/2}(\Gamma_{ij})}.$$

The seminorm of $H^{1/2}$ is estimated by C while the L_2-norm by $C(1+\log\frac{H}{h_i})$; this follows from straightforward computations. Hence,

$$||\Pi_m(\varphi_{im}^{(m)} - \varphi_A)||^2_{H_{00}^{1/2}(\Gamma_{ij})} \le C(1+\log\frac{H}{h_i}).$$

Substituting this and (43) into (42), we get (41). □

Lemma 4

$$|\varphi_{ilm}|^2_{H^{1/2}(\partial\Omega_i)} \le C\Big(1+\log\frac{H}{h_i}\Big). \tag{44}$$

Proof. Here there are three possibilities of the form φ_{ilm} on $\partial\Omega_i$. If Γ_{im} and Γ_{il} are master sides of $\partial\Omega_i$, the estimate is obvious and without the logarithmic factor.

When the sides Γ_{il} and Γ_{im} are both slaves, the inequality follows from the inverse inequality, and Lemma 2 and 3.

Let Γ_{im} be a master side while Γ_{il} is a slave side. Let $I_H\varphi_{ilm}$ be the linear function with the same values as φ_{ilm} at the vertices of Ω_i. We have $|\varphi_{ilm}|^2_{H^{1/2}(\partial\Omega_i)} \le 2||\varphi_{ilm} - I_H\varphi_{ilm}||^2_{H_{00}^{1/2}(\Gamma_{il})} + C$.

Note that $\varphi_{ilm} = \varphi_{il}^{(s)} + \varphi_{li}^{(ms)}$ on Γ_{il}. The term with $\varphi_{il}^{(s)} - I_H\varphi_{il}^{(s)}$ is estimated straightforwardly by $C(1+\log\frac{H}{h_i})$ using the inverse inequality and (40) while the term with $\varphi_{li}^{(ms)}$ is handled by Lemma 3. Thus, $|\varphi_{ilm}|^2_{H^{1/2}(\partial\Omega_i)} \le C\Big(1+\log\frac{H}{h_i}\Big)$ which shows (44). □

Lemma 5 *For any* $u_{i,h} \in W_{ij}^{h_i}(\Gamma_{ij})$

$$||u_{i,h}||^2_{L^\infty(\Gamma_{ij})} \leq C\Big(1 + \log\frac{H}{h_i}\Big)||u_{i,h}||^2_{H^{1/2}(\Gamma_{ij})}.$$

A proof of this fact is given in Dryja (1987)

5 PROOF OF THEOREM 1

For the proof of Theorem 1, we use the lemmas of Section 4. According to the general ASM theory, see Dryja, Smith, and Widlund (1994), it reduces to checking three key assumptions.

Assumption (iii)

It is obvious in our case since the space V_1^h is decomposed into two only subspaces and $C_0^2 = 1$.

Assumption (ii)

We need to show that for $k = 0, 1$,

$$a(u_h, u_h) \leq \omega b_{1k}(u_h, u_h), \quad u_h \in V_{1k}. \tag{45}$$

For $k = 1$, this follows from Lemma 1 and well known results on the matrix $K_{ij}^{1/2}$. Indeed, using an extension theorem, we have for $u_h \in V_{11}(\Omega)$

$$\rho_i|u_{i,h}|^2_{H^1(\Omega_i)} \leq C \sum_{\Gamma_{ij}\subset\partial\Omega_i} \rho_i||u_{i,h}||^2_{H_{00}^{1/2}(\Gamma_{ij})}. \tag{46}$$

Let Γ_{ij} be the master side. Then,

$$\rho_i||u_{i,h}||^2_{H_{00}^{1/2}(\Gamma_{ij})} \leq C\rho_i\Big(K_{ij}^{1/2}\underline{u}_{i,h}, \underline{u}_{i,h}\Big)_{R^{N_{ij}}}, \tag{47}$$

see Dryja (1982). Let Γ_{ij} be a slave side of Ω_i, a side common with Ω_j. Using Lemma 1, we get

$$\rho_i||u_{i,h}||^2_{H_{00}^{1/2}(\Gamma_{ij})} \leq C\rho_i||u_{j,h}||^2_{H_{00}^{1/2}(\Gamma_{ij})} \leq C\rho_i\Big(K_{ij}^{1/2}\underline{u}_{j,h}, \underline{u}_{j,k}\Big)_{R^{N_{ij}}} \tag{48}$$

since $u_{i,h}$ and $u_{h,j}$ satisfy the mortar condition (2). By the selection of the master side, we have $\rho_i \leq \rho_j$. Hence,

$$\rho_i||u_{i,h}||^2_{H_{00}^{1/2}(\Gamma_{ij})} \leq C\rho_j\Big(K_{ij}^{1/2}\underline{u}_{j,h}, \underline{u}_{j,h}\Big)_{R^{N_{ji}}}, \tag{49}$$

which proves (45) for $k = 1$ with ω= constant.

To show (45) for $k = 0$, we first note that

$$|u_{i,h}|^2_{H^1(\Omega_i)} \leq C|u_{i,h} - \overline{u}_{i,h}|^2_{H^{1/2}(\partial\Omega_i)}. \tag{50}$$

Any function $u_h \in V_{10}$ on $\partial\Omega_i$ can be represented as follows. Let x_{ilm}, x_{ilp}, x_{ipm} be vertices of Ω_i, denoted also by A, B, C, and let Γ_{im}, Γ_{il}, and Γ_{ip} be the sides of Ω_i common with Ω_m, Ω_l, and Ω_p. Using the functions introduced in Section 3, we see that $u_{i,h} \in V_{10}$ on $\partial\Omega_i$ is of the form

$$u_{i,h}(x) = u_{i,h}(A)\varphi_{ilm}(x) + u_{i,h}(B)\varphi_{ilp}(x) + u_{i,h}(C)\varphi_{ipm}(x). \tag{51}$$

Note that $u_{i,h}$ on Γ_{il}, the master side, is of the form

$$u_{i,h}(x) = u_{i,h}(A)\varphi^{(m)}_{il,A} + u_{i,h}(B)\varphi^{(m)}_{il,B}, \tag{52}$$

while on Γ_{im}, the slave side

$$u_{i,h}(x) = u_{i,h}(A)\varphi^{(s)}_{im,A} + u_{i,h}(C)\varphi^{(s)}_{im,C} + u_{m,h}(A)\varphi^{(ms)}_{im,A}(x) + u_{m,h}(C)\varphi^{(ms)}_{im,C}(x), \tag{53}$$

where $\varphi^{(m)}_{il,A}$, $\varphi^{(s)}_{im,A}$, and $\varphi^{(ms)}_{im,A}$, are $\varphi^{(m)}_{il}$, $\varphi^{(s)}_{im}$ and $\varphi^{(ms)}_{im}$ associated with the vertex A etc. Here we consider the configuration where only Γ_{il} is a master side. A similar representation exists for $u_{i,h}$ for other configurations of master and slave sides of Ω_i and when Ω_i is a rectangle.

Note that the sum of the basis functions which appear in (51) is one. Using this and (52) and (53) in (50), and applying Lemmas 1 to 4, we get

$$\begin{aligned} \rho_i\Big|u_{i,h} - \overline{u}_{i,h}\Big|^2_{H^{1/2}(\partial\Omega_i)} \leq\ & C\Big\{\sum_{x\in\nu_i}\Big(1 + \log\frac{H}{h_i}\Big)\rho_i\Big(u_{i,h}(x) - \overline{u}_{i,h}\Big)^2 \\ & + \sum_{x\in\nu_{im}}\Big(1 + \log\frac{H}{h_m}\Big)\rho_m\Big(u^{(m)}_{m,h}(x) - \overline{u}^{(s)}_{i,h}\Big)^2 \\ & + \sum_{x\in\nu_{ip}}\Big(1 + \log\frac{H}{h_p}\Big)\rho_p\Big(u^{(m)}_{p,h}(x) - \overline{u}^{(s)}_{i,h}\Big)\Big\}, \end{aligned} \tag{54}$$

where ν_{im} and ν_{ip} are the ends of Γ_{im} and Γ_{ip}, the slave sides of Ω_i and $\overline{u}^{(s)}_{i,h} = \overline{u}_{i,h}$. Taking a sum in (54) with respect to i, we get the estimate (45) for $k = 1$ with ω=constant. The proof of Assumption (ii) is complete.

Assumption (i)

We need to show that for any $u_h \in V_1^h$, there exists

$$u_h = u_h^{(0)} + u_h^{(1)}, \quad u_h^{(0)} \in V_{10}, \quad u_h^{(1)} \in V_{11} \tag{55}$$

such that

$$\sum_{k=0}^{1} b_k\Big(u_h^{(k)}, u_h^{(k)}\Big) \le C\delta^2 a(u_h, u_h). \tag{56}$$

Let $u_h^{(0)}$ be the interpolant of $u_h \in V_1^h$ belonging to V_{10}. Its representation on $\partial\Omega_i$ is given by (51) - (53). Let $u_h^{(1)} = u_h - u_h^{(0)}$; of course, $u_h^{(1)} \in V_{11}$. Let us first discuss $k = 1$ and the form b_{11} given by (13). It is known, see e.g. Dryja (1982), that

$$\rho_i\Big(K_{ij}^{1/2}\,\underline{u}_{i,h}^{(1)}, \underline{u}_{i,h}^{(1)}\Big)_{R^{N_{ij}}} \le C\rho_i\Big|\Big|u_{i,h}^{(1)}\Big|\Big|^2_{H_{00}^{1/2}(\Gamma_{ij})} \le C\Big(1+\log\frac{H}{h_i}\Big)^2 \rho_i|u_{i,h}|^2_{H^1(\Omega_i)} \tag{57}$$

since $u_h^{(0)}$ on Γ_{ij}, as the master side, is a linear function. Summing this over the master sides, we get

$$b_{11}(u_h, u_h) \le C\delta^2\, a(u_h, u_h). \tag{58}$$

To check this estimate for the form b_{11} given by (14) and (15), it is enough to see that

$$b_{\Gamma_{ij}}\Big(u_{i,h}^{(1)}, u_{i,h}^{(1)}\Big) \le C\rho_i||u_{i,h}^{(1)}||^2_{H_{00}^{1/2}(\Gamma_{ij})}.$$

We now show that, see (16),

$$b_{10}\Big(u_h^{(0)}, u_h^{(0)}\Big) \le C\delta^2 a(u_h, u_h). \tag{59}$$

The first term of (16) is estimated by

$$\sum_{x\in\nu_i} \rho_i(u_{i,h}(x) - \overline{u}_{i,h}) \le C\Big(1+\log\frac{H}{h_i}\Big)\rho_i|u_{i,h}|^2_{H^{1/2}(\partial\Omega_i)}$$

in view of Lemma 5 and Poincaré's inequality. To estimate the second term of (16), we again apply Lemma (5) and use the fact that the average values of $u_{i,h}^{(m)}$ and $u_{j,h}^{(s)}$ on Γ_{ij} are equal to each other which follows from the mortar condition. Thus,

$$\rho_j\Big(u_{i,h}^{(m)}(x) - \overline{u}_{j,h}^{(s)}\Big)^2 \;\le\; C\Big\{\Big(1+\log\frac{H}{h_i}\Big)\rho_i|u_{i,h}|^2_{H^{1/2}(\partial\Omega_i)} + \Big(1+\log\frac{H}{h_j}\Big)\rho_j|u_{j,h}|^2_{H^{1/2}(\partial\Omega_j)}\Big\}.$$

Adding these two estimates and taking a sum with respect to i, we get (59). In turn, adding the estimates (58) and (59), we get (56). *Assumption (i)* have been checked. The proof of Theorem 1 is complete.

6 THE GEOMETRICALLY NONCONFORMING CASE

In this section, we generalize the considerations from the previous sections to the case when the coarse partition is not triangulation. In the mortar element method this is called a geometrically nonconforming case. Unfortunately, it is not clear how to analyze ASM considered in Section 3 for the problem with the jump coefficients. Therefore here we discuss only elliptic problems with continuous coefficients. To simplify our presentation, we consider only the following problem:

Find $u \in H_0^1(\Omega)$ such that

$$a(u,v) = f(v), \quad v \in H_0^1(\Omega), \tag{60}$$

$$a(u,v) = \int_\Omega \nabla u \cdot \nabla v dx, \quad f(v) = \int_\Omega f v dx. \tag{61}$$

The problem (60) is discretized by the mortar element method. For that let us, as in Section 2, assume that Ω is polygonal region divided into subregions Ω_i, which are triangles or rectangles. We make no assumption that this partition is a finite element triangulation. Let $H = \max_i H_i$ where H_i is the diameter of Ω_i. We now introduce a triangulation of each Ω_i with triangular elements and a parameter h_i. In general, this triangulation is nonmatching but we assume that the triangulation of each Ω_i is quasi-uniform, see Ciarlet (1978). Let Γ_{ij} denoted the intersection of closures of Ω_i and Ω_j; now this can be a part of sides of Ω_i and Ω_j. The sides of Ω_i are denoted by γ_{ij}.

Let $X^h(\Omega)$ and $W_{ij}^{h_i}(\gamma_{ij})$ be defined as in Section 2. Let us choose the master and slave sides γ_{ij} of Ω_i. Here the rule is that the mortar sides are entire sides of substructures, that they are disjoint, and that their union is equal to $\Gamma = \cup \partial\Omega_i \backslash \partial\Omega$. From this follows that the γ_{ij}, across the interface, is a union of Γ_{ik}, intersections of the closures of Ω_k and Ω_i.

Let γ_{ij} be the slave side of $\partial\Omega_i$ and let the test space, $M_{ij}^{h_i}(\gamma_{ij})$, be defined as in Section 2. We say that $u_h \in X^h$ satisfies the mortar condition on the slave side γ_{ij} if

$$\int_{\gamma_{ij}} \left(u_{i,h_{|\gamma_{ij}}} - \sum_k u_{k,h_{|\Gamma_{ik}}}\right)\Psi = 0, \qquad \Psi \in M_{ij}^{h_i}(\gamma_{ij}), \tag{62}$$

where the sum is taken over the Ω_k with nonzero intersection with γ_{ij}, i.e. over Γ_{ik} such that $\gamma_{ij} = \cup \Gamma_{ik}$ (intersections with one point are excluded).

The space V^h is defined as a space of functions from X^h which satisfy condition (62) for each slave side γ_{ij} belonging to interior of Ω.

The discrete problem for (60) is of the form:

Find $u_h \in V^h$ such that

$$a(u_h, v_h) = f(v_h). \tag{63}$$

The problem has a unique solution and an estimate of the error is as in (4), see Bernardi, Maday and Patera (1990).

Our goal is to design and analyze an ASM like that in Section 3 for problem (63). The first question is that of the condition number of A, the matrix of the system resulting from (63). In Dryja and Widlund (1995), it has been established that also in the geometrically nonconforming case, the condition number of A is as in (5) provided that the h_i are of the order h.

Problem (63), as in Section 3, is reduced to a problem on the interfaces. Let V_1^h be a space of functions belonging to V^h, which are discrete harmonic in each Ω_i, cf. (8):

Find $u_h \in V_1^h$ such that

$$s(u_h, v_h) = \tilde{f}(v_h), \quad v_h \in V_1^h, \tag{64}$$

where $s(v_h, w_h)$ and $\tilde{f}(v_h)$ are the same as in Section 3 with the form $a(u,v)$ defined by (61).

We now design ASM for solving (64) follows the scheme of Section 3.

The decomposition of V_1^h is (cf. (9))

$$V_1^h = V_{10} + V_{11}. \tag{65}$$

Here V_{11}, as in Section 3, is a space of functions from V_1^h which vanish at vertices of Ω_i while V_{10} is the generalization of the space from Section 3. It is defined as follows. Let x_{ikl} be a vertex of Ω_i that is in common with the substructure Ω_k and others, that is an end point of γ_{ij}, a side of Ω_i. The functions $\varphi_{ij}^{(s)}$ and $\varphi_{ij}^{(m)}$ are associated with x_{ikl} and correspond to the slave and mortar sides γ_{ij}, respectively, are defined as in Section 3, i.e. $\varphi_{ij}^{(s)}$ is given by (11) while $\varphi_{ij}^{(m)}$ is linear with value one at x_{ikl} and zero at the other end of γ_{ij}. The function $\varphi_{ij}^{(ms)}$ is defined by (10) where $\varphi_{ij}^{(m)}$ now is replaced by a sum of $\varphi_{ik}^{(m)}$ on the mortar sides of Ω_k which intersect γ_{ij}.

Using these functions, we define φ_{ikl} which is associated with x_{ikl} as follows. It is discrete harmonic in each Ω_i with data on $\partial\Omega_i$ which are zero except on the sides with the common vertex x_{ikl}. On these sides φ_{ikl} is defined by: on the master side it equals $\varphi_{ik}^{(m)}$ while on the slave side it is equal to $\varphi_{ij}^{(s)} + \varphi_{ij}^{(ms)}$. Thus,

$$V_{10} = \text{span}\ \{\varphi_{ikl}\}. \tag{66}$$

It is easy to see that $V_{10} \subset V_1^h$.

The bilinear forms b_{1k}, $k = 0, 1$, defined over $V_{1k} \times V_{1k}$, are of the form:

For k=0

$$\begin{aligned} b_{10}(u_h, u_h) \;=\; & \sum_{\Omega_i}\Big\{\sum_{x\in\nu_i}\left(u_{i,h}(x) - \overline{u}_{i,h}\right)^2 \\ & + \sum_{\gamma_{ij}\subset\partial\Omega_i}\Big[\sum_{x\in\nu_{li}}\left(u_{l,h}^{(m)} - \overline{u}_{i,h}^{(s)}\right)^2 + \ldots + \sum_{x\in\nu_{pi}}\left(u_{p,h}^{(m)} - \overline{u}_{i,h}^{(s)}\right)^2\Big]\Big\}. \end{aligned} \tag{67}$$

Here $\Gamma_{li}, \ldots, \Gamma_{pi}$ are parts of mortar sides given as intersections of $\Omega_l, \ldots, \Omega_p$ with γ_{ij}, the slave side of Ω_i and, as in Section 3, the ν_{li} are the end points of the mortar side of Ω_l which contains Γ_{li} etc.

The form b_{11} is similar to the one given by (13), i.e.

$$b_{11}(u_h, v_h) = \sum_{\Omega_i}\sum_{\gamma_{ij}\subset\partial\Omega_i}\left(K_{ij}^{1/2}\,\underline{u}_{i,h}^{(m)}, \underline{v}_{i,h}^{(m)}\right)_{R^{N_{ij}}}, \tag{68}$$

where the sum is taken over the master sides γ_{ij} of Ω_i.

Remark 3 *The form $b_{11}(u_h, v_h)$ can be also defined by (14) where $b_{\Gamma_{ij}}$ is replaced by $b_{\gamma_{ij}}$.*

Let

$$T_1 = T_{10} + T_{11}, \tag{69}$$

where the T_{1k}, $k = 0, 1$, are defined by (18) with the forms b_{1k} introduced above.

Theorem 2 *For any $v_h \in V_1^h$*

$$\delta^{-2} ca(v_h, v_h) \le a(T_1 v_h, v_h) \le C\delta a(v_h, v_h), \tag{70}$$

where c and C are positive constants independent of H and h_i, $i = 1, \ldots, N$, and $\delta = (1 + \log \frac{H}{h})$ and $h = \inf_i h_i$.

The method can be implemented as described in Section 3.

7 PROOF OF THEOREM 2

We first formulate auxiliary results, which are needed to prove Theorem 2.

The following lemma is the generalization of Lemma 1 to the geometrically nonconforming case.

Lemma 6 *Let $u_{i,h}$ and $u_{l,h}, \ldots, u_{p,h}$ vanish at the vertices of $\Omega_i, \Omega_l, \ldots, \Omega_p$. Let γ_{ij} be a slave side of Ω_i while $\gamma_{lj}, \ldots, \gamma_{pj}$ are master sides of $\Omega_l, \ldots, \Omega_p$ which intersect γ_{ij}. Under the condition (62), assuming that $h_i, h_l, \ldots, h_p$ are of the same order, it holds*

$$\begin{aligned} ||u_{i,h}||^2_{H^{1/2}_{00}(\gamma_{ij})} &\le C\Big\{\Big(1 + \log \frac{H}{h_l}\Big)||u_{l,h}||^2_{H^{1/2}_{00}(\gamma_{lj})} + \ldots + \\ &\quad + \Big(1 + \log \frac{H}{h_p}\Big)||u_{p,h}||^2_{H^{1/2}_{00}(\gamma_{pj})}\Big\}. \end{aligned} \tag{71}$$

Proof. To simplify our presentation, let γ_{ij} be a sum of only two Γ_{ki} with $k = l$ and $k = p$, i.e. $\gamma_{ij} = \Gamma_{li} \cup \Gamma_{pi}$ where $\Gamma_{li} \subset \gamma_{lj}$ and $\Gamma_{pi} \subset \gamma_{pj}$. Let u_{lp} be equal to $u_{l,h}$ on γ_{li} and be equal to $u_{p,h}$ on Γ_{pi}. Note that these functions do not vanish at the ends of γ_{ij}. Using the projection Π_i introduced in Section 4, see (28), we see that $u_{i,h} = \Pi_i u_{lp}$ on γ_{ij} since $u_{i,h}$ and $u_{l,p}$ satisfy the mortar condition (62). We now show that

$$||\Pi_i u_{lp}||^2_{H^{1/2}_{00}(\gamma_{ij})} \le C\Big\{\Big(1 + \log \frac{H}{h_l}\Big)||u_{l,h}||^2_{H^{1/2}_{00}(\gamma_{lj})} + \Big(1 + \log \frac{H}{h_p}\Big)||u_{p,h}||^2_{H^{1/2}_{00}(\gamma_{pj})}\Big\}. \tag{72}$$

To see this, we introduce functions z_l and z_p defined as follows. Let the ends of γ_{ij} be denoted by A and B; they are in general not nodal points of the h_l- and h_p-triangulations on γ_{lj} and γ_{pj}. Let A lie between C and D, two neighboring nodal points of the h_l-triangulation. The function z_l is equal to $u_{l,h}$ at C and D and zero at the remaining nodal points. In a similar way, we define z_p. Let $z_{lp} = z_l + z_p$. We have

$$||\Pi_i u_{lp}||^2_{H^{1/2}_{00}(\gamma_{ij})} \leq 2\Big(||\Pi_i(u_{lp} - z_{lp})||^2_{H^{1/2}_{00}} + ||\Pi_i z_{lp}||^2_{H^{1/2}_{00}}\Big). \tag{73}$$

Using the inverse inequality and the fact that Π_i is L_2 - stable, see (32), we get

$$||\Pi_i z_{lp}||^2_{H^{1/2}_{00}(\gamma_{ij})} \leq \frac{C}{h_i}\Big|\Big|z_{lp}\Big|\Big|^2_{L_2(\gamma_{ij})} \leq C\Big(||u_{l,h}||^2_{L^\infty(\gamma_{lj})} + ||u_{p,h}||^2_{L^\infty(\gamma_{pj})}\Big). \tag{74}$$

provided that h_i, h_l, and h_p are of the same order. Using Lemma 5 and the fact that $u_{l,h}$ and $u_{p,h}$ vanish at some points, we get

$$||\Pi_i z_{lp}||^2_{H^{1/2}_{00}(\gamma_{ij})} \leq C\Big\{\Big(1 + \log\frac{H}{h_l}\Big)|u_{l,h}|^2_{H^{1/2}(\gamma_{lj})} + \Big(1 + \log\frac{H}{h_p}\Big)|u_{p,h}|^2_{H^{1/2}(\gamma_{pj})}\Big\}. \tag{75}$$

To estimate the first term of (73), we can use (31) since $u_{lp} - z_{lp}$ is a continuous function and vanishes at the ends of γ_{ij}. Thus,

$$\begin{aligned} ||\Pi_i(u_{lp} - z_{lp})||^2_{H^{1/2}_{00}(\gamma_{ij})} &\leq C||u_{lp} - z_{lp}||^2_{H^{1/2}_{00}(\gamma_{ij})} \leq \\ &\leq C\Big(||u_{l,h} - z_l||^2_{H^{1/2}_{00}(\Gamma_{li})} + ||u_{p,h} - z_p||^2_{H^{1/2}_{00}(\Gamma_{pi})}\Big). \end{aligned} \tag{76}$$

It is easy to see that

$$||u_{l,h} - z_l||^2_{H^{1/2}_{00}(\Gamma_{li})} \leq C\Big(1 + \log\frac{H}{h_l}\Big)||u_{l,h}||^2_{H^{1/2}_{00}(\gamma_{lj})},$$

since

$$||z_l||^2_{H^{1/2}(\Gamma_{li})} \leq C||u_{l,h}||^2_{L^\infty(\gamma_{lj})} \leq C\Big(1 + \log\frac{H}{h_l}\Big)||u_{l,h}||^2_{H^{1/2}_{00}(\gamma_{lj})}$$

in view of Lemma 5 and Poincaré's' inequality. In a similar way, we estimate $u_{p,h} - z_p$. Substituting these estimates into (76), we get

$$\begin{aligned} ||\Pi_i(u_{lp} - z_{lp})||^2_{H^{1/2}_{00}(\gamma_{ij})} &\leq C\Big\{\Big(1 + \log\frac{H}{h_l}\Big)||u_{l,h}||^2_{H^{1/2}_{00}(\gamma_{lj})} + \\ &+\Big(1 + \log\frac{H}{h_p}\Big)||u_{p,h}||^2_{H^{1/2}_{00}(\gamma_{pj})}\Big\}. \end{aligned} \tag{77}$$

In turn, substituting (75) and (77) into (73), we get (72). □

Proofs of the two following lemmas are almost the same as the proofs of Lemmas 3 and 4, respectively, therefore they are not given here.

Lemma 7

$$\|\varphi_{il}^{(ms)}\|^2_{H_{00}^{1/2}(\gamma_{ij})} \le C\Big(1+\log\frac{H}{h_i}\Big). \tag{78}$$

Lemma 8

$$|\varphi_{ilp}|^2_{H^{1/2}(\partial\Omega_i)} \le C\Big(1+\log\frac{H}{h_i}\Big). \tag{79}$$

Proof of Theorem 2 We must check the three key assumptions as in Section 5.
Assumption (iii) obviously holds.
Assumption (ii). We need to show that for $k=0,1$, and $u_h \in V_{1k}$

$$a(u_h,u_h) \le \omega b_{1k}(u_h,u_h). \tag{80}$$

For $k=1$, it follows from Lemma 6. Let $a_i(u_h,u_h)$ be the restriction of $a(u_h,v_h)$ to Ω_i. We have

$$a_i(u_h,u_h) \le C \sum_{\gamma_{ij}\subset\partial\Omega_i} |u_{i,h}|^2_{H_{00}^{1/2}(\gamma_{ij})}. \tag{81}$$

For the mortar side, we use (49) while for the slave side we use Lemma 6. Substituting these estimates into (81) and summing the resulting inequality with respect to i, we get (80) with $\omega = C\delta$. To show (80) for $k=0$ and $u_h \in V_{10}$, we use a representation of u_h on $\partial\Omega_i$ given by formulas similar to (51) - (53). Note that this time $u_{i,h}$, on the slave side γ_{ij}, is of the form

$$\begin{aligned} u_{i,h}(x) &= u_{i,h}(A)\varphi_{ij,A}^{(s)} + u_{i,h}(B)\varphi_{ij,B}^{(s)} + \\ &+ u_{l,h}(C)\varphi_{li,C}^{(ms)} + u_{l,h}(D)\varphi_{li,D}^{(ms)} + \ldots + u_{p,h}(E)\varphi_{pi,E}^{(ms)} + u_{p,h}(F)\varphi_{pi,F}^{(ms)}, \end{aligned} \tag{82}$$

where A and B are the ends of γ_{ij}, the slave side of Ω_i, while C and D are the ends of γ_{li}, the mortar side of Ω_l which intersects Ω_i, etc.

The function $u_{i,h}$ on the mortar side γ_{ij} is linear, see (52).

Note also that the sum of basis functions used in the representation of $u_{i,h}$ on $\partial\Omega_i$ is equal to one. Using that and Lemmas 2, 6 to 8, we show straightforwardly that

$$\begin{aligned} |u_{i,h}-\overline{u}_{i,h}|^2_{H^{1/2}(\partial\Omega_i)} \le\; & C\Big\{\Big(1+\log\frac{H}{h_i}\Big)\sum_{x\in\nu_i}\Big(u_{i,h}-\overline{u}_{i,h}\Big)^2 + \\ &+ \sum_{\gamma_{ij}\subset\partial\Omega_i}\Big[\sum_{x\in\nu_{li}}\Big(1+\log\frac{H}{h_l}\Big)\Big(u_{l,h}^{(m)}-\overline{u}_{l,h}^{(s)}\Big)^2 + \ldots + \\ &+ \sum_{x\in\nu_{pi}}\Big(1+\log\frac{H}{h_p}\Big)\Big(u_{p,h}^{(m)}-\overline{u}_{i,h}^{(s)}\Big)^2\Big]\Big\}. \end{aligned} \tag{83}$$

Summing this up with respect to i, we get (80) for $k = 0$ with $\omega = C\delta$. The proof of Assumption (ii) is complete.

Assumption (i). We need to show (56) for the form $b_{1k}(u_h, v_h)$ defined in Section 6, see (67) and (68). Let $u_h^{(0)} \in V_{10}$ be the interpolant of $u_h \in V_1^h$ and $u_h^{(1)} = u_h - u_h^{(0)}$. Of course, $u_h^{(1)} \in V_{11}$. The estimate

$$b_{11}(u_h^{(1)}, u_h^{(1)}) \leq C\delta^2 a(u_h, u_h) \tag{84}$$

is proved as in Section 5, using Lemma 6.

To show that

$$b_{10}\left(u_h^{(0)}, u_h^{(0)}\right) \leq C\delta a(u_h, u_h), \tag{85}$$

we proceed in a way similar to that in Section 5. Adding (84) and (85), we get the estimate by checking Assumption (i). The proof of Theorem 2 is complete.

8 REFERENCES

Achdou, Y., Maday, Y., and Widlund, O. (1995) Substructuring preconditioners for the mortar method in two dimensions. *Tech. Rep., Courant Institute* (to appear).

Belgacem, F. (1995) The mortar finite element method with Lagrange multipliers. *Methods in Applied Mechanics and Engineering* (to appear).

Bernardi, Chr. and Maday, Y. (1995) Mesh adaptivity in finite elements by the mortar method. *Tech. Rep.* R 94029, *Université Pierre et Marie Curie and CNRS.*

Bernardi, Chr., Maday, Y., and Patera, A. (1994) A new nonconforming approach to domain decomposition: the mortar element method. *Pitman*, Brezis, H. and Lions J.L., eds.

Ciarlet, P.G. (1978) *The finite element method for elliptic problems.* North-Holland, Amsterdam.

Dryja, M. (1982) A capacitance matrix method for Dirichlet problem on polygonal region. *Numer. Math.* 39, 51–64.

Dryja, M. (1988) A method of domain decomposition for 3-D finite element problems. *In First International Symposium on Domain Decomposition Methods for PDEs*, Glowinski, A., Golub, G.H., Meurant, G.A. and Périaux, J., eds., SIAM, Philadelphia, PA, 43–61.

Dryja, M., Smith, B., and Widlund, O. (1994) Schwarz analysis of iterative substructuring algorithms for elliptic problems in three dimensions. SIAM *J. Numer. Anal.* 31, 1662–1694.

Dryja, M. and Widlund, O. (1995) A bound of the condition number of mortar finite element stiffness matrices (in preparation).

Widlund, O. (1995) Domain decomposition methods for spectral and mortar finite element approximation of elliptic problems. *Proceedings of the Third International Congress on Industrial and Applied Mathematics*, Hamburg, June 1995 (to appear).

4

Soil venting

U. Hornung, Y. Kelanemer and M. Slodička
Department of Computer Science
University of the Federal Armed Forces Munich
D-85577 Neubiberg, Germany. Fax: +49-89-6004-3560
E-mail: ulrich@informatik.unibw-muenchen.de
youcef@informatik.unibw-muenchen.de
marian@informatik.unibw-muenchen.de

Abstract

Soil venting is a technique for remediation of soils contaminated by NAPLs (non aqueous phase liquids, often called VOCs, i.e., volatile organic compounds). During the operation of soil venting systems the vapor phase of the nonaqueous phase liquid NAPL is being removed by imposing a convective air stream and thus enhancing the evaporation of the fluid phase of the NAPL in contaminated regions of the soil.

In mathematical terms, this multi-phase multi-component process is described by a system of nonlinear parabolic equations. The variables therein are mass and volume fractions of the chemical components involved and the pressures of the different phases. Assuming that the mathematical model has been chosen appropriately, two major steps have to be performed:

"Calibration" of the model. This can be done only for each case separately, since the geological structure of the soil and the spatial distribution of the contaminants have to be taken into account. In addition, due to lack of data and to measurement errors, simulations are possible only in a statistical sense, i.e., conditional Monte-Carlo simulations have to be performed.

"Optimization" of the remediation procedure. The control variables are the number and position of extraction wells (and/or injection wells) and the extraction rates. Since the whole process may take many years, even "sub-optimal" strategies could be of great value.

Keywords

Multi-phase and multi-component transport, porous media, soil venting, remediation

1 INTRODUCTION

In recent years, contamination of ground water by industrial waste has become a serious problem. Benzene, mineral oil, solvents, or other organic compounds entering the subsurface may be a serious and potentially long-term hazard for the environment, e.g., soils and

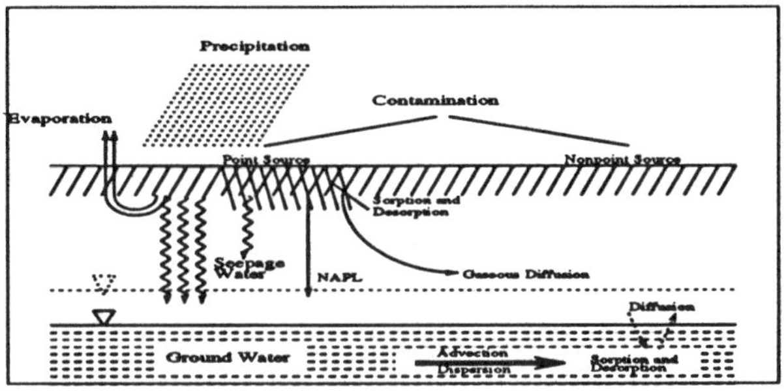

Figure 1 Contaminated soil

ground water systems, see figure 1. For removing volatile organic compounds (VOCs) from the water-unsaturated zone of contaminated soils commonly the vapor extraction method is used. During the operation of soil venting systems the vapor phase of the nonaqueous phase liquid VOC is being removed by imposing a convective air stream and enhancing the evaporation of the fluid phase of the VOCs in contaminated regions of the soil. Frequently, soil venting systems are designed empirically, using extensive experimental studies and observations.

The major problems that the *engineer* has to solve are these:

1. *Exploration and Installation:* This means, he has to find out and describe the spatial distribution of the pollution and the geology, i.e., the soil properties of the site. And he has to decide which type of the clean-up procedure should be chosen.
2. *Monitoring and Maintenance:* He has to monitor the extraction rates and maintain the machinery and equipment. And he has to decide on modifications, if necessary.
3. *Stopping Criteria:* He has to decide at what time the remediation process may be stopped.

The major problems that the *scientist*, i.e. the mathematician, soil physicist, or soil chemist has to solve are these:

1. *General Model:* An appropriate mathematical physical chemical model has to be found.
2. *Model Calibration:* For each individual remediation site the model has to be calibrated taking the geology and the distribution of the pollution into account. If necessary, modifications may have to be made.
3. *Optimal Control:* The choice and the design of the procedure have to be optimized considering duration and costs of the remediation process.

2 A REMEDIATION SITE

On one of the remediation sites in Germany an estimated 100 tons of VOCs have seeped to the upper three to four meters of the soil in an area of about $20,000\ m^2$. This area contains several buildings with basements. Its surface is partially sealed by an asphalt layer of 5...10 cm, see figure 2. The measured concentrations (in units of mg/m^3) of two

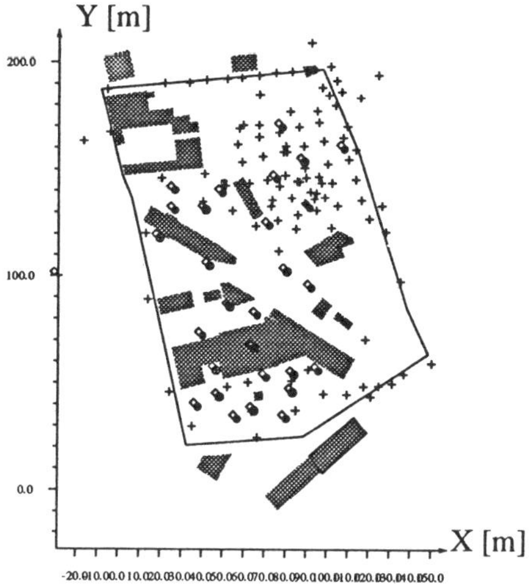

Figure 2 Boreholes, i.e., points where measurements are taken (in the shaded domains there are buildings with basements)

selected VOCs, namely Tetrachlorethen and 1,1,1-Trichlorethan, are shown in figure 3. These plots show only parts of the remediation site, since probes were made only at the points indicated by black dots.

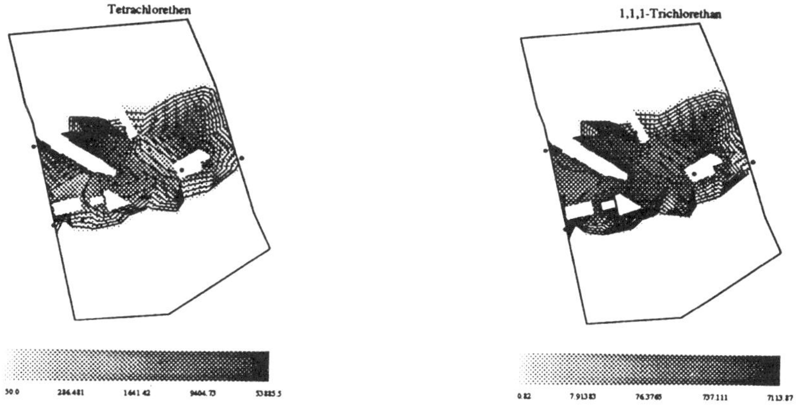

Figure 3 Concentrations of VOCs: Tetrachlorethen and 1,1,1-Trichlorethan

3 A GENERAL PHYSICAL MODEL

Theoretical approaches and simulation models for optimization of the design and the operation of venting systems are often based on narrow physical and simplified mathematical descriptions of the key processes and, furthermore, relatively inefficient numerical methods are frequently applied for the solutions of systems of differential equations.

Most models have been restricted to two-dimensional vertical sections (Kaluarachchi and Parker (1989, 1990), Kuppusamy, Sheng, Parker and Lenhard (1987)) and vertically integrated models (Hochmuth and Sunada (1985), Kaluarachchi, Parker and Lenhard (1990)). Abriola and Pinder (1985) presented a computational model which became a basis for many other papers. Baehr and Corapcioglu (1987) presented a multi-component transport model including oxygen-limited biodegradation. Most papers are based on the assumption of local equilibrium between various phases when describing contaminant volatilization during the soil vapor extraction (Baehr, Hoag and Marley (1989); Rathfelder, Yeh and Mackay (1991)). Non-equilibrium phase models have been presented by Sleep and Sykes (1989), Wilkins, Abriola and Pennell (1995). Mayer, Miller, Poirer-McNeill (1990) presented a comparison between equilibrium and non-equilibrium oil-water models.

3.1 Balance laws

Multi-component and multi-phase transport in porous media is considered (Russell (1995)). Four different phases are taken into consideration: solid *(s)*, liquid (water) *(l)*, gas *(g)* and contaminant *(o)*. The following four components are considered: soil *(s)*, air *(a)*, water *(w)* and volatile contaminant *(v)* capable of crossing phase boundaries.

The macroscopic mass balance for component i in phase α is written as

$$\partial_t(\rho^\alpha \varepsilon^\alpha \omega_i^\alpha) + \nabla \cdot (\rho^\alpha \mathbf{q}^\alpha \omega_i^\alpha) - \nabla \cdot \mathbf{J}_i^\alpha = \rho^\alpha \varepsilon^\alpha \left[f_i^\alpha + e_i^\alpha\right] \tag{1}$$

where ρ^α $\left[\frac{kg}{m^3}\right]$ is the mass density of the phase α; ε^α [1] is the volume fraction occupied by the phase α; $\mathbf{q}^\alpha$ $\left[\frac{m}{s}\right]$ is Darcy's velocity of the phase α; ω_i^α [1] is the mass fraction of component i in the α phase; $\mathbf{J}_i^\alpha$ $\left[\frac{kg}{m^2 s}\right]$ is the flux vector representing the diffusive flux of component i in the phase α; f_i^α $\left[\frac{1}{s}\right]$ is the source of component i in the phase α; e_i^α $\left[\frac{1}{s}\right]$ is the gain of mass of component i due to phase change.

Equation (1) is written under the following constraints:

$$\sum_i \omega_i^\alpha = 1 \;, \quad \sum_\alpha \varepsilon^\alpha = 1 \;, \quad \sum_\alpha \rho^\alpha \varepsilon^\alpha e_i^\alpha = 0. \tag{2}$$

Each component could appear in different phases at the same time, thus the macroscopic mass balance for component i is derived from (1) by summing over all phases, taking into account the last equation of (2)

$$\sum_\alpha \left[\partial_t(\rho^\alpha \varepsilon^\alpha \omega_i^\alpha) + \nabla \cdot (\rho^\alpha \mathbf{q}^\alpha \omega_i^\alpha) - \nabla \cdot \mathbf{J}_i^\alpha\right] = \sum_\alpha \rho^\alpha \varepsilon^\alpha f_i^\alpha. \tag{3}$$

Analogously, one can derive the macroscopic mass balance for phase α

$$\sum_i [\partial_t(\rho^\alpha \varepsilon^\alpha \omega_i^\alpha) + \nabla \cdot (\rho^\alpha \mathbf{q}^\alpha \omega_i^\alpha) - \nabla \cdot \mathbf{J}_i^\alpha] = \rho^\alpha \varepsilon^\alpha \sum_i [f_i^\alpha + e_i^\alpha]. \tag{4}$$

This is the general framework of multi-phase multi-component transport. To make this operational, special assumptions have to be introduced.

3.2 Special assumptions

We adopt the following assumptions in order to simplify the calculations.

- The soil matrix is rigid, i.e. incompressible and immobile. There is no mass exchange across a solid-fluid interface.
- The temperature field is given; no freezing occurs.
- Water appears in the liquid phase l only.
- Air may be present in phase g only.
- The volatile organic component appears in two phases o, g.
- There are no distributed sources for water and for the organic component.
- Darcy's law for phase α is adopted in the following form (Forchheimer's law is compared with Darcy's law below in the text)

  $$\mathbf{q}^\alpha = -\frac{\mathbf{k} k_r^\alpha}{\mu^\alpha} \cdot (\nabla p^\alpha - \rho^\alpha \mathbf{g})$$

  where $\mathbf{k}$ $[m^2]$ is the permeability; k_r^α [1] - relative permeability of phase α; μ^α $\left[\frac{kg}{ms}\right]$ - dynamic viscosity of phase α; p^α $\left[\frac{N}{m^2}\right]$ - pressure of phase α.

The local non-equilibrium is assumed to obey a linear phase change law of volatile contaminant between phases o and g, namely

$$e_v^g = \frac{\varepsilon^o}{\varepsilon^g} \lambda_v^{og} (\omega_{v,sat}^g - \omega_v^g) \text{ and } e_v^o = -\frac{\rho^g}{\rho^o} \lambda_v^{og} (\omega_{v,sat}^g - \omega_v^g)$$

where λ_v^{og} $\left[\frac{1}{s}\right]$ is the phase change rate between both phases, $\omega_{v,sat}^g$ [1] - mass fraction of volatile contaminant v in the phase g at saturation. The continuity equation

$$\rho^g \varepsilon^g e_v^g + \rho^o \varepsilon^o e_v^o = 0$$

is satisfied.

3.3 Equation summary

The general laws of section 3.1 and the assumptions of section 3.2 lead to the following eight equations with eight unknowns: mass fractions (ω_v^g, ω_v^o), volume fractions (ε^l, ε^g, ε^o), and pressures (p^l, p^g, p^o).

1. Water conservation

$$\partial_t(\rho^l\varepsilon^l) + \nabla \cdot (\rho^l \mathbf{q}^l) - \nabla \cdot \mathbf{J}^l_w = 0$$

2. Volatile component conservation for the gaseous phase

$$\partial_t(\rho^g\varepsilon^g\omega^g_v) + \nabla \cdot (\rho^g \mathbf{q}^g \omega^g_v) - \nabla \cdot \mathbf{J}^g_v = \varepsilon^o \lambda^{og}_v \rho^g (\omega^g_{v,sat} - \omega^g_v)$$

3. Volatile component conservation for the organic phase

$$\partial_t(\rho^o\varepsilon^o\omega^o_v) + \nabla \cdot (\rho^o \mathbf{q}^o \omega^o_v) - \nabla \cdot \mathbf{J}^o_v = -\varepsilon^o \lambda^{og}_v \rho^g (\omega^g_{v,sat} - \omega^g_v)$$

4. Air conservation

$$\partial_t(\rho^g\varepsilon^g) + \nabla \cdot (\rho^g \mathbf{q}^g) - \nabla \cdot \mathbf{J}^g_a = 0$$

5. $\omega^g_v + \omega^g_a = 1$
6. $\varepsilon^s + \varepsilon^l + \varepsilon^g + \varepsilon^o = 1$
7. $p^l - p^g = p^{lg}(\varepsilon^l, \varepsilon^g)$
8. $p^o - p^g = p^{og}(\varepsilon^o, \varepsilon^g)$

If the full problem of coupled water, NAPL, and air transport is to be modelled, this nonlinear system of four differential and four algebraic equations has to be solved simultaneously, together with appropriate initial and boundary conditions. At the moment, it is a challenging task to solve this system in 2D or 3D.

3.4 Turbulent flow

In general, Darcy's law is valid for laminar flow (the Reynolds number $Re < 1 \ldots 10$). Here the viscous forces are predominant. As the velocity of flow increases, Darcy's law is no longer valid; at higher values of Re (say $Re > 150 \ldots 300$) the flow becomes turbulent. Forchheimer's law

$$-\nabla p = \frac{\mu}{k}\mathbf{q} + \frac{\rho\alpha}{k^{1/2}}|\mathbf{q}|\mathbf{q}$$

presents one of the generalizations of Darcy's law (α is the viscosity resistance coefficient), where gravity effects for gas have been omitted. Putting this into the mass balance equation for air

$$\partial_t(\rho\theta) + \nabla \cdot (\rho \mathbf{q}) = 0$$

and using a gas law of the form $\rho = \beta p^\gamma$ (where β is the inertia coefficient, and γ is the adiabatic coefficient $0 \leq \gamma \leq 1$), one gets

$$\partial_t(\beta\theta u^{\gamma/(1+\gamma)}) - \nabla \cdot (F(|\nabla u|)\nabla u) = 0$$

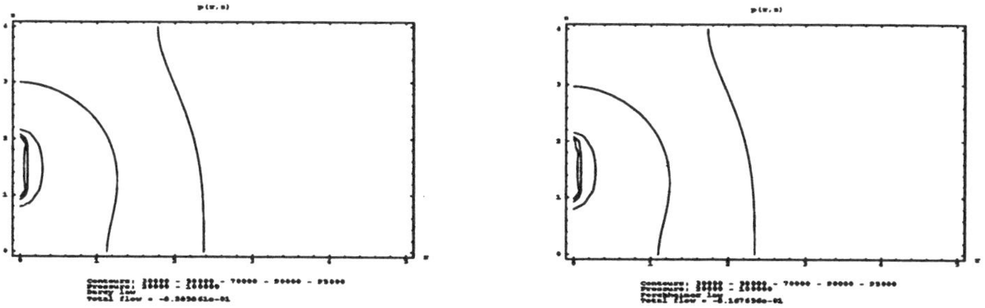

Figure 4 Darcy - Forchheimer

with

$$F(s) = \frac{\mu}{2\alpha k^{1/2}} \left[\left(1 + \frac{4\alpha\beta k^{3/2}}{(1+\gamma)\mu^2} s \right)^{1/2} - 1 \right] \frac{1}{s}.$$

This differential equation is degenerate for $u \to 0$ and for $u \to \infty$; there are special numerical tools for dealing with such degeneracies (cf. Jäger and Kačur (1994)); in practical soil remediation problems the regime of these degeneracies is not reached. We have compared Darcy and Forchheimer's law for a steady state, one well, radially symmetric case. The equipotentials are drawn in figure 4. In spite of high discharge of the well the difference between both cases is small and thus this kind of nonlinearity is practically not significant.

4 THE GAS FLOW FIELD

In practical applications it has turned out that an important piece of information is the gas flow field. Even if this is not being used as part of the full system of section 3.3, this knowledge as such is relevant to estimate the operation of the pumping strategy. If the spatial distribution of the pollutant has been found out, one obviously should try to generate a high gas stream in those domains where the NAPL concentration is large. In this section we describe an approach for calculating the air flow field for the remediation site described in section 2.

4.1 Air permeability

Compressible flow in porous media has been a subject of investigations for many years. Mathematical models of air movement in unsaturated porous media have been calibrated using air pressure data in several investigations in order to determine the air permeability. When considering the standard two-phase model of flow in unsaturated porous media (air and water), the soil consists of the porous matrix, the water and the air phase. Thus one

Layer	$\theta_r[-]$	$\theta_s[-]$	K_s $[10^{-12}m^2]$	$\alpha[cm^{-1}]$	$l[-]$	$n[-]$
Coarse Sand	0	0.449	655.87	0.4976	0.5	3.551
Middle Sand	0	0.426	10.28	0.2789	0.5	1.596
Fine Sand 1	0	0.457	6.57	0.2501	0.5	1.422
Fine Sand 2	0	0.419	6.00	0.1729	0.5	1.912

Table 1 Van-Genuchten parameters for water ($m = 1 - \frac{1}{n}$)

can write

$$\theta_{gas} + \theta_{water} + \theta_{matrix} = 1$$

for the saturations. We suppose that the domain is insulated on all sides and there exist active and passive air-wells; this means pumps generating a certain under-pressure are used at the active wells, and the passive wells are allowed to function as sources of air. In this situation the flux of air will be determined by the pumping rates at the active wells. There are no distributed sink/source for water and we consider the water saturation as a given function of position (x, y, z). We have used van-Genuchten formula for describing the relation between pressure head h and the saturation θ_{water}

$$\theta(h) = \theta_r + \frac{\theta_s - \theta_r}{(1 + (\alpha h)^n)^m}.$$

The parameters are given in the table 1. The relationships $h - \theta$ are plotted in figure 5 for different layers. The air permeability k_{gas} varies in the different layers and it is given by the formula

$$k_{gas} = k_{water}\frac{\mu_{water}}{\mu_{gas}},$$

where μ_{water} and μ_{gas} are the dynamic viscosities of water and air, respectively. In this way one obtains K_{gas} as a function of the position (x, y, z).

4.2 Horizontal flow in 2D

In many cases the domain for which the model has to be developed is essentially two-dimensional. Here we consider a situation in which the vertical thickness is small compared with the horizontal lengths. Then the numerical calculations can be significantly simplified by averaging over the vertical axis (see Gilding (1988)). This methodology is explained here for steady-state compressible gas flow. In three-dimensional space, conservation of mass is given by

$$\nabla \cdot (\rho_{gas}\mathbf{q}_{gas}) = f_{gas}$$

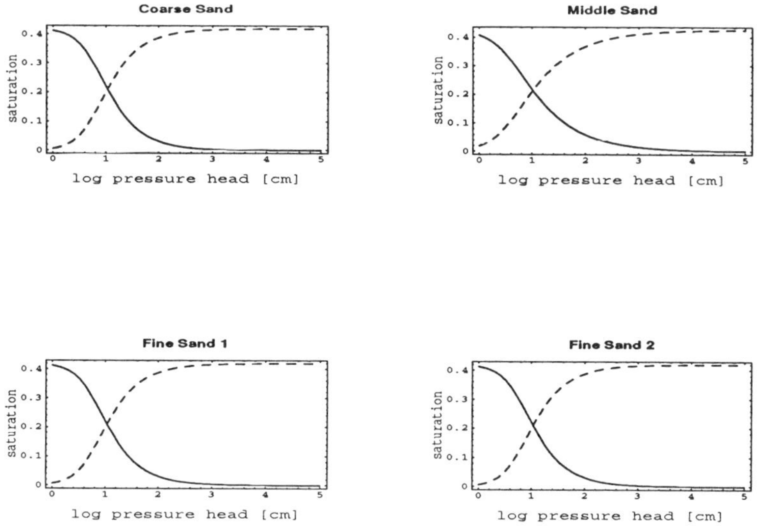

Figure 5 $h - \theta$ relationships for different layers (the full line corrsponds to air and the dashed line to water)

where ρ_{gas} is the gas density, $\mathbf{q}_{gas}$ denotes the specific discharge vector, and f_{gas} a distributed source, if there is any. According to the ideal gas law, Darcy's law and for constant temperature, one obtains

$$\nabla \cdot (\tilde{k}_{gas} \nabla p_{gas}{}^2) = f_{gas}.$$

Under the assumption that the gas flow is essentially horizontal, one integrates between the soil surface at $z = z_2$ and the water table at $z = z_1$ (below $z = z_1$ the soil is water saturated, thus there is no gas flow) and gets

$$\nabla \cdot (T_{gas} \nabla u) = \overline{f_{gas}} \tag{5}$$

where u is the two-dimensional average of $p_{gas}{}^2$, T_{gas} is the transmissivity

$$T_{gas}(x,y) = \int_{z_1}^{z_2} \tilde{k}_{gas}(x,y,z)dz$$

and $\overline{f_{gas}}$ is the average of the source

$$\overline{f_{gas}}(x,y) = \int_{z_1}^{z_2} f_{gas}(x,y,z)dz.$$

Equation (5) has to be considered in a two-dimensional domain Ω. In principle, this is a straightforward and well known method. In practical applications - of course - there is only a limited number of measurements available. In most cases the variable T_{gas} is known at a relatively small number of points that are non-uniformly distributed in the domain Ω. A method that is used very frequently in the geosciences is "Kriging" (see, e.g., Journel and Huijbregts (1978)). The basic idea is that one considers the variable in question - here $T = T_{gas}$ - as a random variable in space and assumes that the measurements T^i taken are realizations of this random field. In a first step one assumes that the field is stationary and estimates its spatial auto-covariance structure. In a second step an "interpolation" $\tilde{T}(\mathbf{X})$ of the T^i at an arbitrary point $\mathbf{X}$ is calculated as a convex combination

$$\tilde{T}(\mathbf{X}) = \sum_i \lambda_i(\mathbf{X})T^i.$$

of the T^i. The weights $\lambda_i(\mathbf{X})$ are determined by the requirement that the variance of the deviation $T(\mathbf{X}) - \tilde{T}(\mathbf{X})$ is as small as possible. As a consequence, for each point $\mathbf{X}$ where the "interpolant" $\tilde{T}(\mathbf{X})$ is needed a linear system with a symmetric matrix has to be solved for the weights $\lambda_i(\mathbf{X})$. A modified conjugate gradient (CG) methods proves to be useful in this context.

4.3 Wells in 2D

Wells are used for extracting gas from the soil during the venting process. For the mathematical model special care has to be taken when incorporating them into the equations. There are two different cases:

1. The radius of the wells is sufficiently **large**. Then one may consider the boundary of the well as part of the boundary of the domain Ω. In this case it may be crucial to use local refinements of a finite element grid in the vicinity of the well, because one expects large gradients of the pressure and thus large discharges there.
2. The radius of the wells is **small** compared to the diameter of the domain Ω. In this case, it is reasonable to approximate the wells as point sources (or sinks).

Here we study the second case. Assuming that the discharge of each well is known, one considers the boundary value problem

$$\begin{cases} -\nabla \cdot (T(\mathbf{X})\nabla u) = \sum_{j=1}^{N} s_j \delta(\mathbf{X} - \mathbf{X}_j) & \text{in } \Omega \\ u = u_D & \text{on } \Gamma_D \\ -T(\mathbf{X})\nabla u \cdot \nu = q_N & \text{on } \Gamma_N. \end{cases} \tag{6}$$

where $\mathbf{X} = (x,y)$ is the space variable, $\mathbf{X}_j \in \Omega$; $j = 1,\ldots,N$ are the locations of the sources/sinks in the domain Ω, δ is the Dirac delta-function, s_j is the specific discharge of the well at the point $\mathbf{X}_j$, and u_D and q_N are given boundary data on Γ_D and Γ_N, resp. Assuming that the transmissivity T is constant $T_j = T(\mathbf{X}_j)$ in the vicinity of each of the wells, the local behavior of the solution u is given by a multiple of the fundamental solution

$$\omega_j(\mathbf{X}) = -\frac{1}{2\pi} \ln |\mathbf{X} - \mathbf{X}_j|.$$

The idea is to synthesize the solution u of the boundary value problem (6) in the form

$$u = \tilde{u} + \sum_j \frac{s_j}{T_j} \omega_j.$$

The function $\tilde{u}$ solves the equation

$$-\nabla \cdot (T(\mathbf{X}) \nabla \tilde{u}) = \sum_j \nabla \cdot \left([T(\mathbf{X}) - T_j] \nabla (\frac{s_j}{T_j} \omega_j) \right).$$

Therefore, $\tilde{u}$ is determined as the solution of the following boundary value problem.

$$\begin{cases} -\nabla \cdot (T(\mathbf{X}) \nabla \tilde{u}) = -\sum_j \frac{s_j}{2\pi T_j} \nabla \cdot \left[(T(\mathbf{X}) - T_j) \frac{\mathbf{X}-\mathbf{X}_j}{|\mathbf{X}-\mathbf{X}_j|^2} \right] & \text{in } \Omega \\ \tilde{u} = u_D + \sum_j \frac{s_j}{2\pi T_j} \ln \mid \mathbf{X} - \mathbf{X}_j \mid & \text{on } \Gamma_D \\ -T(\mathbf{X}) \nabla \tilde{u} \cdot \nu = q_N - T(\mathbf{X}) \sum_j \frac{s_j (\mathbf{X}-\mathbf{X}_j) \cdot \nu}{2\pi T_j |\mathbf{X}-\mathbf{X}_j|^2} & \text{on } \Gamma_N. \end{cases}$$

Since the singularities at the wells are subtracted from the function u, the auxiliary function $\tilde{u}$ is smooth.

If the discharges s_j of the wells are unknown, a modification of the approach described so far has to be made. Here we assume that values u_j for u at the boundaries of the wells are prescribed. Then we impose the condition

$$u = \frac{1}{2\pi R_j} \int_{|\mathbf{X}-\mathbf{X}_j|=R_j} u \, d\sigma = u_j$$

where R_j is the "effective radius" of the well at $\mathbf{X}_j$ and σ is the arc length. Now the discharges s_j have to be determined such that this condition is satisfied; in other words, we have to consider an augmented system of equations where the s_j are additional unknowns.

4.4 Numerical examples

Several test calculations were carried out for the remediation site shown in figure 2. First we have chosen homogeneous Neumann boundary conditions at the boundaries of the domain Ω and the same discharges for all active wells. The resulting pressure field (in units of Pa) is shown in the figure 6. The pressure together with the flow field (the lengths of the flux vectors are taken to be proportional to the logarithm of their moduli) can be found in figure 7. A scenario with a different number and modified locations of the active and passive wells is shown in figure 8.

As a comparison we have also taken Dirichlet boundary conditions at the boundary of

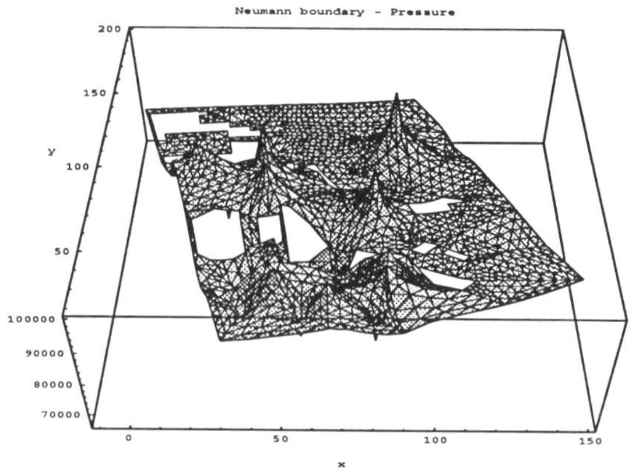

Figure 6 Zero Neumann boundary conditions: Pressure

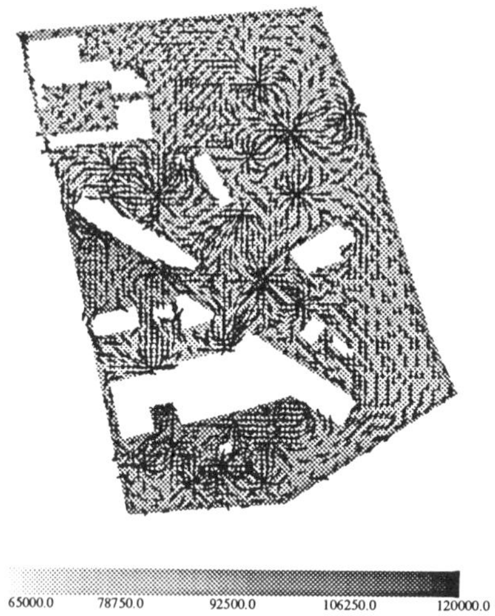

Figure 7 Zero Neumann boundary conditions: Pressure & Log Mass Flux

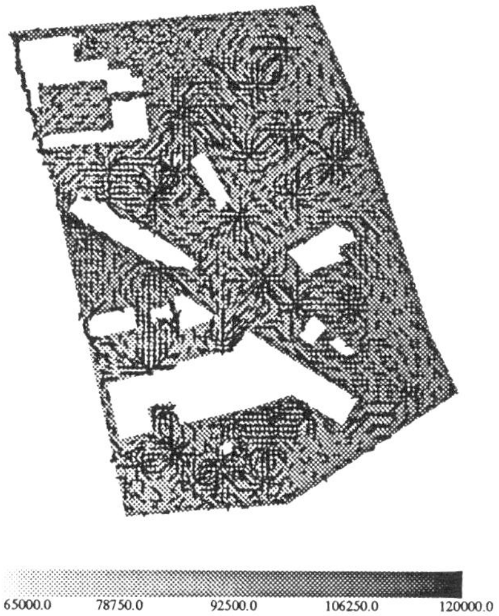

Figure 8 Modified location of wells: Zero Neumann boundary conditions: Pressure & Log Mass Flux

Ω, namely atmospheric pressure. Again, the discharges are the same for all active wells. The resulting fields are shown in figures 9 and 10.

5 OPTIMIZATION

The optimal control aspect of soil venting consists of optimizing the remediation process with respect to its efficiency. There are several conflicting goals: (i) One wants to extract as much VOCs as possible, (ii) the duration of the process should be short, (iii) the costs should be low, and (iv) restrictions with respect to the available technical equipment have to be fulfilled. It is by no means obvious how to formulate an appropriate mathematical objective/cost functional that adequately takes all these points into account (the choice of the "objective" function is highly "subjective").

Another basic difficulty arises when applying the tools from mathematical control theory. The distribution of the pollution and the geological properties of the soil are only partly known. This means that there is quite an uncertainty in an optimal control approach. Recently Unger, Sudicky, and Forsyth P. A. (1995) have addressed the problem of robustness of a remediation strategy with respect to spatial heterogeneities of the soil. Here we describe another methodology, namely we study - as a model problem - a random elliptic control problem. This means, we consider the coefficient (here this is the transmissivity) of the elliptic equation as a random field.

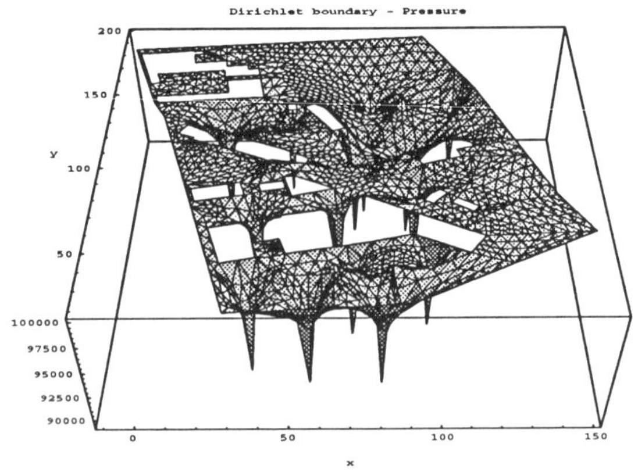

Figure 9 Atmospheric Dirichlet boundary conditions: Pressure

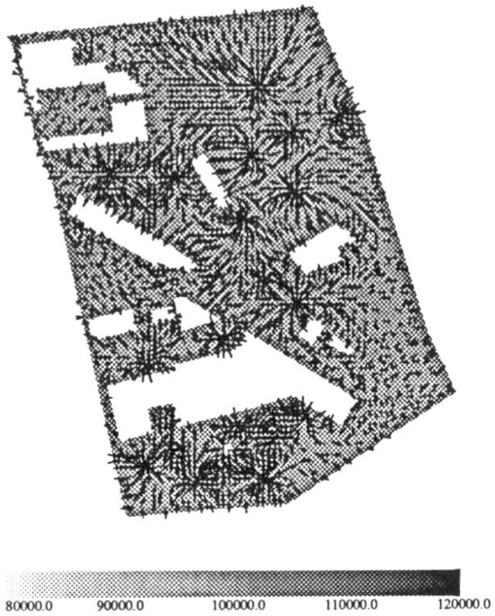

Figure 10 Atmospheric Dirichlet boundary conditions: Pressure & Log Mass Flux

5.1 A model problem

We study a typical problem where the control variables u_j are the pressures that are generated at some part of the boundary of a remediation site Ω, and the cost functional $J(u)$ simultaneously measures the normal flux (using Φ in equation (8)) that is produced at some other part of the boundary and also the costs (using Ψ in equation (8)) that are caused when generating the pressures u_j. In principle, such a problem is in the framework of well known theory for optimal control of partial differential equations. First we describe the problem in its deterministic version. We consider an elliptic boundary value problem of the form

$$\begin{cases} -\nabla\cdot(K\nabla y) & = f, & x \in \Omega \\ y & = g + Bu, & x \in \Gamma_D \\ q_\nu & = h, & x \in \Gamma_N \end{cases} \tag{7}$$

Here K is a strictly positive bounded function in the bounded domain $\Omega \subset \mathbb{R}^2$ the piecewise smooth boundary of which is $\Gamma = \Gamma_D \cup \Gamma_N$ with $\Gamma_D \cap \Gamma_N = \emptyset$. The functions f, g, and h are given boundary data, and the quantity $q_\nu = -\nu \cdot K\nabla y$ is the flux in normal direction on Γ. The *state* variable is $y(u)$, and the *control* variable is $u \in U$. The operator $B : U \to H^{1/2}(\Gamma)$ is assumed to be linear from the Hilbert-space U into $H^{1/2}(\Gamma)$. Its adjoint $B^* : H^{-1/2}(\Gamma) \to U$ is given by

$$\langle B^* r, v\rangle = \langle r, Bv\rangle \; \forall r \in H^{-1/2}(\Gamma), \; v \in U.$$

The *cost functional* is chosen as

$$\boxed{J(u) = \int_{\Gamma_D} \Phi(q_\nu(y(u))\mu)\, d\Gamma + \Psi(u)} \tag{8}$$

where Φ is a differentiable convex function on $\mathbb{R}$, $\mu \in L^\infty(\Gamma)$ is a weight function on Γ, and Ψ is a differentiable convex functional on U.

We introduce the *adjoint* variable p as the solution of the elliptic problem

$$\begin{cases} -\nabla\cdot(K\nabla p(u)) & = 0, & x \in \Omega \\ p(u) & = \Phi'(q_\nu(y(u))\mu)\mu, & x \in \Gamma_D \\ r_\nu & = 0, & x \in \Gamma_N \end{cases}$$

where $r_\nu = -\nu \cdot K\nabla p(u)$ is the normal boundary flux belonging to the adjoint p. We get the following formula for the differential of the cost functional.

$$\boxed{\partial_u J(u) = B^* r_\nu(p(u)) + \partial_u \Psi(u).}$$

Proof: Following the methodology of Lions (1971), chapter II.4, we get

$$\langle \partial_u J(u), v - u\rangle = \int_{\Gamma_D} \Phi'(q_\nu(y(u))\mu)(q_\nu(y(v)) - q_\nu(y(u)))\mu \, d\Gamma + \langle \partial_u \Psi(u), v - u\rangle$$

for any $v \in U$. Green's formula gives

$$\int_\Omega \nabla\cdot(K\nabla p(u))(y(v)-y(u))\,d\Omega - \int_\Gamma \nu\cdot K\nabla p(u)(y(v)-y(u))\,d\Gamma$$
$$= \int_\Omega \nabla\cdot(K\nabla(y(v)-y(u))p(u)\,d\Omega - \int_\Gamma \nu\cdot K\nabla(y(v)-y(u))p(u)\,d\Gamma$$

Since $\nabla\cdot(K\nabla p(u)) = 0$ in Ω, $y(v)-y(u) = Bv - Bu$ on Γ_D, $r_\nu = 0$ on Γ_N, $\nabla\cdot(K\nabla(y(v)-y(u)) = 0$ on Ω, and $q_\nu(y(v)) - q_\nu(y(u)) = 0$ on Γ_N, we get

$$\int_{\Gamma_D} p(u)(q_\nu(y(v)) - q_\nu(y(u)))\,d\Gamma = \int_{\Gamma_D} r_\nu B(v-u)\,d\Gamma,$$

and from this we obtain

$$\langle \partial_u J(u), v-u\rangle = \langle B^* r_\nu, v-u\rangle + \langle \partial_u \Psi(u), v-u\rangle$$

and thus the conclusion. Q.E.D.

Now we specialize more and assume $\Gamma_D = \Gamma_I \cup \Gamma_O$ with $\Gamma_I \cap \Gamma_O = \emptyset$. Further, we assume $B(U) \subset H^{1/2}(\Gamma_I)$. Let $U = I\!R^k$, and $\Gamma_j \subset \Gamma_I$ pairwise disjoint open subsets; for $u = (u_1, \ldots, u_k) \in U$ let $B(u)(x) = u_j$ whenever $x \in \Gamma_j$. In this case we get $(B^* r)_j = \langle r, 1_j\rangle$ for $r \in H^{-1/2}(\Gamma_I)$, where $1_j(x) = 1$ for $x \in \Gamma_j$ and $1_j(x) = 0$ for $x \in \Gamma_I \setminus \Gamma_j$. Let $\mu = 0$ on Γ_I. In addition, let $f = 0$, $g = 0$, and $h = 0$. For $\Phi(z) = \frac{z}{2}(z-2)$ we get $\Phi'(z) = z - 1$. If $\Psi(u) = \frac{1}{2}|u|^2$, we have $\partial_u \Psi = id$. Finally, we take $\mu = 1$ in the lower half of Γ_O and $\mu = 0$ in the upper half.

For an unconstrained control problem, i.e.

$$\inf_U J(u),$$

the necessary and sufficient optimality condition of first order is $\partial_u J(u) = 0$, i.e., $u = -B^* r_\nu(p(u))$. Since in our situation the problem is quadratic with respect to the control variable u, one efficient way of solving this numerically is the CG method. We have applied this technique to a problem in the unit square of $I\!R^2$; here Γ_I is the left vertical boundary, Γ_O is the right vertical boundary, and Γ_N consists of the two - lower and upper - horizontal boundaries. Results that were obtained using a hybrid mixed finite element method can be seen in figures 11 and 12. A transmissivity field K was chosen such that its values are either 1 (light areas) or 100 (dark areas). The meaning of this is that the soil described here contains very highly conducting "holes".

The main point of this numerical study is a Monte-Carlo experiment where we have used 100 realizations of a transmissivity field each of which had a similar geometric structure as the one of figure 11; the sizes and locations of the "holes" were chosen randomly. Now we have made the following comparison: (a) We have determined the optimal solution for each individual realization; the resulting optimal controls u (as functions Bu on the left boundary of the unit square) are plotted in the left part of figure 13. As one can see, the optimal u has quite a large variation. The right part of this figure shows the control for the example of figures 11 and 12 (zig-zag solid line) and also the arithmetic average (dotted line) of all 100 optimal controls u, together with (b) the solution of the "averaged" optimal control problem (dashed line)

$$\inf_U \mathrm{E}\ J(u), \tag{9}$$

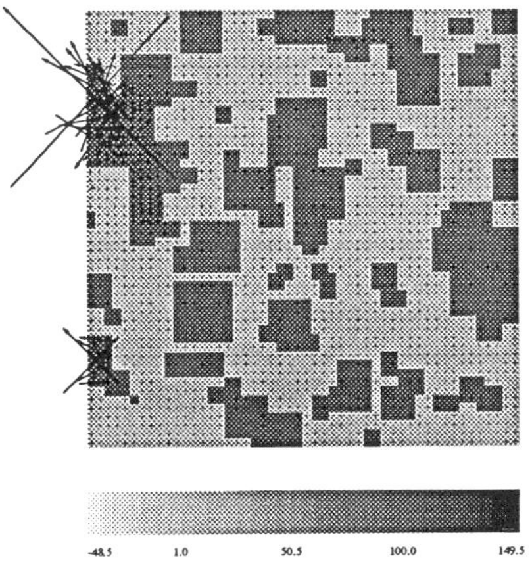

Figure 11 Transmissivity and flux of the state variable for one of the optimal control problems

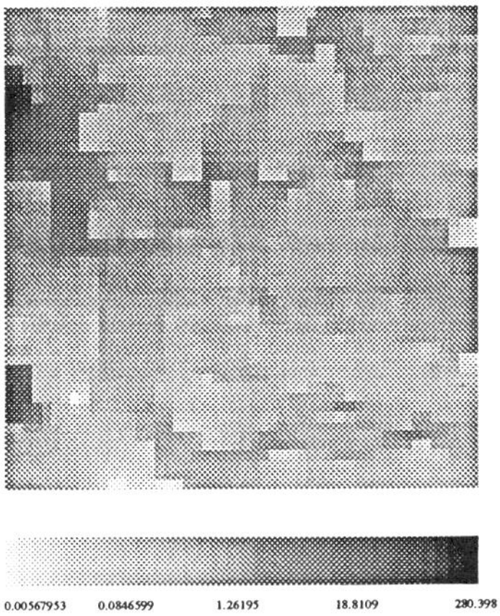

Figure 12 Discharge (on a logarithmic scale) of the state variable for one of the optimal control problems

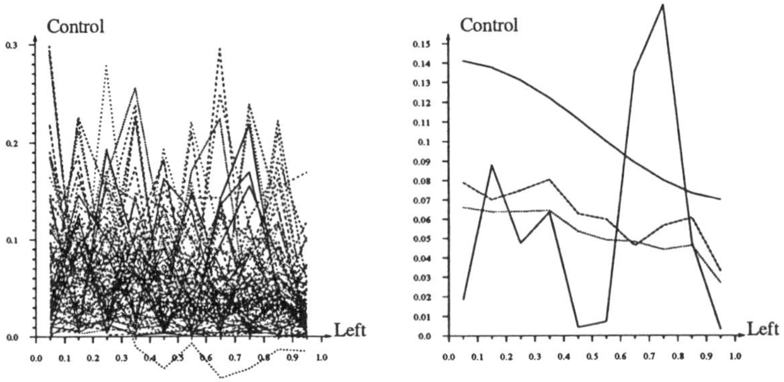

Figure 13 Optimal controls for 100 samples of a Monte-Carlo simulation experiment and averages

where E denotes the *expectation* (in the sense of Monte-Carlo simulations). The monotone curve in the right part of figure 13 shows (c) the optimal solution of the deterministic problem (smooth solid line) for the "effective" (constant) transmissivity K which was determined by solving 100 boundary problems (with constant boundary values) for the state equation.

The left part of figure 14 shows the cumulative distribution function of the objective function; the small vertical line indicates the value of the objective function for the problem with the effective transmissivity. The right part shows the normal flux on the lower half of the right boundary Γ_O for the case of figures 11 and 12 (zig-zag solid line), the arithmetic average (dotted line) of all 100 fluxes of the Monte-Carlo experiment, and the flux of the problem with the effective transmissivity (smooth solid line).

The fact that the three "averages" differ significantly is of great importance for practical applications. Care must be taken when one tries to deal with the influence of the stochastic aspects of optimal control problems. In particular, the often used "effective" parameters prove to be of only limited value for this type of problems.

6 CONCLUSION

The problem of soil remediation presents many difficulties: mathematical, physical, numerical, statistical, and experimental. We can say that the first two are almost solved by many works during the last 40 years. In this paper we deal with the third and the fourth

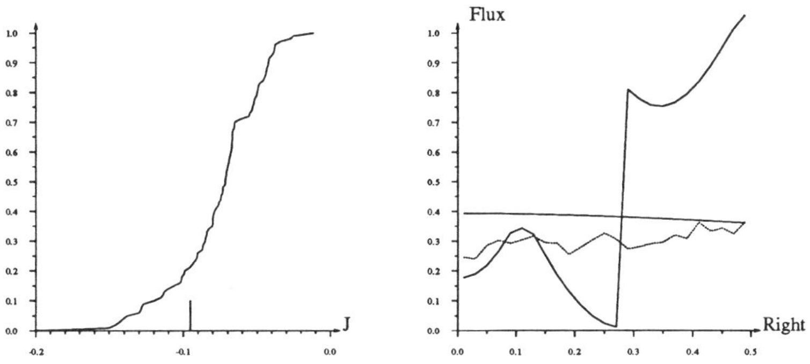

Figure 14 Distribution of the objective function and fluxes

aspect, we use efficient numerical methods such as the "Mixed Finite Element Method", and we show how the random aspect of the problem can make the numerical results - in some cases - useless. The experimental part and the model calibration are still extremely difficult.

For projects of the type described here all important areas of Applied Analysis come into play, namely Partial Differential Equations, Numerical Analysis, Stochastics, Geostatistics, Inverse Problems, and Optimal Control. In addition, methods from Soil Physics, Soil Chemistry, Thermodynamics, Parallel Computing, and Computer Graphics are being used extensively. The consequence is that projects of this kind require interdisciplinary cooperation and teams of researchers that are large enough to deal with very complex problems.

REFERENCES

Abriola L. M. and Pinder G. F. (1985) A multi-phase approach to the modeling of porous media contamination by organic compounds 1. Equation development *Water Resour. Res.* **21.1**, 11-18

Abriola L. M. and Pinder G. F. (1985) A multi-phase approach to the modeling of porous media contamination by organic compounds 2. Numerical simulation *Water Resour. Res.* **21.1**, 19-26

Baehr A. L. and Corapcioglu M. Y. (1987) A compositional multi-phase model for ground water contamination by petroleum products 1. Theoretical consideration *Water Resour. Res.* **23.1**, 191-200

Baehr A. L. and Corapcioglu M. Y. (1987) A compositional multi-phase model for ground water contamination by petroleum products 2. Numerical solution *Water Resour. Res.* **23.1**, 201-213

Baehr A. L., Hoag G. E., and Marley M. C. (1989) Removing volatile contaminants from the unsaturated zone by inducing advective air-phase transport *Journal of Contaminant Hydrology* **4**, 1-26

Gilding B. H. (1988) Mathematical Modelling of Saturated and Unsaturated Ground water Flow *Xiao Shutie: Flow and Transport in Porous Media, World Scientific*

Hochmuth D. P. and Sunada D. K. (1985) Ground water model of two phase immiscible flow in coarse material *Ground Water* **23**, 617-626

Jäger W. and Kačur J. (1994) Solution of doubly nonlinear and degenerate parabolic problems by relaxation schemes *Preprint, Comenius University, Faculty of Mathematics and Physics, Bratislava,* **M2-94**

Journel A. G. and Huijbregts Ch. J (1978) Mining Geostatistics *Academic Press*, London

Kaluarachchi J. J. and Parker J. C. (1989) An efficient finite element method for modeling multi-phase flow *Water Resour. Res.* **25.1**, 43-54

Kaluarachchi J. J. and Parker J. C. (1990) Modeling multi-component organic chemical transport in three-fluid-phase porous media *Journal of Contam. Hydrol.* **5**, 349-374

Kaluarachchi J. J., Parker J. C., and Lenhard R. J. (1990) A numerical model for water for areal migration of water and light hydrocarbon in unconfined aquifers *Adv. Wat. Res.* **13.1**, 29-40

Kuppusamy T., Sheng J., Parker J. C., and Lenhard R. J. (1987) Finite-Element analysis of multi-phase immiscible flow through soils *Water Resour. Res.* **23.4**, 625-631

Lions J.-L. (1971) Optimal Control of Systems Governed by Partial Differential Equations *Springer*, Berlin

Mayer A. S., Miller C. T., and Poirer-McNeill M. M. (1990) Dissolution of trapped non-aqueous phase liquids: mass transfer characteristics *Water Resour. Res.* **26.11**, 2783-2796

Rathfelder K. W., Yeh W. G., and Mackay D. (1991) Mathematical simulation of soil vapor extraction systems: model development and numerical examples *J. Contam. Hydrol.* **8**, 263-297

Russell T. (1995) Modeling of multi-phase multi-contaminant transport, *Rev. Geophys.* **33** Supplement, to appear

Sleep B. E. and Sykes J. F. (1989) Modeling the transport of volatile organics in variably saturated media *Water Resour. Res.* **25.1**, 81-92

Unger A. J. A., Sudicky E. A., and Forsyth P. A. (1995) Mechanisms controlling vacuum extraction coupled with air sparging for remediation of heterogeneous formations contaminated by dense nonaqueous liquids *Water Resour. Res.* **31.8**, 1913-1925

Wilkins M. D., Abriola L. M., and Pennell K. D. (1995) An experimental investigation of rate-limited nonaqueous phase liquid volatization in unsaturated porous media: Steady state mass transfer, *Water Resour. Res.*, to appear

Acknowledgment: This work was supported by the BMBF (German Federal Ministry for Education, Science, Research and Technology) Grant Number 03-HO7BWM. The authors thank Horst Gerke, Eberswalde, and Dirk Stegemann, Garbsen, for their support and generous help.

5

Mathematical models of hysteresis

Augusto Visintin
Dipartimento di Matematica dell'Università di Trento
via Sommarive 14, 38050 Povo (Trento), Italia
Phone: +39-461-881635. Fax: +39-461-881624.
E–mail: visintin@science.unitn.it

Abstract
The concept of *hysteresis operator* is outlined, and some simple models are illustrated. Some differential equations with hysteresis are also briefly discussed.

Keywords
Hysteresis, hysteresis operators, differential equations

1 HYSTERESIS

Hysteresis appears in several phenomena, in physics, engineering, chemistry, biology, economics, and others. Typical examples are plasticity, ferromagnetism, ferroelectricity, and so on.

A Hysteresis Loop. Let us consider the continuous hysteresis loop outlined in Figure 1.

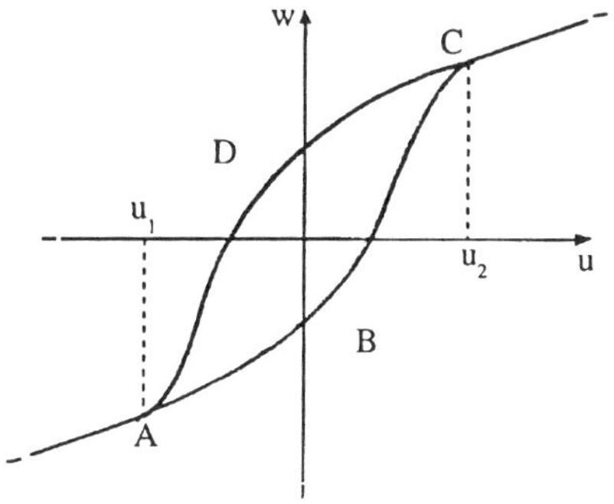

Figure 1 A continuous hysteresis loop.

If u increases from u_1 to u_2, then the pair (u, w) moves along the curve ABC; conversely, if u decreases from u_2 to u_1, then (u, w) moves along the path CDA. Moreover, if u inverts its motion when $u_1 < u(t) < u_2$, then (u, w) moves into the interior of $\mathcal{L}$ (the region bounded by $ABCDA$) along a curve which must be prescribed by the specific hysteresis model.

At any instant t, $w(t)$ depends on the previous evolution of u (*memory effect*), and on the initial state of the system. In the most simple setting one assumes a dependence of the form

$$w(t) = [\mathcal{F}(u)](t) \qquad \forall t \in [0, T]. \tag{1}$$

Here $\mathcal{F}$ represents an operator acting in an appropriate space of time dependent functions, e.g. $C^0([0,T])$. Obviously $\mathcal{F}$ must be *causal:* the output $w(t)$ may not depend on $u_{|]t,T]}$. The definition of $\mathcal{F}$ must include information about a *desired* initial state, which may then be modified because of the value of $v(0)$. For instance, an initial value w^0 such that $(v(0), w^0) \notin \mathcal{L}$ cannot be attained; one may then set $[\mathcal{F}(v)](0)$ equal to the projection of w^0 onto $\{(v(0), w) \in \mathcal{L} : w \in \mathbf{R}\}$.

Here it is implicitly assumed that the pair $(u(t), w(t))$ characterizes the initial state of the system at any instant t. However in several cases the state also depends on *inner variables,* whose initial value must then be specified.

Rate Independence. This means that the pair $(u(t), w(t))$ is invariant with respect to any increasing C^∞-diffeomorphism $\varphi : [0,T] \to [0,T]$:

$$\mathcal{F}(u \circ \varphi) = \mathcal{F}(u) \circ \varphi \qquad \text{in } [0, T]. \tag{2}$$

This property is the characteristic feature of hysteresis. Any rate independent and causal operator will be named a *hysteresis operator.*

This concept raises several problems: the formulation of examples, their adequacy to represent specific applicative phenomena, the analysis of their properties (continuity in various functional spaces, construction of the closure of the graph of discontinuous operators), their characterization, the identification of parameters, and so on.

Hysteresis is often associated with *irreversibility* and *dissipation.* As we shall see, it may have a regularizing effect.

2 CONTINUOUS HYSTERESIS

Here we review the main features of some simple hysteresis models.

Stop and Play. The classical *Prandtl model* of elasto-plasticity, also named *stop,* can be represented by the following variational inequality

$$|w| \le 1, \qquad \left(\frac{du}{dt} - \frac{dw}{dt}\right)(w - v) \ge 0 \qquad \forall v, |v| \le 1, \tag{3}$$

cf. Figure 2(a). For any $u \in W^{1,1}(0,T)$ and $w^0 \in [-1, 1]$, there exists one and only one $w := \mathcal{G}(u)$ which fulfils (3) and the initial condition $w(0) = w^0$.

The *play* operator can be also represented by a variational inequality:

$$|u-w| \leq 1, \qquad \frac{dw}{dt}(u-w-v) \geq 0 \qquad \forall v, |v| \leq 1, \tag{4}$$

cf. Figure 2(b). For any $u \in W^{1,1}(0,T)$ and $w^0 \in \mathbf{R}$, such that $|w^0 - u(0)| \leq 1$, there exists one and only one $w := \mathcal{E}(u)$ which fulfils (4) and such that $w(0) = w^0$.

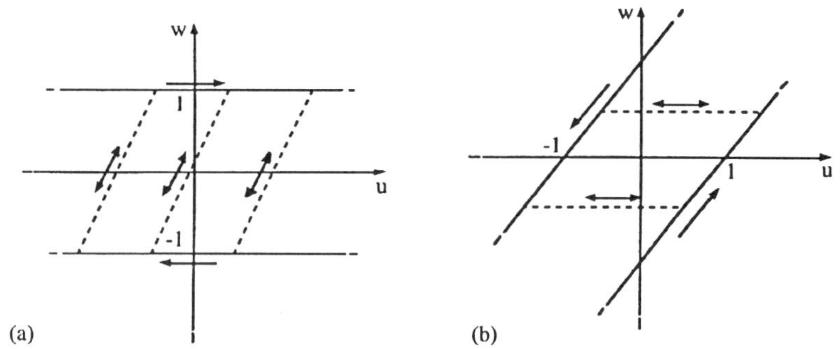

Figure 2 Prandtl's model (or stop) (a), and play (b).

Theorem 1 *$\mathcal{G}$, $\mathcal{E}$ are continuous hysteresis operator acting in $C^0([0,T])$. Moreover*

$$\mathcal{E} : C^0([0,T]) \to C^0([0,T]) \cap BV(0,T). \tag{5}$$

In fact $\mathcal{E}(u)$ is piecewise monotone, since the uniformly continuous function u cannot have an infinite number of oscillations of amplitude larger than 1.

Plays and stops are closely related. Denoting the identity by Id, it is easy to see that

$$\mathcal{E} + \mathcal{G} = \mathrm{Id}, \qquad 2(\mathrm{Id} + \mathcal{E})^{-1} = \mathrm{Id} + \mathcal{G}. \tag{6}$$

The Duhem Model. A model of hysteresis was proposed by Duhem about a century ago. In a simplified form, it reads as follows. Let g_1 and g_2 be given nonnegative continuous functions. For any $u \in W^{1,1}(0,T)$ and $w^0 \in \mathbf{R}$, consider the following Cauchy problem

$$\begin{cases} \dfrac{dw}{dt} = g_1(u,w)\left(\dfrac{du}{dt}\right)^+ - g_2(u,w)\left(\dfrac{du}{dt}\right)^- & \text{in }]0,T[, \\ w(0) = w^0. \end{cases} \tag{7}$$

Here it is implicitly assumed that $dt > 0$: as we said, hysteresis is irreversible. Hence dividing both members by $\frac{du}{dt}$, *formally* we get the equivalent condition

$$\frac{dw}{du} = \begin{cases} g_1(u,w) & \text{if } du > 0, \\ g_2(u,w) & \text{if } du < 0, \end{cases} \qquad \text{in }]0,T[. \tag{8}$$

Under appropriate regularity conditions, two systems of curves in the (u,w)-plane are then obtained by integrating the fields g_1 and g_2. They represent the paths of evolution of the pair (u,w) for increasing and decreasing u, respectively, and may span the whole plane $\mathbf{R}^2$.

A more interesting setting is obtained when (u,w) is confined to a region $\mathcal{L} \subset \mathbf{R}^2$, bounded by the graphs of two nondecreasing continuous functions $\gamma_\ell, \gamma_r : \mathbf{R} \to [-\infty,+\infty]$ $(\gamma_r \leq \gamma_\ell)$.

Theorem 2 *Assume that g_1, g_2 are continuous, and*

$$|g_i(u,w_1) - g_i(u,w_2)| \leq L(u)|w_1 - w_2| \qquad \forall u, w_1, w_2 \in \mathbf{R}\,(i = 1,2), \tag{9}$$

with $L : \mathbf{R} \to \mathbf{R}^+$ continuous. Then for any $u \in W^{1,1}(0,T)$ and any $w^0 \in \mathbf{R}$, there exists a unique solution $w \in W^{1,1}(0,T)$ of (7), such that $w(0) = w^0$.

The hysteresis operator $\mathcal{D} : u \mapsto w$ is strongly continuous in $W^{1,p}(0,T)$, for any $p \in [1,+\infty]$.

This operator is piecewise monotone; in general it is not continuous with respect to the *weak* topology of $W^{1,p}(0,T)$, for any $p \in [1,+\infty]$, and has no continuous extension to $C^0([0,T])$.

Note that in this model at any instant t the state is characterized by the pair $(u(t), w(t))$, with no inner variable. This is not always satisfactory for applications, for ferromagnetism for instance.

Further features of these models and other examples of hysteresis operators are discussed in Visintin (1994).

3 DISCONTINUOUS HYSTERESIS

Relay Operator. For any pair $\rho := (\rho_1, \rho_2) \in \mathbf{R}^2$ $(\rho_1 < \rho_2)$, we introduce the *(delayed) relay operator*

$$h_\rho : C^0([0,T]) \times \{-1,1\} \to BV(0,T). \tag{10}$$

For any $u \in C^0([0,T])$ and any $\xi = \pm 1$, the function $w = h_\rho(u,\xi) : [0,T] \to \{-1,1\}$ is defined as follows:

$$w(0) := \begin{cases} -1 & \text{if } u(0) \leq \rho_1, \\ \xi & \text{if } \rho_1 < u(0) < \rho_2, \\ 1 & \text{if } u(0) \geq \rho_2; \end{cases} \tag{11}$$

for any $t \in]0,T]$, setting $X_t := \{\tau \in]0,t] : u(\tau) = \rho_1 \text{ or } \rho_2\}$,

$$w(t) := \begin{cases} w(0) & \text{if } X_t = \emptyset, \\ -1 & \text{if } X_t \neq \emptyset \text{ and } u(\max X_t) = \rho_1, \\ 1 & \text{if } X_t \neq \emptyset \text{ and } u(\max X_t) = \rho_2. \end{cases} \tag{12}$$

Thus w is uniquely defined in $[0,T]$. For instance, let $u(0) < \rho_1$; then $w(0) = -1$, and $w(t) = -1$ as long as $u(t) < \rho_2$; if at some instant u reaches ρ_2, then w jumps up to 1, where it remains as long as $u(t) > \rho_1$; if later u reaches ρ_1, then w jumps down to -1; and so on, cf. Figure 3.

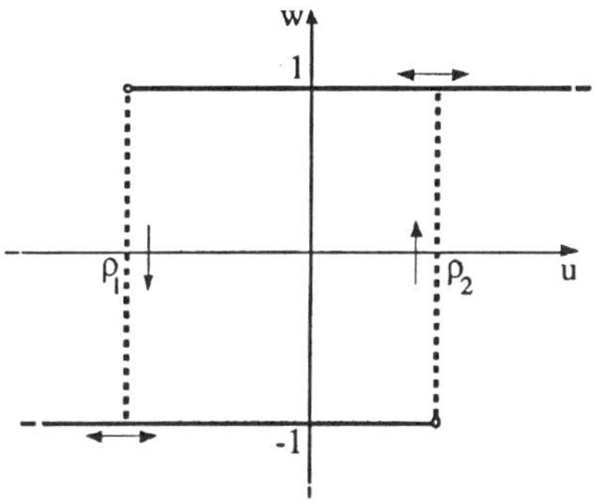

Figure 3 Relay operator.

For any function $u \in C^0([0,T])$, the number of oscillations of u between ρ_1 and ρ_2 is necessarily finite, because of the uniform continuity; hence w can just have a finite number of jumps between -1 and 1, if any. Therefore w is piecewise constant and its total variation in $[0,T]$ is finite.

It is straightforward to check that w is also continuous on the right in $[0,T[$, and that h_ρ is rate independent. Thus h_ρ is a (discontinuous) hysteresis operator.

Remark 1 Generally speaking, hysteresis is strictly related to *multistability:* it can be regarded as a rule for selecting the actual value on the basis of previous evolution. In the case of the relay w tends to stay constant, as long as the value of u allows $w = \pm 1$.

Closure of the Relay Operator. For any $\xi \in \{-1,1\}$, the graph of the operator $h_\rho(\cdot,\xi)$ is not closed with respect to the strong topology of $C^0([0,T])$ for the input u, and the weak star topology of $BV(0,T)$ for the output w. As a counterexample it suffices to take $\xi = -1$, $\{u_n = \rho_2 - \frac{1}{n}\}_{n \in \mathbf{N}}$.

On the basis of the latter remark, we introduce the *multivalued* operator

$$\bar{h}_\rho : C^0([0,T]) \times \{-1,1\} \to \mathcal{P}(BV(0,T)) \text{ (power set)}, \tag{13}$$

defined as follows. For any $u \in C^0([0,T])$ and any $\xi \in \{-1,1\}$, $w \in \bar{h}_\rho(u,\xi)$ if and only if

$$w(0) \in \begin{cases} \{-1\} & \text{if } u(0) < \rho_1, \\ \{-1,\xi\} & \text{if } u(0) = \rho_1, \\ \{\xi\} & \text{if } \rho_1 < u(0) < \rho_2, \\ \{\xi,1\} & \text{if } u(0) = \rho_2, \\ \{1\} & \text{if } u(0) > \rho_2 \end{cases} \tag{14}$$

and, for any $t \in [0,T]$,

$$w(t) \in \begin{cases} \{-1\} & \text{if } u(t) < \rho_1 \\ \{-1,1\} & \text{if } \rho_1 \le u(t) \le \rho_2 \\ \{1\} & \text{if } u(t) > \rho_2 \end{cases} \tag{15}$$

$$\begin{cases} \text{if } u(t) \neq \rho_1,\rho_2, & \text{then } w \text{ is constant in a neighbourhood of } t, \\ \text{if } u(t) = \rho_1, & \text{then } w \text{ is nonincreasing in a neighbourhood of } t, \\ \text{if } u(t) = \rho_2, & \text{then } w \text{ is nondecreasing in a neighbourhood of } t. \end{cases} \tag{16}$$

Such a function w exists, is measurable, and belongs to $BV(0,T)$.

Loosely speaking, the *graph* of $\bar{h}_\rho$ in the (u,w)-plane is obtained by adding the points $(\rho_1,1)$ and $(\rho_2,-1)$ to the *graph* of h_ρ, and then imposing the restrictions $(16)_2$, $(16)_3$.

Theorem 3 *(Closure) For any $\xi \in \{-1,1\}$, $\bar{h}_\rho(\cdot,\xi)$ is the closure of $h_\rho(\cdot,\xi)$ with respect to the strong topology of $C^0([0,T])$ for the input u, and the weak star topology of $BV(0,T)$ for the output w.*

4 HYSTERESIS, REGULARIZATION FOR O.D.E.S

A Simple O.D.E. without Hysteresis. Ordinary and partial differential equations (O.D.E.s and P.D.E.s) can be coupled with hysteresis laws, which can be represented by means of hysteresis operators.

Let us first consider a simple O.D.E. without hysteresis:

$$\frac{dy}{dt} + \mathrm{sign}_0(y) = f(t) \qquad \text{in }]0,T[, \tag{17}$$

where $f \in L^1(0,T)$, and $\mathrm{sign}_0(\xi) := -1$ if $\xi < 0$, $\mathrm{sign}_0(0) := 0$, $\mathrm{sign}_0(\xi) := 1$ if $\xi > 0$.

This equation can represent evolution of temperature (without diffusion), in presence of a source of intensity f and of a thermostat (without hysteresis), which tends to force the temperature $y = 0$. Here $\mathrm{sign}_0(y) = -1$ ($\mathrm{sign}_0(y) = 1$, respect.) means that the thermostat is providing (subtracting, respect.) heat, and $\mathrm{sign}_0(y) = 0$ that it is switched off.

Now for a large class of functions f, e.g. for $f = \text{Constant} \in]-1,0[\cup]0,1[$, the problem obtained coupling (17) with the initial condition $y(0) = 0$ has no solution $y \in W^{1,1}(0,T)$.

This formulation can be easily modified, to allow for existence of a solution; but modifications must be justified from the modelling viewpoint. For instance one might replace the single-valued function sign_0 by the maximal monotone multivalued function sign (with $\text{sign}(0) := [-1,1]$, the whole interval). However values $w \notin \{-1,0,1\}$ require an interpretation.

Hysteresis Regularization. It looks more sensible to account for some hysteresis in the behavior of the thermostat, as it is often evident in practice. Let us fix any $\rho := (\rho_1, \rho_2)$ $(\rho_1 < 0 < \rho_2)$, define $\bar{h}_\rho$ as in Sect. 3, and replace (17) by

$$\begin{cases} \dfrac{dy}{dt} + w = f(t) \\ w \in \bar{h}_\rho(y, w^0) \end{cases} \qquad \text{in }]0, T[. \tag{18}$$

By the closure property stated Theorem 3, for any $f \in L^1(0,T)$ the Cauchy problem governed by (18) has one (and only one) solution, which also depends continuously on the data. Note that after a transient y periodically oscillates between the values ρ_1, ρ_2, and that $dy/dt \in BV(0,T)$ if $f \in L^1(0,T)$.

Generalization. Let us consider an equation of the form

$$\frac{dy}{dt} + \varphi(y) = f(t) \qquad \text{in }]0, T[. \tag{19}$$

This equation has a solution whenever φ is bounded and is of the form $\varphi = \sum_{i=0}^{4} \varphi_i$. Here φ_0 is a continuous function, φ_1 is a maximal monotone graph, φ_2 is an antimonotone function (taking two values at the point of discontinuity), φ_3 and $-\varphi_4$ are relay operators. It can also be shown that if $\varphi_3 \equiv 0$ then there exist a maximal and a minimal solution.

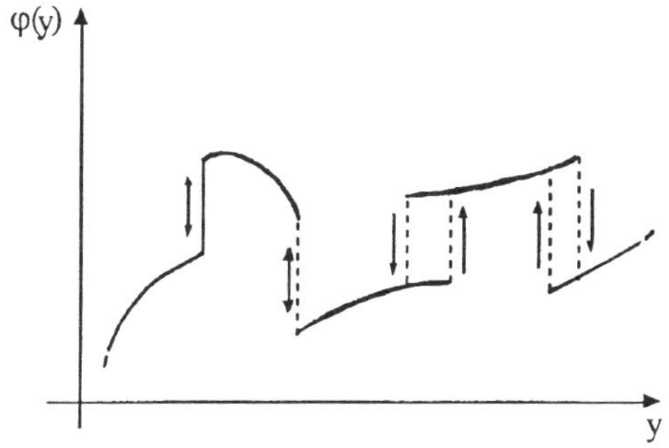

Figure 4 Example of function φ such that equation (19) has a solution.

5 P.D.E.S WITH HYSTERESIS

Two Classes of P.D.E.s with Hysteresis. Let Ω be a domain of $\mathbf{R}^N$, A an elliptic operator, $\mathcal{F}$ a (continuous) hysteresis operator, and consider the equations

$$\frac{\partial}{\partial t}[u + \mathcal{F}(u)] + Au = f \qquad \text{in } \Omega \times]0, T[, \tag{20}$$

$$\frac{\partial}{\partial t}u + Au + \mathcal{F}(u) = f \qquad \text{in } \Omega \times]0, T[, \tag{21}$$

where f is a given function. Of course each of these equations must be coupled with suitable limit conditions.

For $A = \text{curl}^2$ if $N = 3$ ($A = -\frac{d^2}{dx^2}$ if $N = 1$), (20) can represent the evolution of a ferromagnetic system. (21) is a model of heat diffusion in presence of a distribution of thermostats.

Hysteresis and Monotonicity. A natural question concerns the classification of equation (20). If the operator $\mathcal{F}$ fulfils some monotonicity property, then (21) can be regarded as forward parabolic, and it is then natural to impose an initial condition on $u + \mathcal{F}(u)$.

The standard *L^2-monotonicity* is too strong a requirement for hysteresis operators. In fact it is easy to find a counterexample whenever a rate independent loop can occur.

In several cases, but not always, *order preservation* is fulfilled:

$$\begin{aligned}
&\forall (u_1, w_1^0), (u_2, w_2^0) \in \text{Dom}(\mathcal{F}), \forall t \in]0, T], \\
&\text{if } u_1 \leq u_2 \text{ in } [0, t] \text{ and } w_1^0 \leq w_2^0, \text{ then} \\
&[\mathcal{F}(u_1, w_1^0)](t) \leq [\mathcal{F}(u_2, w_2^0)](t).
\end{aligned} \tag{22}$$

The following property of *piecewise monotonicity preservation* (more briefly, *piecewise monotonicity)* seems to be especially appropriate for hysteresis operators:

$$\begin{cases}
\forall v \in C^0([0, T]), \forall [t_1, t_2] \subset [0, T], \\
\text{if } v \text{ is either nondecreasing or nonincreasing in } [t_1, t_2], \text{ then} \\
\{[\mathcal{F}(v)](t_2) - [\mathcal{F}(v)](t_1)\} [v(t_2) - v(t_1)] \geq 0.
\end{cases} \tag{23}$$

That is, $\frac{dw}{dt}\frac{du}{dt} \geq 0$ a.e. in $]0, T[$, whenever $u, w := \mathcal{F}(u) \in W^{1,1}(0, T)$. This means that *hysteresis branches* are nondecreasing.

6 AN EXISTENCE RESULT

Let $\Omega \subset \mathbf{R}^N$ ($N \geq 1$) be a bounded domain of Lipschitz class, $T > 0$, and set $Q := \Omega \times]0, T[$. Let

$$\mathcal{F} : C^0([0, T]) \to C^0([0, T]) \tag{24}$$

be a *causal* and continuous operator, that is,

$$\forall v_1, v_2 \in C^0([0,T]), \forall t \in [0,T], \text{ if } v_1 = v_2 \text{ in } [0,t] \text{ then } [\mathcal{F}(v_1)](t) = [\mathcal{F}(v_2)](t), \tag{25}$$

$$\begin{aligned}&\forall \{v_n \in C^0([0,T])\}_{n\in\mathbf{N}}, \text{ if } v_n \to v \text{ uniformly in } [0,T]\\ &\text{then } \mathcal{F}(v_n) \to \mathcal{F}(v) \text{ uniformly in } [0,T].\end{aligned} \tag{26}$$

Model Problem. We set $V := H_0^1(\Omega)$, identify the space $L^2(\Omega)$ to its dual $L^2(\Omega)'$, and the latter to a subspace of V'. This yields the Hilbert triplet $V \subset L^2(\Omega) = L^2(\Omega)' \subset V'$, with continuous, dense and compact injections. We define the operator $A : V \to V'$ by

$$_{V'}\langle Au, v\rangle_V := \int_\Omega \nabla u \cdot \nabla v dx \qquad \forall u, v \in V,$$

so that $Au = -\Delta u$ in $\mathcal{D}'(\Omega)$. We assume that

$$u^0, w^0 \in L^2(\Omega), \qquad (u^0, w^0) \in \mathcal{L} \text{ a.e. in } \Omega, \qquad f \in L^2(0,T;V'). \tag{27}$$

Problem 1. *To find $u : \Omega \to C^0([0,T])$ measurable, such that*

$$u \in L^2(0,T;V), \qquad \mathcal{F}(u) \in L^2(Q), \qquad u + \mathcal{F}(u) \in H^1(0,T;V'), \tag{28}$$

$$\frac{\partial}{\partial t}[u + \mathcal{F}(u)] + Au = f \qquad \text{in } V', \text{ a.e. in }]0,T[, \tag{29}$$

$$[u + \mathcal{F}(u)]|_{t=0} = u^0 + w^0 \qquad \text{in } V'. \tag{30}$$

Obviously (29) is a weak formulation of (20).

Theorem 4 *(Existence) Assume that (24) – (27) hold. Let $\mathcal{F}$ be piecewise monotone, cf. (23), and affinely bounded, i.e.,*

$$\exists L, M \in \mathbf{R}^+ : \forall v \in C^0([0,T]), \|\mathcal{F}(v)\|_{C^0([0,T])} \le L\|v\|_{C^0([0,T])} + M. \tag{31}$$

Moreover let

$$u^0 \in V, \qquad f \in L^2(Q) \cap W^{1,1}(0,T;V'). \tag{32}$$

Then Problem 1 has at least one solution such that

$$u \in H^1\left(0,T;L^2(\Omega)\right) \cap L^\infty(0,T;V), \qquad \mathcal{F}(u) \in L^2\left(\Omega; C^0([0,T])\right). \tag{33}$$

For the proof we refer to Visintin (1994; Chap. IX).

For a large class of hysteresis operators (including plays, for instance) the previous assumptions are fulfilled, and continuous and monotone dependence on the data (whence uniqueness of the solution) can be proved.

In case of *discontinuous* hysteresis operators some modifications of the setting of Problem 1 are needed. In that case however well-posedness can also be proved.

REFERENCES

Here are some basic references to the mathematics of hysteresis.

Bossavit, A., C. Emson, C. and Mayergoyz, I.D. (1991) *Géométrie différentielle, éléments finis, modèles d' hystérésis.* Eyrolles, Paris.

Brokate, M. (1987) Optimale Steuerung von gewöhnlichen Differentialgleichungen mit Nichtlinearitäten vom Hysteresis-Typ. Lang, Frankfurt am Main. English translation: Optimal control of ordinary differential equations with nonlinearities of hysteresis type. In: *Automation and Remote Control*, **52** (1991), 53.

Brokate, M., et al. Contributions to the session on "Problems in hysteresis". In: *Proceedings of World Congress of Nonlinear Analysts*, Tampa, August 1992 (to appear).

Brokate, M., et al. (Visintin, A. ed.) (1994) *Phase Transitions and Hysteresis.* Lecture Notes in Mathematics, vol. 1584. Springer, Berlin.

Brokate, M. and Sprekels, J. *Hysteresis phenomena in phase transitions* (Monograph in preparation).

Krasnosel'skiĭ, M.A. and Pokrovskiĭ, A.V. (1989) *Systems with hysteresis.* Springer, Berlin (Russian edition: Nauka, Moscow 1983).

Krejčí, P. *Convexity, hysteresis and dissipation in hyperbolic equations.* (Monograph in preparation).

Mayergoyz, I.D. (1991) *Mathematical models of hysteresis.* Springer, New York.

Visintin, A. (ed.) (1993) *Models of hysteresis.* Longman, Harlow.

Visintin, A. (1994) *Differential models of hysteresis.* Springer, Berlin.

PART TWO

Properties of Solutions to PDE

6

Blow-up points to one phase Stefan problems with Dirichlet boundary conditions

Toyohiko Aiki
Department of Mathematics, Faculty of Education, Gifu University, Yanagido, Gifu, 501-11, Japan. Phone: 81-58-293-2239.
Fax: 81-58-293-2243. E–mail: aiki@cc.gifu-u.ac.jp
and
Hitoshi Imai
Faculty of Engineering, University of Tokushima, 2-1 Minami-josanjima, Tokushima 770, Japan. Phone: 81-886-56-7541.
Fax: 81-886-24-6799. E–mail: imai@pm.tokushima-u.ac.jp

Abstract

We study one-phase Stefan problems for semilinear parabolic equations with Dirichlet boundary conditions in one-dimensional space. We show behavior of free boundaries of blow-up solutions at finite blow-up time and numerical experiments for our problem.

Keywords

One-phase Stefan problem, blow-up solution, behavior of free boundary

1 INTRODUCTION

Let us consider the following one-phase Stefan problem DP (resp. NP) with homogeneous Dirichlet (resp. Neumann) boundary condition in one-dimensional space: The problem is to find a curve $x = \ell(t) > 0$ on $[0,T]$, $(0 < T < \infty)$, and a function $u = u(t,x)$ on $Q_\ell(T) := \{(t,x); 0 < t < T, 0 < x < \ell(t)\}$ satisfying that

$$u_t = u_{xx} + u^{1+\alpha} \quad \text{in} \quad Q_\ell(T), \tag{1}$$

$$u(0,x) = u_0(x) \quad \text{for} \quad 0 \le x \le \ell_0, \tag{2}$$

$$u(t,0) = 0 \quad \text{for} \quad 0 < t < T, \tag{3}$$

$$(\text{resp. } \frac{\partial}{\partial x} u(t,0) = 0 \quad \text{for} \quad 0 < t < T,) \tag{4}$$

$$u(t,\ell(t)) = 0 \quad \text{for} \quad 0 < t < T, \tag{5}$$

$$\frac{d}{dt}\ell(t) = -u_x(t,\ell(t)) \quad \text{for} \quad 0 < t < T, \tag{6}$$

$$\ell(0) = \ell_0 \tag{7}$$

where α and ℓ_0 are given positive constants and u_0 is a given initial function on $[0, \ell_0]$.

In Fasano-Primicerio(1979) they established the local existence in time and the uniqueness for solutions to the above DP and NP in the classical formulation (which means that u_t and u_{xx} are continuous functions). Besides, for solutions of DP and NP in the distribution sense (which means that u_t and u_{xx} belong to L^2-class) the existence, the comparison and the behavior were studied by Aiki(1990), Kenmochi(1990) and Aiki-Kenmochi(1991).

It is well known that there are blow-up solutions of the usual initial boundary value problem for semilinear equation (1) in a bounded domain (cf. Tsutsumi(1972)). Accordingly, by using comparison principle it is clear that $DP(NP)$ has a blow-up solution for a sufficiently large initial data. Here, we note the the following global existence result of a solution: Let $[0, T^*)$ be the maximal interval of existence of the solution to DP and NP, we see (cf. Aiki-Kenmochi(1991)) that the following cases (a) or (b) must occur:

(a) $T^* = +\infty$;
(b) $T^* < +\infty$ and $|u|_{L^\infty(Q_\ell(t))} \to +\infty$ as $t \uparrow T^*$.

However, from the above result we can get no information for the behavior of free boundary ℓ at blow-up time. In the present paper we shall show the behavior of blow-up solutions to DP and NP at finite blow-up time (see Theorems 2 and 3). Our proofs to Theorems 2 and 3 are essentially due to Friedman-McLeod(1985). Also, we get an estimates for the time-derivative of free boundary, ℓ' (see Theorem 4).

In the final section we shall show numerical experiments to DP. For the investigation of the influence of a free boundary, numerical results are compared with those of the normal blow-up problem with the fixed boundary and the homogeneous Dirichlet boundary conditions.

We begin with the precise definition of a solution to DP and NP. In this paper we consider classical solutions to our problems since we shall apply the strong maximum principle for the proofs to our theorems. Let $C^{1,0}(\overline{Q_\ell(T)})$ be the set of functions which are continuous on $\overline{Q_\ell(T)}$ with their $x-$derivatives.

Definition 1 *A couple $\{u, \ell\}$ of functions $u = u(t, x)$ and $x = \ell(t)$ is said to be a solution of DP (resp. NP) on a compact interval $[0, T]$, $0 < T < +\infty$, if the following conditions (S1) and (S2) are satisfied:*

(S1) $\ell \in C^1([0, T])$, and $u \in C^{1,0}(\overline{Q_\ell(T)})$, u_{xx} and u_t are continuous in $Q_\ell(T)$;
(S2) (1) $\sim$ (3) and (5) $\sim$ (7) (resp. (1) $\sim$ (2) and (4) $\sim$ (7)) hold in the classical sense.

Also, we call a couple $\{u, \ell\}$ is a solution of DP(resp. NP) on an interval $[0, T')$, $0 < T' \leq \infty$, if it is a solution of DP(resp. NP) on $[0, T]$ in the above sense for any $0 < T < T'$.

First, we recall the theorem concerned with local existence of solutions to the above DP and NP.

Theorem 1 *(cf. Fasano-Primicerio(1979)) We assume that $u_0 \in C^1([0, \ell_0])$, $u_0 \geq 0$ on $[0, \ell_0]$, $u_0(\ell_0) = 0$ and $u_0(0) = 0$(resp. $(u_{0,x}(0) = 0)$. Then there exists a positive number T_0 depending only on $|u_0|_{C^1([0,\ell_0])}$, ℓ_0 and α such that problem DP(resp. NP) has a unique solution $\{u, \ell\}$ on $[0, T_0]$.*

For the problems DP and NP, we say that $[0,T)$, $0 < T \leq +\infty$, is the maximal interval of existence of the solution, if the problem has a solution on time-interval $[0,T']$, for every T' with $0 < T' < T$ and the solution can not be extended in time beyond T.

2 MAIN RESULTS

In order to establish results concerned with the behavior of blow-up solutions we give assumptions (H1) $\sim$ (H4) for initial data ℓ_0 and u_0.

(H1) $\ell_0 > 0$ and $u_0 \in C^2((0,\ell_0)) \cap C^1([0,\ell_0])$ and $u_0(x) > 0$ for $x \in (0,\ell_0)$,
(H2) $u_{0,xx}(x) + u_0^{1+\alpha}(x) \geq 0$ for $x \in (0,\ell_0)$,
(H3) $u_0(\ell_0) = 0$, $u_{0,x} < 0$ for $x \in (0,\ell_0)$ and $u_{0,x}(0) = 0$,
(H4) $u_0(0) = u_0(\ell_0) = 0$, $u_{0,x} > 0$ on $[0,x_0)$ and $u_{0,x} < 0$ on $(x_0,\ell_0]$ for some $x_0 \in (0,\ell_0)$.

Theorem 2 *(cf. Aiki(to appear)) Assume that (H1) $\sim$ (H3) hold. Let $\{u,\ell\}$ be a solution of NP. If $T^* < \infty$, then $\ell(t) \uparrow L < +\infty$ as $t \uparrow T^*$, $u(t,0) \to +\infty$ as $t \uparrow T^*$, and for any $x \in (0,L)$ there exists a positive number $M(x)$ such that*

$$|u(t,x)| \leq M(x) \quad \textit{for any } t \textit{ with } (t,x) \in Q_\ell(T).$$

In Fujita and Chen(1988) they studied the following initial boundary value problem.

$$u_t = u_{xx} + u^{1+\alpha} \quad \text{in } (0,T) \times (0,1), \tag{8}$$
$$u_x(t,0) = 0 \quad \text{for } t \in (0,T], \tag{9}$$
$$u(t,1) = 0 \quad \text{for } t \in (0,T], \tag{10}$$
$$u(0,x) = u_0(x) \quad \text{for } x \in [0,1]. \tag{11}$$

They showed that under the similar assumptions for u_0 to (H1) $\sim$ (H3) if the solution u blows up then blow-up point is one and only one point $x = 0$. In the proof of Theorem 2 we done with help of the idea in Fujita-Chen(1988).

In case the homogeneous Neumann boundary condition the maximum point is always the point $x = 0$ for some initial data, so we get Theorem 2. However, in case Dirichlet boundary condition the maximum point may move so that we can not estimate the blow-up point. Hence, we conclude the following theorem.

Theorem 3 *(cf. Aiki-Imai(submitted)) Assume that (H1),(H2) and (H4) hold. Let $[0,T^*)$ be the maximal interval of existence of the solution $\{u,\ell\}$ to DP. If T^* is finite, then either the following cases (A) or (B) always happens:*
(A) $\ell(t) \to \ell_\infty$ as $t \uparrow T^*$ where ℓ_∞ is some positive number, there exists one and only one point $x^* \in (0,\ell_\infty)$ such that $u(t,x) \to +\infty$ as $t \uparrow T^*$ and for $x \in (0,\ell_\infty)$ with $x \neq x^*$ there is a positive constant $M_1(x)$ such that $|u(t,x)| \leq M_1(x)$ for t with $(t,x) \in Q_\ell(T^*)$;
(B) $\ell(t) \to +\infty$ as $t \uparrow T^*$ and for any $x > 0$ there is a positive number $M_2(x)$ satisfying that $|u(t,\xi)| \leq M_2(x)$ for $(t,\xi) \in Q_\ell(T^*) \cap \{\xi < x\}$.

The proof of Theorem 3 is done in the following way. We assume that (B) does not hold, that is, the following cases (A1) or (A2) is valid:

(A1) There is a number $x_1 \in (0,\infty)$ satisfying that for some sequence $\{t_n, \xi_n\} \subset Q_\ell(T^*)$, $t_n \uparrow T^*$ and $u(t_n,\xi_n) \to \infty$ as $n \to$;

(A2) there is a positive constant L_0 such that $\ell(t) \leq L_0$ for any $t \in [0,T^*)$.

Next, by using the similar argument to those of Friedman-McLeod(1985) we conclude that (A1) is a sufficient condition for (A). Similarly, we can prove that (A2) implies (A).

Remark 1 *By the numerical experiments it seems that the case (B) in the statement of Theorem 3 does not occur, however, we can not prove it, theoretically.*

Theorem 4 *Under the same assumptions as in Theorem 3 we assume that the case(A) in the statement of Theorem 3 occurs. Then, there is a positive constant C such that*

$$|\ell'(t)| \leq C \quad \textit{for } t \in [0,T^*).$$

Proof. First, from Lemma 5.5 in Aiki-Imai(submitted) we observe that there are $x^* < a < \ell_\infty$ and $t_0 \in [0,T^*)$ such that $a < \ell(t_0)$ and

$$|u(t,x)| \leq M_1(a) := M - 1 \quad \text{for } (t,x) \in \hat{Q}$$

where $\hat{Q} = \{(t,x) \in Q_\ell(T^*); x > a, t > t_0\}$

Here, we denote by v a solution of the following initial boundary value problem in non-cylindrical domain $\hat{Q}$;

$$\begin{aligned} &v_t - v_{xx} = M^{1+\alpha} \quad \text{in } \hat{Q},\\ &v(t_0,x) = v_0(x) \quad \text{for } x \in (a,\ell(t_0)),\\ &v(t,a) = M \quad \text{for } t \in (t_0,T^*),\\ &v(t,\ell(t)) = 0 \quad \text{for } t \in (t_0,T^*) \end{aligned}$$

where $v_0 \in C^1([a,\ell(t_0)]$ with $u(t_0,\cdot) \leq v(t_0,\cdot) \leq M$ on $[a,\ell(t_0)]$, $v_0(a) = M$ and $v_0(\ell(t_0)) = 0$.

By using comparison principle we see that $u \leq v$ on $\hat{Q}$ and $v_x(t,\ell(t)) \leq u_x(t,\ell(t)) \leq 0$ for $t \in [t_0,T^*)$.

Next, let w be a solution to the following problem:

$$\begin{aligned} &w_t - w_{xx} = 0 \quad \text{in } (t_0,T^*) \times (a,\ell(t_0)) := \tilde{Q},\\ &w(t_0,x) = w_0(x) \quad \text{for } x \in (a,\ell(t_0)),\\ &w(t,a) = M \quad \text{for } t \in (t_0,T^*),\\ &w(t,\ell(t_0)) = 0 \quad \text{for } t \in (t_0,T^*) \end{aligned}$$

where $w_0 \in C^1([a,\ell(t_0)])$ with $0 \leq w_0 \leq v_0$ on $[a,\ell(t_0)]$, $w_0(a) = M$ and $w_0(\ell(t_0)) = 0$. It is clear that $|w_x(t,x)| \leq C_1$ on $\tilde{Q}$ for some positive constant C_1, $w \leq v$ on $\tilde{Q}$ and $w_x(t,a) \leq v_x(t,a)$ for $t \in [t_0,T^*)$ so that

$$v_x(t,a) \geq -C_1 \quad \text{for } t \in [t_0,T^*).$$

Putting $z(t,x) = v(t,x) + K\exp(x - \ell(t))$ where $K = (M^{1+\alpha} + |v_x(t_0)|_{L^\infty(a,\ell(t_0))} + C_1)\exp(\ell_\infty)$, z satisfies that $z(t,\ell(t)) = K$ for $t_0 \leq t < T^*$ and

$$\begin{aligned}
&z_t - z_{xx} < 0 \quad \text{in } \hat{Q},\\
&z_x(t_0, x) > 0 \quad \text{for } a \le x \le \ell(t_0),\\
&z_x(t, a) > \quad \text{for } t_0 < t < T^*.
\end{aligned}$$

Then, we conclude that z takes its maximum value on any point in the curve $\{(t, x); x = \ell(t), t_0 \le t < T^*\}$. Hence, $z_x(t, \ell(t)) \le 0$ for $t_0 \le t < T^*$. By the definition of z, we have

$$v_x(t, \ell(t)) \ge -K \quad \text{for } t \in [t_0, T^*).$$

Therefore, we obtain that $|\ell'(t)| = |u_x(t, \ell(t))| \le K$ for $t_0 \le t < T^*$. Thus this theorem is proved. □

We refer to Lemma 6.3.1 in Ladyzenskaja-Solonnikov-Ural'ceva (1968) and Lemma 4.2 in Aiki(to appear) for the proof of Theorem 4 and to Aiki-Kenmochi(1991) for comparison principle.

3 NUMERICAL RESULTS

We carried out numerical computations to DP with $\ell_0 = 1$ and $u_0(x) = u_0^A(x) \equiv Ax^2(x-1)^2$. We note (H2) is satisfied for large A.

Numerical computations need some additional techniques due to treatment of the unknown boundary. Therefore, we used the fixed domain method using the following mapping function

$$x(t, \xi) = \frac{\ell(t)}{2}(\xi + 1), \quad 0 \le t, \quad -1 \le \xi \le 1. \tag{12}$$

The free boundary problem is transformed to the following equivalent problem in the fixed domain using the variable transform.

$$u_t(t, \xi) = \frac{4}{\ell^2(t)} u_{\xi\xi}(t, \xi) - \frac{2(\xi + 1)}{\ell^2(t)} u_\xi(t, 1) u_\xi(t, \xi) + u^{1+\alpha}(t, \xi), \quad 0 < t, -1 < \xi < 1, \tag{13}$$

$$u(0, \xi) = u_0^A\left(\frac{\xi + 1}{2}\right), \quad -1 < \xi < 1, \tag{14}$$

$$u(t, -1) = 0, \qquad 0 \le t, \tag{15}$$

$$u(t, 1) = 0, \qquad 0 \le t, \tag{16}$$

$$\ell'(t) = -\frac{2}{\ell(t)} u_\xi(t, 1), \quad 0 < t, \tag{17}$$

$$\ell(0) = 1. \tag{18}$$

Here we should remark this problem becomes the normal blow-up problem with the fixed boundary((8), $u(t, 0) = 0$ for $t \in (0, T]$, (10) $\sim$ (11)) by setting $\ell(t) \equiv 1$ and neglecting the 2nd term in (13) and (17).

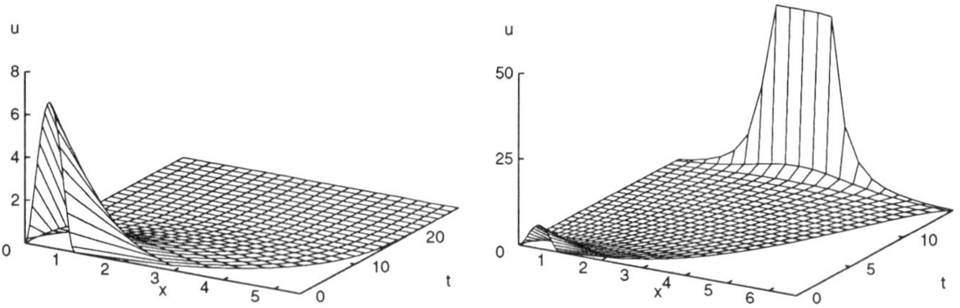

(a) (b)

Figure 1 $u(t,x)$ to DP for $\alpha = 1$. (a):A=108.8, $u(t,x) \downarrow 0$. (b):A=108.9, $u(t,x) \uparrow +\infty$ as $t \uparrow 14.6$.

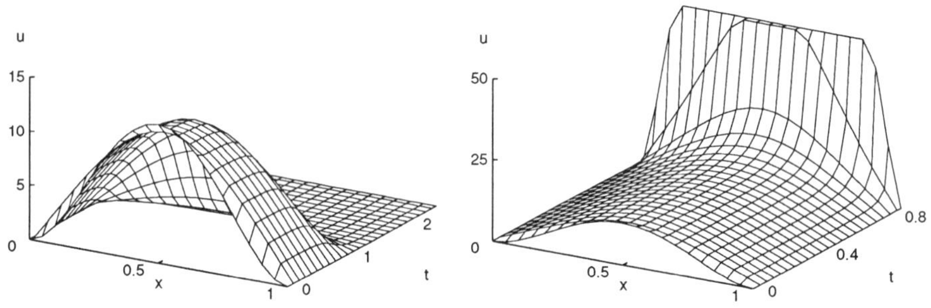

(a) (b)

Figure 2 $u(t,x)$ to the normal blow-up problem for $\alpha = 1$. (a):A=205.05, $u(t,x) \downarrow 0$. (b):A=205.1, $u(t,x) \uparrow +\infty$ as $t \uparrow 0.86$.

To these transformed problems we applied the 2nd order finite difference method in space and the 4th order Runge-Kutta method in time. We used 201 grid points in space. The basic time increment is 10^{-4}. It is controlled adaptively as in Chen (1986). Figure 1 shows numerical results to DP. Figure 2 shows numerical results to the normal blow-up problem with the fixed boundary ((8) $\sim$ (10)).

From comparison principle mentioned in the previous section it is expected the free boundary has the influence of the enhancement of the blow-up phenomena. This expectation is confirmed by these numerical results. However, the blow-up phenomena of the free boundary are not observed. This is because as Figure 1 shows blow-up points do not sufficiently get toward the free boundary.

REFERENCES

Aiki, T. (1990) The existence of solutions to two-phase Stefan problems for nonlinear parabolic equations, *Control and Cybernetics*, **19**, 41–62.

Aiki, T. (to appear) Behavior of free boundaries blow-up solutions to one-phase Stefan problems, Nonlinear Anal. TMA.

Aiki, T. and Imai, H. (submitted) Behavior of blow-up solutions to one-phase Stefan problems with Dirichlet boundary conditions.

Aiki, T. and Kenmochi, N. (1991) Behavior of solutions to two-phase Stefan problems for nonlinear parabolic equations, *Bull. Fac. Education, Chiba Univ.*, **39**, 15–62.

Chen, Y.-G. (1986) Asymptotic behaviours of blowing-up solutions for finite difference analogue of $u_t = u_{xx} + u^{1+\alpha}$, *J. Fac. Sci. Univ. Tokyo, Sect. IA, Math.*, **33**, 541–574.

Fasano, A. and Primicerio, M. (1979) Free boundary problems for nonlinear parabolic equations with nonlinear free boundary conditions, *J. Math. Anal. Appl.*, **72**, 247–273.

Friedman, A. and McLeod, B. (1985) Blow-up of positive solutions of semilinear heat equations, *Indiana Univ. Math. J.*, **34**, 425–447.

Fujita, H. and Chen, Y.-G. (1988) On the set of blow-up points and asymptotic behaviours of blow-up solutions to a semilinear parabolic equation, in *Analyse Mathématique et Applications*, Gauthier-Villars, Paris, pp. 181–201.

Kenmochi, N. (1990) A new proof of the uniqueness of solutions to two-phase Stefan problems for nonlinear parabolic equations, *Free Boundary Problems*, ISNM., Vol. 95, Birkhäuser, Basel, pp. 101–126.

Ladyzenskaja, O. A., Solonnikov, V. A. and Ural'ceva, N. N. (1968) , *Linear and Quasi-Linear Equations of Parabolic Type,* Transl. Math. Monograph 23, Amer. Math. Soc., Providence R. I.

Tsutsumi, M. (1972) Existence and nonexistence of global solutions for nonlinear parabolic equations, *Publ. Res. Inst. Math. Sci.*, **8**, 211–229.

7

Some new ideas for a Schiffer's conjecture

T. Chatelain , M. Choulli and A. Henrot
Équipe de Mathématiques – URA CNRS 741
Université de Franche-Comté, F-25030 Besançon Cedex France

Abstract

We discuss a Schiffer's conjecture which is a symmetry problem for an overdetermined spectral p.d.e.. We show the connection between this problem and the critical points of the eigenvalue with a volume constraint as well as the Faber-Krahn inequality. We give two original proofs of these symmetry results in the case of the first eigenvalue.

Keywords

Domain derivative, first eigenvalue, Faber-Krahn inequality, continuous Steiner symmetrization

1 INTRODUCTION

The motivation of this work is a conjecture of M. Schiffer in spectral theory: let Ω be a connected regular domain in $\mathbf{R}^N$ such that there exists an eigenvalue λ and a Dirichlet-eigenfunction $u \neq 0$ satisfying

$$(SC) \quad \begin{cases} -\Delta u &= \lambda u \quad \text{in} \quad \Omega, \\ u &= 0 \quad \text{on} \quad \partial\Omega, \\ \dfrac{\partial u}{\partial n} &= cst \quad \text{on} \quad \partial\Omega \end{cases}$$

then does it imply that Ω is a ball ?
A domain Ω for which (SC) has a non trivial solution will be called in the following a solution of the (SC) problem.
This conjecture, for the Dirichlet case, can also, of course, be considered for the Neumann case : if $\lambda > 0$ and $u \neq 0$ are such that

$$(SC)' \quad \begin{cases} -\Delta u &= \lambda u \quad \text{in} \quad \Omega, \\ \dfrac{\partial u}{\partial n} &= 0 \quad \text{on} \quad \partial\Omega, \\ u &= cst \quad \text{on} \quad \partial\Omega \end{cases}$$

can we prove that Ω is a ball ?
This conjecture $(SC)'$ has been more intensively studied than the previous one, because

Williams in a paper of 1976, proves that it was equivalent to the famous Pompeiu problem. In spite of many efforts these twenty last years, only partial results are available for $(SC)'$, a good reference is Zalcman(1992), and (to our knowledge), very few is known for (SC). So these two conjectures (SC) and $(SC)'$ remain open even in 2-dimensions.

In this paper, we are interested in (SC) for the case $\lambda = \lambda_1$ the first Dirichlet eigenvalue. Of course, in this case the result is already known, since it is a consequence for example of the classical result of Serrin (1971). Nevertheless, our aim is to give another original proof of this result which does not use maximum principle.
This fact is of importance since maximum principle is unavailable for (SC) when $\lambda \geq \lambda_2$ and for $(SC)'$. Unfortunately, we apparently need in our approach the positivity of the eigenfunction. We also link this Schiffer's conjecture, using domain derivative, to the classical problem of minimizing the first eigenvalue with a volume constraint (see also Cox (1994) for similar ideas) and we give another proof of the fact that equality in the Faber-Krahn inequality is achieved only for the ball.
Some of the ideas developed in this paper could be used for the conjecture (SC) or $(SC)'$ in the general case. A work in this direction is in preparation (see Chatelain and Henrot(work in progress)).
Let us describe now briefly the content of this paper.

In the second part, we recall the classical formula of derivative of the first eigenvalue with respect to the domain. This formula allows us to point out the link which exists between the Schiffer's conjecture (SC), the isoperimetric Faber-Krahn inequality and the critical point of the first eigenvalue, considered as a domain functional, under a volume constraint. Indeed we make the simple observation that Ω is a solution of the (SC) problem if and only if there exists a constant c such that

$$d\lambda(\Omega; V) = c\, dVol(\Omega; V) \quad \text{for all displacement field } V, \tag{1}$$

where $d\lambda(\Omega; V)$ and $dVol(\Omega; V)$ denote respectively the derivative of λ and the derivative of the volume at the point Ω in the direction of the displacement field V.

In the third part, we prove that equality in the Faber-Krahn inequality is realized only for balls. In other words, the only minimum for the first eigenvalue with a volume constraint is the ball. This result was, of course, already known ; but as pointed out in Kawohl (1985), most of the proofs were done in the analytic case and used the analyticity of the eigenfunction.

At last, in the fourth part, we prove, using a continuous Steiner symmetrization, that, if Ω is a convex domain which is not a ball, there exists at least one displacement field V such that (1) fails. This will imply the desired result that the only solution (at least in the convex case) of (SC) are the balls. The same idea is used in a similar context in Brock and Henrot (work in progress) to prove a generalization of the Serrin's result to some nonlinear elliptic problems.

2 DERIVATIVE OF THE FIRST EIGENVALUE AND SOME CONSEQUENCES

Domain derivative is now a very usual tool in field such as shape optimization. From a mathematical point of view, it goes from Hadamard in 1905 and Garabedian-Schiffer

in 1953. Good references are the works of Murat-Simon in 1975 which is summed up in a paper of Simon (1980), or the more recent book of Sokolowski and Zolesio (1992).
The first eigenvalue of the Laplacien-Dirichlet, $\lambda_1(\Omega)$ which is characterized as

$$\lambda_1(\Omega) = \inf_{\substack{v \in H_0^1(\Omega) \\ v \neq 0}} \frac{\int_\Omega |\nabla v|^2}{\int_\Omega v^2} \tag{2}$$

can be regarded as a domain functional since it depends obviously on the domain Ω. Moreover, for a connected domain, $\lambda_1(\Omega)$ is a simple eigenvalue so it is possible to prove that $\Omega \mapsto \lambda_1(\Omega)$ is differentiable. This is proved, for example in Rousselet and Chesnais (1990), for a similar problem in Mignot, Murat and Puel (1979) and also formally (from a "mechanist" point of view) in Sanchez-Hubert and Sanchez-Palencia. We do not want, here, to give the proof of the differentiability. It is technical and it uses the Fredholm alternative and an implicit function theorem. Let us just establish here the formula of the derivative of the first eigenvalue by a formal calculation : we assume in the following that Ω is a C^2 connected bounded domain and let u $(= u(\Omega))$ be a positive normalized eigenfunction associated to λ_1. In the sequel we will drop out the subscript 1 since we are always working with the first eigenvalue, so $\lambda = \lambda_1$ and :

$$\begin{cases} -\Delta u &= \lambda u \quad \text{in} \quad \Omega \\ u &= 0 \quad \text{on} \quad \partial\Omega \\ \int_\Omega u^2(x)dx &= 1. \end{cases} \tag{3}$$

Let us choose a field of deformation $V \in C^2(\mathbf{R}^N; \mathbf{R}^N)$. By differentiating (3), it is well known that u' (the derivative of u with respect to the domain in the direction of V) satisfies

$$\begin{cases} -\Delta u' &= \lambda u' + \lambda' u \quad \text{in} \quad \Omega \\ u' &= -\frac{\partial u}{\partial n} V.n \quad \text{on} \quad \partial\Omega, \end{cases} \tag{4}$$

where λ' is the derivative of λ with respect to the deformation of Ω by V. Multiplying the first equation by u and integrating on Ω yields

$$-\int_\Omega u\Delta u'\, dx = \lambda \int_\Omega uu'\, dx + \lambda' \int_\Omega u^2\, dx = \lambda \int_\Omega uu'\, dx + \lambda', \tag{5}$$

but the Green formula gives

$$-\int_\Omega u\Delta u'\, dx = -\int_{\partial\Omega} u\frac{\partial u'}{\partial n} d\sigma + \int_{\partial\Omega} u'\frac{\partial u}{\partial n} d\sigma + \lambda \int_\Omega u'u\, dx, \tag{6}$$

so using (3) and (4) and replacing in (5) yields the desired formula :

$$\lambda' = d\lambda(\Omega; V) = -\int_{\partial\Omega} \left(\frac{\partial u}{\partial n}\right)^2 V.n\, d\sigma. \tag{7}$$

Since the derivative of the volume is given by

$$dVol(\Omega; V) = \int_{\partial\Omega} V.n \, d\sigma, \tag{8}$$

we have immediately

Proposition 1 *Ω is a solution of the problem (SC) if and only if there exists a constant c such that*

$$d\lambda(\Omega; V) = -c^2 \, dVol(\Omega; V). \tag{9}$$

In other words, the solutions of the problem (SC) can be considered as critical points of the functional $\lambda(\Omega)$ with a volume constraint, the constant $-c^2$ being a Lagrange multiplier.

Another consequence of the proposition 1 is the following : the well known Faber-Krahn tells us that for all domains Ω with $\text{Vol}(\Omega) = V_0$ given, we have

$$\lambda_1(\Omega^*) \leq \lambda_1(\Omega), \tag{10}$$

where Ω^* is the ball of volume V_0. Now, thanks to proposition 1, any such minimum satisfies (9) and then

Corollary 1 *Ω realizes the equality in the Faber-Krahn inequality if and only if Ω is a solution of the problem (SC).*

As an application of the Serrin's result (Serrin, 1971) we obtain immediately that :

Theorem 1 (equality in the Faber-Krahn inequality) *Let*

$$\Theta = \{\Omega \textit{ bounded, connected } C^2 \textit{ domains in } \mathbf{R}^N, \; Vol(\Omega) = V_0\}$$

and Ω^* be the ball of volume V_0 then

$\lambda_1(\Omega) = \lambda_1(\Omega^*)$ and $\Omega \in \Theta$ if and only if $\Omega = \Omega^*$.

3 ANOTHER (ELEMENTARY) PROOF OF THEOREM 1

The aim of this section is to give another very simple proof of the theorem 1 :
let Ω be a minimum of λ_1 with $\text{Vol}(\Omega) = V_0$ (that is to say a domain for which we have equality in the Faber-Krahn inequality). According to corollary 1, we have

$$\frac{\partial u}{\partial n} = \text{constant } = c \qquad on \ \partial\Omega \tag{11}$$

and then, integrating over $\partial\Omega$ gives :

$$c \, P(\partial\Omega) = \int_{\partial\Omega} \frac{\partial u}{\partial n} \, d\sigma = \int_{\Omega} \Delta u \, dx = -\lambda(\Omega) \int_{\Omega} u \, dx, \tag{12}$$

where $P(\Omega)$ is the perimeter of $\partial\Omega$.
Now we use the well known Rellich identity (see Rellich (1940)) :

$$2\int_{\partial\Omega}(x.\nabla u)\frac{\partial u}{\partial n}\,d\sigma - \int_{\partial\Omega}(x.n)\,|\nabla u|^2\,d\sigma = 2\int_{\Omega}(x.\nabla u)\Delta u\,dx + (2-N)\int_{\Omega}|\nabla u|^2\,dx. \quad (13)$$

In our case, this identity gives :

$$2\lambda(\Omega) = 2\lambda(\Omega)\int_{\Omega}u^2\,dx = Nc^2 Vol(\Omega) \quad (14)$$

and then (12) and (14) yield

$$\lambda(\Omega)\left(\int_{\Omega}u\,dx\right)^2 = \frac{2P\,(\partial\Omega)^2}{N\,Vol(\Omega)}. \quad (15)$$

Now, let us introduce u^* the (Schwarz) spherical decreasing rearrangement of u (see Kawohl (1985)) which is defined on the ball Ω^*. We know that

$$\int_{\Omega^*}u^{*^2}(x)\,dx = \int_{\Omega}u^2(x)\,dx = 1$$

that

$$\int_{\Omega^*}u^*(x)\,dx = \int_{\Omega}u(x)\,dx$$

and also that (Polya-Szegö theorem)

$$\int_{\Omega^*}|\nabla u^*(x)|^2\,dx \leq \int_{\Omega}|\nabla u(x)|^2\,dx = \lambda_1(\Omega). \quad (16)$$

Now, by (2) and (16) :

$$\lambda(\Omega) = \lambda(\Omega^*) \leq \int_{\Omega^*}|\nabla u^*(x)|^2\,dx \leq \int_{\Omega}|\nabla u(x)|^2\,dx = \lambda(\Omega). \quad (17)$$

Then, the inequality chain in (17) is an equality chain and

$$\lambda(\Omega^*) = \int_{\Omega^*}|\nabla u^*(x)|^2\,dx,$$

which implies that u^* is a (normalized) eigenfunction associated to $\lambda(\Omega^*)$. Now identity (15) remains true for u^* and Ω^* (u^* being radial, its normal derivative is constant and then all the calculations are valid). So as a consequence :

$$\frac{2P\,(\partial\Omega)^2}{N\,Vol(\Omega)} = \lambda(\Omega)\left(\int_{\Omega}u\,dx\right)^2 = \lambda(\Omega^*)\left(\int_{\Omega^*}u^*\,dx\right)^2 = \frac{2P\,(\partial\Omega^*)^2}{N\,Vol(\Omega^*)}. \quad (18)$$

Now, since $\mathrm{Vol}(\Omega^*) = \mathrm{Vol}(\Omega)$, (18) implies that $P(\partial\Omega^*) = P(\partial\Omega)$ and the classical isoperimetric inequality gives us the desired result.

4 CONTINUOUS STEINER SYMMETRIZATION

In this section, we want to give another proof (in a particular case) of the Serrin's theorem for the problem (SC), which is based on the proposition 1 and which does not use maximum principle.

Theorem 2 *The only convex solutions of the problem* (SC) *(for* $\lambda = \lambda_1$*) are the balls.*

Our proof is based on the following very simple idea. Since each solution of the problem (SC) satisfies (9), to prove theorem 2 it is sufficient to prove that for a domain Ω which is not a ball, there exists (at least) one vector field of displacement V such that (9) is false. More precisely, we are going to exhibit a vector field of displacement V (given by the Continuous Steiner symmetrization) which leaves the volume unchanged, then $dVol(\Omega; V) = 0$, and such that

$$d\lambda(\Omega; V) < 0, \tag{19}$$

which will give the theorem 2. The use of a continuous Steiner symmetrization field of displacement can also be found in Brock and Henrot (to appear) to prove a more general result.

Let us assume that Ω is not a ball : so there exists a (direction of) hyperplane with respect to which Ω is not symmetric. Let us denote by $(x', y), x' \in \mathbf{R}^{N-1}, y \in \mathbf{R}$, the points in $\mathbf{R}^N$, and we can always assume that the hyperplane is $y = 0$. Since Ω is convex, the intersection of Ω with every line $x' =$ constant is a segment say $y_1(x') \le y_2(x')$.

Then we define a displacement field $V : \overline{\Omega} \to \mathbf{R}^N$ by :

$$V(x', y) = (0, -\frac{1}{2}(y_1(x') + y_2(x'))) \tag{20}$$

(where 0 is the zero of $\mathbf{R}^{N-1}$) which is continuously extended to $\mathbf{R}^N$.

In this way $\Omega_t = (\mathrm{Id} + tV)(\Omega)$ is the classical continuous Steiner symmetrization of Ω (see Brock(1995)) and we know that, for every positive continuous function f in the Sobolev space $H_0^1(\Omega)$, we are able to define f^t, its continuous Steiner symmetrization and we have

$$\int_{\Omega_t} f^{t^2}(x)\, dx = \int_\Omega f^2(x)\, dx, \tag{21}$$

$$\int_{\Omega_t} |\nabla f^t(x)|^2\, dx \le \int_\Omega |\nabla f(x)|^2\, dx. \tag{22}$$

The following lemma gives a more precise statement of the inequality (22) in the case of a quasiconcave function f (i.e. a function f whose level sets are convex) :

Lemma 1 *Let* u *be a quasiconcave function as above defined on the convex set* Ω *and* u^t *its continuous Steiner symmetrization ; then when* t *goes to zero we have the expansion*

$$\int_{\Omega_t} |\nabla u^t(x)|^2\, dx = \int_\Omega |\nabla u(x)|^2\, dx + \alpha(u, \Omega)\, t + o(t) \tag{23}$$

with $\alpha(u, \Omega)$ *(which is a number depending on* u *and* Ω*) strictly negative as soon as* Ω *is not symmetric with respect to* $\{y = 0\}$.

Proof. Let us denote by

$$\Delta = \{(x',u) \in \mathbf{R}^{N-1} \times \mathbf{R}_+ \ / \ \exists (x',y) \in \Omega \ such\ that\ u = u(x',y)\}$$

The level sets of u being convex, for each $(x',u) \in \Delta$ the equation $u = u(x',y)$ has exactly two solutions $y = y_1$ and $y = y_2$ with $y_1 \leq y_2$. Thus u can be represented in Δ by the inverse function $y = y_k(x',u)$, $k = 1,2$ and moreover we have

$$u_y(x',y_k) = \left(\frac{\partial y_k}{\partial u}\right)^{-1} \begin{cases} > 0 & for\ k = 1 \text{ (almost everywhere)} \\ < 0 & for\ k = 2 \text{ (almost everywhere)} \end{cases} \tag{24}$$

$$u_{x_i}(x',y_k) = -\frac{\partial y_k}{\partial x_i}\left(\frac{\partial y_k}{\partial u}\right)^{-1} \qquad (i = 1,...,N-1). \tag{25}$$

At last, by definition of the continuous symmetrization (see Brock, 1995), u^t is defined by the inverse functions y_k^t, $k = 1,2$ given by

$$y_k^t = y_k - \frac{t}{2}(y_1 + y_2) \qquad k = 1,2. \tag{26}$$

By the change of variable induced by y_k, we have

$$\int_{\Omega_t} |\nabla u^t(x)|^2\,dx = \int_\Delta \sum_{k=1}^{2} \left|\frac{\partial y_k^t}{\partial u}\right|^{-2} \left(1 + \sum_{i=1}^{N-1}\left(\frac{\partial y_k^t}{\partial x_i}\right)^2\right) \left|\frac{\partial y_k^t}{\partial u}\right| dx'du \tag{27}$$

this formula remaining valid for t = 0. Using (26) and a straightforward calculation, we obtain immediately

$$\int_{\Omega_t} |\nabla u^t(x)|^2 dx = \int_\Omega |\nabla u(x)|^2\,dx$$
$$+ t\sum_{k=1}^{2}\int_\Delta \left[\frac{1}{2}\left(1 + \Sigma_{i=1}^{N-1}\left(\tfrac{\partial y_k}{\partial x_i}\right)^2\right)\frac{\left(\frac{\partial y_1}{\partial u}+\frac{\partial y_2}{\partial u}\right)}{\frac{\partial y_k}{\partial u}} - \Sigma_{i=1}^{N-1}\frac{\partial y_k}{\partial x_i}\left(\frac{\partial y_1}{\partial x_i}+\frac{\partial y_2}{\partial x_i}\right)\right]\left|\frac{\partial y_k}{\partial u}\right|^{-1} dx'du \tag{28}$$
$$+ o(t).$$

Using (24), we deduce of (28) that

$$\alpha(u,\Omega) = \frac{1}{2}\int_\Delta \left(\tfrac{\partial y_1}{\partial u} + \tfrac{\partial y_2}{\partial u}\right)\left(\left(\tfrac{\partial y_1}{\partial u}\right)^{-2} - \left(\tfrac{\partial y_2}{\partial u}\right)^{-2}\right)$$
$$+\left(\frac{\partial y_2}{\partial u} - \frac{\partial y_1}{\partial u}\right)\left(\sum_{i=1}^{N-1}\left(\frac{\frac{\partial y_1}{\partial x_i}}{\frac{\partial y_1}{\partial u}} - \frac{\frac{\partial y_2}{\partial x_i}}{\frac{\partial y_2}{\partial u}}\right)^2\right) dx'du. \tag{29}$$

Since each above quantity is non positive, $\alpha(u,\Omega)$ would be zero if and only if $\dfrac{\partial(y_1+y_2)}{\partial u} = 0$ and $\dfrac{\partial(y_1+y_2)}{\partial x_i} = 0$ a.e. $\forall i = 1,..,N-1$, this is only possible if Ω is symmetric which completes the proof. □

Now, we can conclude for the proof of theorem 2 :
let us denote by u_t the first (normalized) eigenfunction on Ω_t (which is - a priori - different from u^t). Since u^t belongs to $H_0^1(\Omega_t)$, the definition of $\lambda_1(\Omega_t)$ gives :

$$\lambda(\Omega_t) = \int_{\Omega_t} |\nabla u_t|^2 \, dx. \leq \int_{\Omega_t} |\nabla u^t|^2 \, dx. \tag{30}$$

Now we can apply lemma 1 to u since the level sets of the first eigenfunction of a convex domain are convex (see Kawohl (1985)), so (30) and (23) yield

$$\lambda(\Omega_t) \leq \int_{\Omega} |\nabla u|^2 \, dx + \alpha(u, \Omega)\, t + o(t). \tag{31}$$

So, we obtain

$$\frac{\lambda(\Omega_t) - \lambda(\Omega)}{t} \leq \alpha(u, \Omega) + o(1) \tag{32}$$

and passing to the limit $d\lambda(\Omega; V) \leq \alpha(u, \Omega) < 0$ what is the desired result since the Steiner symmetrization field preserves the volume.

REFERENCES

Brock, F. (1995) Continuous Steiner Symmetrization. *Math. Nachrichten*, 172, 25-48.

Brock, F. and Henrot, A. (to appear) A symmetry result for overdetermined boundary value problem using domain derivative and continuous Steiner symmetrization.

Chatelain, T. and Henrot, A. (work in progress).

Cox, S.J. (1994) Extremal eigenvalue problems for the Laplacian. M. A. Herrero, E. Zuazua eds, *R.A.M.* J. Wiley and Masson, 1994.

Kawohl, B. (1985) *Rearrangements and Convexity of level sets in Partial Differential Equations.* Springer Lecture Notes in Math., 1150.

Mignot, F., Murat, and Puel, J.P. (1979) Variation d'un point de retournement par rapport au domaine. *Comm. in p.d.e.*, 4, 11, 1263-1297.

Rellich, F. (1940) Darstellung der eigenwerk $\Delta u + \lambda u$ durch ein randintegral. *Math. Z.*, 46, 635-646.

Rousselet, B. and Chesnais, D. (1990) Continuité et différentiabilité d'éléments propres: application à l'optimisation de structures. *Appl. Math. and Optim.*, 22, 27-59.

Sanchez-Hubert, J. and Sanchez-Palencia, E. *Vibration and coupling of continuous systems: asymptotic methods.* Springer-Verlag.

J. Serrin, J. (1971) A symmetry problem in potential theory, *Arch. Rat. Mech. Anal.*, 43, 304-318.

Simon, J. (1980) Differentiation with respect to the domain in boundary value problems. *Num. Funct. Anal. Optimiz.*, 2 (7,8), 649-689.

Sokolowski, J. and Zolesio, J.P. (1992) *Introduction to shape optimization: shape sensitivity analysis.* Springer Series in Computational Mathematics, vol 10, Springer.

L. Zalcman, L. (1992) A bibliographical survey of the Pompeiu problem, in *Approximation by solutions of p.d.e.*, B. Fuglede et al. eds. , Kluwer.

8

A frequency method for H^∞ operators

Richard Datko
Georgetown University
Washington, D.C. 20057, U.S.A.

Abstract

We present numerical methods for determining the stability or instability of autonomous linear functional differential equations. These are based on some qualitative properties of analytic mappings and the ability to use standard mathematical packages to evaluate transcendental functions and approximate complex integrals.

Keywords

Functional differential equations, Poisson integral, H^∞ spaces, uniform exponential stability

1 INTRODUCTION

The use of quadratic Lyapunov functionals in the investigation of the stability of linear autonomous dynamical systems is largely passé. With the exception of finite dimensional linear differential equations their construction presents formidable computational difficulties. Consequently, when used to study infinite dimensional systems, Lyapunov functionals are often introduced in an ad hoc manner. In this paper we outline a general approach for studying the stability of a class of linear dynamical systems using what might be loosely described as a frequency domain approach. In reality it is a return to a complex analysis approach in the study of linear systems which is enhanced by the use of computational techniques such as MATLAB. The fundamental idea is quite straight-forward and is based on the following observation. The majority of linear dynamical systems one encounters in practice are uniformly exponentially stable (u.e.s. for short) if and only if their Laplace transforms are H^∞ operators in some Banach space. How does one determines whether a given analytic operator is in H^∞? Certainly not by direct methods, such as locating poles in the right half plane. However by reversing certain arguments used to study H^∞ functions and using standard numerical packages it is possible in many instances to determine whether or not a given analytic operator valued function is in H^∞.

Due to space limitations we shall state our results either without proof or with a brief sketch. However the details can easily be filled in by recourse to the associated references. All computations in this paper were carried out using the Texas Instruments TI-85 Scientific Calculator.

2 STABILITY AND H^∞ OPERATORS

Definition 2.1

(i) An operator valued function F from the complex plane, C, is in H^∞ if $F(\lambda)$ is analytic and uniformly bounded in $Re\lambda > 0$.
(ii) A function F which is in H^∞ will be said to be in $H^\infty(B)$ if it is in H^∞, $\lim_{|w|\to\infty} |F(iw)| = 0$, where $|\cdot|$ denotes a suitable norm, and if F is analytic in a neighborhood of $Re\lambda = 0$.

Property 2.2 Let $F : C \to H$, a Hilbert space, be in $H^\infty(B)$. Then if $\lambda = x + iy$, F has the following integral representation.

$$F(x+iy) = \frac{x}{\pi}\int_{-\infty}^{\infty} \frac{F(it)}{(t-y)^2 + x^2} dt; \tag{1}$$

$$F(x+iy) = \frac{1}{\pi}\int_{-\pi/2}^{\pi/2} F[i(x\tan\sigma + y)]d\sigma \tag{2}$$

The formula (1) is given in Hille (1973) and (2) is obtained from (1) by the change of variable

$$x\tan\sigma = t - y. \tag{3}$$

Property 2.3 If F is in $H^\infty(B)$, then

$$\int_{-\pi/2}^{\pi/2} e^{-2i\sigma} F[i(x\tan\sigma + y)]d\sigma = 0 \tag{4}$$

for all $\lambda = x + iy, \;\; x > 0$. Formula (4) is obtained from formula (2) and the requirement that $F_x = \frac{1}{i}F_y$.

Property 2.4 Let F be holomorphic in a neighborhood of $Re\lambda = 0$. Then F is in $H^\infty(B)$ if and only if equation (4) is satisfied for all $x+iy, \;\; x > 0$. If F is not analytic in $Re\lambda > 0$, but analytic and uniformly bounded in some neighborhood of $Re\lambda = 0$, then equation (2) is a harmonic function which is not analytic.

Example 1

Let $F(\lambda) = \frac{1}{\lambda - 2}$. Then, if $x > 0$,

$$\frac{x}{\pi}\int_{-\infty}^{\infty} \frac{F(it)\,dt}{x^2 + (t-y)^2} = \frac{-1}{\overline{\lambda} + 2},$$

where $\lambda = x + iy$. Also $\int_{-\pi/2}^{\pi/2} e^{-2\lambda i\sigma} F(i\tan\sigma)d\sigma = -\frac{2}{9} \neq 0$.

Theorem 2.5 *The linear dynamical system described by*

$$\frac{d}{dt}[x(t) - \sum_{j=1}^{m} D_j x(t-h_j)] = A_0 x(t) + \sum_{j=1}^{m} A_j x(t-h_j), \tag{5}$$

where $\{A_j\}$, $0 \leq j \leq m$, *and* $\{D_j\}$, $1 \leq j \leq m$, *are real* $n \times n$ *matrices and* $\{h_j\}$, $1 \leq j \leq m$, *are positive constants, is u.e.s. if and only if the scalar function*

$$F(\lambda) = [\det(\lambda(I - \Sigma D_j e^{-\lambda h_j}) - A_0 - \Sigma A_j e^{-\lambda h_j})]^{-1} \tag{6}$$

is in $H^\infty(B)$. *That is, equation (6) satisfies Property 2.3 for all* $\lambda = x + iy$ *in some neighborhood,* $N(\lambda_0)$, *of* $\lambda_0 = x_0 + iy_0$, $x_0 > 0$.

Sketch of the Proof It can be shown that if (4) is satisfied at some point $\lambda_1 = x_1 + iy$, $x_1 > 0$ then $F(\lambda_1)$ satisfies the Cauchy-Riemann equations at λ_1. Thus, if (4) is satisfied in an open neighborhood of λ_0, F is analytic in that neighborhood. But F must also be analytic is a neighborhood of $Re\lambda = 0$ and uniformly bounded there (otherwise the integral (4) would not exist). This means, using an analytic continuation argument, that the Poisson integral (2) represents F in $Re\lambda > 0$. This also means F has no poles in $Re\lambda \geq 0$ and since it is uniformly bounded in some neighborhood of $Re\lambda = 0$ all the poles of F must lie in $Re\lambda \leq -\alpha$, $\alpha > 0$ (see e.g. Hale and Lunel (1993) or Henry (1974)). Hence the system (5) is u.e.s. (see e.g. Henry (1974)).

Remark 2.6 The practical implementation of Theorem 2.5 would be to evaluate $F(\lambda)$ at a finite number of points and then check the Poisson integrals (2) against these points using some numerical method. If these are consistent then the system is probably u.e.s. We indicate this procedure in the following example.

Example 2

Consider the system

$$\frac{d}{dt}\left[x(t) - \begin{pmatrix} 0 & -\frac{1}{2} \\ \frac{1}{2} & 0 \end{pmatrix} x(t-h)\right] = \begin{pmatrix} 0 & 1 \\ -1 & 0 \end{pmatrix} x(t) + \begin{pmatrix} 0 & 0 \\ 0 & -1 \end{pmatrix} x(t-h), \tag{7}$$

where $h > 0$. The corresponding function is

$$F(\lambda, h) = \left[\lambda^2\left(1 + \frac{e^{-2\lambda h}}{4}\right) + 1\right]^{-1} \tag{8}$$

Let $h = 1$. Then the values of $F(1,1), F(.5,1)$ and $F(.25,1)$ and their corresponding Poisson integral approximations are given in the following table.

$$\begin{aligned} F(1,1) &= .491682255339, & I(1,1) &= .491716045511 \\ F(.5,1) &= .78550604137, & I(.5,1) &= .785504502663 \\ F(.25,1) &= .932855799362, & I(.25,1) &= .932948880576. \end{aligned}$$

The integral approximations were obtained using Simpson's Rule with upper limit and lower limit respectively $\pm$ 1.57069632679 and the number of partitions $N = 100$.

We now let $h = 2$ in (8). The corresponding values of $F(1,2), F(.5,2)$ and $F(.25,2)$ and the approximate integrals are

$$\begin{array}{llll} F(1,2) & = .498857887385, & I(1,2) & = .559635925368 \\ F(.5,2) & = .794622973739, & I(.5,2) & = .815297457756 \\ F(.25,2) & = .936112111757, & I(.25,2) & = .941191927459. \end{array}$$

We conclude that for $h = 1$ (7) is u.e.s. but not for $h = 2$. Indeed when $h = 2$ the integral (4) in Property 2.3 yields $I(1,2) \doteq .659367589698$, but for $h = 1$ the integral (4) is $I(1,1) \doteq -3.48519770543 \times 10^{-5}$

3 STABILITY ESTIMATES USING LYAPUNOV FUNCTIONALS

Although we stated in the introduction that Lyapunov functionals were passé, they still have their uses in studying certain types of dynamical systems. One such class is linear autonomous functional differential equations with multiple discrete delays described by equation (5).

Assumption 3.1 The poles of the function

$$G(\lambda) = [\det(I - \sum_{j=1}^{m} D_j e^{-\lambda h_j})]^{-1} \tag{9}$$

lie in $Re\lambda \leq -\beta, \;\; \beta > 0$.

Assumption 3.1 is known as the D-stability condition of Cruz and Hale (1969). Henry (1974) has shown that it is a necessary condition for the system (5) to be u.e.s.

The solutions of (5) must satisfy an initial condition on the interval $[-h, 0]$, where

$$h = \max_j h_j. \tag{10}$$

If this initial condition is a continuously differentiable vector function,

$$\phi : [-h, 0] \to C^n, \tag{11}$$

then the solution has for $t \geq 0$ the representation.

$$\begin{aligned} x(t) &= [S(t) - \sum_{j=1}^{m} S(t - h_j) D_j]\phi(0) \\ &+ \sum_{j=1}^{m} \int_{-h_j}^{0} S(t - \sigma - h_j)[A_j\phi(\sigma) + D_j\dot{\phi}(\sigma)]d\sigma, \end{aligned} \tag{12}$$

where $S(t)$ is the inverse Laplace transform of the analytic matrix valued function

$$\hat{S}(\lambda) = [\lambda(I - \sum_{j=1}^{m} D_j e^{-\lambda h_j}) - A_0 - \sum A_j e^{-\lambda h_j}]^{-1}. \tag{13}$$

Definition 3.2

(i) Let $A = (a_{ij})$ be an $n \times n$ matrix. The norm of A, $|A|$, is defined by
$|A| = \sup_j \sum_{i=1}^{n} |a_{ij}|$

(ii) The norm of any n-vector $x = \{x_j\}$ is
$|x| = \sup_j |x_j|$

(iii) The inner product in C^n is denoted by $(\cdot,\cdot)$ with the conventions that for any λ in C and x, y in C^n
$(\lambda x, y) = \overline{\lambda}(x,y), \quad (x, \lambda y) = \lambda(x,y).$

(iv) The adjoint of any $n \times n$ matrix A is denoted by A^*.

Theorem 3.3 *Let the system (5) be u.e.s. with a continuously differentiable initial condition ϕ. Then there exists a continuously differentiable mapping $q : [0,\infty] \to C^n$ such that:*

(i)
$$q(t) = \int_0^\infty S^*(\sigma)x(\sigma + t)d\sigma, \tag{14}$$

(ii)
$$\frac{d}{dt}[q(t) - \sum_{j=1}^{m} D_j^* q(t+h_j)] = -x(t) - A_0^* q(t) - \sum_{j=1}^{m} A_j^* q(t+h_j) \tag{15}$$

and

(iii)
$$(q(0) - \sum_{j=1}^{m} D_j^* q(h_j),\ \phi(0)) + \sum_{j=1}^{m} \int_{h_j}^{0} (q(\sigma + h_j),\ A_j\phi(\sigma) + D_j\dot{\phi}(\sigma))d\sigma \tag{16}$$
$$= \int_0^\infty (x(\sigma), x(\sigma))d\sigma.$$

The proof of Theorem 3.3 is an extrapolation of the results in Datko (1974). In particular (14) and (16) induce a bilinear functional on the space of continuous vector valued functions from $[-h, 0]$ into C^n. The problem is that an explicit form for this functional is impossible to obtain. However as we shall see equations (14) and (16) have their uses. These are shown in the following theorem whose proof is given in Datko (1995).

Theorem 3.4 *Let the system (5) be u.e.s. and let*

$$M_0 \geq \sup_w |\hat{S}(iw)^*| \tag{17}$$

and

$$M_1 \geq \sup_w |wS(iw)^*|. \tag{18}$$

Then all poles of the matrix operator function $\hat{S}(\lambda)$ lie in

$$Re\lambda < -\alpha, \quad \alpha > 0, \tag{19}$$

where α is the unique real solution of the equation

$$2(1 + \sum |D_j|)\, M_0\alpha + 2\sum_{j=1}^{m} [M_0|A_j| + M_1|D_j|] \sinh\alpha h_j = 1 \tag{20}$$

Remark 3.5 One consequence of Theorem 3.4 is that all solutions of the system (5) have a decay rate $e^{-\alpha t}$ (see e.g. Henry (1974)).

Example 3

Let us return to example 2. When $h = 1$ the system is u.e.s. and unstable when $h = 2$. In this case $|D| = \frac{1}{2}$, $|A_1| = 1$. When $h = 1$ the quantities M_0 and M_1 in (17) and (18) satisfy

$$M_0 < 7, \quad M_1 < 8 \tag{21}$$

(see Datko (1995)). Since the real solution of (20) decreases as M_0 and M_1 increase the solution of that equation for the values $M_0 = 7$ and $M_1 = 8$ yield an upper bound on the poles of $\hat{S}(\lambda)$. The value, $-\alpha$, in this case is

$$-\alpha = -.02354741522. \tag{22}$$

When $h = 2$ in the system of example 2 the values of M_0 and M_1 respectively satisfy

$$M_0 < 2.5, \quad M_1 < 2.3. \tag{23}$$

We substitute these values into equation (20) to obtain

$$\alpha = .045208159134. \tag{24}$$

We now estimate $\max_w |\hat{S}(-\alpha + iw)^*|$ for this value of α. We obtain $\max_w |\hat{S}(-\alpha + iw)^*| > 2.7 > \max_w |S(iw)^*|$.

This implies that $\hat{S}(-\overline{\lambda})^*$ is not analytic in the entire left half plane $Re\lambda < 0$. Hence $\hat{S}(\lambda)$ cannot be in H^∞, which means that for $h = 2$ the system is unstable.

Another method for arriving at the same conclusion is to observe that $F(\lambda, 2)$ in (8) satisfies the inequalities $\sup_w |F(iw, 2)| < 4.3 < 8.9 < \sup_w |F(\alpha + iw, 2)|$, where α is given by (24). Thus if $F(\lambda, 2)$ is not in H^∞ neither is $\hat{S}(\lambda)$.

REFERENCES

Cruz, M. A. and Hale, J. K. (1969) Asymptotic behavior of neutral functional differential equations. *Arch. Rat. Mech. Anal.* **34**, 331-53.

Datko, R. (1974) Neutral autonomous functional equations with quadratic cost. *SIAM J. Control* **12**, 70-81.

Datko, R. (1995) Some numerical methods for studying stability of linear delay systems. To appear.

Hale, J. K. and Lunel, S. M. V. (1993) *Introduction to Functional Differential Equations.* Springer-Verlag, New York.

Henry, D. (1974) Linear autonomous neural functional differential equation. *J. Diff. Eqs.*, **15**, 106-28.

Hille, E. (1973) *Analytic Function Theory, Vol. 2*, Chelsea, New York.

9

Local smoothing properties of a Schrödinger equation with nonconstant principal part

*Mary Ann Horn**
Department of Mathematics, Vanderbilt University
Nashville, Tennessee 37240, U.S.A.
and
Walter Littman
School of Mathematics, University of Minnesota
Minneapolis, Minnesota 55455, U.S.A.

Abstract
A Schrödinger equation with continuous, nonconstant coefficients appearing in the principal part of the differential operator is considered. Through the use of Littman and Taylor's general technique for proving boundary controllability of evolution equations, a simple proof of exact boundary controllability for the Schrödinger equation is obtained. Proof of the necessary regularity for the solution relies on an approximation argument and the spectral results for elliptic operators of Birman and Solomyak.

Keywords
Schrödinger equation, exact controllability, regularity.

1 INTRODUCTION

In the context of controllability, local smoothing properties for the Schrödinger equation were first addressed in Littman and Taylor (1992) where the Schrödinger equation with nonsmooth potential was considered. In this work, a very general technique for proving boundary controllability was developed which is applicable to many classes of equations. It can be summed up in the following way:

local smoothing + uniqueness + reversibility $\Longrightarrow$ *exact boundary controllability.*

Here "uniqueness" is the uniqueness property implying approximate controllability by duality. If control is to be exercised on the whole lateral boundary of a cylindrical domain,

*This material is based upon work partially supported under a National Science Foundation Mathematical Sciences Postdoctoral Research Fellowship.

$\Omega \times (0, T)$, in space-time, this means that every solution (in an appropriate space) of the homogeneous linear evolution equation in this domain having zero Chauchy data on the lateral boundary *must vanish identically.* "Reversibility" means that the backward problem (in time) is wellposed. For the Schrödinger equation, the difficulty of applying Littman and Taylor's approach lies in proving the necessary smoothing properties.

Although various systems of equations have been considered and a variety of techniques developed in the context of boundary control, the appeal of this method is its generality and avoidance of many of the technicalities of other related works. In comparison, Lasiecka and Triggiani (1992) use the method of multipliers to obtain boundary controllability for the Schrödinger equation with constant coefficients. However, the estimates involved are very sensitive to the lower order differential terms which will be inevitably produced by the variations of the coefficients. While some oscillations of the coefficients can still be handled by the existing techniques, to treat the general case, new approaches need to be developed.

In Lebeau (1992), the results of his work with Bardos and Rauch (1989) on exact controllability for hyperbolic equations are modified and applied to Schrödinger type equations. However, in the statement of his results, Lebeau assumes constant coefficients and an analytic boundary. Although his work appears to extend to the case of nonconstant coefficients, our goal, as already stated, is to avoid the technical difficulties of applying his technique to the more general case.

We note that this paper is a preliminary report. A more detailed and comprehensive treatment, including regions other than $\mathbf{R}^n$, is planned for the future.

2 STATEMENT OF THE PROBLEM

In this paper, if our system exhibits local smoothing properties, the result is that the solution belongs to the class Gevrey-δ, where this Gevrey class is defined with respect to the time variable t. We formulation this more precisely in the definition below.

Definition *A function $f(x,t)$ belongs to the space γ^δ (Gevrey-δ class) uniformly for (x,t) in a compact set K if, for every $(x,t) \in K$ and for every $\theta > 0$,*

$$|\frac{\partial^j}{\partial t^j} f(x,t)| \leq C_{K,\theta}\theta^j (j!)^\delta.$$

Our study of Schrödinger equations with nonconstant coefficients in the principal part of the differential operator begins with the following system.

$$\begin{array}{ll} i\frac{\partial u}{\partial t} + \mathcal{A}u = 0, & x \in \mathbf{R}^n, \quad t > 0, \\ \quad u(x,0) = \phi(x), & \end{array} \tag{1}$$

where

$$\mathcal{A} \equiv \sum_{i,j=1}^{n} \frac{\partial}{\partial x_i} a_{ij}(x) \frac{\partial}{\partial x_j} + c(x). \tag{2}$$

Note that, by definition, any potential is included in the operator $\mathcal{A}$.

2.1 Main results

In Horn and Littman (1995), the authors prove the following regularity theorem.

Theorem 1 *Assume the coefficients of $\mathcal{A}$ are real and satisfy the following conditions:*

$$\begin{array}{lll} i) & a_{ij}(x) = a_{ji}(x) & \text{for all } i,j, \\ ii) & \sum_{i,j=1}^{n} a_{ij}(x)\xi_i\xi_j \geq \alpha|\xi|^2 & \text{for some } \alpha > 0, \\ iii) & a_{ij}(x) - \delta_{ij}, c(x) \in C_0^\infty(\mathbf{R}^n), & \end{array} \tag{3}$$

where δ_{ij} is the Kronecker delta. Without loss of generality, we assume that the support of these functions is contained in the ball $B_a \equiv \{x : |x| < a\}$.

Assume $\phi(x) \in L^2(\mathbf{R}^n)$ and has compact support. Then the solution $u(x,t)$ to the initial value problem (1) is of class Gevrey-2 with respect to t uniformly in compact sets of $\mathbf{R}^n \times \{t > 0\}$.

Our goal is to reduce the smoothness assumptions on the coefficients of $\mathcal{A}$. In Littman and Taylor (1992), the potential, $c(x)$, needed only to be bounded to obtain regularity results for the solution when $\mathcal{A} \equiv \Delta$. However, we also wish to reduce assumption *iii)* of Theorem 1 from $a_{ij}(x) - \delta_{ij} \in C_0^\infty(\mathbf{R}^n)$ to $a_{ij}(x) - \delta_{ij} \in C_0(\mathbf{R}^n)$, thus proving the following theorem.

Theorem 2 *Assume the coefficients of $\mathcal{A}$ are real-valued, satisfy conditions i.) and ii.) of Theorem 1, and*

$$iii)' \quad a_{ij}(x) - \delta_{ij}, c(x) \in C_0(\mathbf{R}^n) \quad \text{for all } i,j. \tag{4}$$

Assume $\phi(x) \in L^2(\mathbf{R}^n)$ and has compact support. Then the solution $u(x,t)$ to the initial value problem (1) is of class Gevrey-2 with respect to t uniformly in compact sets of $\mathbf{R}^n \times \{t > 0\}$.

The proof of the Gevrey regularity in time (as given in Littman and Taylor (1992) and Horn and Littman (1995)) consists of solving the Schrödinger equation by the Laplace transform and showing that the inversion integral can be transformed into a contour integral along the boundary of what is essentially a union of two parabolas in the left hand plan of the transformed time variable λ.

The justification of this contour change consists of studying the modified resolvent, $\tilde{R}(k^2;\mathcal{A}) = \chi(x)(k^2 + \mathcal{A})^{-1}\chi(x)$, (where $\chi(x)$ is a cutoff function) and to show that $\tilde{R}(k^2;\mathcal{A})$ can be continued as a bounded analytic operator function into a set of the type

$$\{k : \Im m k \geq -b, \Re e k > a\}.$$

This has been done in Horn and Littman (1995) for C^∞ coefficients. To study nonsmooth coefficients, a perturbation argument is used below. The function $\tilde{u}$ introduced below is essentially the continuation of the modified resolvent acting on f,

$$\tilde{u} = \tilde{R}(k^2;\mathcal{A})f.$$

Thus, in this paper, we will focus on proving the modified resolvent exists and is bounded in this extended region. Once these bounds have been established, proof of the Gevrey regularity of the solution follows as in Horn and Littman (1995) and combines straightforward estimation with the semigroup theory of Pazy (1983).

3 PROOF OF GEVREY REGULARITY OF THE SOLUTION

In a sense, we will treat the case of $C_0(\mathbf{R}^n)$ coefficients as a perturbation of the $C_0^\infty(\mathbf{R}^n)$ case. However, this does not allow us to apply a standard perturbation argument because the perturbation occurs in the principal part of the differential operator. Similarly to the proof of Theorem 1, we begin with the Laplace transform of the system but after this, we quickly diverge from the proof found in Horn and Littman (1995).

3.1 Step 1: Laplace transform

Taking the Laplace transform of (1) with respect to t, we find

$$\begin{aligned} i\lambda\tilde{u} - \mathcal{A}\tilde{u} &= i\phi(x) \\ \text{or } \mathcal{A}\tilde{u} + k^2\tilde{u} &= i\phi(x) \equiv f(x), \end{aligned} \tag{5}$$

where $\lambda = -ik^2$. Define $\mathcal{L} \equiv \mathcal{A} + k^2$, $\mathcal{L}_0 \equiv \Delta + k^2$. As in Theorem 1, without loss of generality, we assume $a_{ij}(x) \equiv 1$ outside a ball of radius a, $\mathcal{B}_a \equiv \{x : |x| < a\}$, and that the supports of both $c(x)$ and $\phi(x)$ are contained in $\mathcal{B}_a$.

3.2 Step 2: Representation of solutions

To facilitate the proof, we impose an "artificial" boundary condition, $\mathcal{B}u|_{\partial\mathcal{B}_{a+3\epsilon}} = 0$, on the system 5. This boundary condition could be, for example, either a Dirichlet or Neumann boundary condition. In order to localize the solution inside and outside of $\mathcal{B}_a$, we define the following two cutoff functions, $\psi(x), \zeta(x) \in C^\infty(\mathbf{R}^n)$, $0 \leq \psi(x), \zeta(x) \leq 1\ \forall x \in \mathbf{R}^n$, as follows.

$$\psi(x) = \begin{cases} 0 & |x| \leq a \\ 1 & |x| \geq a + \epsilon \end{cases} \qquad \zeta(x) = \begin{cases} 1 & |x| \leq a + 2\epsilon \\ 0 & |x| \geq a + 3\epsilon \end{cases} \tag{6}$$

and let $w = \psi\tilde{u}$ (outer solution) and $v = \zeta\tilde{u}$ (inner solution). Then $\tilde{u}$ can be written in terms of w and v.

$$\begin{aligned} \tilde{u} &= (1-\psi)\tilde{u} + \psi\tilde{u} = (1-\psi)\tilde{u} + w \\ &= (1-\psi)(1-\zeta)\tilde{u} + (1-\psi)\zeta\tilde{u} + w \\ &= (1-\psi)v + w, \end{aligned} \tag{7}$$

since $(1-\psi)(1-\zeta) \equiv 0$ for all $x \in \mathbf{R}^n$.

Applying $\mathcal{L}$ to w and v, we find

$$\mathcal{L}w = \mathcal{L}_0 w \quad = \psi\mathcal{L}\tilde{u} + [\mathcal{L},\psi]\tilde{u} \quad = [\mathcal{L},\psi]v$$

$$\mathcal{L}v \quad = \zeta\mathcal{L}\tilde{u} + [\mathcal{L},\zeta]\tilde{u} \quad = f + [\mathcal{L},\zeta]w,$$

where $[\cdot,\cdot]$ denotes the commutator of the two operators and where ψ and ζ also represent multiplication operators. Solving for w, we find

$$w = \mathcal{L}^{-1}[\mathcal{L},\psi]v = \mathcal{L}_{ext}^{-1}[\mathcal{L}_{int},\psi]v \tag{8}$$
$$\Longrightarrow v = \mathcal{L}_{int}^{-1}\{f + [\mathcal{L}_{ext},\zeta]\mathcal{L}_{ext}^{-1}[\mathcal{L}_{int},\psi]v\}, \tag{9}$$

where

$$\mathcal{L}_{int} \equiv \mathcal{L}, \quad \mathcal{D}(\mathcal{L}_{int}) \equiv H_0^1(\mathcal{B}_{a+3\epsilon}), \quad \mathcal{L}_{ext} \equiv \mathcal{L}_0, \quad \mathcal{D}(\mathcal{L}_{ext}) \equiv H^1(\mathbf{R}^n). \tag{10}$$

We note for future reference that, with the above definition, $\mathcal{L}_{int}$ is constant outside of $\mathcal{B}_a$.

3.3 Approximation of continuous coefficient problem

At this point, in order to differentiate between the problem with C^∞ coefficients and continuous coefficients, we define the following notation.

- $\mathcal{A}_\infty$ and $\mathcal{A}_c$ are defined as in (2) with C^∞ and continuous coefficients, respectively.
- $\mathcal{L}_\infty \equiv \mathcal{A}_\infty + k^2$, $\mathcal{L}_c \equiv \mathcal{A}_c + k^2$.

We choose the coefficients of $\mathcal{A}_\infty$ such that given $\epsilon > 0$, $\max_x |a_{\infty,ij}(x) - a_{c,ij}(x)| < \epsilon$. Thus, $\mathcal{A}_\infty$ can be considered to be a perturbation of the operator $\mathcal{A}_c$.

3.4 Estimation of the inner solution

Define

$$M_\infty \equiv [\mathcal{L}_0,\zeta]\mathcal{L}_0^{-1}[\mathcal{L}_{\infty,int},\psi] \qquad M_c \equiv [\mathcal{L}_0,\zeta]\mathcal{L}_0^{-1}[\mathcal{L}_{c,int},\psi]. \tag{11}$$

Then

$$v_\infty \equiv \mathcal{L}_{\infty,int}^{-1}\{f + M_\infty v_\infty\} \qquad v_c \equiv \mathcal{L}_{c,int}^{-1}\{f + M_c v_c\}, \tag{12}$$

$$\begin{aligned}\Longrightarrow \quad v_c - v_\infty &= (\mathcal{L}_{c,int}^{-1} - \mathcal{L}_{\infty,int}^{-1})f + \mathcal{L}_{c,int}^{-1}M_c v_c + \mathcal{L}_{\infty,int}^{-1}M_\infty v_\infty \\ &= \delta\mathcal{L}_{int}^{-1}f + \delta\mathcal{L}_{int}^{-1}M_c v_c + \mathcal{L}_{\infty,int}^{-1}\delta M v_c + \mathcal{L}_{\infty,int}^{-1}M_\infty \delta v\end{aligned} \tag{13}$$

$$\Longrightarrow \quad (I - \mathcal{L}_{\infty,int}^{-1}M_\infty)\delta v = \delta\mathcal{L}_{int}^{-1}f + \delta\mathcal{L}_{int}^{-1}M_c v_c + \mathcal{L}_{\infty,int}^{-1}\delta M v_c,$$

where

$$\delta v \equiv v_c - v_\infty, \quad \delta M \equiv M_c - M_\infty, \quad \delta\mathcal{L}_{int}^{-1} \equiv \mathcal{L}_{c,int}^{-1} - \mathcal{L}_{\infty,int}^{-1}. \tag{14}$$

By definition, we have chosen $\mathcal{L}_{\infty,int}$ such that

$$\mathcal{L}_{\infty,int} \equiv \mathcal{L}_{c,int} + \mathcal{P},$$

where $\mathcal{P}$ is a small perturbation of the differential operator. Thus,

$$\begin{aligned} \mathcal{L}_{\infty,int}^{-1} - \mathcal{L}_{c,int}^{-1} &= \mathcal{L}_{\infty,int}^{-1} - (\mathcal{L}_{\infty,int} + \mathcal{P})^{-1} = \mathcal{L}_{\infty,int}^{-1} - (\mathcal{L}_{\infty,int} + \mathcal{P}\mathcal{L}_{\infty,int}^{-1}\mathcal{L}_{\infty,int})^{-1} \\ &= \mathcal{L}_{\infty,int}^{-1} - [(I + \mathcal{P}\mathcal{L}_{\infty,int}^{-1})\mathcal{L}_{\infty,int}]^{-1} \\ &= \mathcal{L}_{\infty,int}^{-1} - \mathcal{L}_{\infty,int}^{-1}(I + \mathcal{P}\mathcal{L}_{\infty,int}^{-1})^{-1} \\ &= \mathcal{L}_{\infty,int}^{-1}\{I - (I + \mathcal{P}\mathcal{L}_{\infty,int}^{-1})^{-1}\}. \end{aligned} \tag{15}$$

Since $\|\mathcal{P}\|_{H^1 \to H^{-1}} < \epsilon$, $I + \mathcal{P}\mathcal{L}_{\infty,int}^{-1} \sim I$, therefore,

$$\|\mathcal{L}_{c,int}^{-1} - \mathcal{L}_{\infty,int}^{-1}\|_{H^{-1} \to H^1} < \epsilon \|\mathcal{L}_{\infty,int}^{-1}\|_{H^{-1} \to H^1}. \tag{16}$$

Next, we need to bound δM. Rewriting δM, we find

$$\begin{aligned} \delta M \equiv M_c - M_\infty &= [\mathcal{L}_0, \zeta]\mathcal{L}_0^{-1}[\mathcal{L}_{c,int}, \psi] - [\mathcal{L}_0, \zeta]\mathcal{L}_0^{-1}[\mathcal{L}_{\infty,int}, \psi] \\ &= [\mathcal{L}_0, \zeta]\mathcal{L}_0^{-1}[\mathcal{L}_{c,int} - \mathcal{L}_{\infty,int}, \psi]. \end{aligned} \tag{17}$$

Note that in the above equation,

$$\begin{aligned} &\mathrm{supp}[\mathcal{L}_{c,int}, \zeta] \subset \mathcal{B}_{a+3\epsilon} \backslash \mathcal{B}_{a+2\epsilon} \\ &\mathrm{supp}[\mathcal{L}_{c,int}, \psi] \subset \mathcal{B}_{a+\epsilon} \backslash \mathcal{B}_a \\ &\mathrm{supp}[\mathcal{L}_{c,int} - \mathcal{L}_{\infty,int}, \psi] \subset \mathcal{B}_{a+\epsilon} \backslash \mathcal{B}_a, \end{aligned} \tag{18}$$

and we know, by definition, that $\mathcal{L}_c \equiv \mathcal{L}_\infty$ in $\mathcal{B}_{a+3\epsilon} \backslash \mathcal{B}_a$. Therefore, $\delta M \equiv 0$.

Combining the results for $\delta\mathcal{L}_{int}^{-1}$ and δM, we find that for some constant $C_v > 0$,

$$\|v_c - v_\infty\|_{L^2} < \epsilon C_v \|\mathcal{L}_{\infty,int}^{-1}\|_{H^{-1} \to H^1}. \tag{19}$$

We note that the constant C_v (and C_w, C_u that follow in the next section) is notational only and does not imply any dependence on v.

3.5 Estimates for $\tilde{u}_c - \tilde{u}_\infty$

Using equation (8), we find

$$\begin{aligned} w_c - w_\infty &= \mathcal{L}_0^{-1}[\mathcal{L}_{c,int}, \psi]v_c - \mathcal{L}_0^{-1}[\mathcal{L}_{\infty,int}, \psi]v_\infty \\ &= \mathcal{L}_0^{-1}\{[\mathcal{L}_{c,int} - \mathcal{L}_{\infty,int}, \psi]v_c - [\mathcal{L}_{\infty,int}, \psi](v_c - v_\infty)\}. \end{aligned} \tag{20}$$

$$\Longrightarrow \|w_c - w_\infty\|_{L^2} < \epsilon C_w \|\mathcal{L}_{\infty,int}^{-1}\|_{H^{-1} \to H^1} \qquad \text{for some constant, } C_w > 0.$$

Therefore, from (7), there exists a constant, $C_u > 0$, such that

$$\|\tilde{u}_c - \tilde{u}_\infty\|_{L^2} < \epsilon C_u \|\mathcal{L}_{\infty,int}^{-1}\|_{H^{-1} \to H^1}. \tag{21}$$

3.6 Step 6: Phragmén-Lindelöf theorem

From the proof of Theorem 1 in Horn and Littman (1995), if the coefficients of problem (1) are C^∞, we know that the modified resolvent corresponding to this problem satisfies

$$\|\tilde{R}(k^2; \mathcal{A}_\infty)\|_{H^{-1} \to H^1} < C \qquad \forall k \in \mathcal{T}_{a,b} \equiv \{k : \Im mk \geq -b, |k| \geq a\}. \tag{22}$$

Hence, in particular, the inequality in (21) implies

$$\|\tilde{R}(k^2; \mathcal{A}_c)\|_{H^{-1} \to H^1} < C \qquad \forall k \in \{k : |\Im mk| = b, \Re ek = a\}, \tag{23}$$

i.e., the modified resolvent associated with $\mathcal{A}_c$ is uniformly bounded with respect to k on the boundary of the strip

$$\{k : |\Im mk| \leq b, \Re ek > a\}.$$

In deriving the boundedness of v as a function of k, the boundedness of $\mathcal{L}_c^{-1}$ as a function of k is needed. Unfortunately, there is a sequence of poles (corresponding to the eigenvalues of the Dirichlet problem for $\mathcal{A}_c$) which prevents the direct estimation of the modified resolvent in the strip. From the results of Birman and Solomjak (1972) on the distribution of eigenvalues, it follows that we can find a sequence of vertical line segments in the strip on which bounds for the modified resolvent will increase no faster than a power of $\Re ek$. Then an application of the Phragmén-Lindelöf Theorem (see Titchmarsh (1939), section 5.65) gives the boundedness of $\tilde{R}(k^2; \mathcal{A})$ in the entire strip.

REFERENCES

Bardos, C., Lebeau, G. and Rauch, J. (1989) Contrôle et stabilisation dans les problèmes hyperboliques. In *Contrôlabilité exacte, perturbations et stabilization de systèmes distribués*, volume 2, Paris. Masson. Appendix 2.

Birman, M. S. and Solomjak, M. Z. (1972) Spectral asymptotics of nonsmooth elliptic operators. *Soviet Math. Dokl.*, 13(2):906–910.

Horn, M. A. and Littman, W. (1995) Boundary control of a Schrödinger equation with nonconstant principal part. In *Control of Partial Differential Equations and Applications*, Lecture Notes in Pure and Applied Mathematics, Marcel Dekker, New York.

Lasiecka, I. and Triggiani, R. (1992) Optimal regularity, exact controllability and uniform stabilization of Schrödinger equations with Dirichlet control. *Differential and Integral Equations*, 5:521-535.

Lebeau, G. (1992) Contrôle de l'equation de Schrödinger. *Journal Math. Pures Appl.*, 71:267–291.

Littman, W. and Taylor, S. (1992) Smoothing evolution equations and boundary control theory. *Journal D'Analyse Mathématique*, 59:117–131.

Pazy, A. (1983) *Semigroups of Linear Operators and Applications to Partial Differential Equations.* Springer-Verlag, New York.

Titchmarsh, E. C. (1939) *The Theory of Functions.* Oxford University Press, London.

10

Regularity results for multiphase Stefan-like equations

Giulia Sargenti
Dipartimento di Matematica, Università di Roma La Sapienza
Piazzale Aldo Moro 2, 00185 Roma, Italy. Phone: 06 49913.
Fax: 44701007. E-mail: sargenti@mat.uniroma1.it
and
Vincenzo Vespri
Dipartimento di Matematica Pura e Applicata, Università dell'Aquila
Via Vetoio, 67010 Coppito (AQ), Italy. Phone: 0862 43311.
Fax: 0862 433180. E-mail: vespri@vxscaq.aquila.infn.it

Abstract

In this note the local continuity of any bounded local weak solution of degenerate multiphase Stefan problem is proved. Moreover the modulus of continuity can be determined a priori only in terms of the data.

Keywords

Continuity, multiphase Stefan problem, p-Laplacian

1 INTRODUCTION

In this note we study the partial regularity of bounded weak solutions of nonlinear parabolic equations of the type

$$(\beta(u))_t = \text{ Div } (|\nabla u|^{p-2}\nabla u) \quad \text{in} \quad D'(\Omega_T) \tag{1}$$

$$u \in L^p_{loc}(0,T; W^{1,p}_{loc}(\Omega)) \text{ and } \ p \geq 2 \tag{2}$$

Here Ω is a domain in R^N, $T > 0$ and Ω_T is the cylindrical domain $\Omega_T = \Omega \times (0,T)$.
Assume $\beta(\cdot)$ be a maximal monotone graph in $R \times R$ satisfying:

$$\beta(s_1) - \beta(s_2) \geq c_0(s_1 - s_2) \quad \forall s_1, s_2 \in R \tag{3}$$

for some given constant $c_0 > 0$,

$$\sup_{|u| \leq M} |\beta(u)| < \infty \tag{4}$$

for every $M > 0$. Lastly we assume that u is the limit of a sequence of local smooth solutions of approximating problems.

Let K be a compact set contained in Ω. In what follows we say that $c = c(data)$ if c is a constant that can be determined a priori only in terms of N, M, p and the distance between $K \times (\varepsilon, T)$ and the parabolic boundary of Ω_T. Now we can state our main result.

Main Theorem 1 *Let (3)–(4) hold and let u be a locally bounded weak solution of (1), (2). Then there exists a continuous nonnegative increasing function ω_{data}, $\omega(0) = 0$ such that*

$|u(P_1) - u(P_2)| \leq \omega(|x_1 - x_2| + |t_1 - t_2|^{\frac{1}{p}})$
for every $P_i = (x_i, t_i) \in K \times (\varepsilon, T)$, $i = 1, 2$.

In the last years several authors studied equations of the type (1) not only for their mathematical interest, but also for their application in many physical phenomena like the transition of phase or the Buckeley Leverett model of two immiscible fluids in a porous media (see Alt and Di Benedetto (1985), Caffarelli and Evans (1983), Chavent and Jaffre (1986), Di Benedetto (1982), Di Benedetto (1993)). In Caffareli and Evans (1983), Di Benedetto (1982) and Ziemer (1982) the case of a β with only a singular point is studied . In Di Benedetto and Vespri (to appear) was faced the case of a general β but the continuity theorem was proved only for local solution of equation of the type (1) with $p = 2$. In this paper we will apply the technique developed in Di Benedetto and Vespri (to appear) ; therefore we will point out only what is really new from this work.

2 NOTATION AND LOCAL ESTIMATES

We denote with B_R a ball of radius R and center in the origin of R^N, with K_ρ a cube centered in the origin of wedge 2ρ, $\rho > 0$ and with $Q(\rho, \rho^p) = K_\rho \times (-\rho^p, 0)$ the cylinder of vertex at the origin, height ρ^p and cross section K_ρ, while for a point $(x_o, t_o) \in R^{N+1}$ we let $(x_o, t_o) + Q(\rho, \rho^p)$ be the cylinder of vertex at (x_o, t_o) and congruent to $Q(\rho, \rho^p)$. In what follows we always assume that u is a locally bounded weak solution of (1) and that ρ is so small that $(x_o, t_o) + Q(\rho, \theta\rho^p)$ is contained in Ω_T, where $\theta > 0$ will be chosen later. For $k \in R$ we define the truncations

$$(u - k)_+ = \max(u - k, 0) \quad (u - k)_- = \max(-u + k, 0),$$

the numbers

$$H_k^\pm = ||(u - k)_\pm||_{\infty, \Omega_T}$$

and the function

$$\Psi(H_k^\pm, (u - k)_\pm, a) = \ln^+\left(\frac{H_k^\pm}{H_k^\pm - (u - k)_\pm + a}\right) \quad a \in (0, H_k^\pm)$$

simply denoted by Ψ. Further we assume ξ is a smooth cut off function defined in $(x_o, t_o) + Q(\rho, \rho^p)$ such that $\xi \in [0, 1]$, $|\nabla \xi| < \infty$, $0 \leq \xi_t < \infty$. We will be working with the smooth approximations to derive a modulus of continuity for smooth solutions of (1). The following two propositions can be proved as in Di Benedetto and Vespri (to appear).

Proposition 2.1 *There exists a constant* $c = c(data)$ *such that*

$$\begin{cases} sup_{t_o-\theta\rho^p\le t<t_o} \int_{[x_o+K_\rho]}(u-k)_\pm^2\xi^p(x,t)\,dx \\ +c\int\int_{[(x_o,t_o)+Q(\rho,\theta\rho^p)]}|\nabla(u-k)_\pm\xi|^p\,dx\,dt \\ \le \int\int_{[(x_o,t_o)+Q(\rho,\theta\rho^p)]}(u-k)_\pm\xi^{p-1}\xi_t\,dx\,dt \\ +\int\int_{[(x_o,t_o)+Q(\rho,\theta\rho^p)]}(u-k)_\pm^p|\nabla\xi|^p\,dx\,dt \\ +\int_{[x_o+K_\rho]}(u-k)_\pm\xi^p(x,t_o-\theta\rho^p)\,dx \end{cases} \tag{5}$$

Proposition 2.2 *There exists a constant c=c(data) such that*

$$\begin{cases} sup_{t_o-\theta\rho^p\le t<t_o}\int_{[x_o+K_\rho]}\Psi^2(x,t)\xi(x)^p\,dx \le c\int_{[x_o+K_\rho]}\Psi(x,t_o-\theta\rho^p)\xi^p(x)\,dx \\ +c\int\int_{[(x_o,t_o)+Q(\rho,\theta\rho^p)]}|\Psi'|^{p-2}\Psi|\nabla\xi|^p\,dx\,dt \end{cases} \tag{6}$$

3 SOME BASIC RESULTS

After a translation we may take $(x_o,t_o)=(0,0)$. Set

$\mu^+ = \sup u\ \mu^- = \inf u$ in $Q(\rho,\theta\rho^p)$, $\omega = \mu^+ - \mu^-$

For $\xi^\pm \in (0,1)$ define

$$A^+_{\xi^+,\rho}(t) = \{x\in K_\rho\ :\ u(x,t) > \mu^+ - \xi^+\omega\}$$
$$A^-_{\xi^-,\rho}(t) = \{x\in K_\rho\ :\ u(x,t) < \mu^- + \xi^-\omega\}$$

and set $A^\pm_{\xi^\pm,\rho} = \int_{-\theta\rho^p}^0 |A^\pm_{\xi^\pm,\rho}(t)|\,dt$.

Proposition 3.1 *There exists a number* $\nu^+\in(0,1)$ *depending upon* ω, ρ, *data such that if*

$$|A^+_{\xi^+,\rho}| < \nu^+|Q(\rho,\theta\rho^p)| \tag{7}$$

then

$$u(x,t) < \mu^+ - \frac{2}{3}\xi^+\omega\ \forall(x,t)\in Q\Big(\frac{\rho}{2},\theta\Big(\frac{\rho}{2}\Big)^p\Big). \tag{8}$$

If also

$$u(x,-\theta\rho^p) < \mu^+ - \xi^+\omega\ \forall x\in K_\rho \tag{9}$$

there exists a number $\nu^+\in(0,1)$ *depending upon* θ, *data, independent of* ω,ξ^+, *such that (8) implies*

$$u(x,t) < \mu^+ - \frac{2}{3}\xi^+\omega\ \forall(x,t)\in Q\Big(\frac{\rho}{2},\theta\rho^p\Big). \tag{10}$$

Remark 3.1 *By the proof we get the number* ν^+ *given by*

$$\nu^+ = c(data)\Big(\theta\Big(1+\frac{(\xi^+\omega)^{1-p}}{\theta}\Big)^{\frac{N+p}{p}}\Big)^{-1} \tag{11}$$

Sketch of the proof

For $n = 0, 1, 2, \ldots$ consider the sequence of radii

$$\rho_n = \frac{\rho}{2} + \frac{\rho}{2^{n+1}} \quad \tilde{\rho}_n = \frac{\rho}{2} + \frac{3}{2}\frac{\rho}{2^{n+2}}$$

and the sequence of numbers

$$K_n = \mu^+ - \frac{2}{3}\xi^+\omega - \frac{1}{3}\frac{\xi^+}{2^n}\omega$$

We rewrite (5) over the pair of cylinder $\tilde{Q}_n = K_{\tilde{\rho}_n} \times (-\theta\tilde{\rho}_n^p, 0)$ and $Q_n = K_{\rho_n} \times (-\theta\rho_n^p, 0)$ and using the sequence of cutoff functions ξ_n defined over $Q_n = K_{\rho_n}$ such that $\xi_n = 1$ over $\tilde{Q}_n$, $|\nabla\xi_n| \le \frac{2^{n+1}}{\rho}$, $\partial_t\xi_n \in [0, 2^{p(n+1)}/\theta\rho^p]$. After some calculations (see Di Benedetto and Vespri (to appear) we get

$$Y_{n+1} \le c(data)\Big(1+\frac{(\xi^+\omega)^{1-p}}{\theta}\Big)\theta^{\frac{N+p}{p}} Y_n^{1+\frac{p}{N+p}} \tag{12}$$

where $Y_n = \frac{|A_{\xi_n,\rho_n}|}{|Q_n|}$. To prove (9) we apply Lemma 4.1 of Di Benedetto (1993). For (11) we argue in a similar way using a sequence ξ_n independent of t.

We can state an analogous result for $A_{\xi^-,\rho^-(t)}$, with $\mu^+ - \frac{2}{3}\xi^+\omega$ replaced by $\mu^- + \frac{2}{3}\xi^-\omega$. As the proof is similar to the one in Di Benedetto and Vespri (to appear) we omit it.

Proposition 3.2 *Suppose that for some* $\xi_o^+ \in (0,1)$ *the number* θ *satisfies*

$$\theta \ge (\xi_o^+\omega)^{p-2} \tag{13}$$

and that

$$u(x, -\theta\rho^p) \le \mu^+ - \xi_o^+ \ \forall x \in K_\rho \tag{14}$$

then for any $\nu^+ \in (0,1)$ *there exists a number* $\xi^+ \in (0, \xi_o^+/4)$ *depending upon* θ, ξ_o^+, $c(data)$ *such that*

$$|A_{\xi^+,\frac{\rho}{2}}(t)| \le \nu^+|K_{\frac{\rho}{2}}| \ \forall t \in (-\theta\rho^p, 0) \tag{15}$$

An analogous proof enables us to get the logarithmic estimate near μ^-.

If (7) holds, then arguing as in Di Benedetto and Vespri (to appear, the first alternative) we get:

$$\text{osc u} \le \eta_o(\omega)\omega \ \forall (x,t) \in Q\Big(\frac{\rho}{8}, \theta\Big(\frac{\rho}{8}\Big)^p\Big) \tag{16}$$

where $\eta_o(\omega) \in (0,1)$.

The unfavourable case, is when both conditions (7) and the analogous one for $A^{-}_{\xi-,\rho}$ are violated in the fixed cylinder $Q(\rho,\theta\rho^p)$, thus in particular in the coaxial boxes

$$(0,\tau)+Q(r,\theta r^p)$$

where $\tau\in[-\theta(\rho^p-r^p),0]$, $r\in(\delta\rho,\rho)$ and $\delta\in(0,1)$ is to be chosen.

Proposition 3.3 *There exists a constant $c=c(data,\omega)$ such that*

$$\theta\omega^p\rho^p\le c\int_{-\theta\rho^p}^{0}\int_{\{\delta\rho<||x||<\rho\}}|\nabla u|^p\,dx\,dt \tag{17}$$

Proof

Arguing as in Di Benedetto and Vespri (to appear), the second alternative] we get the existence of two cylinders Q_i both contained in $\{\delta\rho<||x||<\rho\}\times[-\theta\rho^p,0)$ such that u is near μ^+ in Q_1, u is near μ^- in Q_2 (see Di Benedetto and Vespri (to appear), Proposition 8.1]). Let $\Gamma_{1,2}$ be a path piecewise parallel to the coordinates axes joining P_1 to P_2, P_i belonging to the cross section of Q_i, $i=1,2$. Then we obtain

$$\omega<\int_{\Gamma_{1,2}}|\nabla u|\,d\Gamma \tag{18}$$

By integration we get (17).

4 PROOF OF THE MAIN THEOREM

Consider first the case $N=p$. Rewrite (17) for the cylinders

$$Q_n^j=[(0,t_n^j)+Q(\delta^n\rho,\theta\delta^n\rho)]\subset Q(\rho,\theta\rho^p), \tag{19}$$

where $t_n^j=-j\theta\delta^{np}\rho^p$, $j=0,1,\dots,\delta^{-np}-1$.

We suppose δ^{-1} be an integer number. After adding over n, $n=0,1,..,n_o-1$, with n_o to be chosen, we obtain

$$n_o\theta\rho^p\omega^p\le\int\int_{Q(\rho,\theta\rho^p)}|\nabla u|^p\,dx\,dt \tag{20}$$

Inequality (5) written for the value $k=\mu^+-\frac{\omega}{2}$ and using a cutoff function $\xi=1$ on $Q(\rho/2,\theta\rho^p/2)$, $|\nabla\xi|\le 2^p/\rho$, $\xi_t\in[0,2^{p+1}/\rho^p]$ gives

$$\int\int_{Q(\rho,\theta\rho^p)}|\nabla u|^p\,dx\,dt\le c\omega\theta\rho^N(1+\omega^{p-1}) \tag{21}$$

Equations (20) and (21) imply

$$n_o\le c(data,\omega) \tag{22}$$

with $c(data, \omega)$. We choose an n_o so large that

$$n_o > c(data, \omega) \tag{23}$$

This implies a contradiction. Therefore (7) or the analogous one for $A^-_{\xi^-,\rho}$ must hold for the cylinders Q^j_n. If we start from (7), we fix a value $\xi^\pm = \frac{1}{12}$, we use proposition 3.1 to arrive at

$$u(x,t) < \mu^+ - \frac{\omega}{18} \; \forall (x,t) \in [(0, t^j_n) + Q(\delta^n \rho/2, \theta(\delta^n \rho/2)^p] \tag{24}$$

Setting $\delta^n \rho = 8r$, $\sigma = \frac{8^p}{4}(j + 1/2^p)\theta$

$$u(x, -4\sigma r^p) < \mu^+ - \frac{\omega}{18} \forall x \in K_r.$$

We choose ν^+ for the largest value of σ i.e. $\nu^+ = c(data)/\theta\delta^{n_o p}$, n_o as in (23). From Proposition 3.1 we get

$$u(x,t) < \mu^+ - \frac{2}{3}\xi^+ \omega \; \forall (x,t) \in Q(\delta^{n_o}\rho/8, \theta(\delta^{n_o}\rho/8)^p)$$

We can repeat the method of the first alternative: there exists $\eta_1(\omega) \in (0,1)$ such that

$$\operatorname{osc} u \leq \eta_1(\omega)\omega \text{ in } Q(\delta^{n_o}\rho/8, \theta(\delta^{n_o}\rho/8)^p) \tag{25}$$

Therefore arguing as in Di Benedetto and Vespri (to appear), Corollary 12.1, we have that

$$\operatorname{osc} u \leq \omega_n \text{ in } Q(\rho_n, \theta\rho^p), \; \omega_n \to 0, \text{ as } n \to \infty.$$

The case $N > p$ is more complicated. We follow sections 13-29 of Di Benedetto and Vespri (to appear). We partition the original set into disjointed boxes of the type $(0, t_i) + Q(\rho, \rho^p)$, with $t_i = 0, -\rho^p, .., -(\theta - 1)\rho^p$ and divide each of these boxes in the disjoint cylinders

$$(x_l, t_h) + Q(\rho/m, \rho^p/m), \; l = 1, \ldots, m^N, \; h = 1, \ldots, m^p.$$

Either the first or the second alternative holds for these sets. Since they are not centered in the origin, we prove that an estimate similar to (24) holds by means of a comparison function. Propositions 24.2, 24.3 of Di Benedetto and Vespri (to appear) show that a number $\theta > 1$ $\lim_{\omega \downarrow 0} \theta(\omega) = \infty$, which can be determined a priori in terms of the data and ω, implies that the first alternative holds.

REFERENCES

Alt, H.M. and Di Benedetto, E. (1985) Non steady flow of water and oil through inhomogeneous porous media. *Ann. Sc. Norm. Sup. Pisa Ser IV*, **12** 335-392

Caffarelli, L. and Evans, L.C.(1983) Continuity of the temperature in the two Stefan problem. *Arch. Rat. Mech. Anal.*, **81**, 199–220.

Chavent, G. and Jaffrè, J.(1986) *Mathematical models and finite elements methods for reservoir simulation.* North-Holland.

Di Benedetto, E.(1982) Continuity of weak solutions to certain singular parabolic equations *Ann. Mat. Pura Appl.*, (4) **121** , 131–176.

Di Benedetto, E. (1993) *Degenerate parabolic equations.* Springer Verlag Series Universitext, New York.

E.Di Benedetto, and V.Vespri, V. On the singular equation $(\beta(u))_t = \Delta u$, *Arch. Rat. Mech.* to appear.

Ziemer, W. (1982) Interior and boundary continuity of weak solutions of degenerate parabolic equations. *Trans. Amer. Math. Soc.*, **271** (2) , 733–747.

PART THREE

Control and Optimization

Linear Systems

11

Optimal control of variational inequalities: A mathematical programming approach

M. Bergounioux
URA-CNRS 1803, Université d'Orléans
U.F.R. Sciences, B.P. 6759, F-45067 Orléans Cedex 2, France
Phone: +33 38 41 73 16. Fax: +33 38 41 71 93.
E-mail: maitine@talcy.univ-orleans.fr

Abstract

We investigate optimal control problems governed by variational inequalities. and more precisely the obstacle problem. Since we adopt a numerical point of view, we first relax the feasible domain of the problem; then using both mathematical programming methods and penalization methods we get optimality conditions with smooth Lagrange multipliers.

Keywords

Optimal control, optimality conditions, variational inequalities, mathematical programming

1 INTRODUCTION

In this paper we are going to investigate optimal control problems where the state is described by variational inequalities. Moreover, we consider constraints on both the control and the state. Our purpose is to give some optimality conditions that can be easily exploited numerically. These kind of problems have been extensively studied by many authors, as for example Barbu (1984), Mignot and Puel (1984) or Friedman (1982,1986). Let us present the problem we are interested in more precisely now.

Let Ω be an open, bounded subset of $\mathbb{R}^n$ $(n \leq 3)$ with a smooth boundary $\partial\Omega$. We consider a continuous, coercive bilinear form $a(.,.)$ defined on $\mathrm{H}_0^1(\Omega) \times \mathrm{H}_0^1(\Omega)$; the "associate" partial derivative operator is $A \in \mathcal{L}(\mathrm{H}_0^1(\Omega), H^{-1}(\Omega))$ and we assume that the coefficients are smooth enough.(We shall denote $\| \ \|_V$, the norm in the Banach space V, and more precisely $\| \ \|$ the $\mathrm{L}^2(\Omega)$-norm. In the same way, $\langle \ , \ \rangle$ denotes the duality product between $H^{-1}(\Omega)$ and $\mathrm{H}_0^1(\Omega)$; we shall denote similarly the $\mathrm{L}^2(\Omega)$-scalar product when there is no ambiguity.)

Before we describe the optimal control problem itself (cost functional and constraints)

we present the state-"equation" which is a variational inequality : let U be a non empty, closed, convex subset of $L^2(\Omega)$. For each v in U we define $y = y(v)$ (the state function of the system) as the solution of the variational inequality :

$$y \in K \text{ and } a(y, \varphi - y) \geq \langle v, \varphi - y \rangle \quad \forall \varphi \in K . \tag{1}$$

Following the paper of Mignot and Puel (1984) we may interpret this variational inequality as a state equation, introducing another control function, so that (1) is equivalent to:

$$Ay = v + \xi \text{ in } \Omega,\; y = 0 \text{ on } \Gamma,\; y \geq 0 \;,\; \xi \geq 0 \;,\; \langle \xi, y \rangle = 0 \; . \tag{2}$$

It is well known that the system (2) has a unique solution (Barbu 1984, Friedman 1982). Moreover, as $v \in L^2(\Omega)$, then $y(v)$ belongs to $H^2(\Omega)$ (Friedman 1982 p. 29) so that $\xi \in L^2(\Omega)$. We set also

$$\tilde{K} = \{\xi \mid \xi \in L^2(\Omega) \;,\; \xi \geq 0 \text{ a.e. in } \Omega\} \quad \text{and } K = \tilde{K} \cap H^1_0(\Omega). \tag{3}$$

The set K is a non empty, closed, convex subset of $H^1_0(\Omega)$; now, let us consider the optimal control problem defined as follows :

$$\min \left\{ J(y, v) = \frac{1}{2} \int_\Omega (y - z_d)^2 \, dx + \frac{\nu}{2} \int_\Omega v^2 \, dx \right\}, \tag{$\mathcal{P}_0$}$$

$$a(y, \varphi - y) \geq \langle v, \varphi - y \rangle \qquad \forall \varphi \in K,$$

$$v \in U \;,\; y \in K \;,$$

where $z_d \in L^2(\Omega)$, $v \in L^2(\Omega)$, and $\nu > 0$.

This optimal control problem appears as a problem governed by a state **equation** (instead of inequality) with mixed state and control constraints :

$$\min \left\{ J(y, v) = \frac{1}{2} \int_\Omega (y - z_d)^2 \, dx + \frac{\nu}{2} \int_\Omega v^2 \, dx \right\}, \tag{$\mathcal{P}$}$$

$$Ay = f + v + \xi \;\text{ in }\; \Omega \;,\; y = 0 \;\text{ on }\; \Gamma \;, \tag{4}$$

$$(y, v, \xi) \in \mathcal{D} \;,$$

where

$$\mathcal{D} = \{(y, v, \xi) \in H^1_0(\Omega) \times L^2(\Omega) \times L^2(\Omega) \mid v \in U,\; y \geq 0,\; \xi \geq 0, \langle y, \xi \rangle = 0\}.$$

We assume that the set $\tilde{\mathcal{D}} = \{ \, (y, v, \xi) \in \mathcal{D} \mid \text{relation (4) is satisfied} \, \}$ is non empty; we know, then that problem $(\mathcal{P})$ has at least an optimal solution (not necessarily unique) that we shall denote $(\bar{y}, \bar{v}, \bar{\xi})$ (see Mignot and Puel (1984) for instance).

Similar problems have been studied also in Bergounioux and Tiba (1994) when the set $\mathcal{D}$ is convex. Here, the main difficulty comes from the fact that the feasible domain $\mathcal{D}$ is not convex because of the bilinear constraint "$\langle y, \xi \rangle = 0$". So, we cannot use directly the convex analysis methods that have been used for instance in Bergounioux and Tiba (1994).

To derive optimality conditions in this case, we are going to use methods adapted to quite general mathematical programming problems. Unfortunately, the domain $\mathcal{D}$ (i.e. the constraints set) does not satisfy the usual (quite weak) assumptions of mathematical programming theory. This comes essentially from the fact that the L^∞-interior of $\mathcal{D}$ is empty. Nevertheless, our aim is to compute the solutions of the original problem. As we have a numerical purpose, we are going to consider the domain $\mathcal{D}_\alpha$ instead of $\mathcal{D}$, with $\alpha > 0$ and

$$\mathcal{D}_\alpha = \{(y, v, \xi) \in \mathrm{H}_0^1(\Omega) \times \mathrm{L}^2(\Omega) \times \mathrm{L}^2(\Omega) \mid v \in U,\ y \geq 0,\ \xi \geq 0, \langle y, \xi \rangle \leq \alpha\}. \tag{5}$$

This point of view is motived and justified numerically, since it is not possible to ensure "$\langle y, \xi \rangle = 0$" during a calculus with a computer but rather "$\langle y, \xi \rangle \leq \alpha$" where α may be chosen as small as wanted, but strictly positive.
First, we define the "relaxed" problem.

2 A RELAXED PROBLEM

As we mentioned it in the previous section, we consider the problem $(\mathcal{P})$ with $\mathcal{D}_\alpha$ instead of $\mathcal{D}$. More precisely, we investigate the following problem :

$$\begin{aligned} &\min J(y, v), && (\mathcal{P}^\alpha)\\ &Ay = v + \xi \text{ in } \Omega,\ y \in \mathrm{H}_0^1(\Omega), \\ &(y, v, \xi) \in \mathcal{D}_\alpha\ . \end{aligned}$$

Remark 2.1 *In fact, we are also obliged to bound ξ by a constant $R \geq \|\tilde{\xi}\|$ as well, to ensure the existence of a solution of the relaxed problem, but this constraint is not very restrictive. We don't mention it to make the presentation of the method more clear.*

Then we may say that problem $(\mathcal{P}^\alpha)$ is a "good" approximation of problem $(\mathcal{P})$ in the following sense :

Theorem 2.1 *Let $\alpha > 0$. Then problem $(\mathcal{P}^\alpha)$ has at least one optimal solution that we call $(y_\alpha, v_\alpha, \xi_\alpha)$. Moreover, when α tends to 0, y_α converges to $\tilde{y}$ weakly in $H_0^1(\Omega)$ (and strongly in $L^2(\Omega)$), v_α converges to $\tilde{v}$ strongly in $L^2(\Omega)$ and ξ_α converges to $\tilde{\xi}$ weakly in $L^2(\Omega)$, where $(\tilde{y}, \tilde{v}, \tilde{\xi})$ is a solution of $(\mathcal{P})$.*

Now, we would like to derive optimality conditions for problem $(\mathcal{P}^\alpha)$.

3 THE MATHEMATICAL PROGRAMMING POINT OF VIEW

As we have already mentioned it, the non convexity of the feasible domain, does not allow to use convex analysis to get the existence of Lagrange multipliers. So we are going to use quite general mathematical programming methods in Banach spaces and adapt them to our framework. The following results are mainly due to Zowe and Kurcyusz (1979) and Tröltzsch (1983,1984) and we briefly present them.

Let us consider real Banach spaces $\mathcal{X}, \mathcal{V}, \mathcal{Z}_1, \mathcal{Z}_2$ and a convex closed "admissible" set $\mathcal{U} \subseteq \mathcal{V}$. In $\mathcal{Z}_2$ a convex closed cone $\mathbf{P}$ is given so that $\mathcal{Z}_2$ is partially ordered by $x \geq y \Leftrightarrow x - y \in \mathbf{P}$. We deal also with :

$f : \mathcal{X} \times \mathcal{V} \rightarrow \mathbb{R}$, Fréchet-differentiable functional ,

$T : \mathcal{X} \times \mathcal{V} \rightarrow \mathcal{Z}_1$ and $G : \mathcal{X} \times \mathcal{V} \rightarrow \mathcal{Z}_2$ continuously Fréchet-differentiable operators .

Now, let be the mathematical programming problem defined by :

$$\min \{ \ f(x,u) \mid T(x,u) = 0, \ G(x,u) \leq 0 \ , u \in \mathcal{U} \ \} \ . \tag{6}$$

We denote the partial Fréchet-derivative of f, T, and G with respect to x and u by a corresponding index x or u. Assume that the problem (6) has an optimal solution that we call (x_o, u_o); we introduce the sets :

$$\mathcal{U}(u_o) = \{ \ u \in \mathcal{V} \mid \exists \lambda \geq 0, \exists u^* \in \mathcal{U}, \ u = \lambda(u^* - u_o) \ \}, \tag{7}$$

$$\mathbf{P}(G(x_o, u_o)) = \{ \ z \in \mathcal{Z}_2 \mid \exists \lambda \geq 0, \exists p \in -\mathbf{P}, \ z = p - \lambda G(x_o, u_o) \ \}, \tag{8}$$

$$\mathbf{P}^+ = \{ \ y \in \mathcal{Z}_2^* \mid \langle y, p \rangle \geq 0 \ , \forall p \in \mathbf{P} \ \}. \tag{9}$$

One may now announce the main result about the existence of optimality conditions.

Theorem 3.1 *Let u_o be an optimal control with corresponding optimal state x_o and suppose that the following regularity condition is fulfilled :*

$$\forall (z_1, z_2) \in \mathcal{Z}_1 \times \mathcal{Z}_2 \text{ the system } \left\{ \begin{array}{lcl} T'(x_o, u_o)(x, u) & = & z_1 \\ G'(x_o, u_o)(x, u) - p & = & z_2 \end{array} \right. \tag{10}$$

is solvable with $(x, u, p) \in \mathcal{X} \times \mathcal{U}(u_o) \times \mathbf{P}(G(x_o, u_o))$.
Then a Lagrange multiplier $(y_1, y_2) \in \mathcal{Z}_1^* \times \mathcal{Z}_2^*$ exists such that

$$f'_x(x_o, u_o) + T'_x(x_o, u_o)^* y_1 + G'_x(x_o, u_o)^* y_2 = 0 \ , \tag{11}$$

$$\langle f'_u(x_o, u_o) + T'_u(x_o, u_o)^* y_1 + G'_u(x_o, u_o)^* y_2, u - u_o \rangle \geq 0 \ , \quad \forall \ u \in \mathcal{U} \ , \tag{12}$$

$$y_2 \in \mathbf{P}^+ \ , \ \langle y_2, G(x_o, u_o) \rangle = 0 \ . \tag{13}$$

The mathematical programming theory in Banach spaces allows to study problems where the feasible domain is not convex : this is precisely our case (and we cannot use the classical convex theory and the Gâteaux differentiability to derive some optimality conditions). The Zowe and Kurcyusz condition is a very weak condition to ensure the existence of Lagrange multipliers. It is natural to try to see if this condition is satisfied for the original problem $(\mathcal{P})$: unfortunately, it is impossible (see Bergounioux 1995) and this is another justification (from a theoretical point of view) of the fact that we have to take $\mathcal{D}_\alpha$ instead of $\mathcal{D}$.

On the other hand, if we apply the previous general result "directly" to $(\mathcal{P}^\alpha)$ we obtain a qualification condition a little complicated which seems difficult to ensure. So we would rather mix these "mathematical-programming methods" with a penalization method in order to "relax" the state-equation as well and make the qualification condition weaker.

4 PENALIZATION OF THE PROBLEM

Now we are going to "decouple" the different constraints : the (linear) state-equation is going to be penalized in a standard way, and the non-convex (state) constraints are to be treated via the Zowe and Kurcyusz condition and with the methods of the previous section.

First we penalize the state equation to obtain an optimization problem with non convex constraints. Moreover, as we want to focus on the solution $(y_\alpha, v_\alpha, \xi_\alpha)$; so, following Barbu (1984), we add some adapted penalization terms to the functional J.

From now, $\alpha > 0$ is fixed; so we omit the index α for convenience (when no confusion is possible). For any ε we define a penalized functional $J_\varepsilon > 0$ on $\mathrm{H}_0^1(\Omega) \times \mathrm{L}^2(\Omega) \times \mathrm{L}^2(\Omega)$ as following :

$$J_\varepsilon(y,v,\xi) = \begin{cases} J(y,v) + \dfrac{1}{2\varepsilon}\|Ay - v - \xi\|^2_{L^2(\Omega)} & \\ \qquad + \dfrac{1}{2}\|y - y_\alpha\|^2_{H_0^1(\Omega)} + \dfrac{1}{2}\|v - v_\alpha\|^2_{L^2(\Omega)} & \text{if } Ay - v - \xi \in \mathrm{L}^2(\Omega) \\ +\infty & \text{else} \end{cases} \tag{14}$$

and we consider the penalized optimization problem

$$\min \{\ J_\varepsilon(y,v,\xi) \mid (y,v,\xi) \in \mathcal{D}_\alpha\} \tag{$\mathcal{P}^\alpha_\varepsilon$}$$

Theorem 4.1 *The penalized problem $(\mathcal{P}^\alpha_\varepsilon)$ has at least a solution $(y_\varepsilon, v_\varepsilon, \xi_\varepsilon) \in (H^2(\Omega) \cap H_0^1(\Omega)) \times L^2(\Omega) \times L^2(\Omega)$. Moreover, when ε tends to 0, $(y_\varepsilon, v_\varepsilon, \xi_\varepsilon)$ strongly converges to $(y_\alpha, v_\alpha, \xi_\alpha)$ in $H_0^1(\Omega) \times L^2(\Omega) \times L^2(\Omega)$.*

Now, we apply the general theorem of the previous section to the penalized problem $(\mathcal{P}^\alpha_\varepsilon)$. We set

$$\mathcal{X} = H^2(\Omega) \cap \mathrm{H}_0^1(\Omega)\ ,\ \mathcal{Z}_2 = \mathcal{X}\ ;$$

$$u = (v,\xi)\ ,\ \mathcal{V} = \mathrm{L}^2(\Omega) \times \mathrm{L}^2(\Omega)\quad ,\quad \mathcal{U} = U \times \tilde{K}\ ,$$

$$\mathbf{P} = \{\ y \in \mathcal{X} \mid y \geq 0\} \times \mathbb{R}^+\ ,$$

$$G(y,v,\xi) = (-y\ ,\langle y,\xi\rangle - \alpha) \ \text{ and } f(x) = J_\varepsilon(y,v)\ ,(x_o,u_o) = (y_\varepsilon, v_\varepsilon, \xi_\varepsilon)\ .$$

There is no equality operator, G is $\mathcal{C}^1$ and $G'(y_\varepsilon, v_\varepsilon, \xi_\varepsilon)(y,v,\xi) = (-y, \langle y_\varepsilon, \xi\rangle + \langle y, \xi_\varepsilon\rangle)$. Here

$$\mathcal{U}(u_o) = \{ (\lambda(v - v_\varepsilon), \mu(\xi - \xi_\varepsilon)) \mid \lambda \geq 0, \mu \geq 0, v \in U, \xi \geq 0 \},$$

$$\mathbf{P}(G(x_o, u_o)) = \{(-p + \lambda y_\varepsilon, -\gamma - \lambda(\langle y_\varepsilon, \xi_\varepsilon \rangle - \alpha)) \in H^2(\Omega) \cap \mathrm{H}_0^1(\Omega) \times \mathbb{R} | \gamma, \lambda \geq 0, p \geq 0\}.$$

Let us write the condition (10) : for any (z, β) in $\mathcal{X} \times \mathbb{R}$ we must solve the system :

$$-y + p - \lambda y_\varepsilon = z$$

$$\langle y_\varepsilon, \mu(\xi - \xi_\varepsilon) \rangle + \langle y, \xi_\varepsilon \rangle + \gamma + \lambda(\langle y_\varepsilon, \xi_\varepsilon \rangle - \alpha) = \beta ,$$

with $\mu, \gamma, \lambda \geq 0$, $p \geq 0$, $\xi \in V$, $v \in U$ and $y \in \mathcal{X}$.

It is not difficult to see that the condition (10) is always satisfied and that we may apply theorem 3.1, since J_ε is Fréchet-differentiable. So we obtain some optimality conditions on the penalized system, without any further assumption:

Theorem 4.2 *The solution $(y_\varepsilon, v_\varepsilon, \xi_\varepsilon)$ of problem $(\mathcal{P}_\varepsilon^\alpha)$ satisfies the following optimality system :*

$$\langle p_\varepsilon + q_\varepsilon, A(y - y_\varepsilon) \rangle + \langle r_\varepsilon \xi_\varepsilon, y - y_\varepsilon \rangle + \langle y_\varepsilon - y_\alpha, y - y_\varepsilon \rangle_{H_0^1(\Omega)} \geq 0 \ , \ \forall\, y \in K \tag{15}$$

$$\langle M v_\varepsilon - q_\varepsilon + (v_\varepsilon - v_\alpha), v - v_\varepsilon \rangle \ \geq 0 \ \ \forall\, v \in U \ , \tag{16}$$

$$\langle r_\varepsilon y_\varepsilon - q_\varepsilon, \xi - \xi_\varepsilon \rangle \ \geq 0 \ \ \forall\, \xi \in V \ , \tag{17}$$

$$r_\varepsilon \in \mathbb{R}^+ , \ r_\varepsilon \left(\langle y_\varepsilon, \xi_\varepsilon \rangle - \alpha \right) = 0 \ , \tag{18}$$

where p_ε is the solution of

$$A^* p_\varepsilon = y_\varepsilon - z_d \ \text{on} \ \Omega, \quad p_\varepsilon \in H_0^1(\Omega) \ , \tag{19}$$

and $q_\varepsilon = \dfrac{1}{\varepsilon}(A y_\varepsilon - v_\varepsilon - \xi_\varepsilon) \ \in L^2(\Omega)$ *(A^* is the adjoint operator of A).*

Now we would like to study the asymptotic behaviour of these relations when ε tends to 0, and we need estimations on q_ε and r_ε. So we have to assume some qualification conditions to pass to the limit in the penalized optimality system; first, to get an estimation for r_ε we assume :

$$\forall \alpha \text{ such that } \langle y_\alpha, \xi_\alpha \rangle = \alpha \ , \exists (\tilde{y}, \tilde{v}, \tilde{\xi}) \in K \times U_{ad} \times V_{ad} \quad \text{such that}$$
$$A\tilde{y} = \tilde{v} + \tilde{\xi} \quad \text{and} \quad \langle \tilde{y}, \xi_\alpha \rangle + \left\langle y_\alpha, \tilde{\xi} \right\rangle < 2\alpha \ . \tag{20}$$

and an estimation on q_ε is given with the following assumption :

$$\exists\, \rho > 0 \ , \ \ \forall\, \chi \in \mathrm{L}^2(\Omega), \ \|\chi\|_{L^2(\Omega)} \leq 1 \ , \exists (y_\chi, v_\chi, \xi_\chi) \text{ uniformly bounded}$$
$$\text{with respect to } \chi, \text{ such that } A y_\chi = v_\chi + \xi_\chi + \rho \chi \ \text{ in } \Omega. \tag{21}$$

Then we may obtain some optimality conditions :

Theorem 4.3 *Assume (20) and (21) and let* $(y_\alpha, v_\alpha, \xi_\alpha)$ *be a solution of* $(\mathcal{P}^\alpha)$*; then there exists a Lagrange multiplier* $(q_\alpha, r_\alpha) \in L^2(\Omega) \times I\!R^+$ *, such that*

$$\langle p_\alpha + q_\alpha, A(y - y_\alpha) + r_\alpha \langle \xi_\alpha, y - y_\alpha \rangle \geq 0 \quad \forall\, y \in K \ , \tag{22}$$

$$\langle M v_\alpha - q_\alpha, v - v_\alpha \rangle \ \geq 0 \quad \forall\, v \in U \ , \tag{23}$$

$$\langle r_\alpha y_\alpha - q_\alpha, \xi - \xi_\alpha \rangle \ \geq 0 \quad \forall\, \xi \in \tilde{K} \ , \tag{24}$$

$$r_\alpha \left(\langle y_\alpha, \xi_\alpha \rangle - \alpha \right) = 0 \ , \tag{25}$$

where p_α *is the solution* $(\in \mathcal{X})$ *of*

$$A^* p_\alpha = y_\alpha - z_d \quad \text{on} \quad \Omega, \ p_\alpha \in H_0^1(\Omega) \ . \tag{26}$$

Example 4.1 *One can prove that assumption (20) is fulfilled if*

$$0 \in U. \tag{27}$$

So the previous theorem is valid when $U = L^2(\Omega)$ *since assumptions (27) and (21) are obviously satisfied.*

When $U = \{\, v \in L^2(\Omega) \mid v \geq \psi \geq 0$ *a.e. on* $\Omega\}$ *(21) is satisfied as well, but (27) may not be true any longer. Anyway, we can prove that the general form of the qualification condition (20) is satisfied (see Bergounioux 1995); this allows to claim that the result of the previous theorem is still valid.*

5 CONCLUSION

We have established optimality conditions for the relaxed problem only. Nevertheless, as we have already mentioned it, it is good enough for the numerical experimentation. The optimality system may be solved with usual algorithms of optimization (see Glowinski et al. (1976) and Fortin-Glowinski (1982) for instance). In conclusion, we shall add that this kind of "mixed" method should be efficient to study more general variational inequalities for the state equation. It seems that it could be also used for optimal control problems where the state equation is nonlinear. In addition this method allows to give some regularity results for the multiplier, or conversly gives the existence of less regular multipliers with a some weaker qualification condition.

REFERENCES

Barbu V. (1984) *Optimal Control of Variational Inequalities* , Research Notes in Mathematics, **100**, Pitman, Boston.

Bergounioux M. (1993) An Augmented Lagrangian Algorithm for Distributed Optimal Control Problems with State Constraints, *Journal of Optimization Theory and Applications*, **78**, no. 3.

Bergounioux M. (1995) Optimality Conditions For Optimal control of Elliptic Problems Governed by Variational Inequalities, Rapport de Recherche, **95-1**, Université d'Orléans, submitted.

Bergounioux M. and Tiba D. (1994) General Optimality Conditions for Constrained Convex Control Problems, *SIAM Journal on Control and Optimization,* to appear.

Fortin M. and Glowinski R. (1982) *Méthodes de Lagrangien Augmenté - Applications à la Résolution de Problèmes aux Limites*, Méthodes Mathématiques pour l'Informatique, Dunod, Paris, France.

Friedman A. (1982) *Variational Principles and Free-Boundary Problems*, Wiley, New York.

Friedman A. (1986) Optimal Control for Variational Inequalities, *SIAM Journal on Control and Optimization*, **24**, no. 3, 439-451.

Glowinski R. , Lions J.L. and Trémolières R. (1976) *Analyse Numérique des Inéquations Variationnelles* , Méthodes Mathématiques pour l'Informatique, Dunod, Paris, France.

Mignot F. and Puel J.P. (1984) Optimal Control in Some Variational Inequalities, *SIAM Journal on Control and Optimization*, **22**, no. 3, 466–476.

Tröltzsch F. (1983) A modification of the Zowe and Kurcyusz regularity condition with application to the optimal control of Noether operator equations with constraints on the control and the state, *Math. Operationforsch. Statist., Ser. Optimization*, **14**, no. 2, 245-253.

Tröltzsch F. (1984) *Optimality conditions for parabolic control problems and applications*, Teubner Texte, Leipzig.

Zowe J. and Kurcyusz S. (1979) Regularity and stability for the mathematical programming problem in Banach spaces, *Applied mathematics and Optimization*, **5**, 49-62.

12

Finite horizon regulator problem: the non–standard case

Francesca Bucci[*]
Università di Firenze, Dipartimento di Matematica Applicata
Via S. Marta 3, 50132 Firenze, Italy. Phone: +39-11-4796248.
E–mail fbucci@ing.unifi.it
and
Luciano Pandolfi[†]
Politecnico di Torino, Dipartimento di Matematica
Corso Duca degli Abruzzi, 24, 10129 Torino, Italy.
Phone: +39-11-5647516.
E–mail lucipan@polito.it

Abstract

We present several regularity properties of the value function of a quadratic control problem. The system is distributed, with distributed control action while the quadratic functional is not coercive; even it may not be positive.

Keywords

Quadratic regulator problem, Dissipative system, Singular control

1 INTRODUCTION

We present in this talk several new results about the *finite horizon* non–standard quadratic regulator problem for a distributed system with bounded input operator.

In this introduction we recall the main known results for finite dimensional systems, a case in which the theory is complete; and we mention the few known extensions to the distributed case.

The system that we intend to study is described by the equation

$$\dot{x} = Ax + Bu, \qquad x(\tau) = x_0 \,. \tag{1}$$

[*]Partially supported by the Italian Ministero della Ricerca Scientifica e Technologica within the program of GNAFA–CNR.

[†]Partially supported by the Italian Ministero della Ricerca Scientifica e Technologica within the program of GNAFA–CNR and by HCM network CEC n. ERB–CHRX–CT93–0402.

Here τ is a point which belongs to the interval $[0,T)$ and we solve equation (1) on the interval $[\tau, T]$.

The (mild) solution to Eq. (1) is given by

$$x(t) = x(t;\tau,x_0,u) = e^{A(t-\tau)}\, x_0 + \int_\tau^t e^{A(t-s)}\, Bu(s) \; = e^{A(t-\tau)}x_0 + (L_\tau u)(t) \tag{2}$$

for each square integrable control $u(\cdot)$. Here e^{As} is a C_0–semigroup on a Hilbert space X whose generator is A; $B \in \mathcal{L}(U,X)$. The space U is a second Hilbert space.

We associate the following quadratic functional to the system described by (1):

$$J_\tau(x_0,u) = \int_\tau^T F(x(t),u(t))\, dt + \langle x(T), P_0 x(T)\rangle\,, \tag{3}$$

where $x = x(t) = x(t;\tau,x_0,u)$ and $F(x,u) = \langle x, Qx\rangle + \langle u, Ru\rangle$. The operators Q, R and P_0 are linear bounded operators in the proper spaces, R and P_0 are selfadjoint. *We do not make any assumption on their definiteness,* in contrast with the so called *standard* case, when

$$Q \geq 0, \quad R \geq \alpha > 0, \quad P_0 \geq 0\,. \tag{4}$$

The standard regulator problem is completely understood and the solution is reduced to the study of an operator Riccati equation. Its solution provides the synthesis of the unique optimal control. In contrast with this, there will be no optimal control in the non standard case and, correspondingly, no Riccati equation.
We note that the same results we are going to describe below hold even if F presents a mixed term in x and u.

Our goal is to characterize the following property

Property 1 $V_\tau(x_0) := \inf_{u\in L^2(\tau,T;U)} J_\tau(x_0,u) > -\infty \qquad \forall x_0 \in X \quad \forall \tau \in [0,T].$

If this condition holds, we want to study the regularity properties of the function $\tau \to V_\tau(x_0)$ for any specified $x_0 \in X$.

The non–standard - or singular - LQR problem was completely solved in finite dimension (see in particular Clements and Anderson (1978), Molinari (1975)). These authors were stimulated mostly by the analysis of the second variations of non–linear optimization problems. These problems are obtained computing the system along a candidate optimal solution, so that in general the matrices A, B, Q, R are time–dependent. Here we confine ourselves to the time–invariant case.

The main results that have been obtained are as follows:

When Property 1 holds, then for each τ the functional $x_0 \to V_\tau(x_0)$ is a (continuous) quadratic functional, namely

$$V_\tau(x_0) = \langle x_0, W(\tau)x_0\rangle,$$

where $W(\tau)$ is a matrix of the same dimension as x. In general *the dependence on τ is not "regular"* as the following example from Clements and Anderson (1978) shows.

Consider the dynamic $\dot{x} = u$, with cost given by

$$J_\tau(x_0, u) = -\int_\tau^1 k(t)x(t)u(t)\,dt + x^2(1)/2,$$

where $k(t) = 0$ on $[0, 1/2)$, $k(t) = 1$ on $[1/2, 1]$. The value function $V_\tau(x_0)$ is equal to 0 for $\tau < 1/2$ and to $x_0^2/2$ for $t \geq 1/2$.
In this example the discontinuity is produced by the coefficient $k(\cdot)$. However *we shall see* that *(for distributed parameter systems) we can have discontinuities even for time–invariant systems.*

A second property that it is proved for finite dimensional systems is the *Dissipation Inequality*, a certain inequality satisfied by $W(\cdot)$, *which also holds for distributed parameter systems* (see below). Actually, systems which enjoy this dissipation inequality have importance in physical applications since they correspond to systems which "dissipate energy" (compare Brune (1931), Willems (1972)). In fact the study of such dissipative systems–on $[0, +\infty)$–is a second problem which stimulated the analysis of the non–standard regulator problem. A third problem is the stability of Lur'e type feedback control systems. A very nice overview of the finite dimensional theory can be found in Bittanti, Laub, Willems (1991).

We stress that the previous results do not require that the cost $J_\tau(x_0, u)$ reaches a minimum value. We shall see that when a minimum is reached, more precise regularity properties on the function $\tau \to V_\tau(x_0)$ can be proved, see below.

Most of the proofs of the cited results, which all concern systems in R^n, are not readily adaptable to the case of distributed systems, since they often rely on special features of finite dimensional systems, such as group property of the evolution operator, exact controllability, etc.

For completeness we add that the distributed parameters time varying case was studied by Jacob (1995), under the restriction that the operator R is the identity operator. This paper shows, in particular, that the existence of the optimal control and the existence of solutions to the Riccati equation are related but not equivalent properties. In contrast with this paper by Jacob (1995) we are interested in the time–invariant case, with special emphasis in non–coercive quadratic functional.

The existing results for finite dimensional systems - infinite horizon case - were extended to the distributed control case in Louis and Wexler (1991).

2 MAIN RESULTS

In this section we describe our results. Some examples and proofs are sketched in the next section.

Theorem 1 *Let* $\tau \in [0, T]$. *If*

$$V_\tau(x_0) > -\infty \quad \forall x_0 \in X, \tag{5}$$

there exists a selfadjoint operator $W(\cdot) \in \mathcal{L}(X)$ such that $W(T) = P_0$ and

$$V_\tau(x_0) = \langle x_0, W(\tau)x_0 \rangle. \tag{6}$$

We consider now the *dissipation inequality.*

Theorem 2 *Property 1 holds if and only if for each $\tau \in [0,T]$ there exist a linear bounded operator $W(\tau) \in \mathcal{L}(X)$ which satisfies, for any $t \geq \tau$, the following* **dissipation inequality**

$$\langle x(t), W(t)x(t) \rangle - \langle x(\tau), W(\tau)x(\tau) \rangle + \int_\tau^t F(x(s), u(s)) \geq 0, \qquad W(T) \leq P_0. \tag{7}$$

In the above inequality $u(\cdot)$ is any admissible control, $x(t) = x(t;\tau,x_0,u)$. Moreover, the inequality becomes an equality if and only if the control u in (7) is optimal.

If the previous inequality holds true, then there exists a solution $P(t)$ of (7) such that $V_\tau(x_0) = \langle x_0, P(\tau)x_0 \rangle$.

Corollary 3 *If for some number τ we have $V_\tau(x_0) > -\infty$ then we have also $V_t(x(t)) > -\infty$ for each $t \in [\tau, T]$. Here, $x(t)$ denotes the value at time t of the function in (2), for any fixed control $u(\cdot)$ on $[\tau, t]$.*

Examples in the next section will show that the function $\tau \to V_\tau(x_0)$ may not be continuous. ¿From this point of view, the following result have an interest.

Theorem 4 *Let us assume that the point x_0 has the following property: $\tau \to V_\tau(x_0)$ is finite on $[0,T]$. Then*

- *the function $\tau \to V_\tau(x_0)$ is upper semicontinuous on $[0,T]$;*
- *the function $\tau \to V_\tau(x_0)$ is lower semicontinuous at τ_0 provided that for each sequence τ_n which tends monotonically to τ_0 there exists a sequence $\{u_n\} \in L^2(\tau_n, T; U)$ such that*

$$\begin{array}{ll} i) & V_{\tau_n}(x_0) \leq J_{\tau_n}(x_0; u_n) \leq V_{\tau_n}(x_0) + \frac{1}{n}; \\ ii) & \exists \gamma_0 > 0: \qquad ||u_n(\cdot)||_{L^2(\tau_n,T;U)} \leq \gamma_0. \end{array} \tag{8}$$

Of course the previous choice for the properties of $u_n(\cdot)$ is suggested by the construction of minimizing sequences. And it may be easier to test the required condition in the case that an optimal control exists for each τ near τ_0.
We shall see that lower semicontinuity needs not hold in all cases.

We now recall the following general result:

Lemma 5 *Consider the functional*

$$J_y(\zeta) = 2Re\langle \mathcal{N}y, \zeta \rangle + \langle \zeta, \mathcal{R}\zeta \rangle \tag{9}$$

over a Hilbert space Z (it will be $Z = L^2(\tau, T; U)$ in the following). If there exists y such that $\inf_\zeta J_y(\zeta) > -\infty$, then $\mathcal{R} \geq 0$; moreover the functional $J_y(\zeta)$ admits at least one *minimum on Z for any fixed y if and only if the following conditions are satisfied:*

$$i) \quad \mathcal{R} \geq 0; \qquad ii) \quad im\, \mathcal{N} \subseteq im\, \mathcal{R}.$$

If the previous conditions hold, then any ζ such that $\mathcal{N}y + \mathcal{R}\zeta = 0$ is a minimum.

In our application the operators $\mathcal{N}$ and $\mathcal{R}$ will be the operators given by

$$(\mathcal{N}_\tau x)(t) = (L_\tau^* Q\, e^{A(\cdot-\tau)}x)(t) + S^* e^{A(t-\tau)}x + (L_{\tau,T}^* P_0\, e^{A(T-\tau)}x)(t), \tag{10}$$

$$(\mathcal{R}_\tau u)(t) = (L_\tau^* Q\,(L_\tau u))(\cdot) + S^*(L_\tau u)(t) + (L_\tau^* S u)(t) + Ru(t) + (L_{\tau,T}^* P_0\, L_{\tau,T} u)(t), \tag{11}$$

respectively.

The following result is crucial:

Theorem 6 *If there exists $\tau_0 \in [0,T]$ such that*

$$\mathcal{R}_{\tau_0} \geq \gamma, \tag{12}$$

then $R \geq \gamma$.

Of course the most important cases are $\gamma = 0$ and $\gamma > 0$ (coercive operator).

We note that this result is well known in the case that $T = +\infty$. But in this last case it is proved thanks to the *frequency domain inequality.* The frequency domain inequality is essentially related to the Parseval equality for Fourier transform so that it has no simple correspondence in the finite–horizon case.

If $\mathcal{R}_{\tau_0}$ is coercive, then an optimal control exists, and the control is unique (even if the quadratic form F of the cost is not positive). Thanks to this property, the value function $V_{\tau_0}(x_0)$ displays better regularity properties. More precisely, we have the following

Theorem 7 *Assume that there exists τ_0 and $\gamma > 0$ such that $\mathcal{R}_{\tau_0} \geq \gamma$. Then, for any $\tau \in [\tau_0, T]$,*

- $\tau \to V_\tau(x)$ *is continuous for any $x \in X$;*
- $\tau \to V_\tau(x)$ *is differentiable for any $x \in D(A)$.*

In this case it is possible to deduce a *differential form* of the dissipation inequality.

Theorem 8 *Assume that $\mathcal{R}_{\tau_0} \geq \gamma > 0$ holds true. Then there exists a selfadjoint operator $W(\cdot) \in L(X)$ such that*

i) $W(T) = P_0$*;*
ii) $W(\cdot)$ *satisfies*

$$\frac{d}{d\tau}\langle a, W(\tau)\, a\rangle + 2Re\langle Aa + Bu, W(\tau)a\rangle + F(a,u) \geq 0, \tag{13}$$

for any $(a,u) \in D(A) \times U$, for any $\tau \in (\tau_0, T)$.

3 EXAMPLES AND SKETCH OF THE PROOFS

3.1 Examples

We present a first example which shows that the value function $V_\tau(x_0)$ may not be continuous at the final time T.

Example 1 Minimize the functional

$$J_\tau(x_0, u) = \langle x(T), P_0 x(T)\rangle, \qquad P_0 > 0,$$

with $x(\cdot)$ subject to any system (finite or infinite dimensional) which is null–controllable from any state in an arbitrary time. It is readily verified that $V_T(x_0) = \langle x_0, P_0 x_0\rangle$, while $V_\tau(x_0) = 0$ for any $\tau < T$, hence $V_\tau(x_0)$ is discontinuous in T except when $x_0 = 0$.

Example 2 Consider the dynamic

$$\begin{cases} \dot{x}(t) &= \xi(t,-1) + u(t) \\ \frac{d}{dt}\xi(t) &= A\xi - w(t) \\ \dot{u}(t) &= w(t) \end{cases} \tag{14}$$

on the state space $\mathrm{R} \times L^2(-1,0) \times \mathrm{R}$. The operator A is defined by

$$D(A) = \{\xi \in H^1([-1,0]),\, \xi(0) = 0\} \qquad (A\xi)(\theta) = \frac{\mathrm{d}}{\mathrm{d}\theta}\xi(\theta).$$

It is easily seen that any solution $t \to (x(t), \xi(t,\theta), u(t))$ is equal to $t \to (x(t), u(t+\theta) - u(t), u(t))$, where $x(t)$ satisfies the *input delay equation*

$$\dot{x} = u(t-1), \tag{15}$$

provided that $u(\cdot)$ is differentiable. This shows that system (14) defines a C_0–semigroup on the space $\mathrm{R} \times L^2(-1,0) \times \mathrm{R}$.

We now study system (14) on the interval $[\tau, 2]$ and we associate to it the following cost:

$$J(x_0, \xi_0, u_0; w) = |x(2)|^2.$$

Consider the problem with $x_0 = x(0) \neq 0$, $\xi_0(\theta) \equiv 0$, and $u_0 = 0$. Controls $u(\cdot)$ are taken to be zero out of $[\tau, T]$.

It is seen from (15) that if $1 < \tau < 2$ then $x(\cdot)$ is not affected by $u(\cdot)$ (i.e. by $w(\cdot)$). Hence, $x(t) \equiv x_0$ and $V_\tau(x_0) = \inf_w J(x_0, 0, 0; w) \equiv |x_0|^2$ on $[1, 2]$.

If $\tau \in (0,1)$ then the function $x(\cdot)$ is affected by the restriction of $u(\cdot)$ to the interval $[\tau, 1]$ and this function can be arbitrarily selected, within the set of H^1 functions which are zero at τ, by a suitable choice of $w(\cdot)$. Hence we can use the control on $[\tau, 1]$ in order to have $x(1) = 0$ i.e. $x(t) \equiv 0$ on $[1, 2]$: $V_\tau(x_0) = 0$ for $\tau < 1$. In conclusion, the function $\tau \to V_\tau(x_0)$ is discontinuous at $\tau_0 = 1$.

Remark *The previous example shows that in the statement of Theorem 4 concerning lower semicontinuity of $V_\tau(x_0)$, assumption ii) cannot be dispensed with. Notice that in fact ii) holds for $\tau \to 1^+$, but not for $\tau \to 1^-$.*

3.2 Sketch of some proofs

Proof. of Theorem 4: Fix $x_0 \in X$, and let $\tau_0 \in (0,T)$. In order to show upper semicontinuity, we shall show that for any real number $\alpha > V_{\tau_0}(x_0)$ we have $\alpha > V_\tau(x_0)$, if $|\tau - \tau_0|$ is taken small enough. Consider the case when $\tau > \tau_0$. Let u be an admissible control such that $J_{\tau_0}(x_0,u) < \alpha$, and define $x_\tau(t) = e^{A(t-\tau)}x_0 + \int_\tau^t e^{A(t-s)}Bu(s)$. It is readily verified that

1. $\lim_{\tau\to\tau_0} x_\tau(T) = x_{\tau_0}(T)$,
2. $\lim_{\tau\to\tau_0} \int_\tau^T F(x_\tau(s),u(s)) - \int_{\tau_0}^T F(x_{\tau_0}(s),u(s)) = 0$,

so that if $|\tau - \tau_0|$ is small enough, $V_\tau(x_0) \le J_\tau(x_0,u) < \alpha$. The case $\tau < \tau_0$ is similar.

As to lower semicontinuity, we consider, for $\tau_n \downarrow \tau_0$, the sequence

$$\hat{u}_n(s) := \begin{cases} 0 & \tau_0 < s < \tau_n \\ u_n(s) & s \ge \tau_n. \end{cases}$$

By assumption ($8ii$) it follows that $\hat{u}_n(\cdot) \rightharpoonup v(\cdot) \in L^2(\tau_0,T;U)$, and moreover

$$\int_{\tau_0}^T F(x(t;\tau_0,x_0,\hat{u}_n),\hat{u}_n(t))\,dt - \int_{\tau_n}^T F(x(t;\tau_n,x_0,u_n),u_n(t))\,dt \to 0.$$

Therefore, by convexity of the cost, and due to the dominated convergence theorem,

$$V_{\tau_0}(x_0) \le J_{\tau_0}(x_0,v) \le \liminf_{n\to\infty} J_{\tau_0}(x_0,\hat{u}_{\tau_n}) = \liminf_{n\to\infty} V_{\tau_n}(x_0). \tag{16}$$

If $\tau \uparrow \tau_0$, define instead $\hat{u}_n(s) = u_n(s)_{|[\tau_0,T]}$. We omit details. □

Proof. of Theorem 6 We note preliminarily that if $\tau > \tau_0$, we can write

$$\langle \mathcal{R}_\tau u, u\rangle_{L^2(\tau,T;U)} = \langle \mathcal{R}_{\tau_0} v, v\rangle_{L^2(\tau_0,T;U)},$$

where $v(\cdot)$ is

$$v(t) = \begin{cases} 0 & \tau \le t < \tau \\ u(t) & t \ge \tau \end{cases}$$

Hence, from (12) it follows that

$$\mathcal{R}_\tau \ge \gamma \qquad \forall \tau \in [\tau_0, T[.$$

We first consider the case $\alpha = 0$, hence by assumption $\mathcal{R}_{\tau_0} \geq 0$. By contradiction, suppose that there exists a control $u_0 \in U$ and a constant $\alpha > 0$ such that $\langle Ru_0, u_0 \rangle = -\alpha$. Given a small $\epsilon > 0$, choose a control u as follows:

$$u(t) = \begin{cases} 0 & \tau_0 \leq t < T - \epsilon \\ u_0 & T - \epsilon \leq t \leq T \end{cases}$$

and compute

$$\begin{aligned} \langle \mathcal{R}_\tau u, u \rangle &= \int_{T-\epsilon}^{T} \langle Q \int_{T-\epsilon}^{t} e^{A(t-s)} Bu_0 \, ds, \int_{T-\epsilon}^{t} e^{A(t-s)} Bu_0 \, ds \rangle \, dt + \epsilon \langle Ru_0, u_0 \rangle + \\ &+ \int_{T-\epsilon}^{T} \langle P_0 \int_{T-\epsilon}^{T} e^{A(T-s)} Bu_0 \, ds, \int_{T-\epsilon}^{T} e^{A(T-s)} Bu_0 \, ds \rangle \, dt = -\epsilon\alpha + o(\epsilon^2) + o(\epsilon^3) \quad (17) \end{aligned}$$

Since ϵ can be taken arbitrarily small, (17) yields $\langle \mathcal{R}_{\tau_0} u, u \rangle < 0$, and this contradicts the statement of Lemma 5.

Assume instead (12) with $\gamma > 0$. By choosing $u(t) = 0$ for $t \in [\tau_0, T - \epsilon[$, $u(t) = u_0 \in U$ arbitrary when $t \in [T - \epsilon, T]$, a direct computation yields

$$\gamma\epsilon ||u_0||^2 \leq \epsilon \langle Ru_0, u_0 \rangle + o(\epsilon^3) ||u_0||^2,$$

which implies $\langle Ru_0, u_0 \rangle \geq \gamma ||u_0||^2$ for any $u_0 \in U$. □

REFERENCES

Bittanti, S., Laub, A.J., Willems, J.C. (1991) *The Riccati equation,* Springer–Verlag, Berlin.

Brune, O. (1931) Synthesis of a finite two terminal network whose driving point impedance is a prescribed function of impedancy, *Journal of Mathematics & Physics,* **10** 191–236.

Clements, D.J. and Anderson, B.D.O. (1978) *Singular optimal control: the linear–quadratic problem,* Lecture Notes LNCIS no.5, Springer Verlag, Berlin.

Jacob, B. (1995) Linear quadratic optimal control of time–varying systems with indefinite costs on Hilbert spaces: the finite horizon problem, *Journal of Mathematical Systems, Estimation & Control,* **5** 1–28.

Louis, J-Cl. and Wexler, D. (1991) The Hilbert space regulator problem and operator Riccati equation under stabilizability, *Annales de la Société Scientifique de Bruxelles,* **105** 137–65.

Molinari, B.P. (1975) Nonnegativity of a quadratic functional, *SIAM Journal Control & Optimization,* **13** 792-806.

Willems, J.C. (1972) Dissipative dynamical systems, Part I: general theory, Part II: Linear systems with quadratic supply rates, *Arch. Rational Mechanics and Analysis,* **45** 321–351 and 352–392.

13

On abstract boundary control problems for vibrations

W. Krabs
Fachbereich Mathematik der Technischen Hochschule Darmstadt
Schlossgartenstrasse 7, 64289 Darmstadt, Germany.
Phone: 06151/162487. Fax: 06151/164011.
E-mail: krabs@mathematik.th-darmstadt.de

Abstract

We consider abstract control problems which are governed by a homogeneous linear wave equation for Hilbert space valued functions under nonhomogeneous boundary conditions which contain the controls that have values in an abstract Hilbert space. These problems are transformed into distributed control problems under homogeneous boundary conditions which have the controls on the right-hand side of a nonhomogeneous wave equation.
After this transformation the problem of null-controllability is studied.

Keywords

Abstract boundary control, distributed control, problem of null-controllability

0 INTRODUCTION

In (Fattorini, 1968) boundary control problems are considered for abstract differential equations of first and second order with respect to the time which in the case of second order are slightly more general than the abstract wave equation (1.1) and under boundary conditions of which (1.2) is a special case. Among others an assumption is made there which in the case of the abstract wave equation (1.1) and the boundary conditions (1.2) reads as follows: There exists a bounded linear operator $G : H_B \to D(A)$ such that

$$B(G(\psi)) = \psi \quad \text{for all} \quad \psi \in H_B$$

(which, of course, requires that $D(A) \subseteq D(B)$).

As control space $C^\infty([0,T], H_B)$ is considered. If then, for some $\varphi \in C^\infty([0,T], H_B)$, $y \in C^2([0,T], D(A)) \cap C^1([0,T], H)$ is a solution (1.1), (1.2), (1.3), we put $\tilde{y} = y - G(\varphi)$ and conclude that

$$B(\tilde{y}(t)) = \Theta_H \quad \text{for all} \quad t \in [0,T]\,, \tag{0.1}$$

$$\ddot{\tilde{y}}(t) + \tilde{A}\tilde{y}(t) = -G(\ddot{\varphi})(t) - A(G(\varphi))(t) \quad \text{for all} \quad t \in (0,T) \quad \text{and} \tag{0.2}$$

$$\tilde{y}(0) = y_0 - G(\varphi)(0) \ , \ \dot{\tilde{y}}(0) = \dot{y}_0 - G(\dot{\varphi})(0) \ . \tag{0.3}$$

Conversely, if $\tilde{y} \in C^2([0,T], D(\tilde{A})) \cap C^1([0,T], H)$ is a solution of (0.1), (0.2), (0.3) for some $\varphi \in C^\infty([0,T], H_B)$, then $y = \tilde{y} + G(\varphi)$ is in $C^2([0,T], D(A)) \cap C^1([0,T], H)$ and solves (1.1), (1.2), and (1.3).

This approach is taken in (Fattorini, 1968) in order to reduce a boundary control problem to a problem of distributed control. Based on this reduction then problems of approximate controllability are investigated.

In this paper we assume in addition that the bounded linear operator $G : H_B \to D(A)$ is such that $A(G(\psi)) = \Theta_H$ for all $\psi \in H_B$. Further we replace the control space $C^\infty([0,T], H_B)$ by $H_0^1([0,T], H_B) = \{\varphi \in H^1([0,T], H_B) | \ \varphi(0) = \varphi(T) = \Theta_{H_B}\}$ and define a weak solution of (1.1), (1.2), (1.3) in such a way that it can be explicitly represented in terms of an infinite series (see (1.7)).

Then we formulate the problem of (exact) null-controllability and investigate under which conditions it can be solved. Finally, we illustrate the abstract results by an example.

1 REDUCTION OF BOUNDARY CONTROL TO DISTRIBUTED CONTROL

We consider a vibrating system whose states are Hilbert space valued functions $y : [0,T] \to H$ for some $T > 0$ (H is a real Hilbert space) which satisfy an abstract wave equation

$$\ddot{y}(t) + Ay(t) = \Theta_H \ , \ t \in (0,T) \ , \tag{1.1}$$

and boundary conditions

$$By(t) = \varphi(t) \ , \quad t \in [0,T] \ , \tag{1.2}$$

where A and B are linear operators on dense linear subspaces $D(A)$ and $D(B)$ of H, respectively, mapping $D(A)$ and $D(B)$ into H and H_B, respectively, with H_B being a second Hilbert space. We assume that $D(A) \subseteq D(B)$. The control function $\varphi : [0,T] \to H_B$ on the right-hand side of (1.2) is chosen in the control space

$$H_0^2([0,T], H_B) = \{\varphi \in H^2([0,T], H_B) | \ \varphi(0) = \varphi(T) = \dot{\varphi}(0) = \dot{\varphi}(T) = \Theta_{H_B}\} \ .$$

Let

$$D_0 = \{z \in D(A) | \ Bz = \Theta_{H_B}\} \quad \text{and} \quad \tilde{A} = A|_{D_0} \ .$$

We assume $\tilde{A} : D_0 \to H$ to be self adjoint and positive definite and we also assume that $E = D(\tilde{A}^{\frac{1}{2}}) \subseteq D(B)$. For $y_0 \in E$ and $\dot{y}_0 \in H$ being given we further consider initial conditions

$$y(0) = y_0 \quad \text{and} \quad \dot{y}(0) = \dot{y}_0 . \tag{1.3}$$

Assumption 1: For every $\psi \in H_B$ there exists exactly one element $w_\psi \in D(A)$ such that

$$A\,w_\psi = \Theta_H \quad \text{and} \quad B\,w_\psi = \psi$$

and the linear operator $G : H_B \to H$ given by $G(\psi) = w_\psi$ is continuous.

Let $\varphi \in H_0^2([0,T], H_B)$ be given. Then we define

$$r(t) = G(\varphi)(t) \quad \text{for every} \quad t \in [0,T]$$

and conclude that

$$\begin{aligned} r \in H_0^2([0,T], H) = \{w \in H^2([0,T], H) |\, & w(0) \\ = w(T) = \dot{w}(0) = \dot{w}(T) = \Theta_H\} &. \end{aligned}$$

Let, for some $\varphi \in H_0^2([0,T], H_B)$, $y \in C([0,T], D(A)) \cap C^1([0,T], H)$ be a solution of (1.1), (1.2), (1.3). Then we put $r = G(\varphi)$ and define $\tilde{y} = y - r$. Then we conclude

$$B\tilde{y}(t) = By(t) - Br(t) = \varphi(t) - \varphi(t) = \Theta_H ,$$

hence $\tilde{y}(t) \in D_0$ for every $t \in [0,T]$.

From $r \in C([0,T], D(A)) \cap C^1([0,T], H)$ we infer that $\tilde{y} \in C([0,T], D_0) \cap C^1([0,T], H)$. Finally, we obtain

$$\begin{aligned} \ddot{\tilde{y}}(t) &= \ddot{y}(t) - \ddot{r}(t) = -Ay(t) - \ddot{r}(t) \\ &= -A(y(t) - r(t)) - \ddot{r}(t) = -\tilde{A}\tilde{y}(t) - \ddot{r}(t) , \end{aligned}$$

hence

$$\ddot{\tilde{y}}(t) + \tilde{A}\tilde{y}(t) = -\ddot{r}(t) \quad \text{for all} \quad t \in (0,T) . \tag{1.4}$$

It also follows that

$$\tilde{y}(0) = y_0 \quad \text{and} \quad \dot{\tilde{y}}(0) = \dot{y}_0 . \tag{$\widetilde{1.3}$}$$

Conversely, if $\tilde{y} \in C([0,T], D_0) \cap C^1([0,T], H)$ is a solution of (1.4), $(\widetilde{1.3})$ where $r = G(\varphi)$ for some $\varphi \in H_0^2([0,T], H_B)$, then $y = \tilde{y} + r \in C([0,T], D(A)) \cap C^1([0,T], H)$ is a solution of (1.1), (1.2), (1.3).

If we replace (1.4) by the weak equation

$$-\int_0^T \langle \dot{\tilde{y}}(t), \dot{v}(t)\rangle_H \, dt + \int_0^T \langle \tilde{A}^{\frac{1}{2}}\tilde{y}(t),\ \tilde{A}^{\frac{1}{2}}v(t)\rangle_H \, dt = -\int_0^T \langle \ddot{r}(t), v(t)\rangle_H \, dt \tag{1.5}$$

for all $v \in H_0^1([0,T],E) = \{w \in H^1([0,T],E) |\ w(0) = w(T) = \Theta_E\}$, then there is exactly one solution $\tilde{y} \in C([0,T],E) \cap C^1([0,T],H)$ of (1.5), (1.3) which is given by

$$\begin{aligned}\tilde{y}(t) &= \sum_{j=1}^{\infty} (\langle y_0, \varphi_j\rangle_H \cos\sqrt{\lambda_j}t + \tfrac{1}{\sqrt{\lambda_j}} \langle \dot{y}_0, \varphi_j\rangle_H \sin\sqrt{\lambda_j}t)\,\varphi_j \\ &- \sum_{j=1}^{\infty} \tfrac{1}{\sqrt{\lambda_j}} \int_0^t \langle \ddot{r}(s), \varphi_j\rangle_H \ sin\sqrt{\lambda_j}(t-s)\, ds\, \varphi_j\end{aligned} \tag{1.6}$$

for $t \in [0,T]$, if we assume that $\tilde{A} : D_0 \to H$ has a complete orthonormal system of eigenelements $\lambda_j \in D_0$, $j \in I\!N$, and corresponding eigenvalues $\lambda_j \in I\!R$, $j \in I\!N$, of finite multiplicity such that

$$0 < \lambda_1 \leq \lambda_2 \leq \dots \quad \text{and} \quad \lambda_j \to \infty \quad \text{as} \quad j \to \infty\,.$$

The time derivative of $\tilde{y}$ is given by

$$\begin{aligned}\dot{\tilde{y}}(t) &= \sum_{j=1}^{\infty} (-\sqrt{\lambda_j}\,\langle y_0, \varphi_j\rangle_H \sin\sqrt{\lambda_j}t + \langle \dot{y}_0, \varphi_j\rangle \cos\sqrt{\lambda_j}t)\,\varphi_j \\ &- \sum_{j=1}^{\infty} \int_0^t \langle \ddot{r}(s), \varphi_j\rangle_H \cos\sqrt{\lambda_j}\,(t-s)\, ds\, \varphi_j\end{aligned}$$

for $t \in [0,T]$.

We therefore define a weak solution $y \in C([0,T],D(B)) \cap C^1([0,T],H)$ of (1.1), (1.2) for some $\varphi \in H_0^2([0,T],H_B)$ and (1.3) by $y = \tilde{y} + r$ with $r = G(\varphi)$ and $\tilde{y}$ given by (1.6). This can then be represented in the form

$$\begin{aligned}y(t) &= \sum_{j=1}^{\infty} (\langle y_0, \varphi_j\rangle_H \cos\sqrt{\lambda_j}t + \tfrac{1}{\sqrt{\lambda_j}} \langle \dot{y}_0, \varphi_j\rangle_H \sin\sqrt{\lambda_j}t)\,\varphi_j \\ &+ \sum_{j=1}^{\infty} \sqrt{\lambda_j} \int_0^t \langle r(s), \varphi_j\rangle_H \sin\sqrt{\lambda_j}(t-s)\, ds\, \varphi_j\end{aligned} \tag{1.7}$$

for $t \in [0,T]$ and its time derivative reads

$$\begin{aligned}\dot{y}(t) &= \sum_{j=1}^{\infty} (-\sqrt{\lambda_j}\,\langle y_0, \varphi_j\rangle_H \sin\sqrt{\lambda_j}t + \langle \dot{y}_0, \varphi_j\rangle_H \cos\sqrt{\lambda_j}t)\,\varphi_j \\ &+ \sum_{j=1}^{\infty} \lambda_j \int_0^t \langle r(s), \varphi_j\rangle_H \cos\sqrt{\lambda_j}(t-s)\, ds\, \varphi_j\end{aligned} \tag{1.8}$$

for $t \in [0,T]$.

If the control space $H_0^2([0,T],H_B)$ is replaced by

$$H_0^1([0,T],H_B) = \{\varphi \in H^1([0,T],H_B) |\ \varphi(0) = \varphi(T) = \Theta_{H_B}\}\,,$$

then $y = y(t)$, $t \in [0,T]$, defined by (1.7) with $r = G(\varphi)$ for some $\varphi \in H_0^1([0,T], H_B)$ is taken as weak solution of (1.1), (1.2) for this φ, and (1.3) und $\dot{y} = \dot{y}(t)$, $t \in [0,T]$, defined by (1.8) is taken as its time derivative.

2 THE PROBLEM OF NULL-CONTROLLABILITY

Given $T > 0$ and $y_0 \in E$, $\dot{y}_0 \in H$, find $\varphi \in H_0^1([0,T], H_B)$ such that the weak solution $y = y(t)$, $t \in [0,T]$, of (1.1), (1.2), (1.3) which is defined by (1.7) with $r = G(\varphi)$ satisfies the end conditions

$$y(T) = \Theta_H \quad \text{and} \quad \dot{y}(T) = \Theta_H \tag{2.1}$$

where $\dot{y} = \dot{y}(t)$, $t \in [0,T]$, is given by (1.8).

From (1.7) and (1.8) it follows that the end conditions (2.1) are equivalent to

$$\begin{aligned} \lambda_j \int_0^T \langle r(t), \varphi_j \rangle_H \sin \sqrt{\lambda_j}\, t \, dt &= \sqrt{\lambda_j}\, \langle y_0, \varphi_j \rangle_H \,, \\ \lambda_j \int_0^T \langle r(t), \varphi_j \rangle_H \cos \sqrt{\lambda_j}\, t \, dt &= -\langle \dot{y}_0, \varphi_j \rangle_H \end{aligned} \tag{2.2}$$

for $j \in I\!N$.

Assumption 2: There exists a linear operator $\tilde{B} : D_0 \to H_B$ such that

$$\langle r, \tilde{A}\varphi_j \rangle_H = \langle Br, \tilde{B}\varphi_j \rangle_{H_B} \quad \text{for all} \quad j \in I\!N$$

and all $r \in D(A)$.

Then (2.2) can be rewritten as

$$\begin{aligned} \int_0^T \langle \varphi(t), \tilde{B}\varphi_j \rangle_H \sin \sqrt{\lambda_j}\, t \, dt &= \sqrt{\lambda_j}\, \langle y_0, \varphi_j \rangle_H \,, \\ \int_0^T \langle \varphi(t), \tilde{B}\varphi_j \rangle_H \cos \sqrt{\lambda_j}\, t \, dt &= -\langle \dot{y}_0, \varphi_j \rangle_H \end{aligned} \tag{2.3}$$

for $j \in I\!N$. Integration by parts shows that this is equivalent to

$$\begin{aligned} \frac{1}{\sqrt{\lambda_j}} \int_0^T \langle \dot{\varphi}(t), \tilde{B}\varphi_j \rangle_H \cos \sqrt{\lambda_j}\, t \, dt &= \sqrt{\lambda_j}\, \langle y_0, \varphi_j \rangle_H \,, \\ \frac{1}{\sqrt{\lambda_j}} \int_0^T \langle \dot{\varphi}(t), \tilde{B}\varphi_j \rangle_H \sin \sqrt{\lambda_j}\, t \, dt &= \langle \dot{y}_0, \varphi_j \rangle_H \end{aligned} \tag{2.4}$$

for $j \in I\!N$.

For every $u \in L^2([0,T], H_B)$ let us define

$$\varphi(t) = \int_0^t u(s)\, ds \,, \quad t \in [0,T] \,. \tag{2.5}$$

Then it follows that $\dot{\varphi}(t) = u(t)$ for almost all $t \in [0,T]$, i.e., $\varphi \in H^1([0,T], H_B)$ and $\varphi(0) = \Theta_{H_B}$. $\varphi(T) = \Theta_{H_B}$ is equivalent to

$$\int_0^T u(t)\,dt = \Theta_{H_B} \ . \tag{2.6}$$

Result: If $u \in L^2([0,T], H_B)$ satisfies (2.4) with u instead of $\dot{\varphi}$ and (2.6) and if $\varphi \in H^1([0,T], H_B)$ is defined by (2.5), then $\varphi \in H_0^1([0,T], H_B)$ and $y = y(t)$, $t \in [0,T]$, defined by (1.7) with $r = G(\varphi)$ satisfies the end conditions (2.1).

Thus, in order to solve the problem of null-controllability one has to determine some $u \in L^2([0,T], H_B)$ which satisfies (2.4) for all $j \in I\!N$ with u instead of $\dot{\varphi}$ and (2.6).

3 ON THE SOLVABILITY OF THE PROBLEM OF NULL – CONTROLLABILITY

Let us define a linear operator $S : L^2([0,T], H_B) \to (I\!R^2)^{I\!N}$ by $S(u) = (S_j^1(u), S_j^2(u))_{j \in I\!N}$ where

$$\begin{aligned} S_j^1(u) &= \tfrac{1}{\lambda_j} \int_0^T \langle u(t), \tilde{B}\,\varphi_j\rangle_H \cos\sqrt{\lambda_j}\,t\,dt\ , \\ S_j^2(u) &= \tfrac{1}{\lambda_j} \int_0^T \langle u(t), \tilde{B}\,\varphi_j\rangle_H \sin\sqrt{\lambda_j}\,t\,dt \end{aligned} \tag{3.1}$$

for all $j \in I\!N$ and $u \in L^2([0,T], H_B)$.

Further let us define a sequence $c = (c_j^1, c_j^2)_{j \in I\!N} \in \ell^2$ by

$$c_j^1 = \langle y_0, \varphi_j\rangle_H \quad \text{and} \quad c_j^2 = \frac{1}{\sqrt{\lambda_j}} \langle \dot{y}_0, \varphi_j\rangle_H \quad \text{for all} \quad j \in I\!N\ . \tag{3.2}$$

Then (2.4) for all $j \in I\!N$ with u instead of $\dot{\varphi}$ can be rewritten in the form

$$S(u) = c\ . \tag{3.3}$$

By Assumption 1, for every $u \in L^2([0,T], H_B)$ there is exactly one $w_u \in L^2([0,T], H)$ such that $w_u(t) \in D(A)$ for all $t \in [0,T]$ and

$$A\,w_u(t) = \Theta_H\ , \quad B\,w_u(t) = u(t) \quad \text{for almost all} \quad t \in [0,T]\ .$$

By Assumption 2 we obtain

$$\frac{1}{\lambda_j} \langle u(t), \tilde{B}\,\varphi_j\rangle_{H_B} = \langle w_u(t), \varphi_j\rangle_H \quad \text{for all} \quad j \in I\!N\ ,$$

hence

$$\sum_{j=1}^{\infty} \frac{1}{\lambda_j^2} \langle u(t), \tilde{B}\varphi_j\rangle_H^2 = \sum_{j=1}^{\infty} \langle w_u(t), \varphi_j\rangle_H^2 = \|w_u(t)\|_H^2$$

for almost all $t \in [0,T]$. This implies

$$\sum_{j=1}^{\infty} [S_j^1(u)^2 + S_j^2(u)^2] \leq T \sum_{j=1}^{\infty} \frac{1}{\lambda_j^2} \int_0^T \langle u(t), \tilde{B}\varphi_j\rangle_H \, dt = T \int_0^T \|w_u(t)\|_H^2 \, dt$$

and shows that $S(L^2[0,T], H_B) \subseteq \ell^2$. In addition, S is closed. Therefore, by the closed graph theorem, S is continuous. The adjoint operator $S^* : \ell^2 \to L^2([0,T], H_B)$ of S is given by

$$S^*(y)(t) = \sum_{j=1}^{\infty} \frac{1}{\lambda_j} (y_j^1 \cos \sqrt{\lambda_j}\, t + y_j^2 \sin \sqrt{\lambda_j}\, t)\, \tilde{B}\varphi_j$$

for almost all $t \in [0,T]$ and $y = (y_j^1, y_j^2)_{j\in\mathbb{N}} \in \ell^2$.

Let us define a continuous linear operator $G : L^2([0,T], H_B) \to L^2([0,T], H_B)$ by

$$G(u) = u - \frac{1}{T}\int_0^T u(t)\, dt \quad \text{for} \quad u \in L^2([0,T], H_B) \,.$$

Obviously it follows that

$$\int_0^T G(u)(t)\, dt = \Theta_{H_B} \quad \text{for all} \quad u \in L^2([0,T], H_B) \,.$$

Therefore, finding some $u \in L^2([0,T], H_B)$ which satisfies (3.3) and (2.6) is equivalent to finding some $v \in L^2([0,T], H_B)$ which satisfies

$$S(G(v)) = c \,. \tag{3.4}$$

If $u \in L^2([0,T], H_B)$ solves (3.3) and (2.6), then $G(u) = u$ solves (3.4). Conversely, if $v \in L^2([0,T], H_B)$ solves (3.4) then $G(v) \in L^2([0,T], H_B)$ is a solution of (3.3) and (2.6).

Let $u, v \in L^2([0,T], H_B)$ be given arbitrarily. Then one can show that

$$\langle v, G(u)\rangle_{L^2([0,T],H_B)} = \langle G(v), u\rangle_{L^2([0,T],H_B)}$$

which implies that the adjoint operator $G^* : L^2([0,T], H_B) \to L^2([0,T], H_B)$ of G is equal to G and therefore

$$(S \circ G)^*(y) = G(S^*(y)) = S^*(y) - \frac{1}{T}\int_0^T S^*(y)(t)\, dt$$

for all $y \in \ell^2$.

By (Krabs, 1982), Theorem 3.6, the operator equation (3.4) has a solution $v \in L^2([0,T],H_B)$, if and only if there is a constant $\lambda > 0$ such that

$$\begin{aligned}\langle c,y\rangle_{\ell^2} &\le \lambda \,\|(S\circ G)^*(y)\|_{L^2([0,T],H_B)} \\ &= \lambda\,\|S^*(y) - \tfrac{1}{T}\int_0^T S^*(y)(t)\,dt\|_{L^2([0,T],H_B)}\end{aligned}$$

for all $y \in \ell^2$.

This leads to the following

Problem: Minimize

$$\frac{1}{2}\,\|S^*(y) - \frac{1}{T}\int_0^T S^*(y)(t)\,dt\|^2_{L^2([0,T],H_B)}$$

for $y \in \ell^2$ subject to

$$\langle c,y\rangle_{\ell^2} = 1\ .$$

Let $\hat{y} \in \ell^2$ be a solution of this problem. Then we have to distinguish between two cases:

a) $\frac{1}{2}\,\|S^*(\hat{y}) - \frac{1}{T}\int_0^T S^*(\hat{y})(t)\,dt\|^2_{L^2([0,T],H_B)} = 0$, hence

$$S^*(\hat{y}) - \frac{1}{T}\int_0^T S^*(\hat{y})(t)\,dt = \Theta_{H_B} \quad \text{a.e..}$$

Let $u \in L^2([0,T],H_B)$ be any solution of $S(G(u)) = c$. Then it follows that

$$\langle c,\hat{y}\rangle_{\ell^2} = \langle S(G(u)),\hat{y}\rangle_{\ell} = \langle u, \underbrace{G(S^*(\hat{y}))}_{=\Theta_{H_B}\text{ a.e.}}\rangle_{L^2([0,T],H_B)} = 0$$

which contradicts $\langle c,\hat{y}\rangle_{\ell^2} = 1$.

Therefore $S(G(u)) = c$ has no solution $u \in L^2([0,T],H_B)$.

b) $\frac{1}{2}\,\|S^*(\hat{y}) - \frac{1}{T}\int_0^T S^*(\hat{y})(t)\,dt\|_{L^2([0,T],H_B)} > 0$.

Then, by the Lagrangian multiplier rule, there is some $\lambda \in I\!R$ such that

$$\langle (G\circ S^*)(\hat{y}), (G\circ S^*)(y)\rangle_{L^2([0,T],H_B)} = \lambda\,\langle c,y\rangle_{\ell^2}$$

for all $y \in \ell^2$ which implies

$$\lambda = \|(G\circ S^*)(\hat{y})\|^2_{L^2([0,T],H_B)} > 0$$

and

$$\langle (S \circ G \circ G \circ S^*)(\hat{y}), y \rangle_{\ell^2} = \langle c, y \rangle_{\ell^2} \quad \text{for all} \quad y \in \ell^2 ,$$

hence

$$S(G(\hat{u})) = c ,$$

if we put

$$\hat{u} = \frac{1}{\lambda}(G \circ S^*)(\hat{y}) = \frac{1}{\|(G \circ S^*)(\hat{y})\|^2_{L^2([0,T],H_B)}} (G \circ S^*)(\hat{y}) .$$

If we define $y^* = \frac{1}{\lambda}\hat{y}$, then $y^* \in \ell^2$ and solves the operator equation

$$(S \circ G \circ S^*)(y^*) = c \tag{3.5}$$

(because of $G \circ G = G$) which can be rewritten in the form

$$S(S^*(y^*) - \frac{1}{T}\int_0^T S^*(y^*)(t)\,dt) = c$$

where

$$S^*(y^*)(t) = \sum_{j=1}^{\infty} \frac{1}{\lambda_j}(y_j^{1*} \cos\sqrt{\lambda_j}\,t + y_j^{2*} \sin\sqrt{\lambda_j}\,t)\,\tilde{B}\varphi_j .$$

for almost all $t \in [0,T]$.

If $y^* \in \ell^2$ is a solution of (3.5) and if we define $\hat{u} \in L^2([0,T], H_B)$ by

$$\hat{u} = (G \circ S^*)(y^*) = S^*(y^*) - \frac{1}{T}\int_0^T S^*(y^*)(t)\,dt ,$$

then $S(G(\hat{u})) = c$ and $G(\hat{u}) \in L^2([0,T], H_B)$ solves (3.3) and (2.6).

If $S \circ G : L^2([0,T], H_B) \to \ell^2$ is surjective, then one can show that the operator $S \circ G \circ S^* : \ell^2 \to \ell^2$ is positive definite and therefore invertible. As a result we conclude that (3.5) has a unique solution which is given as

$$y^* = (S \circ G \circ S^*)^{-1}\,c .$$

4 AN EXAMPLE

As an example of (1.1) we consider the wave equation

$$y_{tt}(x,t) - \Delta\, y(x,t) = 0 \quad \text{for} \quad x \in \Omega , \quad t \in (0,T)$$

where $\Omega \subseteq I\!R^2$ is a bounded open set with a sufficiently smooth boundary $\partial\Omega = \bar{\Omega}\backslash\Omega$ and $T > 0$ is given. By Δ we denote the Laplacian operator $\frac{\partial^2}{\partial x_1^2} + \frac{\partial^2}{\partial x_2^2}$. As boundary condition we consider

$$y(x,t) = \varphi(x,t) \quad \text{for} \quad x \in \partial\Omega\ , \quad t \in [0,T]\ .$$

We choose $H = L^2(\Omega)$, $A = -\Delta$, $D(A) = H^2(\Omega)$, $B = \cdot|_{\partial\Omega}$, $D(B) = H^1(\Omega)$, $H_B = L^2(\partial\Omega)$.

This implies $D_0 = H^2(\Omega) \cap H_0^1(\Omega)$ and $E = D(\tilde{A}^{\frac{1}{2}}) = H_0^1(\Omega)$. For every $\psi \in L^2(\partial\Omega)$ there exists exactly one $w_\psi \in H^2(\Omega)$ such that

$$\Delta\, w_\psi = 0 \quad \text{and} \quad w_\psi|_{\partial\Omega} = \psi$$

and $\psi \rightarrow w_\psi$ is continuous. Therefore Assumption 1 holds true. Assumption 2 is a consequence of Green's formula

$$\int_\Omega (v\,\Delta\, u - u\,\Delta\, v) dx = \int_{\partial\Omega} (v\, u_\nu - u\, v_\nu)\, ds$$

for all $u, v \in H^2(\Omega)$ where $\frac{\partial}{\partial\nu}$ denotes the normal derivative. If we choose $v = r \in D(A)$ and $u = \varphi_j$, $j \in I\!N$, then it follows

$$\int_\Omega r\,\Delta\,\varphi_j\, dx = \int_{\partial\Omega} (Br)(\varphi_j)_\nu\, ds$$

which is equivalent to Assumption 2, if $\tilde{B} : D_0 \rightarrow H_B$ is defined by $\tilde{B}v = v_\nu$, $v \in D_0$.

REFERENCES

Fattorini, H.O. (1968): Boundary Control Systems. *SIAM Journal of Control*, vol. 6, No.3, 349 - 385.

Krabs, W. (1982): Convex Optimization and Approximation. In: *Modern Applied Mathematics*. Edited by B. Korte, North-Holland Publishing Company: Amsterdam, New York, Oxford, 327 - 357.

14

A frequency domain approach to the boundary control problem for parabolic equations

*L. Pandolfi**
Politecnico di Torino, Dipartimento di Matematica
Corso Duca degli Abruzzi, 24, 10129 Torino, Italy.
Phone: +39-11-5647516. E-mail: lucipan@polito.it

Abstract

The quadratic regulator problem for boundary control systems was studied in many papers, compare Lasiescka and Triggiani (1991) and Bensoussan *et al.* (1993) for an overview of the existing results. In particular, precise results are available in the case of control systems whose free evolution generates a holomorphic semigroup, due to the high regularity. However, the analysis of the regulator problem was carried out completely in the time domain. Hence, we propose to rederive existing results for parabolic boundary control systems, and standard quadratic cost, following a route which is largely (but not completely) in the frequency domain.

Keywords

Quadratic regulator problem, parabolic equations, frequency domain techniques

1 INTRODUCTION

The control system that we are going to study is the system

$$\dot{x} = A(x - Du) \qquad x(0) = x_0 \tag{1}$$

where as usual $x \in X$, a Hilbert space, and the operator A generates a stable holomorphic semigroup on X. The vector u belongs to a Hilbert space U and $D \in \mathcal{L}(U, X)$. Even more, we $\boxed{\text{assume}}$ $\text{im} D \subseteq \text{dom}\,(-A)^{1-\alpha}$, $0 < \alpha < 1$. A crucial consequence of this assumption:

$$||AR(z; A)D|| = ||A(zI - A)^{-1}D|| =\leq \frac{1}{|z|^\alpha} \qquad \forall z \in \{\Re e\, z > 0\}\,.$$

We associate the following cost to the system described by (1)

*Paper partially supported by the Italian Ministero della Ricerca Scientifica e Tecnologica within the program of GNAFA–CNR and by HCM network CEC n. ERB–CHRX–CT93–0402.

$$J(x_0;u)=\int_0^{+\infty}\{||Qx(t)||^2+||u(t)||^2\}\,\mathrm{d}t \tag{2}$$

where $x(t)=x(t;x_0,u)$ solves Eq. (1) so that the cost does depend both on the control $u(\cdot)$ and on the initial datum x_0. We assume that $Q=Q^*\in\mathcal{L}(X)$, $Q\geq 0$.

The cost is finite for each square integrable control since we assume that the semigroup generated by the operator A is exponentially stable.

2 PRELIMINARIES

In this section we set up the preliminary material and the notations that we use in the present paper.

It is well known that any $L^2(\mathbb{R})$–function admits a Laplace transform

$$\hat{f}(z)=\int_0^{+\infty}e^{-zt}f(t)\,\mathrm{d}t$$

which belongs to the Hardy space H^2 of the right half plane; i.e. such that

$$\sup_x\{\int_{-\infty}^{+\infty}|f(x+iy)|^2\,\mathrm{d}y\}^{1/2}<+\infty\,.$$

Any such function has boundary values on the imaginary axis (defined as non–tangential limits) which identify an $L^2(i\mathbb{R})$–function (whose norm is equal to (2)). If $f(\cdot)$ is also in $L^1(-\infty,+\infty)$ we can define its Fourier transform

$$F(\omega)=\hat{f}(z)=\int_{-\infty}^{+\infty}e^{-\omega t}f(t)\,\mathrm{d}t$$

and it is known that a continuous extension exists to all the square integrable functions, thanks to Parseval equality

$$\int_{-\infty}^{+\infty}|f(t)|^2\,\mathrm{d}t=\frac{1}{2\pi}\int_{-\infty}^{+\infty}|F(\omega)|^2\,\mathrm{d}\omega\,.$$

Moreover, $F(\omega)=\hat{f}(i\omega)$. For simplicity we shall always write $f(i\omega)$.

The Fourier transform is a continuous function from $L^2(-\infty,+\infty)$ to itself while the Laplace transform is a continuous transformation from $L^2(0,+\infty)$ to the Hardy space H^2 (of the right half plane). We use these facts in order to express the transformations of solutions of Eq. (1).

Let us assume that the input $u(\cdot)$ is both square integrable and C^1, with square integrable derivative. It is known that this set of functions is dense in $L^2(0,+\infty)$. We introduce the function

$$\xi(t)=e^{At}\xi(0)-\int_0^t e^{A(t-s)}D\dot{u}(s)\,\mathrm{d}s$$

By definition, a solution to Eq. (1) is the function

$$x(t)=\xi(t)-Du(t)$$

and it is possible to prove that

$$x(t) = e^{A(t-s)}x_0 + A\int_0^t e^{A(t-s)}Du(s)\,\mathrm{d}s\,.$$

(A formula that can be extended by continuity to any square integrable input $u(\cdot)$, compare Balakrishnan (1978).) Let us take the Laplace transform of the function $\xi(\cdot)$. We have

$$\hat{\xi}(i\omega) = R(i\omega; A)\xi(0) - R(i\omega; A)D[i\omega\hat{u}(i\omega) - u(0)]$$

and $\xi(0) = x_0 - Du(0)$. from which we obtain

$$\hat{x}(i\omega) = R(i\omega; A)x_0 - i\omega R(i\omega; A)D\hat{u}(i\omega) + D\hat{u}(i\omega) \tag{3}$$
$$= R(i\omega; A)x_0 - AR(i\omega; A)D\hat{u}(i\omega) = R(i\omega; A)x_0 - H(i\omega)\hat{u}(i\omega) \tag{4}$$

Now, this expression was derived under the differentiability assumption on $u(\cdot)$; but, the right hand side depends continuously on $u(\cdot) \in L^2(0, +\infty)$ so that we are justified if we assume the expression (3) as the definition of the solution to Eq. (1) for each input $u(\cdot)$ which is simply square integrable.

The assumption that the semigroup be exponentially stable, can be removed.

3 THE REGULATOR PROBLEM

After this introduction we study explicitly the system described by

$$\dot{\xi} = A\xi - D\dot{u} \qquad \xi(0) = \xi_0 = x(0) - Du(0) \tag{5}$$

under the framed assumptions listed in sect. 1. This system is equivalent to (1) only if the control $u(\cdot)$ is differentiable. Consequently, it is not at all obvious that we shall find a minimum for the cost, when using this system.

Fourier transformation gives

$$(zI - A)(\hat{x} - D\hat{u}) = x_0 - zD\hat{u} \tag{6}$$

i.e.

$$\hat{x} = R(z; A)x_0 + H(z)\hat{u} \qquad H(z) = -AR(z; A)D \tag{7}$$

Parseval equality implies that we must minimize

$$\hat{J}(x_0, \hat{u}) = \int_{-\infty}^{+\infty} \{|Q\hat{x}(i\omega)|^2 + |\hat{u}(i\omega)|^2\}\,\mathrm{d}\omega \tag{8}$$

on the solutions to Eq. (6). We noted already that it is conceivable that we lost the optimal control, if the optimal control is not differentiable. But, in fact, we see directly that Eq. (3) makes sense for every square integrable control $\hat{u}(\cdot)$ and that $\hat{x}(\cdot)$ is a continuous function of $\hat{u}(\cdot)$ (in the L^2 norms). Hence, the minimum of the cost defined in (8) along the solutions of (3) exists and is unique for each given x_0.

We can characterize the optimal control $u^+(\cdot; x_0)$ (and the optimal state $x^+(\cdot; x_0)$) in the frequency domain in two ways, both useful. The first is easier:

$$\langle \hat{v}, \hat{u}^+ + H^* Q \hat{x}^+ \rangle = 0 \qquad \forall \hat{v} \in H^2 . \tag{9}$$

This condition is obtained by computing the derivative at $\rho = 0$ of the function $\rho \to \hat{J}(x_0, \hat{u}^+ + \rho \hat{v})$ for any $\hat{v} \in H^2$. Hence, we find that

$$\hat{u}^+ = -P_+ M_{H^* Q} \hat{x}^+ \tag{10}$$

where P_+ denotes the projection of $L^2(i\mathbb{R})$ onto H^2 and $M_{H^* Q}$ is the multiplication by $H^* Q$.

A second expression for the optimal control is more artificial. We extend the control $u(\cdot)$ to the negative axis. Let us assume for the moment that the extended function is differentiable. We introduce the function

$$\tilde{\xi}(t) = \begin{cases} e^{At}\{\xi_0 + \int_{-\infty}^{0} e^{-As} D\dot{u}(s)\, ds\} & \text{if } t > 0 \\ 0 & \text{if } t < 0 \end{cases} - \int_{-\infty}^{t} e^{A(t-s)} D\dot{u}(s)\, ds .$$

Moreover, we define $\tilde{x}(\cdot) = \tilde{\xi}(\cdot) + Du(\cdot)$ so that

$$\tilde{x}(t) = \begin{cases} x(t; x_0, u) & \text{if } t > 0 \\ -\int_{-\infty}^{t} e^{A(t-s)} D\dot{u}(s)\, ds + Du(t) & \text{if } t < 0 \end{cases}$$

and, we define $\tilde{J}(x_0; u) = \int_{-\infty}^{+\infty} \{\langle \tilde{x}, Q\tilde{x} \rangle + |u(t)|^2\}\, dt$. We note that neither $\tilde{\xi}(\cdot)$ nor $\tilde{x}(\cdot)$ are continuous at $t = 0$; But, the transformations from $u(\cdot)$ to $\tilde{\xi}(\cdot)$ and to $\tilde{x}(\cdot)$ are continuous transformations from $L^2(-\infty, +\infty; U)$ to $L^2(-\infty, +\infty; X)$.

As $\tilde{x}(t) = x(t)$ for $t > 0$ then $J(x_0, u) \le \tilde{J}(x_0, u)$ and equality holds if $u(\cdot)$ is zero for negative times. Hence: $\inf_u \tilde{J}(x_0; u) = \inf_u J(x_0; u)$. We see that if we compute the function which minimize the cost $\tilde{J}(x_0; u)$ we find the optimal control for the original regulator problem (1), (2): The artificial extension of the control $u(\cdot)$ to the negative axis is not going to change either the optimal control or the value of the minimum.

We take the Fourier transforms and we get the following form for of $\hat{\tilde{\xi}}(\cdot)$, when $u(\cdot) \in C^1(-\infty; +\infty)$:

$$\begin{aligned} \hat{\tilde{\xi}}(i\omega) &= R(i\omega; A)\{\xi_0 + \int_{-\infty}^{0} e^{-As} D\dot{u}(s)\, ds\} - i\omega R(i\omega, A) D\hat{u}(i\omega) \\ &= R(i\omega; A)[\xi_0 + Du(0)] - D\hat{u}(i\omega) + H(i\omega)\hat{u}(i\omega) + AR(i\omega; A)\nu \end{aligned}$$

and

$$\nu = \int_{-\infty}^{0} e^{-As} Du(s)\, ds = \frac{1}{2\pi} \int_{-\infty}^{+\infty} R(i\omega; A) D\hat{u}(i\omega)\, d\omega \in \operatorname{dom} A^\gamma \qquad \gamma < 1/2 + \alpha .$$

Consequently, we add $D\hat{u}$ and we get

$$\hat{\tilde{x}}(i\omega) = R(i\omega, A)x_0 + H(i\omega)\hat{u}(i\omega) + AR(i\omega; A)\nu\,. \tag{11}$$

Now we minimize in the frequency domain the functional $\tilde{J}$. We differentiate with respect to ρ the expression

$$\tilde{J}(x_0, \hat{u}^+ + \rho\hat{v}) =$$
$$= (1/2\pi)\int_{-\infty}^{+\infty} \{\langle \hat{x}^+ + \rho H\hat{v} + \rho\Lambda\hat{v}, Q[\hat{x}^+ + \rho H\hat{v} + \rho\Lambda\hat{v}]\rangle + \langle \hat{u}^+ + \rho\hat{v}, \hat{u}^+ + \rho\hat{v}\rangle\}\, \mathrm{d}\omega$$

where

$$\Lambda\hat{v} = AR(i\omega; A)\int_{-\infty}^{0} e^{-As} Dv(s)\, \mathrm{d}s\,. \tag{12}$$

The derivative above must be zero for each $\hat{v} \in L^2(i\mathbb{R}; U)$. Consequently,

$$\hat{u}^+ = -H^* Q\hat{\tilde{x}}^+ - \Lambda^* Q\hat{\tilde{x}}^+ \tag{13}$$

The projection operator is now replaced by the subtraction of $-\Lambda^* Q\hat{\tilde{x}}^+$ (and, we recall, $\tilde{x}^+(\cdot) = \tilde{x}(\cdot)$).

A point should be noted: we are making these computations in the frequency domain; but, the operator Λ acts on the time function $v(\cdot)$. We should take this into account in the computation of the of the adjoint of Λ. Standard calculations give Λ^*:

$$\Lambda^*\hat{z} = D^* A^* R^*(i\omega; A)\int_{-\infty}^{+\infty} R^*(is; A)\hat{z}(is)\, \mathrm{d}s,\ \ ||\Lambda^*\hat{z}|| = O(1/\omega^{1-\alpha})$$

Now we have:

Theorem 1 *The function $t \to x^+(t)$ is continuous for each $t \geq 0$ and each x_0.*

Proof. The Fourier transform of $t \to x^+(t)$ is the function

$$\hat{x}(i\omega) = R(i\omega; A)x_0 + H(i\omega)\hat{u}^+(i\omega) \tag{14}$$

The inverse transform of $R(i\omega; A)x_0$ is the function $e^{At}x_0$ for $t \geq 0$ while it is zero for $t < 0$: it is a continuous function for $t \geq 0$.

It is known that the inverse Fourier transform of an $L^1(i\mathbb{R})$ is continuous on $\mathbb{R}$. Hence, if we can prove that $H(i\omega)\hat{u}(i\omega)$ is an integrable function then the inverse Laplace transform of $H(i\omega)\hat{u}^+(i\omega)$ is even continuous on $\mathbb{R}$. The following bootstrap argument shows that $H(i\omega)\hat{u}^+(i\omega)$ is integrable by proving that $\hat{u}(i\omega) = O(1/\omega^{1-\alpha+\epsilon})$ for large $|\omega|$.

A direct observation of the expression (11) for $\hat{\tilde{x}}(i\omega)$ shows that $\hat{\tilde{x}}(i\omega) = O(1/\omega^{\sigma})$ for some positive σ, $0 < \sigma <$min$\{\alpha, (3/2) - \alpha$.

We use this observation in the expression (13) for $\hat{u}(i\omega) = -H^*Q\hat{\hat{x}}^+ - \Lambda^*Q\hat{\hat{x}}^+$ and we find that the first addendum is of the order $O(1/\omega^{\alpha+\sigma})$ and the second addendum is of the order $O(1/\omega^{\alpha+\sigma-\epsilon})$. In fact,

$$(\Lambda^*\hat{\hat{x}})(i\omega) = D^*(-A)^{*\alpha}(-A)^{*1-\alpha-\sigma+\epsilon}R^*(i\omega)\int_{-\infty}^{+\infty}(-A)^{*\sigma-\epsilon}R^*(is;A)O(1/s^\sigma)\,\mathrm{d}s$$

The integral is convergent and the factor in front of it is of the order of $1/\omega^{\alpha+\sigma-\epsilon}$. Hence, this is also the order of $\hat{u}(i\omega)$ for $|\omega| \to +\infty$.

In the case that $\alpha+\sigma > 1$ we are done; otherwise we repeat this procedure. We replace again in the expression (11) for $\hat{\hat{x}}^+(i\omega)$ and we find that

$$\hat{\hat{x}}^+(i\omega) = O(1/\omega) + O(1/\omega^{2\alpha+\sigma-\epsilon}) + O(1/\omega^{\alpha+\omega-2\epsilon}) :$$

we gained $\alpha - 2\epsilon$ of regularity above the previous estimate as long as $\alpha+\sigma-2\epsilon < 0$.

If we insert this further information in the expression (13) we see that $\hat{u}^+(i\omega) = O(1/\omega^{\alpha+\epsilon+(\alpha+\sigma-2\epsilon)})$.

We iterate these steps till we have

$$\hat{u}(i\omega) = O(1/\omega^{1-\alpha+\epsilon})$$

and consequently $\hat{x}(i\omega)$ is the sum of $R(i\omega;A)x_0$ (whose inverse Fourier transformation is continuous on $t \geq 0$) and of a function which is continuous on the whole real axis. □

Analogously,

Theorem 2 *The function $t \to u^+(t;x_0)$ is a continuous function of t for $t \geq 0$ and moreover $x_0 \to u^+(t;x_0)$ is continuous for each $t \geq 0$; the function $x_0 \to u^+(\cdot;x_0)$ is continuous from X to $L^2(0,+\infty)$.*

Proof. We replace $\hat{u}^+(i\omega)$ in the the expression (11) of $\hat{\hat{x}}(i\omega)$; then we replace the resulting formula in the expression (13) of the optimal control. We note that when $v = u^+$ then $AR(i\omega;A)\nu = \Lambda\hat{u}^+$, compare (12). Hence,

$$\hat{u}^+(i\omega) = -\{I + (H+\Lambda)^*Q(H+\Lambda)\}^{-1}\{H^* + \Lambda^*\}QR(i\omega)x_0\,.$$

The first brace is bounded for large $|\omega|$ and the second is the order of $1/\omega^\alpha$ so that $\omega \to \hat{u}^+(i\omega)$ is integrable, since $R(i\omega;A) = O(1/\omega)$: the optimal control is a time continuous function. Moreover, the expression above depends continuously, as an $L^1(-\infty,+\infty)$ function, on x_0. This shows that $x_0 \to u^+(t;x_0)$ is continuous for each time $t \geq 0$. □

The next crucial step is the following result:

Theorem 3 *The transformation $(t,x_0) \to x^+(t;x_0)$ is a C_0–semigroup.*

Proof. We know that $t \to x(t;x_0)$ is a continuous function from $[0,+\infty)$ to X and that $x_0 \to x(t;x_0)$ is continuous from X to X (compare previous Lemma).

The semigroup property is a consequence of the causality of the problem, and can be proved with a time domain argument for example as in Lasiecka, Lukes, Pandolfi, (1995). □

Let use denote A_+ the infinitesimal generator of the semigroup $T^+(t)x_0 = x^+(t;x_0)$. We note that A_+ generates an **exponentially stable semigroup** because we know the function $t \to x^+(t;x_0)$ is square integrable for each x_0. Hence, exponential stability follows from Datko (1970).

It is also possible to study further regularity property of the optimal control. We already noted continuity. Moreover, we can prove

Theorem 4 *The optimal control belongs to $W^{k,2}$ if $x_0 \in dom(A_+)^{k+1}$.*

Proof. We present first a different proof of the continuity of the optimal control, which however requires the further condition $x_0 \in \text{dom}\, A_+$. We proved that

$$\hat{u}^+(i\omega) = P_+ M_{H^*Q}\hat{x}^+(i\omega) = P_+(H^*(i\omega)QR(i\omega;A_+)x_0)$$

so that $t \to u^+(t)$ is the restriction to $t \geq 0$ of the inverse Fourier transform of $M_{H^*Q}x^+$. We prove that this inverse Fourier transform is continuous on $\mathbb{R}_+$. We note that the inverse Fourier transform of $x_0/(1+i\omega)$ is zero for $t<0$ and it is $e^{-t}x_0$ for $t \geq 0$: its restriction to $\mathbb{R}_+$ is continuous. Now,

$$H^*Q\{R(i\omega,A_+)x_0 - \frac{1}{i\omega+1}x_0\} = H^*Q(A_+ + I)\frac{R(i\omega,A_+)}{i\omega+1}x_0$$
$$= [\frac{1}{i\omega+1}H^*(i\omega)][QR(i\omega,A_+)(A_+ + I)x_0]$$

if $x_0 \in \text{dom}A_+$. The second bracket is bounded on the imaginary axis since A_+ generates an exponentially stable semigroup. The first bracket is $O(1/\omega^{1+\alpha})$ is integrable. Hence, the inverse Fourier transform is continuous.

We can iterate this procedure and we find that the optimal control belongs to $W^{k,2}$ if $x_0 \in \text{dom}(A_+)^{k+1}$. We show this for the case $k=1$. In this case $x_0 \in \text{dom}\, A_+^2$ and we consider

$$H^*Q\{R(i\omega;A_+)x_0 - \frac{1}{i\omega+1}x_0 - \frac{(A_+ + I)}{(i\omega+1)^2}x_0\}$$
$$= \frac{1}{i\omega+1}H^*(i\omega)Q(A_+ + I)\{R(i\omega;A_+) - \frac{1}{i\omega+1}x_0\}$$
$$\frac{1}{(i\omega+1)^2}H^*(i\omega)QR(i\omega;A_+)(A_+ + I)^2x_0 = O(1/\omega^2)$$

i.e. its inverse transform belongs to $W^{1,2}(-\infty,+\infty;U)$. As both the inverse transforms of $\frac{x_0}{i\omega+1}$ and of $\frac{(A_+ + I)x_0}{(i\omega+1)^2}$ belong to $W^{1,2}(0,+\infty;U)$ we have the result. □

Now we consider the value of the optimal cost:

$$\hat{J}(x_0;u^+) = \langle R(i\omega;A)x_0 + H\hat{u}^+, Q\hat{x}^+\rangle_{L^2} + \langle \hat{u},\hat{u}\rangle_{L^2}$$

$$= \langle R(i\omega; A)x_0, Q\hat{x}^+\rangle + \{\langle \hat{u}^+, H^*Q\hat{x}^+ + \hat{u}^+\rangle\} = \langle R(i\omega; A)x_0, Q\hat{x}^+\rangle .$$

In fact, the inner product $\langle \hat{u}^+, H^*Q\hat{x}^+ + \hat{u}^+\rangle$ is zero, compare Eq. (9). Consequently

$$J(x_0, u^+) = \langle x_0, Px_0\rangle = (1/2\pi)\langle x_0, \int_{-\infty}^{+\infty} R^*(i\omega; A)Q\hat{x}^+(i\omega; x_0)\, \mathrm{d}\omega\rangle$$
$$= \langle x_0, \int_0^{+\infty} e^{A^*t}Qe^{A_+t}x_0\, \mathrm{d}t\rangle .$$

It follows from these inequalities that $\mathrm{im}P \subseteq \mathrm{dom}\,(-A^*)^{1-\epsilon}$ for each $\epsilon > 0$ (compare Lasiecka, Triggiani (1991)). In particular, D^*A^*P admits a continuous extension to X. We proved already that $t \to u^+(t; x_0)$ is continuous if $x_0 \in \mathrm{dom}A^+$. Hence we can compute $u^+(0; x_0)$ which is the value of the inverse Fourier transform at 0:

$$u^+(0; x_0) = -(1/2\pi) \int_{-\infty}^{+\infty} H^*(i\omega)Q\hat{x}^+(i\omega; x_0)\, \mathrm{d}\omega$$
$$= -D^*A^*\{\int_{-\infty}^{+\infty} R^*(i\omega; A)Q\hat{x}(i\omega; x_0)\, \mathrm{d}\omega\} = -D^*A^*Px_0 . \qquad (15)$$

(We can use the formula for the inverse Fourier Transform since the integrand belongs to $L^1(i\mathbb{R})$). The previous relation was proved under the assumption that $x_0 \in \mathrm{dom}A^+$. We saw that $x_0 \to u^+(0; x_0)$ is continuous on X so that $x_0 \to D^*A^*Px_0$ admits a continuous extension to X, i.e. equality (15) holds on X.

In particular, **the function $t \to u^+(t; x_0)$ is bounded on $[0, +\infty)$.**

Now,

Theorem 5 *If $x \in domA_+$ then $Px \in domA^*$; if $x \in dom\,A$ then $Px \in domA^*$.*

Proof. We show that $y \to \langle Px, Ay\rangle$ is continuous on X for each $x \in \mathrm{dom}A^+$. In fact, we make use of Parseval equality and we see that

$$\langle Px, Ay\rangle = (1/2\pi) \int_{-\infty}^{+\infty} \langle QR(i\omega; A^+)x, AR(i\omega; A)y\rangle\, \mathrm{d}\omega = \int_0^{+\infty} \langle Qe^{A^+s}x, Ae^{As}y\rangle\, \mathrm{d}s$$
$$= -\langle Qx, y\rangle - \int_0^{+\infty} \langle Qe^{A^+s}A^+x, e^{As}y\rangle\, \mathrm{d}s$$

a continuous function of y.

In order to see the second property, we note that

$$\hat{x}^+(i\omega) = R(i\omega; A)x_0 + H(i\omega)\hat{u}^+(i\omega) = R(i\omega; A)A^{-1}Ax_0 = O(1/\omega^2) + O(1/\omega^\alpha)\hat{u}^+(i\omega) .$$

Hence,

$$A^*Px_0 = \int_{-\infty}^{+\infty} A^*R^*(i\omega; A)x^+(i\omega; x_0)$$

is expressed by a convergent integral. □

Moreover,

Theorem 6 *The operator A^+ generates a holomorphic semigroup.*

Proof. The system is causal. This implies that $u^+(s+\tau;x_0) = u^+(s;x^+(\tau;x_0))\forall s,\tau \geq 0$.
We put $s = 0$ and we get:

$$u^+(\tau;x_0) = u^+(0;x^+(\tau,x_0)) = -D^*A^*Px^+(\tau,x_0) \qquad \forall x_0 \in \mathrm{dom}A^+ .$$

If $x_0 \in \mathrm{dom}A^+$ then $t \to x^+(t;x_0)$ is differentiable so that

$$\dot{x}^+(t) = A[I - DD^*A^*P]x^+(t)$$

i.e. $\mathrm{dom}A^+ = \{x, [I - DD^*A^*P]x \in \mathrm{dom}\,A\}$ $\quad A^+x = A[I - DD^*A^*P]x$. As in Lasiecka, Triggiani (1991) this implies that

$$(zI - A^+)^{-1} = [zI - A(I - DD^*A^*P)]^{-1} = \{I + (-A)^{1-\alpha}R(z)(-A)^\alpha D^*A^*P\}^{-1}(zI - A)^{-1} .$$

The brace is bounded for $\Re e\, z > 0$ since $A^{1-\alpha}R(z)o(1/|z|^\alpha)$. Consequently $|R(z;A^+)| < M/|z|$ for $\Re e\, z > 0$ i.e. A^+ generates a holomorphic semigroup. □

Now we use again Parseval equality and the fact that A, A_+ generates holomorphic semigroups in order to arrive at the usual Riccati equation. We note that the following computations make sense for $x \in\mathrm{dom}\,A_+$, $y \in\mathrm{dom}\,A$:

$$\langle y, PA_+x\rangle = \int_{-\infty}^{+\infty} \langle y, R^*(i\omega;A)QR(i\omega;A_+)A_+x\rangle\,\mathrm{d}\omega$$
$$= \int_0^{+\infty} \langle e^{At}y, Qe^{A_+t}A_+x\rangle\,\mathrm{d}t = -\langle y, Qx\rangle - \int_0^{+\infty} \langle e^{At}Ay, Qe^{A_+t}x\rangle\,\mathrm{d}t =$$
$$-\langle y, Qx\rangle - \langle Ay, \int_0^{+\infty} e^{A^*t}Qe^{A_+t}x\rangle\,\mathrm{d}t =$$
$$-\langle y, Qx\rangle - \langle y, A^* \int_0^{+\infty} e^{A^*t}Qe^{A_+t}x\rangle\,\mathrm{d}t = -\langle y, Qx\rangle - \langle y, A^*Px\rangle .$$

Hence we obtained the following equality which holds for each $y \in\mathrm{dom}\,A$ and $x \in\mathrm{dom}\,A_+$:

$$\langle y, PA[I - DD^*A^*P]x\rangle = -\langle y, Qx\rangle - \langle Ay, Px\rangle$$

Now, $y \in\mathrm{dom}\,A$ so that the first inner product is equal to

$$\langle y, PA[I - DD^*A^*P]x\rangle = \langle A^*Py, x\rangle - \langle D^*A^*Py, D^*A^*Px\rangle .$$

Hence we arrive at the Riccati equation

$$\langle y, PAX\rangle + \langle Ay, Px\rangle - \langle y, (D^*A^*P)(D^*A^*P)x\rangle + \langle y, Qx\rangle = 0 .$$

This equality was derived for any $y \in\mathrm{dom}\,A$ and any $x \in\mathrm{dom}\,A_+$, but we note that each addendum is a continuous function of x so that the equality can be extended to hold for any $x \in X$. In particular, if we choose $x \in \mathrm{dom}\, A$ we get the usual form of the Riccati Equation.

REFERENCES

Balakrishnan AV (1978) Boundary control of parabolic equations: L–Q–R theory. in *Theory of nonlinear operators.* Akademie–Verlag, Berlin.

Bensoussan A., Da Prato G., Delfour M. and Mitter S. (1993) *Representation and control of infinite dimensional systems,* Birkhäuser, Boston.

Datko, R. (1970) Extending a Theorem of Lyapunov to Hilbert spaces *J. Math. Analysis Appl.* **32** 610–616.

Lasiecka, I., Lukes, D. and Pandolfi, L. (1995) Input dynamics and non–standard Riccati equations with applications to boundary control of damped wave and plate equations, *J. Optim. Theory Appl.* **84** 549–574.

Lasiecka I. and Triggiani R (1991) *Differential and algebraic Riccati equations with applications to boundary/point control problems: continuous theory and approximation theory.* Springer–Verlag, Berlin.

15

Diffusion processes: invertibility problem and guaranteed estimation theory

Irina F.Sivergina *
Institute for Mathematics and Mechanics, Urals Branch, RAS
16 S.Kovalevskoj Str., 620219, Ekaterinburg, Russia.
Tel.: 3432-493243. Fax: 3432-442581. E-mail: siv@oou.imm.intec.ru

Abstract

This paper considers a series of problems on state and parameter specifying for distributed parameter systems on the basis of measurements generated by sensors. Observability and invertibility issues, duality relations for problems of control, schemes of sensor's generating to guarantee the observability desired are discussed. These problems are treated within a unified framework of guaranteed estimation procedures for systems with unknown but bounded errors.

Keywords

Inverse problem, distributed parameter system, sensor, invertibility, observability, duality, guaranteed estimation problem, regularization

1 INTRODUCTION

This paper deals with the problem of parameter and state specifying for distributed parameter system on the basis of measurements generated by sensors. The issues treated here are the invertibility and observability problems. The approach leans upon describing the system properties in terms of solutions to some guaranteed estimation problems. They are formulated for the systems with disturbances taken to be unknown but bounded. That provides another interpretation for the properties known and reveals quite new features. The necessary and sufficient conditions to provide them are given. The duality relations for the problems of control are considered.

Another aspect of the invertibility problem considered here is in generating sensors ensuring a desired invertibility property for an evolutionary system. A procedure how to do this for the strong observability is discussed. It makes use of results on optimization of observers for guaranteed estimation problem. Regulating the parameters in that problem enables a sensor with the desired property to be approximated.

*Work supported by the Russian Fund of Fundamental Researches under Projects 94-01-00803 and 95-01-00380

2 BASIC NOTIONS

Let there be the initial-boundary problem

$$\begin{cases} \dfrac{\partial u(t,x)}{\partial t} = \Delta u(t,x) + f(t,x), & t \in (0,T),\ x \in \Omega; \\ u(0,x) = u_0(x), & x \in \Omega; \\ u(t,\zeta) = v(t,\zeta), & t \in (0,T),\ \zeta \in \partial\Omega. \end{cases} \tag{1}$$

Here Ω is an open bounded domain in an n-dimensional space having a piecewise-smooth boundary $\partial\Omega$. A symmetric uniformly elliptic operator with the smooth coefficients may stay instead of Laplacian Δ. Here $f(t,x)$ is the control function, $v(t,x)$ is the boundary condition, $u_0(x)$ is the initial distribution. They are supposed to be square integrable functions on the proper sets. We'll consider $u(t,x)$ as a generalized solution to (1).

There is an evolutionary system having been associated with that initial-boundary problem. The state of the system at a time instant t, $0 \leq t \leq T$, is the function $u(t,\cdot)$ being an element of the space $L_2(\Omega)$. The triple $w = (f(\cdot,\cdot), v(\cdot,\cdot), u_0(\cdot))$ will be referred to as an input for that system.

Further it is assumed that the input is taken to be unknown in advance. It is supposed that all the available information on the solution is given through a finite-dimensional measurement equation

$$y(t) = G(t)u(t,\cdot), \quad 0 \leq t \leq T. \tag{2}$$

$y(t)$ is a measurement data, $y(t) \in R^m$, $y(\cdot) \in L_2^m(0,T)$. $G(t)$ is a linear observation operator, or sensor. Let us start with two traditional notions.

Definition 1 *We will say that the system (1),(2) is invertible if the measurement* $y(t) \equiv 0$, $0 \leq t \leq T$, *yields* $w = 0$.

Consider the initial boundary problem (1) assuming that

$$f(t,x) \equiv 0, \quad v(t,x) \equiv 0. \tag{3}$$

Definition 2 *We will say that the system (1)-(3) is observable if the measurement* $y(t) \equiv 0$, $0 \leq t \leq T$, *yields* $u(T,\cdot) = 0$.

The first definition is equivalent to the fact that the linear mapping C, $Cw = y(\cdot)$ is such that Ker$C = \{0\}$. So, if the system is invertible then two different inputs yield two different measurements. The second one implies the similar property for the mapping $Du(T,\cdot) = y(\cdot)$.

There are papers studying these properties and sensors to assure them for the system. Let us think over these properties from another point of view. From here on we are concentrating on the observability problem. We'll return to the invertibility one later.

Let us consider a guaranteed estimation problem closely related to the observability one. Suppose, that the measurement $y(t)$ is not exact and contains additively an uncertain disturbance $\xi(t)$

$$y(t) = G(t)u(t,\cdot) + \xi(t), \quad 0 \le t \le T, \tag{4}$$

$$\| \xi(\cdot) \|_{L_2^m(0,T)} \le \mu, \quad \mu \ge 0. \tag{5}$$

That inequality gives an a priori restriction on $\xi(t)$. Now, in general, there's no way of determining the state $u(T,\cdot)$ of the system on the basis of the measurement data $y(t)$. This inverse problem, as a rule, has got a non-unique solution. This leads us to the following notion (Kurzhanski, 1977).

Definition 3 *The informational set $\mathcal{U}[y]$ of states $u(T,\cdot)$ of the system (1), (3)-(5) being consistent with measurement data $y(t)$ and the a priori restriction on $\xi(t)$, is the set of all those functions $z(\cdot) \in L_2(\Omega)$ for each of which there exists an input w such that the corresponding solution $u(t,\cdot)$ to the problem (1) at the time instant T will be equal to $z(\cdot)$ and the equation (4) will be valid for some $\xi(t)$ satisfying the a priori restriction (5).*

The linearity of the system and the convexity of the set described by (5) imply the convexity of $\mathcal{U}[y]$. This set is said to be a solution to the guaranteed estimation problem of the state $u(T,\cdot)$ on the basis of $y(\cdot)$. We can state the first obvious statement.

Lemma 1 *The system (1),(3) is observable if and only if $\mathcal{U}[y]$ is one-point set for $\mu = 0$, whatever is $y(t)$.*

If the informational set was bounded, one might be defined the minimax estimate of the state $u(T,\cdot)$, being the Chebyshev centre of $\mathrm{cl}\,\mathcal{U}[y]$. It is determined through the formula $u^0 = \arg\min_{u \in L_2(\Omega)} \sup_{v \in \mathcal{U}[y]} \| u - v \|_{L_2(\Omega)}$. (A solution to that variational problem does exist and is unique under the assumption above.) If our system were finite dimensional, i.e. were described by a system of ordinary differential equations, the observability would guarantee the existence of the minimax estimate u^0. Our infinite dimensional system needs a more strong property to ensure that. Namely,

Definition 4 *(Kurzhanski, Khapalov, 1991) We will say that the system (1),(3)-(5) is strongly observable if $\mathcal{U}[y]$ is bounded, whatever is the measurement $y(t)$.*

This definition is equivalent to the continuous observability of the system implying the existences of a constant $\alpha > 0$ such that the inequality $\| u(T,\cdot) \| < \alpha \| y \|$ is true, whatever is $y = G(t)u(t,\cdot)$, $0 \le t \le T$. That means that the operator D is continuously invertible. It is clear that Definition 4 implies Definition 2. It turns out that there are other substantial properties "between" them.

3 ϵ-OBSERVABILITY

A set in a Hilbert space is bounded if and only if the projections of it onto all the one-dimensional subspaces are bounded. So, the system (1),(3)-(5) is just strongly observable if and only if the projections of $U[y]$ onto all the one-dimensional subspaces are bounded, whatever is $y(t)$. It should be emphasized that if the boundedness were inherent in the intersections of the informational set with all the one-dimensional subspaces, whatever the measurement is, the system would be observable. Let us now give the following

Definition 5 *(Kurzhanski, Sivergina, 1991) We will say that the system (1),(3)-(5) is ϵ-observable if for almost every functions $\varphi \in L_2(\Omega)$ the projection of $\mathcal{U}[y]$ onto a subspace generated by that is bounded, whatever is the measurement $y(t)$.*

Theorem 1 *The system (1),(3)-(5) is ϵ- observable if and only if D is invertible and has a closed extension.*

Theorem 2 *Let for almost every functions from $L_2(\Omega)$ the projection of $\mathcal{U}[0]$ onto the subspace generated by that be bounded, whatever $y(t)$ is. Then the system (1),(3)-(5) is ϵ-observable.*

So, we have three notions related to the observability issue: the observability, the ϵ-observability, and the strong observability. It has to be shown that they are distinct.

Example 1 That is an instance of a system being observable but not ϵ- observable. Suppose the sensor is of spatially averaged type:

$$G(t)u(t,\cdot) = \int_{\Omega} g(t,x)u(t,x)\,dx, \ 0 \le t \le T \tag{6}$$

where $g(t,x) = \varphi_1(x)e^{\lambda_1(t-T)}e^{-t} + \sum_{k=2}^{\infty} \varphi_k(x)e^{\lambda_k(t-T)}(e^{-k^2 t} - \frac{k}{k-1}e^{-k-1)^2 t})$. $\varphi_i(\cdot)$, λ_i are the eigen-functions and the eigen-values for the spectral problem $\Delta\varphi_i(x) = -\lambda_i\varphi_i(x)$, $x \in \Omega$, $\varphi_i(\zeta) = 0$, $\zeta \in \partial\Omega$; $\lambda_i \to \infty$ as $i \to \infty$. $\{\varphi_i\}_{i=1}^{\infty}$ make up an orthonormal base in $L_2(\Omega)$. It is easy to show that the condition $G(t)u(t,\cdot) \equiv 0$, implies $u(T,\cdot) = 0$. So, the system is observable. At the same time, the set $\mathcal{U}[0]$ is unbounded. Moreover, a closure of it contains a whole linear subspace generated by the function

$$\varphi(x) = \sum_{i=1}^{\infty} \frac{1}{i}\varphi_i(x), \quad x \in \Omega. \tag{7}$$

Note, in that Example there is no restriction on n, i.e. on the dimension of the space variable. It might be anyone.

Example 2 The instance of a system being ϵ-observable but not strongly observable gives the heat equation in the domain $\Omega = (0,1)$ with the spatially averaged sensor (6) when $g(t,x)$ doesn't depend on t and is equal to $\varphi(x)$ from (7). Now, because of the minimality property in $L_2(\Omega)$ of the system $\{e^{-\lambda_i t}, \ 0 \le t \le T, i = 1,2,\ldots\}$ we have got $|u_{0i}| \le \alpha_i \, (\| y(\cdot) \| + \mu)$, $i = 1,2,\ldots$ where α_i doesn't depend on $y(\cdot)$. This assures the ϵ-observability for the system. But, according to Dolecki's result (1977), the system won't be strongly observable because of the equality $\sum_{i=1}^{\infty} \frac{i^2}{e^{-\lambda_i T}} = \infty$.

4 DUAL PROBLEM

One way of studying the observability property of a system is to turn to the dual control problem. Let us formulate a property being dual to the ϵ-observability. Consider the system

$$\begin{cases} \dfrac{\partial v(t,x)}{\partial t} = -\Delta v(t,x) + G^*(t)\lambda(t), & t \in (0,T),\ x \in \Omega; \\ v(T,x) = 0, & x \in \Omega; \\ u(t,\zeta) = 0, & t \in (0,T),\ \zeta \in \partial\Omega. \end{cases} \tag{8}$$

Here $\lambda(\cdot) \in L_2^m(0,T)$ is a control. The control problem consists in selecting the control to bring the system to the time instant 0 to the position $S^*(T)\varphi(\cdot)$:

$$u(0,\cdot) = S^*(T)\varphi(\cdot); \tag{9}$$

$\varphi(\cdot) \in L_2(\Omega)$ is a given function, $S(t),\ t \geq 0$ is the semi-group corresponding to the evolutionary system (1). So, the set of those $\varphi(\cdot)$ for which the problem (8),(9) may be solvable is of interest. Suppose, $G(t)$ in (8) is the same as in (2). It is known, that if the system (1),(3)-(5) is strongly observable, the control problem is solvable for any $\varphi(\cdot)$. If the system is only observable, one is solvable for almost every $\varphi(\cdot) \in L_2(\Omega)$. In this case one says about approximate controllability of the system (8),(9):

$$\forall\varphi(\cdot) \in L_2(\Omega)\ \forall\epsilon > 0\ \exists\lambda(*\cdot)\ :\ u(0,\cdot) \in S^*(T)\varphi + \mathcal{O}_\epsilon(0).$$

Theorem 3 *The property of the system (1),(3)-(5) to be ϵ-observable is equivalent to the property of the problem (8),(9) to be solvable in the following sense:*

$$\forall\varphi(\cdot)\ \forall\epsilon > 0\ \exists\lambda(\cdot)\ :\ u(0,\cdot) \in S^*(T)(\varphi + \mathcal{O}_\epsilon(0)).$$

It is easy to note that the key to the approach to the observability problem lies with the property of the system to be observable in a direction. We are illustrating that notion with the invertibility one. But before proceeding to that we'll consider another problem connected with the observability problem.

5 REGULARIZATION OF SENSORS

Consider the system

$$\begin{cases} \dfrac{\partial u(t,x)}{\partial t} = \dfrac{\partial^2 u(t,x)}{\partial x^2}, & t \in (0,T),\ x \in \Omega = (0,1); \\ u(t,\zeta) = 0, & t \in (0,T),\ \zeta \in \partial\Omega; \\ y(t) = \int_\Omega g(x)u(t,x)dx, & 0 \leq t \leq T \end{cases} \tag{10}$$

Let us suppose a possibility for a function $g(x)$ to be chosen. It is well known that there exist the functions $g(\cdot) \in L_2(\Omega)$ for which the system (10) would be strongly observable.

Moreover they would exist to belong to the set $\mathcal{G} = \{g(\cdot) \in C^1(\bar{\Omega}) \mid ||g(\cdot)||_{C^1(\Omega)} \leq 1\}$. How extract them? The method to do this is proposed here.

Consider an auxiliary observation system, in which the initial distribution u_0 is unknown, the disturbance $\xi(t)$ is present in the observation equation, and $g(x)$ is a function form $\mathcal{G}$. One supposes u_0 and $\xi(t)$ to be a priori restricted by the joint quadratic inequality

$$\begin{cases} \dfrac{\partial u(t,x)}{\partial t} = \dfrac{\partial^2 u(t,x)}{\partial x^2}, & t \in (0,T),\ x \in \Omega,\ \Omega = (0,1); \\ u(t,\zeta) = 0, & t \in (0,T),\ \zeta \in \partial\Omega; \\ y(t) = \int_\Omega g(x)u(t,x)dx + \xi(t), & 0 \leq t \leq T; \\ \epsilon \parallel u_0 \parallel^2_{L_2(\Omega)} + \parallel \xi(\cdot) \parallel^2_{L_2(0,T)} \leq \mu^2. \end{cases} \tag{11}$$

Here $\epsilon > 0$ is a parameter. Formally one defines an informational set $\mathcal{U}_\epsilon[y]$ of the states $u(T,\cdot)$ of the system (11) being consistent with the measurement data $y(t)$ and the a priori restriction on u_0 and $\xi(t)$. It is a closed bounded set in $L_2(\Omega)$. We would associate the number $\mathcal{J}_\epsilon(g)$ with a sensor in (11) corresponding to a function $g(\cdot) \in \mathcal{G}$

$$\mathcal{J}_\epsilon(g) = \max_{y(\cdot)} \sup_{||\varphi|| \leq 1} \max_{u(\cdot) \in \mathcal{U}_\epsilon[y]} |<\varphi, u>|. \tag{12}$$

By $<\cdot,\cdot>$ it is denoted a scalar production in $L_2(\Omega)$. $\mathcal{J}_\epsilon(g)$ may be interpreted as a functional on $\mathcal{G}$. One can show that the external maximum in (12) is reached on the function $y(\cdot) = 0$. By applying the methods of paper (Kurzhanski, Khapalov, 1986) we make sure that the variational problem $\min_{g(\cdot)\in\mathcal{G}} \mathcal{J}_\epsilon(g)$ is solvable. Denote its solution by $g_\epsilon(\cdot)$. There exists a subsequence $g_{\epsilon_i}(\cdot)$ being convergent in uniform metric when ϵ_i tends to zero. Let $g_0(\cdot)$ be a limit.

Theorem 4 *For $g(\cdot) = g_0(\cdot)$ the system (10) is strongly observable.*

Remark 1 A similar result is true for the point-wise sensors $G(t)u(t,\cdot) = u(t,x(t))$, $0 < \delta \leq t \leq T$. Here $x(t)$ is a function taking the values in $\bar{\Omega}$. In that case one has to take $\mathcal{G} = \{x(\cdot) \in C^1(\delta,T) \mid x(t) \in \bar{\Omega},\ |\dot{x}(t)| \leq 1\}$.

6 INVERTIBILITY PROBLEM

Now we turn to the invertibility problem. Let us remind the system under consideration. It is (1)-(2). Restrict ourselves to an assumption that the input as well as the sensor have the special representations

$$f(t,x) = \sum_{i=1}^{s} b_i(x)f_i(t),\ v(t,\zeta) = \sum_{i=1}^{k} d_i(\zeta)v_i(t),\ G(t)u(t,\cdot) = \mathrm{col}[\int_\Omega g_i(x)u(t,x)\,dx]_{i=1}^m.$$

The functions $b_i(x)$ and $d_i(x)$ are supposed to be given. So, the invertibility problem is to specify $f_i(\cdot)$, $i = 1,\ldots,s$, $v_i(\cdot)$, $i = 1,\ldots,k$, and $u_0(x)$ on the basis of the measurement data $y(t)$, $0 \leq t \leq T$. Here it is rationally to call the element $\tilde{w} = (f_1,\ldots,f_s,v_1,\ldots,v_k,u_0)$ $\in L_2^s(0,T) \times L_2^k(0,T) \times L_2(\Omega)$ as the input for the system. The following assumptions are supposed to be valid from now on: each of the function systems $\{b_i(\cdot)\}$, $\{d_i(\cdot)\}$, and $\{g_i(\cdot)\}$ are linearly independent, and the functions $\{b_i(\cdot), \sum_{l=1}^{\infty} \int_{\partial\Omega} \frac{\partial\varphi_l(\zeta)}{\partial n} d_j(\zeta)\, d\zeta \varphi_l(\cdot)\}$ are, too.

Formulate some results relative to the invertibility properties of the system. They are presented in (Sivergina, 1995) for a special case.

Theorem 5 *For any functions $\{b_i(\cdot)\}_{i=1}^{s}$, $\{d_i(\cdot)\}_{i=1}^{k}$, and for $m = s + k + 1$ there exist $g_1, g_2, ..., g_m$ such ones that the system will be invertible.*

That estimate for m can't be improved. Really, any system being a finite-dimensional Fourier approximation for the observation system under consideration may be invertible only if $m \geq s + k + 1$. In the following theorem there are formulated the necessary and sufficient conditions for the system to be invertible in the case that the functions $\{b_i(\cdot)\}$, $\{d_i(\cdot)\}$, and $\{g_i(\cdot)\}$ belong to the space $\overset{0}{C}{}^{\infty}(\Omega)$. Here the techniques of conditionally invariant spaces developed by Basile, Marro (1992) is useful. Denote $G \equiv G(t)$ and define the two operators $\bar{B}$ and $\bar{V}$ mapping from R^s and R^k, respectively, into $L_2(\Omega)$: $\bar{B}(f_1, \ldots, f_s) = \sum_{i=1}^{s} b_i(x) f_i$, $\bar{V}(v_1, \ldots, v_k) = \sum_{i=1}^{k} \sum_{j=1}^{\infty} \int_{\partial\Omega} \frac{\partial\varphi_i}{\partial\zeta} d_i(\zeta) d\zeta \varphi_j(x) v_i$.

Theorem 6 *The system is invertible if and only if $\bigcup_{l=0}^{\infty} \mathcal{Y}_l = L_2(\Omega)$, where $\mathcal{Y}_0 = \mathrm{Im}(G^*)$, $\mathcal{Y}_l = \mathcal{Y}_0 + \Delta(\bar{B}^* + \bar{V}^*) \bigcap \mathcal{Y}_{l-1}$.*

7 INVERTIBILITY IN A DIRECTION

Now, following the scheme above, consider an information set of inputs, consistent with the measurement data and the restriction on the "noise" in the measurement equation.

Definition 6 *The information set $\tilde{\mathcal{W}}[y]$ of inputs $\tilde{w}$ of the system (1),(4),(5) being consistent with the measurement data $y(t)$ and the a priori restriction on ξ, is the set of all those inputs $\tilde{w}$ for each of which there exists an element $\xi(\cdot) \in L_2^m(0, T)$ satisfying the conditions (4),(5).*

It is obviously that the strong invertibility property fails. Here the matter of current interest is whether the system is invertible in a direction and if this property is strong. Formulate these notions accurately. Let Λ be an element of the input space.

Definition 7 *We will say that the system (1),(2) is invertible in a direction Λ if $y(t) \equiv 0$ implies that the projection of the input on the subspace generated by Λ is equal to 0.*

Definition 8 *We will say that the system (1),(4),(5) is strongly invertible in a direction Λ if the projection of $\tilde{\mathcal{W}}[y]$ on the subspace generated by Λ is bounded, whatever $y(\cdot)$ is.*

Here it turns it is possible to describe the set of all the directions, in which the system is invertible.

Theorem 7 *The system (1),(2) is invertible in the direction Λ if and only if Λ belongs to a linear space generated by the elements $\Lambda_t^{(j)} = (f_{1t}^{(j)}, \ldots, f_{st}^{(j)}, v_{1t}^{(j)}, \ldots, v_{kt}^{(j)}, u_{0t}^{(j)})$, $j =$*

$1,\ldots,m$, $0\le t\le T$ *of the form:* $u_{0t}^{(j)}(\cdot)=\sum_{l=1}^{\infty}e^{-\lambda_l t}g_{jl}\varphi_l(\cdot)$,

$$f_{it}^{(j)}(\tau)=\begin{cases}\sum\limits_{l=1}^{\infty}e^{-\lambda_l(t-\tau)}g_{jl}b_{il}, & 0\le\tau\le t\\ 0, & t<\tau\le T\end{cases},\quad v_{it}^{(j)}(\tau)=\begin{cases}\sum\limits_{l=1}^{\infty}e^{-\lambda_l(t-\tau)}g_{jl}d_{il}, & 0\le\tau\le t\\ 0, & t<\tau\le T\end{cases}.$$

$$g_{jl}=\int_\Omega g_j(z)\varphi_l(z)dz,\quad b_{jl}=\int_\Omega b_j(z)\varphi_l(z)dz,\quad d_{jl}=\int_{\partial\Omega}\frac{\partial\varphi_l(\zeta)}{\partial n}d_j(\zeta)d\zeta.$$

To conclude, we formulate the conditions guaranteeing the strong invertibility in a direction for the simplest case.

Theorem 8 *Let* $s=k=1$, $m=1$, $y(\cdot)\in C[0,T]$, *the functions* $b_1(\cdot)$, $d_1(\cdot)$, $g_1(\cdot)$ *be twice continuously differentiable, and the system be invertible in the direction* Λ. *Then it is strongly invertible in that direction.*

The author is grateful to Academician A.B.Kurzhanski for helpful discussions and encouragement.

REFERENCES

Basile, F. and Marro, G. (1992) *Controlled and conditioned invariants in linear system theory.* Prentice-Hall Inc., Englewood Cliffs.

Dolecki, S. (1977) Observability for regular processes. *Journal of Mathematical Analysis and Applications*, **58**, 178-188.

Kurzhanski, A.B.(1977) *Control and observation under uncertainty.* Nauka, Moscow.

Kurzhanski, A.B. and Khapalov, A.Yu. (1986) On the state estimation problem for distributed systems. *Lecture Notes in Control and Information Sciences*, **83**, 102-113.

Kurzhanski, A.B. and Khapalov, A.Yu. (1991) An observation theory for distributed parameter systems. *Journal of Mathematical Systems, Estimation, and Control*, **1**, 389-440.

Kurzhanskii, A.B. and Sivergina I.F. (1992) ϵ-observability for distributed parameter systems. *Proceedings of Institute for Mathematics and Mechanics, Urals Branch, RAS.* **1**, 1222-137.

Sivergina, I.F. (1995) The invertibility and observability problems for evolutionary systems. *Doklady of Russian Academy of Sciences.* (to appear).

16
Quadratic optimal control of stable abstract linear systems

Olof J. Staffans
Institute of Mathematics, Helsinki University of Technology, FIN-02150 Espoo, Finland

Abstract
We consider the nonstandard infinite horizon quadratic cost minimization problem for a stable abstract linear control system, and show that it can be reduced to a J-inner coprime factorization problem (or equivalently, to a canonical spectral factorization problem) in the (often finite-dimensional) control space. More precisely, we show that the optimal solution of the quadratic cost minimization problem is of static state feedback type if and only if a certain spectral factorization problem has a solution. If both the system and the spectral factor are regular together with their adjoints, then the feedback operator can be expressed in terms of the Riccati operator, and the Riccati operator is a self-adjoint solution of an algebraic Riccati equation. This Riccati equation is similar to the classical algebraic Riccati equation, but one of its coefficients differs from the expected one. We apply our main theorems to prove the first available versions of the positive real and bounded real lemmas for abstract linear systems. Similar results are true for unstable systems.

Keywords
Spectral factorization, J-inner coprime factorization, positive real lemma, bounded real lemma

1 INTRODUCTION

This work treats a nonstandard infinite horizon quadratic cost minimization problem for a stable causal time-invariant abstract linear control system $\Psi = \begin{bmatrix} \mathcal{A} & \mathcal{B} \\ \mathcal{C} & \mathcal{D} \end{bmatrix}$ on a triple of Hilbert spaces (U, H, Y). Here U is the input space, H the state space, Y the output space, $\mathcal{A}$ the semigroup, $\mathcal{B}$ the controllability map, $\mathcal{C}$ the observability map, and $\mathcal{D}$ is the input/output map of Ψ. The object is to find the control $u \in L^2(\mathbf{R}^+; U)$ that minimizes the nonstandard cost function

$$W(x_0, u) = \int_{\mathbf{R}^+} \langle y(s), Jy(s) \rangle \, ds,$$

where J is a self-adjoint operator in Y, and $y = \mathcal{C}x_0 + \mathcal{D}u$ is the observation of Ψ with initial data $x_0 \in H$ and control u. (In the standard problem J is the identity operator.)

We do not require J to be positive, but we do require the input/output map $\mathcal{D}$ of Ψ to be "J-coercive" the sense that there exists some $\epsilon > 0$ such that $\mathcal{D}^* J \mathcal{D} \geq \epsilon I$.

After a short presentation of stable abstract linear systems in Section 2, we give a more precise definition of the nonstandard quadratic cost minimization problem in Section 3. Here we also define the notion of a J-inner right coprime factorization of $\mathcal{D}$. As described in our first main Theorem 7 in Section 4, the solution of the quadratic cost minimization problem is closely related to such factorizations. In our second main Theorem 8 we derive the algebraic Riccati equation satisfied by the Riccati operator in the case where the system is regular. Finally, in Section 5 we apply the main theorems to get weak versions of the strict bounded and positive real lemmas for abstract linear systems.

One of our most striking findings is that the appropriate algebraic Riccati equation differs significantly from the classical one. Roughly speaking, if the strictly causal part of the input/output map is "smoothing", then we get the classical equation. However, if, for example, the system contains discrete time delays, then, in order to get the correct Riccati equation, we must replace the operator $D^* J D$ in the classical equation by $X^* X$, where D is the feed-through operator of $\mathcal{D}$ and X is the feed-through operator of a spectral factor of $\mathcal{D}^* J \mathcal{D}$. In particular, this means that, in order to write down the correct Riccati equation, we must first solve a spectral factorization problem. This phenomenon was first observed and reported in Staffans (1994a).

For more details and for extensions to unstable systems, see Staffans (1994a, 1994b, 1995a, 1995b, 1995c). Many of the results presented here have been discovered independently by Martin and George Weiss (private communication; to be presented at the ECC conference in Rome, September 1995). See Callier and Winkin (1992) and their reference list for earlier work on spectral factorization results for systems with bounded control and observation operators, and Curtain and Green (1994) for a recent work on coprime factorizations. See Lasiecka and Triggiani (1991) and their reference list for work on Riccati equations for systems governed by partial differential equations.

Most of our notations are self-explanatory. We let $\mathcal{L}(U;Y)$ and $\mathcal{L}(U)$ denote the sets of bounded linear operators from U into Y or from U into itself, respectively, and let $\mathbf{R} = (-\infty, \infty)$, $\mathbf{R}^+ = [0, \infty)$, and $\mathbf{R}^- = (-\infty, 0]$. The set of causal bounded linear shift-invariant operators from $L^2(\mathbf{R}; U)$ into $L^2(\mathbf{R}; Y)$ is denoted by $SIC(U;Y)$ (this is the set of input/output maps whose transfer functions belong to $H^\infty(U;Y)$). Additional notations are explained later as they are introduced.

2 A REVIEW OF ABSTRACT LINEAR SYSTEMS

In order to formulate the axioms satisfied by a stable abstract linear system we introduce the "past time" projection π_-, the "future time" projection π_+, and the "time shift" group $\tau(t)$ that operate on functions $u \in L^2(\mathbf{R}; U)$ in the following way:

$$
\begin{aligned}
(\pi_- u)(s) &= \begin{cases} u(s), & \text{if } s \in \mathbf{R}^-, \\ 0, & \text{if } s \in \mathbf{R}^+, \end{cases} \qquad (\pi_+ u)(s) = \begin{cases} u(s), & \text{if } s \in \mathbf{R}^+, \\ 0, & \text{if } s \in \mathbf{R}^-, \end{cases} \\
(\tau(t)u)(s) &= u(t+s), \quad t, s \in \mathbf{R}.
\end{aligned}
$$

Definition 1 *Let U, H, and Y be Hilbert spaces. A (causal) stable abstract linear system on (U,H,Y) is a quadruple $\Psi = \begin{bmatrix} \mathcal{A} & \mathcal{B} \\ \mathcal{C} & \mathcal{D} \end{bmatrix}$, where $\mathcal{A}$, $\mathcal{B}$, $\mathcal{C}$, and $\mathcal{D}$ are bounded linear operators of the following type:*

(i) $\mathcal{A}(t): H \to H$ *is a bounded strongly continuous semigroup on* H;
(ii) $\mathcal{B}: L^2(\mathbf{R};U) \to H$ *satisfies* $\mathcal{A}(t)\mathcal{B}u = \mathcal{B}\tau(t)\pi_- u$ *for all* $u \in L^2(\mathbf{R};U)$ *and* $t \in \mathbf{R}^+$;
(iii) $\mathcal{C}: H \to L^2(\mathbf{R};Y)$ *satisfies* $\mathcal{C}\mathcal{A}(t)x = \pi_+\tau(t)\mathcal{C}x$ *for all* $x \in H$ *and* $t \in \mathbf{R}^+$;
(iv) $\mathcal{D}: L^2(\mathbf{R};U) \to L^2(\mathbf{R};Y)$ *satisfies* $\tau(t)\mathcal{D}u = \mathcal{D}\tau(t)u$, $\pi_-\mathcal{D}\pi_+u = 0$, *and* $\pi_+\mathcal{D}\pi_-u = \mathcal{C}\mathcal{B}u$ *for all* $u \in L^2(\mathbf{R};U)$ *and* $t \in \mathbf{R}$.

If, in addition, $\mathcal{A}(t)x \to 0$ as $t \to \infty$ for all $x \in H$, then Ψ is called strongly stable.

The different components of Ψ are named as follows: U is the input space, H the state space, Y the output space, $\mathcal{A}$ the semigroup, $\mathcal{B}$ the controllability map, $\mathcal{C}$ the observability map, and $\mathcal{D}$ the input/output map of Ψ. In the initial value setting with initial time zero, initial value $x_0 \in H$, and control $u \in L^2(\mathbf{R};U)$, the controlled state $x(t)$ at time $t \in \mathbf{R}^+$ and the observation y of Ψ are given by

$$\begin{pmatrix} x(t) \\ y \end{pmatrix} = \begin{pmatrix} \mathcal{A}(t) & \mathcal{B}\tau(t) \\ \mathcal{C} & \mathcal{D} \end{pmatrix} \begin{pmatrix} x_0 \\ \pi_+ u \end{pmatrix} = \begin{pmatrix} \mathcal{A}(t)x_0 + \mathcal{B}\tau(t)\pi_+ u \\ \mathcal{C}x_0 + \mathcal{D}\pi_+ u \end{pmatrix}.$$

The condition imposed on $\mathcal{D}$ in Definition 1 requires that $\mathcal{D} \in SIC(U;Y)$ (i.e., $\mathcal{D}$ is shift-invariant and causal), and that the Hankel operator induced by $\mathcal{D}$ is equal to $\mathcal{C}\mathcal{B}$. Intuitively, the controllability map $\mathcal{B}$ takes past controls into the present state, the observability map $\mathcal{C}$ takes the present state into future observations, and the input/output map $\mathcal{D}$ takes inputs into outputs in a causal way.

The axioms listed above describe standard properties of the corresponding operators for exponentially stable systems with bounded control and observation operators B and C. For such systems, we have

$$\begin{aligned}
\mathcal{B}u &= \int_{-\infty}^{0} \mathcal{A}(-s)Bu(s)\mathrm{d}s, \\
\mathcal{C}x &= (t \mapsto C\mathcal{A}(t)x), \\
\mathcal{D}u &= \left(t \mapsto \int_{-\infty}^{t} C\mathcal{A}(t-s)Bu(s)\mathrm{d}s + Du(t)\right), \\
x(t) &= \mathcal{A}(t)x_0 + \int_0^t \mathcal{A}(t-s)Bu(s)\mathrm{d}s, \quad t \in \mathbf{R}^+, \\
y(t) &= C\mathcal{A}(t)x_0 + \int_0^t C\mathcal{A}(t-s)Bu(s)\mathrm{d}s + Du(t), \quad t \in \mathbf{R}^+,
\end{aligned}$$

where D is a given bounded linear feed-through operator.

The sets of axioms used by Salamon (1987, 1989) and Weiss (1994a, 1994b) to define an abstract linear system differ slightly from the one in Definition 1. In particular, Salamon and Weiss do not discuss the *stability* of the system.

3 QUADRATIC COST MINIMIZATION, COERCIVITY, AND J-INNER COPRIME FACTORIZATIONS

We consider the following nonstandard infinite horizon quadratic cost minimization problem:

Definition 2 *Let* $\Psi = \begin{bmatrix} \mathcal{A} & \mathcal{B} \\ \mathcal{C} & \mathcal{D} \end{bmatrix}$ *be an abstract linear system on* (U, H, Y), *and let* $J \in \mathcal{L}(Y)$ *be self-adjoint. The nonstandard quadratic cost minimization problem for* Ψ *consists of finding, for each* $x_0 \in H$, *the minimum over all* $u \in L^2(\mathbf{R}^+; U)$ *of the cost function* $W(x_0, u) = \int_{\mathbf{R}^+} \langle y(s), Jy(s) \rangle \, ds$, *where* $y = \mathcal{C}x_0 + \mathcal{D}\pi_+ u$ *is the observation of* Ψ.

Observe that we do not here require the operator J to be positive. Instead we shall require $\mathcal{D}$ to be J-coercive:

Definition 3 *Let* $J \in \mathcal{L}(Y)$. *The abstract linear system* $\Psi = \begin{bmatrix} \mathcal{A} & \mathcal{B} \\ \mathcal{C} & \mathcal{D} \end{bmatrix}$ *on* (U, H, Y) *is called* J*-coercive if* $\mathcal{D}^* J \mathcal{D} \geq \epsilon I$ *for some* $\epsilon > 0$, *i.e.,*

$$\int_{\mathbf{R}} \langle (\mathcal{D}u)(s), J(\mathcal{D}u)(s) \rangle \, ds \geq \epsilon \, \|u\|^2_{L^2(\mathbf{R};U)}, \qquad u \in L^2(\mathbf{R}; U).$$

The following result is immediate (Staffans, 1995c, Lemma 3.3):

Lemma 4 *Let* $J \in \mathcal{L}(Y)$ *be self-adjoint, and let* $\Psi = \begin{bmatrix} \mathcal{A} & \mathcal{B} \\ \mathcal{C} & \mathcal{D} \end{bmatrix}$ *be a stable* J*-coercive abstract linear system. Then, for each* $x_0 \in H$, *the nonstandard quadratic cost minimization problem for* Ψ *has a unique solution* u_{opt}.

Definition 5 *If* Ψ *is* J*-coercive, then the Riccati operator of* Ψ *(with respect to* J*) is the unique self-adjoint operator* Π *that satisfies* $\langle x_0, \Pi x_0 \rangle = W(x_0, u_{\mathrm{opt}})$ *for all* $x_0 \in H$, *where* u_{opt} *is the unique solution of the nonstandard quadratic cost minimization problem.*

A central role in this work is played by coprime factorizations with a J-*inner* numerator:

Definition 6 *Let* $J \in \mathcal{L}(Y)$ *be self-adjoint.*

(i) An operator $\mathcal{N} \in SIC(U; Y)$ *is called* J*-inner (or more precisely,* (J, I)*-inner) iff* $\mathcal{N}^* J \mathcal{N} = I$, *i.e.,* $\int_{\mathbf{R}} \langle (\mathcal{N}u)(s), J(\mathcal{N}u)(s) \rangle \, ds = \|u\|^2_{L^2(\mathbf{R};U)}$ *for all* $u \in L^2(\mathbf{R}; U)$.

(ii) $(\mathcal{N}, \mathcal{M})$ *is called an* J*-inner right coprime factorization of* $\mathcal{D} = \mathcal{N}\mathcal{M}^{-1} \in SIC(U; Y)$ *if* $\mathcal{N}$ *is* J*-inner and* $(\mathcal{N}, \mathcal{M})$ *is a right coprime factorization of* $\mathcal{D}$ *(Staffans, 1995b, Definition 4.2).*

In the stable case, (ii) is satisfied if and only if $\mathcal{D} = \mathcal{N}\mathcal{M}^{-1}$ and $\mathcal{M}^{-1}$ is a canonical right spectral factor of $\mathcal{D}^* J \mathcal{D}$, i.e., $\mathcal{M}$ is invertible in $SIC(U; U)$, and $\mathcal{D}^* J \mathcal{D} = (\mathcal{M}^{-1})^* \mathcal{M}^{-1}$.

4 THE TWO MAIN THEOREMS

The following two theorems are our main results. These theorems are proved in Staffans (1995a, 1995b, 1995c) (in a more general form).

Theorem 7 *Let* $J \in \mathcal{L}(Y)$ *be self-adjoint, and* $\Psi = \begin{bmatrix} \mathcal{A} & \mathcal{B} \\ \mathcal{C} & \mathcal{D} \end{bmatrix}$ *be a stable* J*-coercive abstract linear system on* (U, H, Y). *Let* y_{opt} *be the optimal observation and* u_{opt} *the optimal control in the nonstandard quadratic cost minimization problem described in Definition 2.*

(i) If $(\mathcal{N}, \mathcal{M})$ *is an* J*-inner right coprime factorization of* $\mathcal{D}$, *and if* E *is an arbitrary invertible operator in* $\mathcal{L}(U)$, *then*

$$\begin{pmatrix} \mathcal{K} & \mathcal{F} \end{pmatrix} = \begin{pmatrix} -\pi_+ E^{-1} \mathcal{N}^* J \mathcal{C} & (I - E^{-1}\mathcal{M}^{-1}) \end{pmatrix}$$

is an admissible stable state feedback pair for Ψ *(Staffans, 1995a, Definition 4.3), and*

$$\begin{pmatrix} y_{\text{opt}} \\ u_{\text{opt}} \end{pmatrix} = \begin{pmatrix} \mathcal{C}_\diamond \\ \mathcal{K}_\diamond \end{pmatrix} x_0 = \begin{pmatrix} \mathcal{C} + \mathcal{N}E\mathcal{K} \\ \mathcal{M}E\mathcal{K} \end{pmatrix} x_0 = \begin{pmatrix} \mathcal{C} \\ 0 \end{pmatrix} x_0 - \begin{pmatrix} \mathcal{N} \\ \mathcal{M} \end{pmatrix} \pi_+ \mathcal{N}^* J \mathcal{C} x_0$$

is equal to the observation of the closed loop system

$$\Psi_\diamond = \begin{bmatrix} \mathcal{A}_\diamond & \mathcal{B}_\diamond \\ \begin{pmatrix} \mathcal{C}_\diamond \\ \mathcal{K}_\diamond \end{pmatrix} & \begin{pmatrix} \mathcal{D}_\diamond \\ \mathcal{F}_\diamond \end{pmatrix} \end{bmatrix} = \begin{bmatrix} \mathcal{A}(\cdot) + \mathcal{B}_\mathcal{T}(\cdot)\mathcal{M}E\mathcal{K} & \mathcal{B}\mathcal{M}E \\ \begin{pmatrix} \mathcal{C} + \mathcal{N}E\mathcal{K} \\ \mathcal{M}E\mathcal{K} \end{pmatrix} & \begin{pmatrix} \mathcal{N}E \\ \mathcal{M}E - I \end{pmatrix} \end{bmatrix}$$

with this feedback pair, initial value x_0*, initial time zero, and zero control* u*. The optimal observability map satisfies* $\pi_+ \mathcal{N}^* J \mathcal{C}_\diamond = 0$*, and the Riccati operator* Π *of* Ψ *can be written in the following alternative forms:*

$$\Pi = \mathcal{C}^* J \mathcal{C} - \mathcal{K}^* E^* E \mathcal{K} = \mathcal{C}^* (J - J\mathcal{N}\pi_+ \mathcal{N}^* J)\mathcal{C} = \mathcal{C}^* J \mathcal{C}_\diamond = \mathcal{C}_\diamond^* J \mathcal{C}_\diamond.$$

(ii) Conversely, if the solution of the quadratic cost minimization problem is of state feedback type in the sense that $\begin{pmatrix} y_{\text{opt}} \\ u_{\text{opt}} \end{pmatrix}$ *is equal to the observation of some state feedback perturbation* $\Psi_\diamond$ *of* Ψ *with initial value* x_0*, initial time* 0*, zero control* u*, and some admissible stable state feedback pair* $(\mathcal{K} \;\; \mathcal{F})$*, then there exists an invertible operator* $E \in \mathcal{L}(U)$ *such that* $(\mathcal{N}, \mathcal{M})$ *is a* J*-inner right coprime factorization of* $\mathcal{D}$*, where* $\mathcal{M} = (I - \mathcal{F})^{-1} E^{-1}$ *and* $\mathcal{N} = \mathcal{D}\mathcal{M}$*. Moreover,* $\mathcal{K}$ *is given by* $\mathcal{K} = -\pi_+ E^{-1} \mathcal{N}^* J \mathcal{C}$*. Thus, part (i) captures all possible state feedback solutions to the nonstandard quadratic cost minimization problem.*

Theorem 8 *Make the same assumptions and introduce the same notations as in Theorem 7. Extend the system* Ψ *into* $\Psi = \begin{bmatrix} \mathcal{A} & \mathcal{B} \\ \begin{pmatrix} \mathcal{C} \\ \mathcal{K} \end{pmatrix} & \begin{pmatrix} \mathcal{D} \\ \mathcal{F} \end{pmatrix} \end{bmatrix}$ *by adding the optimal state feedback pair* $(\mathcal{K}, \mathcal{F})$*. Let* $\Psi_\diamond$ *be the optimally controlled feedback solution to the quadratic cost minimization problem given by Theorem 7. Denote the generating operators of* Ψ *and* $\Psi_\diamond$ *by the same letters as the corresponding operators (Salamon, 1989, Theorem 3.1).*

(i) The Riccati operator Π *of* Ψ *satisfies the Lyapunov equation*

$$\langle Ax_0, \Pi x_1\rangle_H + \langle x_0, \Pi A x_1\rangle_H = -\langle Cx_0, JCx_1\rangle_Y + \langle EKx_0, EKx_1\rangle_U, \quad x_0, x_1 \in \operatorname{dom}(A).$$

(ii) In addition, suppose that the extended system Ψ *is regular together with its adjoint (Weiss, 1994a, Definition 2.2). Denote the feed-through operators with the same letters as their corresponding input/output maps, and let an overline denote the strong Abel extension of an observation map (see Weiss (1994a), for example,* $\overline{C}x = \lim_{\lambda\to\infty} C_\lambda x$*, where* $C_\lambda = \lambda C(\lambda I - A)^{-1}$ *is the Yoshida approximation of* C*). Then*

$$\begin{bmatrix} A_\diamond & B_\diamond \\ \begin{pmatrix} C_\diamond \\ K_\diamond \end{pmatrix} & \begin{pmatrix} D_\diamond \\ F_\diamond \end{pmatrix} \end{bmatrix} = \begin{bmatrix} A + BME\overline{K} & BME \\ \begin{pmatrix} \overline{C} + NE\overline{K} \\ ME\overline{K} \end{pmatrix} & \begin{pmatrix} NE \\ ME - I \end{pmatrix} \end{bmatrix},$$

where the equation for $B_\diamond$ *should be interpreted as* $\overline{B}_\diamond^* = E^* M^* \overline{B}^*$*.*

(iii) In the regular case, the operator $\overline{B}^*\Pi$ *satisfies the equation*

$$EKx = -M^* \left(\overline{B}^*\Pi + D^* JC\right) x, \quad x \in \operatorname{dom}(A).$$

(iv) *In the regular case we can combine (i) and (iii) to get the algebraic Riccati equation*

$$\begin{aligned}&\langle Ax_0, \Pi x_1\rangle_H + \langle x_0, \Pi A x_1\rangle_H \\ &\quad = -\langle Cx_0, JCx_1\rangle_Y + \left\langle M^*\left(\overline{B}^*\Pi + D^*JC\right)x_0, M^*\left(\overline{B}^*\Pi + D^*JC\right)x_1\right\rangle_U, \\ &\qquad x_0, x_1 \in \mathrm{dom}(A).\end{aligned}$$

Part (ii) of this theorem is due to Weiss (1994b).

5 APPLICATIONS: THE BOUNDED AND POSITIVE REAL LEMMAS

The preceding theory gives us the first available versions of the strict bounded and positive (real) lemmas for general abstract linear systems. To see this we have to consider a slightly different setting where instead of minimizing a quadratic functional of the observation y alone we minimize a quadratic functional of both the control u and the observation y. We can reduce this setting to the setting considered above by simply adding a copy of the control to the observation, i.e., we replace $\Psi = \begin{bmatrix}\mathcal{A} & \mathcal{B}\\ \mathcal{C} & \mathcal{D}\end{bmatrix}$ by the extended system $\Psi_{\mathrm{ext}} = \begin{bmatrix}\mathcal{A} & \mathcal{B}\\ \begin{pmatrix}\mathcal{C}\\ 0\end{pmatrix} & \begin{pmatrix}\mathcal{D}\\ I\end{pmatrix}\end{bmatrix}$, and apply Theorems 7 and 8 to Ψ_{ext} instead of to Ψ.

By applying Theorems 7 and 8 to Ψ_{ext} we can obtain one version of the *strict bounded real lemma* as follows: We choose J to be $J = \begin{pmatrix}-I & 0\\ 0 & \gamma^2 I\end{pmatrix}$, where γ is a positive constant. Then

$$W(x_0, u) = \gamma^2 \|u\|^2_{L^2(\mathbf{R}^+;U)} - \|y\|^2_{L^2(\mathbf{R}^+;Y)},$$

and Ψ_{ext} is J-coercive if and only if the operator norm of $\mathcal{D}$ is less than γ, in which case Theorem 7 applies. The central formulas in that theorem become

$$\begin{aligned}\mathcal{D} &= \mathcal{N}\mathcal{M}^{-1}, \qquad \gamma^2\mathcal{M}^*\mathcal{M} - \mathcal{N}^*\mathcal{N} = I, \qquad E\mathcal{K} = \pi_+\mathcal{N}^*\mathcal{C}, \\ \begin{pmatrix}\mathcal{C}_\diamond\\ \mathcal{K}_\diamond\end{pmatrix} &= \begin{pmatrix}\mathcal{C}\\ 0\end{pmatrix} + \begin{pmatrix}\mathcal{N}\\ \mathcal{M}\end{pmatrix} E\mathcal{K} = \begin{pmatrix}\mathcal{C}\\ 0\end{pmatrix} + \begin{pmatrix}\mathcal{N}\\ \mathcal{M}\end{pmatrix}\pi_+\mathcal{N}^*\mathcal{C}, \\ \Pi &= \gamma^2\mathcal{K}_\diamond^*\mathcal{K}_\diamond - \mathcal{C}_\diamond^*\mathcal{C}_\diamond = -\mathcal{C}^*\left(I + \mathcal{N}\pi_+\mathcal{N}^*\right)\mathcal{C}.\end{aligned}$$

The connecting and Lyapunov equations in Theorem 8 become (for x_0 and $x_1 \in \mathrm{dom}(A)$)

$$\begin{aligned}EKx_0 &= -M^*\left(\overline{B}^*\Pi - D^*C\right)x_0, \\ \langle Ax_0, \Pi x_1\rangle_H + \langle x_0, \Pi A x_1\rangle_H &= \langle Cx_0, Cx_1\rangle_Y + \langle EKx_0, EKx_1\rangle_U.\end{aligned}$$

Observe that the parameter γ enters these equations only through the operator M, which in the classical case satisfies $(MM^*)^{-1} = \begin{pmatrix}D^* & I\end{pmatrix} J \begin{pmatrix}D\\ I\end{pmatrix} = \gamma^2 I - D^*D$. In our setting Π is negative definite; to get the standard setting where Π is positive (Green and Limebeer, 1995, Theorem 3.7.1), we must replace J by $-J$ and maximize instead of minimize.

The *strictly positive (real) lemma* is a statement about a stable system $\Psi = \begin{bmatrix}\mathcal{A} & \mathcal{B}\\ \mathcal{C} & \mathcal{D}\end{bmatrix}$ on (U, H, U) (i.e., the output space of this system is equal to its input space). The input/output map $\mathcal{D}$ of Ψ is called strictly positive iff there exists some $\epsilon > 0$ such that $\mathcal{D} + \mathcal{D}^* \geq \epsilon I$, i.e.,

$$\int_{\mathbf{R}} \langle u(s), (\mathcal{D}u)(s) \rangle \, ds + \int_{\mathbf{R}} \langle (\mathcal{D}u)(s), u(s) \rangle \, ds \geq \epsilon \, \|u\|^2_{L^2(\mathbf{R};U)}, \qquad u \in L^2(\mathbf{R}; U).$$

Define $J = \begin{pmatrix} 0 & I \\ I & 0 \end{pmatrix}$. Then $Q(x_0, u) = \int_{\mathbf{R}^+} \langle u(s), y(s) \rangle \, ds + \int_{\mathbf{R}^+} \langle y(s), u(s) \rangle \, ds$, and $\Psi_{\mathbf{ext}}$ is is J-coercive with respect to this operator iff $\mathcal{D}$ is strictly positive. The formulas in Theorem 7 become in this case

$$\begin{aligned}
\mathcal{D} &= \mathcal{N}\mathcal{M}^{-1}, \qquad \mathcal{M}^*\mathcal{N} + \mathcal{N}^*\mathcal{M} = I, \qquad E\mathcal{K} = -\pi_+\mathcal{M}^*\mathcal{C}, \\
\begin{pmatrix} \mathcal{C}_\diamond \\ \mathcal{K}_\diamond \end{pmatrix} &= \begin{pmatrix} \mathcal{C} \\ 0 \end{pmatrix} + \begin{pmatrix} \mathcal{N} \\ \mathcal{M} \end{pmatrix} E\mathcal{K} = \begin{pmatrix} \mathcal{C} \\ 0 \end{pmatrix} - \begin{pmatrix} \mathcal{N} \\ \mathcal{M} \end{pmatrix} \pi_+\mathcal{M}^*\mathcal{C}, \\
\Pi &= \mathcal{K}_\diamond^*\mathcal{C}_\diamond + \mathcal{C}_\diamond^*\mathcal{K}_\diamond = -(E\mathcal{K})^*(E\mathcal{K}) = -\mathcal{C}^*\mathcal{M}\pi_+\mathcal{M}^*\mathcal{C}.
\end{aligned}$$

The connecting and Lyapunov equations in Theorem 8 become (for x_0 and $x_1 \in \mathrm{dom}(A)$)

$$\begin{aligned}
EKx_0 &= -M^*\left(\overline{B}^*\Pi + C\right)x_0, \\
\langle Ax_0, \Pi x_1 \rangle_H + \langle x_0, \Pi A x_1 \rangle_H &= \langle EKx_0, EKx_1 \rangle_U .
\end{aligned}$$

In the classical case M satisfies $(MM^*)^{-1} = D + D^*$. Again Π is negative; to get a positive Π we should change the sign of J and maximize instead of minimize (Green and Limebeer, 1995, Problem 3.25).

In the Pritchard-Salamon case, the applications to the bounded and positive (real) lemmas that we have presented above are found in (Weiss, 1994c, Remark 4.34).

REFERENCES

Callier, F. M. and Winkin, J. (1992). LQ-optimal control of infinite-dimensional systems by spectral factorization. *Automatica*, **28**, 757–770.

Curtain, R. F. and Green, M. (1994). Analytic system problems and J-lossless coprime factorization for infinite-dimensional linear systems. Preprint.

Green, M. and Limebeer, D. L. L. (1995). *Linear Robust Control*. Prentice Hall, Englewood Cliffs, New Jersey.

Lasiecka, I. and Triggiani, R. (1991). *Differential and Algebraic Riccati Equations with Applications to Boundary/Point Control Problems: Continuous Theory and Approximation Theory*, volume 164 of *Lecture Notes in Control and Information Sciences*. Springer-Verlag, Berlin.

Salamon, D. (1987). Infinite dimensional linear systems with unbounded control and observation: a functional analytic approach. *Transactions of the American Mathematical Society*, **300**, 383–431.

Salamon, D. (1989). Realization theory in Hilbert space. *Mathematical Systems Theory*, **21**, 147–164.

Staffans, O. J. (1994a). Quadratic optimal control of stable systems through spectral factorization. To appear in *Mathematics of Control, Signals, and Systems*.

Staffans, O. J. (1994b). Quadratic optimal control of unstable systems through coprime factorization. Submitted.

Staffans, O. J. (1995a). Quadratic optimal control of stable abstract linear systems. Submitted.

Staffans, O. J. (1995b). Coprime factorizations and quadratic optimal control of abstract linear systems. Submitted.

Staffans, O. J. (1995c). The nonstandard quadratic cost minimization problem for abstract linear systems. Submitted.

Weiss, G. (1994a). Transfer functions of regular linear systems. Part I: Characterizations of regularity. *Transactions of American Mathematical Society*, **342**, 827–854.

Weiss, G. (1994b). Regular linear systems with feedback. *Mathematics of Control, Signals, and Systems*, **7**, 23–57.

Weiss, M. (1994c). Riccati equations in Hilbert space: A Popov function approach. Doctoral dissertation, Rijksuniversiteit Gronigen.

17

Linear quadratic optimal control for abstract linear systems

H.J. Zwart
University of Twente, Faculty of Applied Mathematics
PO Box 217, 7500 AE Enschede, The Netherlands

Abstract

In this paper we study the linear quadratic optimal control problem on an infinite horizon for abstract linear systems. We show that under a natural condition this optimal control problem is uniquely solvable. The optimal control is given by a, possible unbounded, state feedback. Furthermore, the optimal cost is a quadratic function of the initial state.

Keywords

linear quadratic, optimal control, abstract linear system

1 INTRODUCTION

Linear quadratic optimal control for infinite-dimensional systems has been studied for a long time. For a historic overview we refer the reader to the notes and references of chapter 6 in Curtain and Zwart (1995). We only mention the papers of Flandoli, Lasiecka and Triggiani (1988), Staffans (1995a, 1995b) and Weiss & Weiss (1995). In Flandoli, Lasiecka and Triggiani (1988) the problem of linear quadratic optimal control is studied for a system with an unbounded input operator but with a bounded output operator. They only assume that the system is optimizable. In Staffans (1995a, 1995b) and Weiss & Weiss (1995) the problem is studied for systems with unbounded input and output operators. However, they assume that the system is stable or stabilizable. Here we allow also for unbounded input and output operators, but we only assume that the system is optimizable.

Following Weiss (1989c) we introduce our class of abstract linear systems. Let U, Y and Z be Hilbert spaces which denote the input space, output space and state space, respectively. By $L_2([0,\infty); X)$ we denote the set of strongly measurable, square integrable functions on $[0,\infty)$ which take values in the Hilbert space X. On our signals we need two operations, namely the truncation and the concatenation. The truncation is simple the projection on a finite time-interval;

$$(P_\tau f)(t) = \begin{cases} f(t) & \text{for} \quad t \in [0,\tau) \\ 0 & \text{for} \quad t \geq \tau, \end{cases} \tag{1}$$

Note that this has meaning for every $f \in L_2([0,\infty);X)$. For any $u,v \in L_2([0,\infty);X)$ we define the τ-*concatenation* of u and v by

$$(u \underset{\tau}{\diamond} v)(t) = \begin{cases} u(t) & \text{for } t \in [0,\tau] \\ v(t-\tau) & \text{for } t > \tau. \end{cases} \tag{2}$$

It is easy to see that for any f and $\tau \geq 0$

$$P_\tau f = f \underset{\tau}{\diamond} 0. \tag{3}$$

Definition 1 *$\Sigma_a(T,\mathcal{B},\mathcal{C},\mathcal{F})$ is an* abstract linear system *on the Hilbert spaces, U, Z, and Y if the following hold:*

1. *$T(t)$ is a C_0-semigroup on Z;*
2. *$\mathcal{B} = (\mathcal{B}^\tau)_{\tau\geq 0}$ is a family of bounded linear operators from $L_2([0,\infty);U)$ to Z such that $\mathcal{B}^0 = 0$ and*

$$\mathcal{B}^{\tau+t}(u \underset{\tau}{\diamond} v) = T(t)\mathcal{B}^\tau u + \mathcal{B}^t v \tag{4}$$

   for any $u,v \in L_2([0,\infty);U)$ and any $\tau,t \geq 0$;
3. *$\mathcal{C} = (\mathcal{C}^\tau)_{\tau\geq 0}$ is a family of bounded linear operators from Z to $L_2([0,\infty);Y)$ such that $\mathcal{C}^0 = 0$ and*

$$\mathcal{C}^{\tau+t}z = \mathcal{C}^\tau z \underset{\tau}{\diamond} \mathcal{C}^t T(\tau)z \tag{5}$$

   for any $z \in Z$ and any $t,\tau \geq 0$;
4. *$\mathcal{F} = (\mathcal{F}^\tau)_{\tau\geq 0}$ is a family of bounded linear operators from $L_2([0,\infty);U)$ to $L_2([0,\infty);Y)$ such that $\mathcal{F}^0 = 0$ and*

$$\mathcal{F}^{\tau+t}(u \underset{\tau}{\diamond} v) = \mathcal{F}^\tau u \underset{\tau}{\diamond}(\mathcal{C}^t\mathcal{B}^\tau u + \mathcal{F}^t v) \tag{6}$$

   for any $u,v \in L_2([0,\infty);U)$ and any $t,\tau \geq 0$.

Following earlier terminology, the operators $\mathcal{B}^\tau$, $\mathcal{C}^\tau$ and $\mathcal{F}^\tau$ are called the controllability, observability, and input-output map, respectively, see Weiss (1989c). We remark that it follows from the assumptions that the operators are causal i.e.,

$$\mathcal{B}^\tau P_\tau = \mathcal{B}^\tau, \quad P_\tau\mathcal{C}^\tau = \mathcal{C}^\tau, \quad \text{and } \mathcal{F}^\tau P_\tau = P_\tau\mathcal{F}^\tau = \mathcal{F}^\tau \quad \text{for all } \tau \geq 0. \tag{7}$$

For completeness we shall prove these equalities.

$$\mathcal{B}^\tau P_\tau u = \mathcal{B}^{0+\tau}(u \underset{\tau}{\diamond} 0) = \mathcal{B}^\tau u,$$

where we have used equation (3) and (4).

$$\mathcal{C}^\tau z_0 = \mathcal{C}^{0+\tau} z_0 = \mathcal{C}^\tau z \underset{\tau}{\diamond} 0 = P_\tau \mathcal{C}^\tau z_0,$$

where we have used equation (5) and (3).

$$\mathcal{F}^\tau P_\tau u = \mathcal{F}^{0+\tau}(u \underset{\tau}{\diamond} 0) = \mathcal{F}^\tau u \underset{\tau}{\diamond} 0 = P_\tau \mathcal{F}^\tau z_0,$$

where we have used equation (6) and (3). Furthermore, let v be such that $u = u \underset{\tau}{\diamond} v$, then we have that

$$\mathcal{F}^\tau u = \mathcal{F}^{0+\tau}(u \underset{\tau}{\diamond} v) = \mathcal{F}^\tau u \underset{\tau}{\diamond} 0 = P_\tau \mathcal{F}^\tau z_0,$$

where we again have used equation (6) and (3).

We say that the triple (u, y, z) is a solution of the abstract linear system $\Sigma_a(T, \mathcal{B}, \mathcal{C}, \mathcal{F})$ with initial condition z_0 if for all $t \geq 0$

$$\begin{cases} z(t) &= T(t)z_0 + \mathcal{B}^t P_t u(\cdot) \\ P_t y &= \mathcal{C}^t z_0 + \mathcal{F}^t P_t u(\cdot). \end{cases} \tag{8}$$

Since $T, \mathcal{B}, \mathcal{C}$ and $\mathcal{F}$ are linear, it is clear that this system is linear. Furthermore, it follows from the properties (4)–(6) that the system is time-invariant. We formalize this to a lemma.

Lemma 2 *The abstract linear system $\Sigma_a(T\mathcal{B}, \mathcal{C}, \mathcal{F})$ is linear and time-invariant. Linearity means that if (u_1, y_1, z_1) and (u_2, y_2, z_2) satisfy (8) with initial conditions z_{10} and z_{20}, respectively, then $(u_1 + u_2, y_1 + y_2, z_1 + z_2)$ satisfies (8) with initial condition $z_{10} + z_{20}$. Time-invariance means that if (u, y, z) satisfies (8) with initial condition z_0, then for every $\tau \geq 0$, $(\sigma_\tau u, \sigma_\tau y, \sigma_\tau z)$ satisfies (8) with initial condition $z(\tau)$, where $(\sigma_\tau f)(t) = f(t + \tau)$.*

Proof. The linearity is obvious, so we shall only prove the time-invariance. Since the result for the output is the hardest, we shall only prove that one. First note that for any f

$$f = f \underset{\tau}{\diamond} \sigma_\tau f. \tag{9}$$

So

$$\begin{aligned} P_{t+\tau} y &= \mathcal{C}^{t+\tau} z_0 + \mathcal{F}^{t+\tau} u = \mathcal{C}^{t+\tau} z_0 + \mathcal{F}^{t+\tau} \left[u \underset{\tau}{\diamond} \sigma_\tau u \right] \quad \text{by (9)} \\ &= \mathcal{C}^\tau z_0 \underset{\tau}{\diamond} \mathcal{C}^t T(\tau) z_0 + \mathcal{F}^\tau u \underset{\tau}{\diamond} \left[\mathcal{C}^t \mathcal{B}^\tau u + \mathcal{F}^t(\sigma_\tau u) \right] \quad \text{by (5) and (6)} \\ &= \left[\mathcal{C}^\tau z_0 + \mathcal{F}^\tau u \right] \underset{\tau}{\diamond} \left[\mathcal{C}^t T(\tau) z_0 + \mathcal{C}^t \mathcal{B}^\tau u + \mathcal{F}^t(\sigma_\tau u) \right] \\ &= P_\tau y \underset{\tau}{\diamond} \left[\mathcal{C}^t z(\tau) + \mathcal{F}^t(\sigma_\tau u) \right]. \end{aligned}$$

Since for any f, $P_{t+\tau} f = P_\tau f \underset{\tau}{\diamond} P_t(\sigma_\tau f)$, we see that

$$P_t(\sigma_\tau y) = \mathcal{C}^t z(\tau) + \mathcal{F}^t(\sigma_\tau u).$$

and so we have shown the time-invariance. □

With the solutions of (8) we associate the cost functional

$$J(z_0, u) = \int_0^\infty \|y(\tau)\|^2 + \|u(\tau)\|^2 d\tau, \tag{10}$$

where u, y satisfy (8).

Note that we have made the cost functional only dependent on u and z_0, since if these are given, then (8) gives $z(\cdot)$ and $y(\cdot)$.

We make the following standard assumption.

Assumption 3 *For all $z_0 \in Z$, there exists an input $u(\cdot)$ such that*

$$J(z_0, u) < \infty \tag{11}$$

This we call optimizable.

Note that the above condition is also known as the Finite Cost Condition.

2 EXISTENCE AND UNIQUENESS

Our problem is to minimize the cost functional $J(z_0, u)$. We shall show that this is possible for all $z_0 \in Z$, and that the optimal control is unique. We introduce the space $\mathcal{X} := L_2([0,\infty); U) \oplus L_2([0,\infty); Y)$ with inner product

$$\left\langle \begin{pmatrix} u_1 \\ y_1 \end{pmatrix}, \begin{pmatrix} u_2 \\ y_2 \end{pmatrix} \right\rangle_{\mathcal{X}} := \int_0^\infty \langle u_1(\tau), u_2(\tau)\rangle_U + \langle y_1(\tau), y_2(\tau)\rangle_Y d\tau. \tag{12}$$

Definition 4 *By $V(z_0)$ we denote the set of all in- and output trajectories which are related by (8) and are square integrable i.e.,*

$$V(z_0) = \left\{ \begin{pmatrix} u \\ y \end{pmatrix} \in \mathcal{X} \text{ such that (8) holds} \right\}$$

In the next lemma we show that the set $V(z_0)$ is a closed, affine linear subset of $\mathcal{X}$.

Lemma 5 *$V(z_0)$ has the following properties:*

1. *For all $z_0 \in Z, V(z_0) \neq \emptyset$;*
2. *$V(z_0) = V(0) + \begin{pmatrix} u \\ y \end{pmatrix}$ for some $\begin{pmatrix} u \\ y \end{pmatrix} \in V(z_0)$*
3. *$V(0)$ is a closed linear subspace of $\mathcal{X}$.*

Proof. 1. This follows from our Assumption 3.

2. Let $\begin{pmatrix} u \\ y \end{pmatrix} \in V(z_0)$, and $\begin{pmatrix} u_0 \\ y_0 \end{pmatrix} \in V(0)$. To $\begin{pmatrix} u \\ y \end{pmatrix}$ there corresponds the state trajectory $z(\cdot)$, and to $\begin{pmatrix} u_0 \\ y_0 \end{pmatrix}$ the state trajectory $z_0(\cdot)$. Now it is easy to see that

$$z(t) + z_0(t) = T(t)z_0 + \mathcal{B}^t P_t u + T(t)0 + \mathcal{B}^t P_t u_0 = T(t)z_0 + \mathcal{B}^t P_t[u + u_0] \quad \text{and}$$

$$P_t y + P_t y_0 = \mathcal{C}^t z_0 + \mathcal{F}^t P_t u + \mathcal{F}^t P_t u_0 = \mathcal{C}^t z_0 + \mathcal{F}^t P_t[u + u_0].$$

Thus $\begin{pmatrix} u + u_0 \\ y + y_0 \end{pmatrix} \in \mathcal{V}(z_0)$, and so $\mathcal{V}(0) + \begin{pmatrix} u \\ y \end{pmatrix} \subset \mathcal{V}(z_0)$. The other inclusion is similar.
3. The linearity of $\mathcal{V}(0)$ is similar to the proof in 2. Thus we must prove the closedness. Let $\begin{pmatrix} u_n \\ y_n \end{pmatrix} \xrightarrow{\mathcal{X}} \begin{pmatrix} u_\infty \\ y_\infty \end{pmatrix}$ for $n \to \infty$. Since $y_n(\cdot) = \mathcal{F}^t u_n(\cdot)$ and since $\mathcal{F}^t$ is bounded, we have that $y_\infty(\cdot) = \mathcal{F}^t u_\infty(\cdot)$ for all $t \in [0, \infty)$. So $\begin{pmatrix} u_\infty \\ y_\infty \end{pmatrix} \in \mathcal{V}(0)$. □

Now we can prove our first main theorem.

Theorem 6 *Consider system (8) with Assumption 3. For every $z_0 \in Z$ there exists a unique input $u^{\min}(\cdot; z_0) \in L_2([0, \infty); U)$ such that*

$$J\left(z_0, u^{\min}\right) \leq J(z_0, u).$$

Let $y^{\min}(\cdot; z_0)$, $z^{\min}(\cdot; z_0)$ denote the corresponding output and state trajectory, respectively. The map $Q : Z \to \mathcal{X}$, given by $Q(z_0) = \begin{pmatrix} u^{\min}(\cdot; z_0) \\ y^{\min}(\cdot; z_0) \end{pmatrix}$ is linear and bounded.

Furthermore, $J(z_0, u^{\min}(\cdot; z_0)) = \langle z_0, Q^* Q z_0 \rangle$.

Proof. We have that $J(z_0, u) = \left\| \begin{pmatrix} u \\ y \end{pmatrix} \right\|_{\mathcal{X}}^2$, where u, y satisfies (8). Hence minimizing J is like minimizing the norm of an affine linear subspace in a Hilbert space. Thus, by the standard orthogonal projection lemma

$$\min_u J(z_0, u) = \min_{\begin{pmatrix} u \\ y \end{pmatrix} \in V(z_0)} \left\| \begin{pmatrix} u \\ y \end{pmatrix} \right\|^2 = \left\| P_{V(0)^\perp} \begin{pmatrix} u \\ y \end{pmatrix} \right\|^2,$$

where $P_{V(0)^\perp}$ is the orthogonal projection on $\mathcal{V}(0)^\perp$. The element $P_{V(0)^\perp} \begin{pmatrix} u \\ y \end{pmatrix}$ is unique in $\mathcal{V}(z_0)$. We call this element $\begin{pmatrix} u^{\min}(\cdot; z_0) \\ y^{\min}(\cdot; z_0) \end{pmatrix}$. Define furthermore $z^{\min}(t; z_0) = T(t)z_0 + \mathcal{B}^t u^{\min}$, i.e., the corresponding state trajectory.

We define Q via $Qz_0 = \begin{pmatrix} u^{\min}(\cdot; z_0) \\ y^{\min}(\cdot; z_0) \end{pmatrix}$. Since $P_{V(0)^\perp}$ is linear, it is clear that this map is linear. We shall show that it is closed.

Let $Qz_n = \begin{pmatrix} u_n \\ y_n \end{pmatrix} \to \begin{pmatrix} u_\infty \\ y_\infty \end{pmatrix}$ and $z_n \to z_\infty$ with $\begin{pmatrix} u_n \\ y_n \end{pmatrix} \in \mathcal{V}(z_n)$.

Similar as in the proof of Lemma 5, we can show that $\begin{pmatrix} u_\infty \\ y_\infty \end{pmatrix} \in \mathcal{V}(z_\infty)$.

The minimal vector in $\mathcal{V}(z_\infty)$ is uniquely characterized by the fact that it is the vector orthogonal to $\mathcal{V}(0)$.

Now $\begin{pmatrix} u_n \\ y_n \end{pmatrix}$ are minimal, so they satisfy that $\begin{pmatrix} u_n \\ y_n \end{pmatrix} \perp \mathcal{V}(0) \cdot \begin{pmatrix} u_n \\ y_n \end{pmatrix} \to \begin{pmatrix} u_\infty \\ y_\infty \end{pmatrix}$ for $n \to \infty$, and so $\begin{pmatrix} u_\infty \\ y_\infty \end{pmatrix} \perp \mathcal{V}(0) \Rightarrow u_\infty(\cdot) = u^{\min}(\cdot, z_\infty)$. Hence Q is a closed operator with domain the whole of Z. This implies that $Q \in \mathcal{L}(Z, \mathcal{X})$.

By the definition of Q we have

$$\min J(z_0, u) = \left\| \begin{pmatrix} u^{\min}(\cdot; z_0) \\ y^{\min}(\cdot; z_0) \end{pmatrix} \right\|_{\mathcal{X}}^2 = \langle Qz_0, Qz_0 \rangle_{\mathcal{X}} = \langle z_0, Q^*Qz_0 \rangle_Z.$$

Thus we have proved the assertions. □

Note that in the standard case Q^*Q is the solution to the algebraic Riccati equation.

3 ANALYSIS OF THE SOLUTION

In this section, we analyze the properties of the optimal solution. We start by deriving the fundamental principle of optimality. Let $t > 0$

$$\begin{aligned} &\int_0^\infty \|y^{\min}(\tau; z_0)\|^2 + \|u^{\min}(\tau; z_0)\|^2 d\tau \\ &= \int_0^t \|y^{\min}(\tau; z_0)\|^2 + \|u^{\min}(\tau; z_0)\|^2 d\tau + \int_t^\infty \|y^{\min}(\tau; z_0)\|^2 + \|u^{\min}(\tau; z_0)\|^2 d\tau \\ &= \int_0^t \|y^{\min}(\tau; z_0)\|^2 + \|u^{\min}(\tau; z_0)\|^2 d\tau + \int_0^\infty \|y^{\min}(\tau + t; z_0)\|^2 + \|u^{\min}(\tau + t; z_0)\|^2 d\tau \\ &= \int_0^t \|y^{\min}(\tau; z_0)\|^2 + \|u^{\min}(\tau; z_0)\|^2 d\tau + J(z^{\min}(t; z_0); u^{\min}(\cdot + t; z_0)), \end{aligned} \tag{13}$$

since the system is time-invariant. Suppose there exists a v such that

$$J(z^{\min}(t; z_0); v) < J(z^{\min}(t; z_0); u^{\min}(\cdot + t; z_0)),$$

then the input

$$\tilde{u}(\tau) = \begin{cases} u^{\min}(\tau; z_0) & 0 \le \tau < t \\ v(\tau) & \tau > t \end{cases}$$

would give a lower cost in (13). But this is in contradiction with the optimality of $u^{\min}(\cdot; z_0)$. Hence by the uniqueness of the optimal input

$$u^{\min}(\cdot; z^{\min}(t; z_0)) = u^{\min}(\cdot + t; z_0), \tag{14}$$

where equality is in $L_2([0,\infty),U)$. ¿From this it follows easily that

$$z^{\min}(\tau; z^{\min}(t;z_0)) = z^{\min}(\tau + t; z_0) \quad \text{for all } \tau > 0, \tag{15}$$

$$y^{\min}(\cdot; z^{\min}(t;z_0)) = y^{\min}(\cdot + t; z_0) \qquad \text{in } L_2([0,\infty),Y), \tag{16}$$

and with (13)

$$\langle z_0, Q^*Qz_0\rangle = \int_0^t \|y^{\min}(\tau;z_0)\|^2 + \|u^{\min}(\tau;z_0)\|^2 d\tau + \langle z^{\min}(t;z_0), Q^*Qz^{\min}(t;z_0)\rangle \tag{17}$$

for all $t \geq 0$.

Lemma 7 *The mapping $z_0 \mapsto z^{\min}(\cdot;z_0)$ is given by a C_0-semigroup $T^{\min}(\cdot)$.*

Proof. Define $T^{\min}(t)z_0 := z^{\min}(t;z_0)$. Then it is trivial that

$$T^{\min}(0)z_0 = z_0$$

for all z_0. For the semigroup property we use (15)

$$\begin{aligned} T^{\min}(\tau + t)z_0 &= z^{\min}(\tau + t; z_0) = z^{\min}(\tau; z^{\min}(t;z_0)) \\ &= T^{\min}(\tau)z^{\min}(t;z_0) = T^{\min}(\tau)T^{\min}(t)z_0. \end{aligned}$$

The strong continuity follows from the fact that $z^{\min}(\cdot;z_0)$ is continuous (Weiss 1989a). Namely, it is the state trajectory for the L_2 input $u^{\min}(\cdot;z_0)$. □

Note that if we measure the state, i.e. $y = z$, then it follows easily that the semigroup $T^{\min}(t)$ is exponentially stable, see for instance Curtain and Zwart (1995). Let $A^{\min}$ denote the infinitesimal generator of $T^{\min}(t)$. The next lemma implies that the optimal control is a static state feedback.

Lemma 8 *The mapping $C_F^t : z_0 \mapsto P_t u^{\min}(\cdot;z_0)$ is an admissible output mapping for $T^{\min}(t)$, i.e. they satisfy (5). Furthermore, there exists a feedback $F^{\min} : D(A^{\min}) \to U$ such that for $z_0 \in D(A^{\min})$, $C_F^t z_0 = P_t F^{\min} T^{\min}(\cdot)z_0$.*

Proof. Define $C_F^t z_0 := P_t u^{\min}(\cdot;z_0)$. Then C_F^t is linear and bounded (see Theorem 6).

$$\begin{aligned} C_F^{t+\tau} z_0 &= P_{t+\tau} u^{\min}(\cdot;z_0) \\ &= P_\tau u^{\min}(\cdot;z_0) \underset{\tau}{\diamond} P_t u^{\min}(\cdot + \tau; z_0) \\ &= C_F^\tau z_0 \underset{\tau}{\diamond} P_t u^{\min}(\cdot; z^{\min}(\tau;z_0)) \quad \text{by (14)} \\ &= C_F^\tau z_0 \underset{\tau}{\diamond} C_F^t T^{\min}(\tau) z_0 \quad \text{by Lemma 7.} \end{aligned}$$

The last assertion of the lemma follows directly from Weiss (1989b). □

Using equations (16), one can similar as in the previous lemma show.

Lemma 9 *The mapping $\mathcal{C}_C^t : z_0 \mapsto P_t y^{\min}(\cdot; z_0)$ is an admissible output mapping for $T^{\min}(t)$, i.e. they satisfy (5). Furthermore, there exists an operator $C^{\min} : D(A^{\min}) \to Y$ such that for $z_0 \in D(A^{\min})$, $\mathcal{C}_C^t z_0 = P_t C^{\min} T^{\min}(\cdot) z_0$.*

Note that in general $\mathcal{C}_C^t \neq \mathcal{C}^t$.

If we now multiply equation (17) by $1/t$, and let $t \to 0$, then it is easy to see from the above lemmas that Q^*Q satisfies the following Lyapunov equation.

$$\langle Q^*Qz_0, A^{\min}z_0\rangle + \langle A^{\min}z_0, Q^*Qz_0\rangle = -\langle C^{\min}z_0, C^{\min}z_0\rangle - \langle F^{\min}z_0, F^{\min}z_0\rangle \tag{18}$$

for all $z_0 \in D(A^{\min})$.

Similar as in Definition 4.1.17 of Curtain and Zwart (1995), we define the system $\Sigma_a(T, \mathcal{B}, \mathcal{C}, \mathcal{F})$ to be approximately observable if

$$\bigcap_{t \geq 0} \ker \mathcal{C}^t = \{0\}.$$

Now it is easy to see that $\ker Q^*Q = \ker Q = \{0\}$ if and only if $\Sigma_a(T, \mathcal{B}, \mathcal{C}, \mathcal{F})$ is approximately observable.

REFERENCES

Curtain, R.F. and Zwart H. (1995) *An Introduction to Infinite Dimensional Linear Systems Theory.* Text in Applied Mathematics, Vol. 21, Springer Verlag, New York.

Flandoli, F., Lasiecka, I. and Triggiani, R. (1988) Algebraic Riccati Equations with Nonsmoothing Observation Arising in Hyperbolic and Euler-Bernoulli Boundary Control Problems. *Annali di Matematica pura ed applicata,* (IV), **CLIII**, pp. 307–382.

Staffans O.J. (1995a) Quadratic Optimal Control of Abstract Linear Systems: The Stable Case. *Research Report A344*, Helsinki University of Technology, Institute of Mathematics.

Staffans O.J. (1995b) Coprime Factorizations and Optimal Control of Abstract Linear Systems. *Research Report A348*, Helsinki University of Technology, Institute of Mathematics.

Weiss, G. (1989a) Admissibility of unbounded control operators. *SIAM J. Control & Optim.*, **27**, pp. 527–545.

Weiss, G. (1989b) Admissible observation operators for linear semigroups. *Israel J. Math.*, **65**, pp. 17–43, 1989.

Weiss, G. (1989c) The representation of regular linear systems on Hilbert spaces, in *Control and Estimation of Distributed Parameter Systems* (eds K. Kappel, K. Kunisch and W. Schappacher), 4th International Conference on Control of Distributed Parameter Systems, Vorau, July 10–16, 1988, ISNM 91, Birkhäuser, pp. 401–416.

Weiss, M. and Weiss, G. (1995) Optimal control of stable weakly regular linear systems, in progress.

18

Relation between invariant subspaces of the Hamiltonian and the algebraic Riccati equation

H.J. Zwart and C.R. Kuiper
University of Twente, Faculty of Applied Mathematics
PO Box 217, 7500 AE Enschede, The Netherlands

Abstract

In this paper we study the relationship between solutions of the algebraic Riccati equation and invariant subspaces of the corresponding Hamiltonian. We show that there is a one-to-one relation between solutions of the algebraic Riccati equation and a class of invariant subspaces of the Hamiltonian. This class of subspaces is shown to be invariant under the inverse of the Hamiltonian as well. With this we generalize the results obtained by Potter (1966), Mårtensson (1971) and Willems (1971) to infinite-dimensional systems. As an example we characterize all solutions of the ARE for a class of unitary systems.

Keywords

Algebraic Riccati equation, Hamiltonian operator, invariant subspace

1 INTRODUCTION

For finite-dimensional systems it is standard to calculate solutions of the algebraic Riccati equation (ARE) via the eigenvectors of the corresponding Hamiltonian. It is well known that for a minimal system the stabilizing solution of the ARE is in one-to-one relationship with the stable eigenvectors of the Hamiltonian. The span of all these stable eigenvectors form the stable subspace of the Hamiltonian, i.e., it is an invariant subspace and the restriction of the Hamiltonian to this subspace is stable. For an overview of the finite-dimensional results we refer to Kučera (1991). In this paper we shall investigate the relation between the ARE and the Hamiltonian for infinite-dimensional systems. We study a general ARE, thus not only the one related to optimal control theory. The organization of the paper is as follow. In the next section, we shall see that there is a one-to-one correspondence between solutions of the ARE and certain invariant subspaces of the Hamiltonian. For infinite dimensional systems the Hamiltonian is an unbounded operator, and it is known that for unbounded operators invariant subspaces are hard to characterize. So the invariance result may seem not that useful. However, for our class of invariant subspaces we can show that they are invariant under spectral projections as well. Hence if the Hamiltonian has compact resolvent, then they are spanned by eigenvectors. With this we can in the last section generalize a famous result of Willems (1971) to infinite dimensional systems.

2 GENERAL RELATIONS

In this section, we shall investigate the relation between the algebraic Riccati equation (ARE) and invariant subspaces of the Hamiltonian. We begin by introducing some notation and definitions.

Let Z be a Hilbert space. By $\mathcal{L}(Z)$ we denote the set of bounded, linear operators on Z. If Q is an linear operator on Z, then Q^* denotes its adjoint. A is a closed operator on Z with domain $D(A) \subset Z$. Furthermore, by Q_1 and Q_2 we denote two bounded, self-adjoint operators on Z. With this notation we can define the ARE and the Hamiltonian operator.

We say that $X \in \mathcal{L}(Z)$ the Algebraic Riccati equation, ARE, if

$$A^*X + XA - XQ_1X + Q_2 = 0 \qquad \text{on } D(A). \tag{1}$$

For the same A, Q_1 and Q_2, the corresponding Hamiltonian operator is given by

$$\mathcal{H} := \begin{pmatrix} A & -Q_1 \\ -Q_2 & -A^* \end{pmatrix} \text{ on } D(\mathcal{H}) = D(A) \oplus D(A^*)$$

In the next lemma, we shall present one of the important relation between solutions of the ARE and invariant subspaces of the Hamiltonian.

Lemma 1 *The following relation holds between the Hamiltonian operator and the ARE. $X \in \mathcal{L}(Z)$ is a solution of the ARE if and only if ran$\begin{pmatrix} I \\ X \end{pmatrix}$ is an invariant subspace of $\mathcal{H}$ and $XD(A) \subset D(A^*)$.*

Proof. Let X be a solution of the ARE, then from (1) it follows easily that $XD(A) \subset D(A^*)$. Hence

$$\text{ran}\begin{pmatrix} I \\ X \end{pmatrix} \cap D(\mathcal{H}) = \left\{ \begin{pmatrix} z \\ Xz \end{pmatrix} \mid z \in D(A) \right\} \tag{2}$$

and

$$\mathcal{H}\begin{pmatrix} z \\ Xz \end{pmatrix} = \begin{pmatrix} A & -Q_1 \\ -Q_2 & A^* \end{pmatrix}\begin{pmatrix} z \\ Xz \end{pmatrix} = \begin{pmatrix} (A - Q_1X)z \\ (-Q_2 - A^*X)z \end{pmatrix}$$
$$\overset{(1)}{=} \begin{pmatrix} (A - Q_1X)z \\ (XA - XQ_1X)z \end{pmatrix} = \begin{pmatrix} (A - Q_1X)z \\ X(A - Q_1X)z \end{pmatrix} \subset \text{ran}\begin{pmatrix} I \\ X \end{pmatrix}.$$

Assume next that

$$\mathcal{H}\left(\begin{pmatrix} z \\ Xz \end{pmatrix} \cap D(\mathcal{H})\right) \subset \text{ran}\begin{pmatrix} I \\ X \end{pmatrix} \tag{3}$$

Since $XD(A) \subset D(A^*)$, we have that (2) holds. This together with (3) implies that for any $z \in D(A)$, there exists an y such that

$$\mathcal{H}\begin{pmatrix} z \\ Xz \end{pmatrix} = \begin{pmatrix} (A - Q_1X)z \\ (-Q_2 - A^*X)z \end{pmatrix} = \begin{pmatrix} y \\ Xy \end{pmatrix}$$

Combining both equalities gives $X(A - Q_1X)z = (-Q_2 - A^*X)z$, or equivalently

$$(XA + A^*X - XQ_1X + Q_2)z = 0. \tag{4}$$

Since this holds for all $z \in D(A)$, we conclude that (1) holds. □

So we see that the ARE has a solution if the Hamiltonian operator has an invariant subspace ran$\begin{pmatrix} I \\ X \end{pmatrix}$. From Example I.6 of Zwart (1989), we know that invariant subspaces of unbounded operator can be very strange. For instance, even for a simple operator as the Laplacian there are invariant subspaces such that the Laplacian restricted to this subspace is nowhere invertible. It also follows from Zwart (1989) that if the subspace is invariant for the inverse of the operator as well, then the situation is more standard. The following theorem shows that our invariant subspaces of the Hamiltonian have this property. To simplify the proof we introduce a transformation of the Hamiltonian. Let $X \in \mathcal{L}(Z)$ be such that $XD(A) \subset D(A^*)$, then $\mathcal{H}_X$ is defined by

$$\mathcal{H}_X := \begin{pmatrix} I & 0 \\ -X & I \end{pmatrix} \mathcal{H} \begin{pmatrix} I & 0 \\ X & I \end{pmatrix} = \begin{pmatrix} A - Q_1X & -Q_1 \\ -XA + XQ_1X - Q_2 - A^*X & -A^* + XQ_1 \end{pmatrix}. \tag{5}$$

Theorem 2 *Define for $X \in \mathcal{L}(Z)$ the subspace $V_X := \operatorname{ran}\begin{pmatrix} I \\ X \end{pmatrix}$. Then the following assertions are equivalent:*

1. *X is a solution of the ARE;*
2. *$XD(A) \subset D(A^*)$, and the subspace V_X is an invariant subspace of $\mathcal{H}$;*
3. *$XD(A) \subset D(A^*)$, and for all $\lambda \in \rho(\mathcal{H}) \cap \rho(A - Q_1X)$, we have that*

$$(\lambda I - \mathcal{H})^{-1} V_X \subset V_X. \tag{6}$$

Proof. The equivalence between 1. and 2. is already proved in Lemma 1. We shall first show that 2. implies 3.

First we remark the following. A subspace W is S-invariant if and only if TW is TST^{-1}-invariant, where T is a boundedly invertible operator. Using this we see that we can replace the condition in 2. by $\mathcal{H}_X\begin{pmatrix} D(A) \\ 0 \end{pmatrix} \subset \begin{pmatrix} Z \\ 0 \end{pmatrix}$. Furthermore it is easy to see that $\rho(\mathcal{H}) = \rho(\mathcal{H}_X)$. So we know that

$$\mathcal{H}_X\begin{pmatrix} D(A) \\ 0 \end{pmatrix} \subset \begin{pmatrix} Z \\ 0 \end{pmatrix} \tag{7}$$

For $\lambda \in \rho(A - Q_1X)$ we have that $(\lambda - A + Q_1X)D(A) = Z$. Hence with (7), it follows that for $\lambda \in \rho(A - Q_1X)$

$$(\lambda I - \mathcal{H}_X)\begin{pmatrix} D(A) \\ 0 \end{pmatrix} = \begin{pmatrix} Z \\ 0 \end{pmatrix} \tag{8}$$

Let $\lambda \in \rho(A - Q_1X) \cap \rho(\mathcal{H}_X)$ and $z \in Z$, then

$$(\lambda I - \mathcal{H}_X)^{-1}\begin{pmatrix} z \\ 0 \end{pmatrix} \overset{(8)}{=} (\lambda I - \mathcal{H}_X)^{-1}(\lambda I - \mathcal{H}_X)\begin{pmatrix} z_A \\ 0 \end{pmatrix} = \begin{pmatrix} z_A \\ 0 \end{pmatrix} \in \begin{pmatrix} Z \\ 0 \end{pmatrix},$$

where $z_A \in D(A)$. So we have shown 3. Now we shall prove that 3. implies 2.

Assume that 3. holds. By the remarks made at the beginning of this proof, we see that we may as well assume that

$$(\lambda I - \mathcal{H}_X)^{-1}\begin{pmatrix} Z \\ 0 \end{pmatrix} \subset \begin{pmatrix} Z \\ 0 \end{pmatrix}.$$

We write $(\lambda I - \mathcal{H}_X)^{-1} = \begin{pmatrix} P_{11} & P_{12} \\ P_{21} & P_{22} \end{pmatrix}$ with $P_{ij} \in \mathcal{L}(Z)$ $i, j = 1, 2$.

Since $(\lambda I - \mathcal{H}_X)^{-1}\begin{pmatrix} Z \\ 0 \end{pmatrix} \subset \begin{pmatrix} Z \\ 0 \end{pmatrix}$ we have that $P_{21} = 0$.

Furthermore, we write $(\lambda I - \mathcal{H}_X) = \begin{pmatrix} \lambda - A + Q_1X & Q_1 \\ Q_3 & \lambda + A^* - XQ_1 \end{pmatrix}$ for some Q_3.

So we have that

$$\begin{pmatrix} \lambda I - A + Q_1X & +Q_1 \\ Q_3 & \lambda + A^* - XQ_1 \end{pmatrix}\begin{pmatrix} P_{11} & P_{12} \\ 0 & P_{22} \end{pmatrix} = \begin{pmatrix} I_Z & 0 \\ 0 & I_Z \end{pmatrix} \tag{9}$$

Thus in particular, there holds $(\lambda I - A + Q_1X)P_{11} = I_Z$ and since $\lambda \in \rho(A - Q_1X)$, this implies that

$$P_{11} = (\lambda I - A + Q_1X)^{-1}$$

Furthermore, we see from (9) that $Q_3P_{11} = 0$, and so $Q_3 = 0$ on $D(A)$. Hence

$$(\lambda I - \mathcal{H}_X) = \begin{pmatrix} \lambda - A + Q_1X & +Q_1 \\ 0 & \lambda + A^* + XQ_1 \end{pmatrix}$$

and so $Z \oplus \{0\}$ is $\mathcal{H}_X$-invariant. This implies that 2. holds. □

It may seem strange that whereas the left-hand side of equation (6) has meaning for every $\lambda \in \rho(\mathcal{H})$, the equality is only valid on a smaller set. In the next example, we shall show that (6) need not to hold on the whole $\rho(\mathcal{H})$. Note that for finite-dimensional systems, $\rho(\mathcal{H}) \subset \rho(A - Q_1X)$. Thus in that case, one can just write $\lambda \in \rho(\mathcal{H})$ in part 3.

Example 3 *We take A to be the shift on $\ell_2(N)$, i.e.*

$$A\begin{pmatrix} z_1 \\ z_2 \\ \vdots \end{pmatrix} = \begin{pmatrix} 0 \\ z_1 \\ \vdots \end{pmatrix} \tag{10}$$

Furthermore, we take $Q_1 = I$, $Q_2 = 0$ and $X = 0$. Then it is clear that X satisfies the ARE. So

$$\mathcal{H}\begin{pmatrix} Z \\ 0 \end{pmatrix} \subset \begin{pmatrix} Z \\ 0 \end{pmatrix}$$

Next we shall show that

$$\mathcal{H}^{-1} = \begin{pmatrix} A^* & -I \\ -P_1 & -A \end{pmatrix}, \tag{11}$$

where $P_1\begin{pmatrix} z_1 \\ z_2 \\ \vdots \end{pmatrix} = \begin{pmatrix} z_1 \\ 0 \\ \vdots \end{pmatrix}$. *From this it is easy to see that $Z \oplus \{0\}$ is not $\mathcal{H}^{-1}$ invariant.*

An easy calculation gives that $A^*\begin{pmatrix} z_1 \\ z_2 \\ \vdots \end{pmatrix} = \begin{pmatrix} z_2 \\ z_3 \\ \vdots \end{pmatrix}$. *Hence*

$$AA^* = I - P_1 \quad \text{and } A^*A = I. \tag{12}$$

Using this we see that

$$\begin{pmatrix} A & -I \\ 0 & -A^* \end{pmatrix}\begin{pmatrix} A^* & -I \\ -P_1 & -A \end{pmatrix} = \begin{pmatrix} AA^* + P_1 & -A + A \\ A^*P_1 & A^*A \end{pmatrix} \overset{(12)}{=} \begin{pmatrix} I & 0 \\ 0 & I \end{pmatrix}$$

Furthermore

$$\begin{pmatrix} A^* & -I \\ -P_1 & -A \end{pmatrix}\begin{pmatrix} A & -I \\ 0 & -A^* \end{pmatrix} = \begin{pmatrix} A^*A & -A^* + A^* \\ -P_1A & P_1 + AA^* \end{pmatrix} = \begin{pmatrix} I & 0 \\ 0 & I \end{pmatrix}.$$

Hence (11) holds. Note that $0 \in \sigma(A) \cap \sigma(A^*)$.

So the previous example shows in particular that in general the resolvent set of $\mathcal{H}$ need not to be equal to the union of the resolvent set of $A - Q_1X$ and of $-A^* + XQ_1$. However, for an important subclass of systems there is equality. Without proof we state the following result.

Lemma 4 *Let X be a solution of the ARE, if the spectra of $\mathcal{H}$, $A - Q_1X$, and $A - Q_1X^*$ are all point spectra, then $\sigma(\mathcal{H}) = \sigma(A - Q_1X) \cup \sigma(-A^* + XQ_1)$.*

3 ALL SOLUTIONS OF THE ARE FOR SKEW-ADJOINT GENERATORS

In this section we shall assume that A is skew-adjoint, i.e., $A^* = -A$, and has compact resolvent. Furthermore, we assume that $Q_2 = Q_1 \geq 0$, and $\Sigma(A, -, Q_1)$ is approximately observable (see Curtain and Zwart 1995). Since A has compact resolvent, so has every bounded perturbation of A, or of A^*, see Kato (1984), Theorem 1.16 of chapter IV. Furthermore, it follows easily that the Hamiltonian operator has compact resolvent as well. Hence, from Theorem 6.29 of chapter III of Kato (1984), $A - Q_1X$, $-A^* + XQ_1$, and $\mathcal{H}$ have pure point spectrum. The following results are standard and easy to prove.

Lemma 5 *Let A, Q_1 and Q_2 have the properties as stated above, then*

1. *I and $-I$ are solutions of the ARE;*
2. *$A - Q_1$ is stable, i.e., all eigenvalues have real part negative;*
3. *$-A^* + Q_1 = A + Q_1$ is completely unstable, i.e., all eigenvalues have real part positive.*

Associated with the solutions I and $-I$ of the ARE, we have the invariant subspaces $V_I := \text{ran}\begin{pmatrix} I \\ I \end{pmatrix}$ and $V_{-I} := \text{ran}\begin{pmatrix} I \\ -I \end{pmatrix}$, respectively. It is easy to see that

$$Z \oplus Z = V_I \oplus V_{-I} \tag{13}$$

Let P_I and P_{-I} denote the projection on V_I and V_{-I}, respectively. Note that (13) is equivalent with $I = P_I + P_{-I}$. From this it is easy to see that

$$P_I\begin{pmatrix} z_1 \\ z_2 \end{pmatrix} = \frac{1}{2}\begin{pmatrix} z_1 + z_2 \\ z_1 + z_2 \end{pmatrix} \quad \text{and } P_{-I}\begin{pmatrix} z_1 \\ z_2 \end{pmatrix} = \frac{1}{2}\begin{pmatrix} z_1 - z_2 \\ -z_1 + z_2 \end{pmatrix} \tag{14}$$

We shall show that these projections can also be seen as spectral projections. We recall that for a closed curve Γ which has empty intersection with the spectrum, the spectral projection is defined as

$$S_\Gamma x = \frac{1}{2\pi i}\int_\Gamma (\gamma I - \mathcal{H})^{-1} x d\gamma.$$

To define the spectral projection associated with the stable eigenvalues of $\mathcal{H}$ we take a sequence of curves, Γ_n, all lying in the left-half plane, such that the interior of Γ_n becomes the whole left half plane for n going to infinity. Now the spectral projection associated with the left half plane is defined as

$$S_{\text{stable}} x = \lim_{n\to\infty} \frac{1}{2\pi i}\int_{\Gamma_n} (\gamma I - \mathcal{H})^{-1} x d\gamma. \tag{15}$$

Similarly we define S_{unstable} as the spectral projection associated with the right-half plane. Now we can prove the following lemmas.

Lemma 6 *Assume S_{stable} and S_{unstable} are well-defined, and that X is a solution of the ARE, then*

$$S_{\text{stable}} V_X \subset V_X \quad \textit{and } S_{\text{unstable}} V_X \subset V_X, \tag{16}$$

where V_X is defined in Theorem 2.

Proof. Let $v \in V_X$, then by equation (6) there holds that $(\gamma I - \mathcal{H})^{-1} v \in V_X$. Thus since V_X is a closed linear subspace, $S_\Gamma v \in V_X$. Since this holds for any closed curve Γ, the result follows. □

Lemma 7 *If S_{stable} and $S_{unstable}$ are well-defined, and $I = S_{\text{stable}} + S_{\text{unstable}}$, then $P_I = S_{\text{stable}}$ and $P_{-I} = S_{\text{unstable}}$.*

Proof. Recall that the range of P_I and P_{-I} are the Hamiltonian invariant subspaces V_I and V_{-I}, respectively. So by Lemma 6 we know that both spectral projections map these subspaces into themselves. We shall show that $S_{\text{stable}} V_{-I} = 0$. Let Γ_n be a closed curve in the left-half plane, and let $(z, -z)^T$ be an element of V_{-I}, then we see that

$$\begin{aligned}
\frac{1}{2\pi i} \int_{\Gamma_n} (\gamma I - \mathcal{H})^{-1} \begin{pmatrix} z \\ -z \end{pmatrix} d\gamma &= \frac{1}{2\pi i} \int_{\Gamma_n} \begin{pmatrix} I & 0 \\ -I & I \end{pmatrix} (\gamma I - \mathcal{H}_{-I})^{-1} \begin{pmatrix} I & 0 \\ I & I \end{pmatrix} \begin{pmatrix} z \\ -z \end{pmatrix} d\gamma \\
&= \begin{pmatrix} I & 0 \\ -I & I \end{pmatrix} \frac{1}{2\pi i} \int_{\Gamma_n} (\gamma I - \mathcal{H}_{-I})^{-1} \begin{pmatrix} z \\ 0 \end{pmatrix} d\gamma \\
&= \begin{pmatrix} I & 0 \\ -I & I \end{pmatrix} \frac{1}{2\pi i} \int_{\Gamma_n} \begin{pmatrix} (\gamma I - A - Q_1)^{-1} z \\ 0 \end{pmatrix} d\gamma = \begin{pmatrix} 0 \\ 0 \end{pmatrix}
\end{aligned}$$

since by Lemma 5 $A + Q_1$ has no poles inside Γ_n. From the definition of S_{stable} it follows directly that $S_{\text{stable}} V_{-I} = 0$. Similarly one can show that $S_{\text{unstable}} V_I = 0$. From these relations it follows easily that $S_{\text{stable}} = S_{\text{stable}} P_I$. Letting P_I operate on this equation gives $P_I S_{\text{stable}} = P_I S_{\text{stable}} P_I = S_{\text{stable}} P_I$, since $\operatorname{ran} S_{\text{stable}} P_I \subset \operatorname{ran} P_I = V_I$. Hence combining these equalities gives

$$S_{\text{stable}} = S_{\text{stable}} P_I = P_I S_{\text{stable}}. \tag{17}$$

Similarly, one can show that

$$S_{\text{unstable}} = S_{\text{unstable}} P_{-I} = P_{-I} S_{\text{unstable}}. \tag{18}$$

For $z \in Z$ we have that

$$\begin{aligned}
z &= P_I z + P_{-I} z = P_I \left[S_{\text{stable}} z + S_{\text{unstable}} z \right] + P_{-I} \left[S_{\text{stable}} z + S_{\text{unstable}} z \right] \\
&= S_{\text{stable}} z + P_I S_{\text{unstable}} z + P_{-I} S_{\text{stable}} z + S_{\text{unstable}} z = z + P_I S_{\text{unstable}} z + P_{-I} S_{\text{stable}} z.
\end{aligned}$$

Since $\operatorname{ran} P_I \cap \operatorname{ran} P_{-I} = \{0\}$, we see that $P_I S_{\text{unstable}} = P_{-I} S_{\text{stable}} = 0$. Now it follows directly that $P_I = S_{\text{stable}}$ and $P_{-I} = S_{\text{unstable}}$. □

With these lemmas we can characterize all solutions of the ARE.

Theorem 8 *Assume that S_{stable} and $S_{unstable}$ are well-defined, and that $I = S_{\mathbf{stable}} + S_{\mathbf{unstable}}$. If X is a solution of the ARE, then X can be written as $X = I - 2\mathcal{P}$, where $\mathcal{P}$ is a projection. If X is self-adjoint, then $-I \leq X \leq I$.*

Proof. From Lemma 7 we know that $P_I = S_{\mathbf{stable}}$ and $P_{-I} = S_{\mathbf{unstable}}$. Let X be a solution of the ARE, and thus V_X is an invariant subspace of the Hamiltonian. Now from Lemma 6 it follows directly that $P_I V_X \subset V_X$. Thus for all $z \in Z$, there exists a $y \in Z$ such that

$$P_I \begin{pmatrix} z \\ Xz \end{pmatrix} = \frac{1}{2} \begin{pmatrix} z + Xz \\ z + Xz \end{pmatrix} = \begin{pmatrix} y \\ Xy \end{pmatrix}$$

Thus for all $z \in Z$, $z + Xz = 2Xy = X(z + Xz)$, and so

$$I = X^2. \tag{19}$$

If we write $X = I - 2\mathcal{P}$, then we see that (19) implies that $\mathcal{P}^2 = \mathcal{P}$. Hence $\mathcal{P}$ is a projection.

It is easy to see that X is self-adjoint if and only if $\mathcal{P}$ is self-adjoint. Since for a self-adjoint projection one always has that $0 \leq \mathcal{P} \leq I$, the assertion follows directly. □

Note that the assumptions in the theorem are always satisfied if $\mathcal{H}$ has a Riesz basis of eigenvectors. One can use this Riesz basis to characterize all solutions of the ARE, even if A is not skew-adjoint, see Kuiper and Zwart (1995). Theorem 8 is a generalization of Theorem 6 of Willems (1971). As in Willems (1971), one can make the result in this theorem if and only if. However, due to space limitations, we cannot present a proof here.

REFERENCES

Curtain, R.F. and Zwart, H.J. (1995) *An Introduction to Infinite Dimensional Linear Systems Theory.* Text in Applied Mathematics, Vol. 21, Springer Verlag, New York.

Kato, T. (1984) *Perturbation Theory for Linear Operators*, Grundlehren der mathematischen Wissenschaften, 132, Springer Verlag, Berlin.

Kučera, V. (1991) Algebraic Riccati Equation: Hermitian and Definite Solutions, in *The Riccati Equation* (ed. S. Bittanti, A.J. Laub, J.C. Willems), Springer Verlag, Berlin.

Kuiper, C.R. and Zwart, H.J. (1995) Relations between the algebraic Riccati equation and the Hamiltonian for Riesz-spectral systems, *Journal of Mathematical Systems, Estimation and Control* (to appear)

Mårtensson, K. (1971) On the Matrix Riccati Equation, *Inform. Sci.* **3**, 17-49.

Potter, J.E. (1966) Matrix quadratic solutions, *SIAM J. Appl. Math.* **14**, 496–501.

Willems, J.C. (1971) Least Squares Stationary Optimal Control and The Algebraic Riccati Equation, *IEEE Trans. Autom. Control*, **16**, 621–634.

Zwart, H.J. (1989) *Geometric Theory for Infinite Dimensional Systems*, Lecture Notes in Control and Information Sciences, **115**, Springer Verlag, Berlin.

Non-linear Systems

19

Strong Pontryagin's principle for state–constrained control problems governed by parabolic equations

Eduardo Casas
Dpto. de Matemática Aplicada y Ciencias de la Computación,
E.T.S.I. Industriales y de Telecomunicación, Universidad de Cantabria,
Av. Los Castros s/n, 39071 Santander, Spain. Phone: 34-42-201427.
Fax: 34-42-201829. E–mail: casas@macc.unican.es

Abstract

This paper deals with state-constrained optimal control problems governed by semilinear parabolic equations. We establish a minimum principle of Pontryagin's type in a qualified form, the so-called strong principle. To achieve this result we formulate two different assumptions, any of them leading to the desired result. One of the assumptions is an extension of Slater's condition. Since no hypothesis of differentiability with respect to the control, even continuity, of the functions involved in the control problem is made, the generalization of Slater's conditions is not obvious. The second assumption is announced in terms of a certain Lipschitz dependence of the optimal cost functional with respect to small perturbations of the set of feasible states.

Keywords

Pontryagin's principle, Slater's condition, state constraints, parabolic equations

1 SETTING OF THE CONTROL PROBLEM

Let $\Omega \subset \mathbb{R}^n$, $n \geq 1$, be an open and bounded set, with Lipschitz boundary Γ. Given $0 < T < +\infty$, we set $\Omega_T = \Omega \times (0, T)$ and $\Sigma_T = \Gamma \times (0, T)$. Let $(\mathcal{K}, d)$ be a metric space and let us consider a function $f : \Sigma_T \times \mathbb{R} \times \mathcal{K} \longrightarrow \mathbb{R}$ of class C^1 with respect to the second variable and satisfying the following assumptions:

$$\begin{cases} \dfrac{\partial f}{\partial y}(x,t,y,u) \leq 0 \quad \forall (x,t,y,u) \in \Sigma_T \times \mathbb{R} \times \mathcal{K}; \\ \forall M > 0 \ \exists C_M > 0 \text{ such that } \forall (x,t,u) \in \Sigma_T \times \mathcal{K} \text{ and } |y| \leq M \\ |f(x,t,0,u)| + \left|\dfrac{\partial f}{\partial y}(x,t,y,u)\right| \leq C_M. \end{cases} \tag{1}$$

The state equation is as follows

$$\begin{cases} \dfrac{\partial y}{\partial t}(x,t) + Ay(x,t) + a_0(x,t,y(x,t)) = 0 & \text{in } \Omega_T, \\ \partial_{\nu_A} y(x,t) = f(x,t,y(x,t),u(x,t)) & \text{on } \Sigma_T, \\ y(x,0) = y_0(x) & \text{in } \Omega, \end{cases} \tag{2}$$

where $y_0 \in C(\bar{\Omega})$, A is a linear operator with bounded coefficients and satisfying the usual ellipticity condition

$$\begin{aligned} Ay = & -\sum_{j=1}^{n} \partial_{x_j} \left\{ \sum_{i=1}^{n} [a_{ij}(x,t)\partial_{x_i} y(x,t)] + b_j(x,t)y(x,t) \right\} \\ & + \sum_{j=1}^{n} d_j(x,t)\partial_{x_j} y(x,t) + c(x,t)y(x,t) \end{aligned} \tag{3}$$

$$\partial_{\nu_A} y(x,t) = \sum_{j=1}^{n} \left\{ \sum_{i=1}^{n} [a_{ij}(x,t)\partial_{x_i} y(x,t)] + b_j(x,t)y(x,t) \right\} \nu_j(x), \tag{4}$$

$\nu(x)$ being the outward unit normal vector to Γ at the point x. Function $a_0 : \Omega_T \times \mathbb{R} \longrightarrow \mathbb{R}$ is a Carathéodory function of class C^1 with respect to the second variable and satisfies the following assumptions

$$\begin{cases} \exists \psi_0 \in L^{\hat{p}}([0,T], L^{\hat{q}}(\Omega)) \text{ and } C_1 > 0 \text{ such that} \\ a_0(x,t,y)y \geq \psi_0(x,t) - C_1 y^2 \quad \forall (x,t,y) \in \Omega_T \times \mathbb{R}; \end{cases} \tag{5}$$

$$\begin{cases} a_0(\cdot,\cdot,0) \in L^{\hat{p}}([0,T], L^{\hat{q}}(\Omega)) \text{ and } \forall M > 0 \ \exists C_M > 0 \text{ such that} \\ \left| \dfrac{\partial a_0}{\partial y}(x,t,y) \right| \leq C_M \ \forall (x,t) \in \Omega_T, |y| \leq M; \end{cases} \tag{6}$$

where $\hat{q}, \hat{p} \in [1, +\infty]$ and $\dfrac{1}{\hat{p}} + \dfrac{n}{2\hat{q}} < 1$.

Once given the sate equation, we introduce the cost functional

$$J(u) = \int_{\Omega_T} L(x,t,y_u(x,t))dxdt + \int_{\Sigma_T} l(x,t,y_u(x,t),u(x,t))d\sigma(x)dt,$$

where y_u is the solution of (2) associated to u; $L : \Omega_T \times \mathbb{R} \longrightarrow \mathbb{R}$ and $l : \Sigma_T \times \mathbb{R} \times \mathcal{K} \longrightarrow \mathbb{R}$ are of class C^1 with respect to the second variable, measurable with respect to the first one, and satisfying

$$\begin{cases} \forall M > 0 \ \exists \psi_{dM} \in L^1(\Omega_T) \text{ such that } \forall (x,t) \in \Omega_T, |y| \leq M \\ |L(x,t,0)| + \left| \dfrac{\partial L}{\partial y}(x,t,y) \right| \leq \psi_{dM}(x,t), \end{cases} \tag{7}$$

$$\begin{cases} \forall M > 0 \ \exists \psi_{bM} \in L^1(\Sigma_T) \text{ such that } \forall (x,t,u) \in \Sigma_T \times \mathcal{K}, |y| \leq M \\ |l(x,t,0,u)| + \left|\dfrac{\partial l}{\partial y}(x,t,y,u)\right| \leq \psi_{bM}(x,t). \end{cases} \tag{8}$$

The space of controls $\mathcal{U}$ is formed by the measurable functions $u : \Sigma_T \longrightarrow \mathcal{K}$ such that the mapping $(x,t) \in \Sigma_T \longrightarrow (f(x,t,y,u(x,t)), l(x,t,y,u(x,t))) \in \mathbb{R}^2$ is measurable for every $y \in \mathbb{R}$. In $\mathcal{U}$ we consider Ekeland's distance

$$d_E(u,v) = m_{\Sigma_T}\left(\{(x,t) \in \Sigma_T : u(x,t) \neq v(x,t)\}\right), \tag{9}$$

where m_{Σ_T} is the measure on Σ_T obtained as the product of σ and the Lebesgue measure in the interval $(0,T)$. It is known (see Ekeland (1979)) that $(\mathcal{U}, d_E)$ is a complete metric space. Finally we formulate the optimal control problem as follows

(P_δ) Minimize $\{J(u) : u \in \mathcal{U}, g(x,t,y_u(x,t)) \leq \delta \ \forall (x,t) \in \bar{\Omega}_T\}$.

Before concluding this section, let us to state some results concerning the state equation.

Theorem 1 *Under assumptions (1)–(6), problem (2) has a unique solution in* $Y = C(\bar{\Omega}_T) \cap L^2([0,T], H^1(\Omega))$ *for every control* $u \in \mathcal{U}$. *Moreover there exists a constant* $M > 0$ *such that*

$$\|y_u\|_\infty + \|y_u\|_{L^2([0,T],H^1(\Omega))} \leq M \quad \forall u \in \mathcal{U}. \tag{10}$$

Finally, if $\{u_k\}_{k=1}^\infty \subset \mathcal{U}$ *is a sequence converging to* u *in* $\mathcal{U}$, *i.e.* $d_E(u_k,u) \to 0$, *then* $\{y_{u_k}\}_{k=1}^\infty$ *converges to* y_u *strongly in* Y.

Theorem 2 *Let* $u, v \in \mathcal{U}$. *Given* $\rho \in (0,1)$, *there exist* m_{Σ_T}*-measurable sets* $E_\rho \subset \Sigma_T$, *with* $m_{\Sigma_T}(E_\rho) = \rho m_{\Sigma_T}(\Sigma_T)$, *such that if we define* $u_\rho = u + [v-u]\chi_{E_\rho}$, χ_{E_ρ} *being the characteristic function on the set* E_ρ, *and if we denote by* y_ρ *and* y *the states corresponding to* u_ρ *and* u, *respectively, then the following equalities hold*

$$y_\rho = y + \rho z + r_\rho, \quad \lim_{\rho \to 0} \frac{1}{\rho}\|r_\rho\|_Y = 0, \tag{11}$$

$$J(u_\rho) = J(u) + \rho z^0 + r_\rho^0, \quad \lim_{\rho \to 0} \frac{1}{\rho} r_\rho^0 = 0, \tag{12}$$

where $z \in Y$ *and* $z_0 \in \mathbb{R}$ *satisfy*

$$\begin{cases} \dfrac{\partial z}{\partial t} + Az + \dfrac{\partial a_0}{\partial y}(x,t,y(x,t))z = 0 \quad \text{in } \Omega_T \\[2ex] \partial_{\nu_A} z = \dfrac{\partial f}{\partial y}(x,t,y(x,t),u(x,t))z \\[2ex] +f(x,t,y(x,t),v(x,t)) - f(x,t,y(x,t),u(x,t)) \text{ on } \Sigma_T, \\[2ex] z(x,0) = 0 \text{ in } \Omega \end{cases} \tag{13}$$

$$z^0 = \int_{\Omega_T} \frac{\partial L}{\partial y}(x,t,y(x,t))z(x,t)dxdt + \int_{\Sigma_T} \frac{\partial l}{\partial y}(x,t,y(x,t),u(x,t))z(x,t)d\sigma(x)dt$$

$$+\int_{\Sigma_T} [l(x,t,y(x,t),v(x,t)) - l(x,t,y(x,t),u(x,t))]d\sigma(x)dt. \tag{14}$$

The reader is referred to Casas (1995) for a proof of these theorems.

2 PONTRYAGIN'S PRINCIPLE

Given $\alpha \geq 0$, we define the Hamiltonian $H_\alpha : \Sigma_T \times \mathbb{R} \times \mathcal{K} \times \mathbb{R} \longrightarrow \mathbb{R}$ as follows

$$H_\alpha(x,t,y,u,\varphi) = \alpha l(x,t,y,u) + \varphi f(x,t,y,u).$$

Theorem 3 *If $\bar{u} \in \mathcal{U}$ is a solution of (P_δ), then there exist $\bar{\alpha} \geq 0$, $\bar{y} \in C(\bar{\Omega}_T) \cap L^2([0,T],H^1(\Omega))$, $\bar{\varphi} \in L^r([0,T],W^{1,p}(\Omega))$, for all $p,r \in [1,2)$ with $(2/r)+(n/p) > n+1$, $\bar{\mu} \in Z'$ and $\bar{\lambda} \in R^l$ such that*

$$\bar{\alpha} + \|\bar{\mu}\|_{Z'} + |\bar{\lambda}| > 0; \tag{15}$$

$$\begin{cases} \dfrac{\partial \bar{y}}{\partial t} + A\bar{y} + a_0(x,t,\bar{y}(x,t)) = 0 \quad in\ \Omega_T, \\ \partial_{\nu_A}\bar{y}(x,t) = f(x,t,\bar{y}(x,t),\bar{u}(x,t)) \ on\ \Sigma_T, \\ \bar{y}(0) = y_0 \quad in\ \Omega; \end{cases} \tag{16}$$

$$\begin{cases} -\dfrac{\partial \bar{\varphi}}{\partial t} + A^*\bar{\varphi} + \dfrac{\partial a_0}{\partial y}(x,t,\bar{y})\bar{\varphi} = \bar{\alpha}\dfrac{\partial L}{\partial y}(x,t,\bar{y}) + \dfrac{\partial g}{\partial y}(x,t,\bar{y})\bar{\mu}|_{\Omega_T} \ \ in\ \Omega_T, \\ \partial_{\nu_{A^*}}\bar{\varphi} = \dfrac{\partial f}{\partial y}(x,t,\bar{y},\bar{u})\bar{\varphi} + \bar{\alpha}\dfrac{\partial l}{\partial y}(x,t,\bar{y},\bar{u}) + \dfrac{\partial g}{\partial y}(x,t,\bar{y})\bar{\mu}|_{\Sigma_T} \ on\ \Sigma_T, \\ \bar{\varphi}(T) = \dfrac{\partial g}{\partial y}(x,t,\bar{y})\bar{\mu}|_{\bar{\Omega}\times\{T\}} \ in\ \bar{\Omega}; \end{cases} \tag{17}$$

$$\int_{\bar{\Omega}_T} [z(x,t) - g(x,t,\bar{y}(x,t))]d\bar{\mu}(x,t) \leq 0 \quad \forall z \in C(\bar{\Omega}_T),\ with\ z(x,t) \leq \delta; \tag{18}$$

$$\begin{aligned} &\int_{\Sigma_T} H_{\bar{\alpha}}(x,t,\bar{y}(x,t),\bar{u}(x,t),\bar{\varphi}(x,t))d\sigma(x)dt \\ &= \min_{u\in\mathcal{U}} \int_{\Sigma_T} H_{\bar{\alpha}}(x,t,\bar{y}(x,t),u(x,t),\bar{\varphi}(x,t))d\sigma(x)dt; \end{aligned} \tag{19}$$

where A^* denotes the formal adjoint operator of A. Moreover, if one of the following assumptions is satisfied:

A1) *Functions f and l are continuous with respect to the third variable on $(\mathcal{K}, d)$ and this space is separable;*
A2) *There exists a set $\Sigma_T^0 \subset \Sigma_T$, with $m_{\Sigma_T}(\Sigma_T^0) = m_{\Sigma_T}(\Sigma_T)$, such that the function*

$$(x,t) \in \Sigma_T \longrightarrow (f(x,t,y,u), l(x,t,y,u)) \in \mathbb{R}^2$$

is continuous in Σ_T^0 for every $(y,u) \in \mathbb{R} \times \mathcal{K}$;

then the following pointwise relation holds a.e.$[\sigma]$ $x \in \Gamma$ and a.e. $t \in [0,T]$

$$H_{\bar\alpha}(x,t,\bar y(x,t),\bar u(x,t),\bar\varphi(x,t)) = \min_{u\in\mathcal{K}} H_{\bar\alpha}(x,t,\bar y(x,t),u,\bar\varphi(x,t)). \tag{20}$$

See Casas (1995) for the proof of this theorem. Our aim is to formulate some conditions under which the theorem remains to be true with $\bar\alpha = 1$. In this section we introduce a Slater's type hypothesis that leads to the desired result. In the next section, we will develop another method to achieve the same goal.

Slater Assumption: There exists $u_0 \in \mathcal{U}$ such that

$$g(x,t,\bar y(x,t)) + \frac{\partial g}{\partial y}(x,t,\bar y(x,t))z_0(x,t) < \delta \ \forall (x,t) \in \bar\Omega_T, \tag{21}$$

$z_0 \in C(\bar\Omega_T) \cap L^2([0,T], H^1(\Omega))$ is the solution of (13), with $u_0 = v$, $\bar u = u$ and $\bar y = y$.

Corollary 1 *If $\bar u$ is a solution of (P_δ) and Slater's assumption holds, then Theorem 3 remains to be true with $\bar\alpha = 1$.*

Proof. If $\bar\alpha > 0$, then we can divide relations (16)–(18) by $\bar\alpha$ and rename $\bar\varphi/\bar\alpha$ and $\bar\mu/\bar\alpha$ as $\bar\varphi$ and $\bar\mu$, respectively, obtaining the searched result. Let us prove that Slater's assumption implies the strict positivity of $\bar\alpha$.

Let us assume that $\bar\alpha = 0$, then (14) implies that $\bar\mu \neq 0$. Since $\bar\mu \neq 0$, the inequality (18) is strict for every $z \in C(\bar\Omega_T)$ with $z(x,t) < \delta \ \forall (x,t) \in \bar\Omega_T$, in particular (21) leads to

$$\int_{\bar\Omega_T} \frac{\partial g}{\partial y}(x,t,\bar y(x,t))z_0(x,t)\,d\bar\mu(x,t) =$$

$$\int_{\bar\Omega_T} \left\{ \left[g(x,t,\bar y(x,t)) + \frac{\partial g}{\partial y}(x,t,\bar y(x,t))z_0(x,t) \right] - g(x,t,\bar y(x,t)) \right\} d\bar\mu(x,t) < 0. \tag{22}$$

On the other hand, (19) along with (13) and (16) gives

$$0 \geq \int_{\Sigma_T} [H_{\bar\alpha}(x,t,\bar y(x,t)u_0(x,t),\bar\varphi(x,t)) - H_{\bar\alpha}(x,t,\bar y(x,t)\bar u(x,t),\bar\varphi(x,t))]\,d\sigma(x)dt =$$

$$\int_{\Sigma_T} \bar\varphi(x,t)[f(x,t,\bar y(x,t),u_0(x,t)) - f(x,t,\bar y(x,t),\bar u(x,t))]\,d\sigma(x)dt =$$

$$\int_{\bar{\Omega}_T} \frac{\partial g}{\partial y}(x,t,\bar{y}(x,t))z_0(x,t)d\bar{\mu}(x,t). \tag{23}$$

Thus (22) and (23) prove that $\bar{\alpha}$ can not be zero. □

A similar assumption to the above one was used by Fattorini and Frankowska (1991) to derive the optimality conditions in a qualified form.

3 THE STABILITY ASSUMPTION

In this section we introduce a stability notion which allows to prove a qualified Pontryagin's principle. Here the qualification is not a consequence of Theorem 3, but it requires a new proof of this theorem with $\bar{\alpha} = 1$.

Definition 1 *We say that* (P_δ) *is strongly stable if there exist* $\epsilon > 0$ *and* $C > 0$ *such that*

$$\inf(\mathrm{P}_\delta) - \inf(\mathrm{P}_{\delta'}) \leq C(\delta' - \delta) \quad \forall \delta' \in [\delta, \delta + \epsilon]. \tag{24}$$

This concept was first introduced in relation with optimal control problems by Bonnans (1991); see also Bonnans and Casas (1995). A weaker stability concept was used by Casas (1992) to analyze the convergence of the numerical discretizations of optimal control problems. The following proposition states that almost all problems (P_δ) are strongly stable.

Proposition 1 *Let* $\delta_0 \geq 0$ *be the smallest number such that* (P_δ) *has feasible controls for every* $\delta > \delta_0$. *Then* (P_δ) *is strongly stable for all* $\delta > \delta_0$ *except at most a zero Lebesgue measure set.*

Proof. It is enough to consider the function $h : (\delta_0, +\infty) \longrightarrow \mathbb{R}$ defined by $h(\delta) = \inf(\mathrm{P}_\delta)$ and remark that it is a nonincreasing monotone function and, consequently, differentiable at every point of $(\delta_0, +\infty)$ except at a zero measure set. Now it is easy to check that (P_δ) is strongly stable at every point where h is differentiable. □

Theorem 4 *If* (P_δ) *is strongly stable and* $\bar{u}$ *is a solution of this problem, then Theorem 3 holds with* $\bar{\alpha} = 1$.

The proof of this theorem is carried out by showing that the stability assumption implies the possibility of an exact penalization of the state constraints. Indeed, let us define

$$Q_\delta = \{z \in C(\bar{\Omega}_T) : z(x,t) \leq \delta \ \forall (x,t) \in \bar{\Omega}_T\}.$$

Since $C(\bar{\Omega}_T)$ is separable, we can take in $C(\bar{\Omega}_T)$ a norm $\|\cdot\|$ equivalent to the usual such that the distance function $d_{Q_\delta} : C(\bar{\Omega}_T) \longrightarrow \mathbb{R}$, $d_{Q_\delta}(z) = \inf\{\|y - z\| : y \in Q_\delta\}$ is convex, Lipschitz and Gâteaux differentiable at every point $z \notin Q_\delta$; see Li and Yong (1995).

Proposition 2 *If* (P_δ) *is strongly stable and* $\bar{u}$ *is a solution of this problem, then there exists* $q_0 > 0$ *such that for every* $q \geq q_0$ $\bar{u}$ *is also a solution of*

$$\inf_{u \in \mathcal{U}} J_q(u) = J(u) + q d_{Q_\delta}(g(\cdot, y_u)) \tag{25}$$

Proof. Let us suppose that it is false. Then there exists a sequence $\{q_k\}_{k=1}^{\infty}$ of real numbers, with $q_k \to +\infty$, and elements $\{u_k\}_{k=1}^{\infty} \subset \mathcal{U}$ such that

$$J(u_k) + q_k d_{Q_\delta}(g(\cdot, y_k)) < J(\bar{u}) \quad \forall k \geq 1,$$

where y_k is the state corresponding to u_k. From here we obtain that

$$d_{Q_\delta}(g(\cdot, y_k)) < \frac{J(\bar{u}) - J(u_k)}{q_k} \longrightarrow 0 \quad \text{when } k \to +\infty$$

and $g(\cdot, y_k) \notin Q_\delta$. Let $\delta_k > \delta$ be the smallest number such that $g(\cdot, y_k) \in Q_{\delta_k}$. Since $\delta_k \to \delta$, we can use (24) to deduce

$$C(\delta_k - \delta) \geq \inf(P_\delta) - \inf(P_{\delta_k}) \geq J(\bar{u}) - J(u_k) > q_k d_{Q_\delta}(g(\cdot, y_k)) = q_k(\delta_k - \delta) \quad \forall k \geq k_\epsilon,$$

which is not possible. □

Since J_q is not Gâteaux differentiable on Q_δ, we are going to modify slightly this functional to attain the differentiability necessary for the proof.

Proposition 3 *Let us take* $q \geq q_0$ *and for every* $\epsilon > 0$ *let us consider the problem*

$$(P_{\delta,\epsilon}) \ \inf_{u \in \mathcal{U}} J_{q,\epsilon}(u) = J(u) + q\left\{d_{Q_\delta}(G(y_u))^2 + \epsilon^2\right\}^{1/2}.$$

Then $\inf(P_{\delta,\epsilon}) \to \inf(P_\delta)$ *when* $\epsilon \to 0$.

Proof. It is an immediate consequence of the inequality $J_q(u) \leq J_{q,\epsilon}(u) \leq J_q(u) + q\epsilon$ $\forall u \in \mathcal{U}$. □

Proof of Theorem 4. Propositions 2 and 3 imply that $\bar{u}$ is a σ_ϵ^2–solution of $(P_{\delta,\epsilon})$, with $\sigma_\epsilon \to 0$ when $\epsilon \to 0$, i.e. $J_{q,\epsilon}(\bar{u}) \leq \inf(P_{\delta,\epsilon}) + \sigma_\epsilon^2$. Then we can apply again Ekeland's principle and deduce the existence of an element $u^\epsilon \in \mathcal{U}$ such that

$$d(u^\epsilon, \bar{u}) \leq \sigma_\epsilon, \quad J_{q,\epsilon}(u^\epsilon) \leq J_{q,\epsilon}(u) + \sigma_\epsilon d_E(u^\epsilon, u) \quad \forall u \in \mathcal{U}.$$

Now we can argue as in the proof of Theorem 3; see Casas (1995). □

ACKNOWLEDGMENT

The author would like to thank Dirección General de Investigación Científica y Técnica (Spain) for its support to this research.

REFERENCES

Bonnans, J. (1991) Pontryagin's principle for the optimal control of semilinear elliptic systems with state constraints, in *30th IEEE Conference on Control and Decision*, Brighton, England, 1976–1979.

Bonnans, J. and Casas, E. (1995) An extension of Pontryagin's principle for state-constrained optimal control of semilinear elliptic equations and variational inequalities. *SIAM J. Control Optim.*, **33**, 274–298.

Casas, E. (1992) Finite element approximations for some state-constrained optimal control problems, in *Mathematics of the Analysis and Design in Process Control* (eds. P. Borne, S. Tzafestas, and N. Radhy), Amsterdam, North Holland, 293–301.

Casas, E. (1995) Pontryagin's principle for state-constrained boundary control problems of semilinear parabolic equations. Technical Report 1299, IMA Preprint Series.

Ekeland, I. (1975) Nonconvex minimization problems. *Bull. Amer. Math. Soc.*, **1**, 76–91.

Fattorini, H. and Frankowska, H. (1991) Necessary conditions for infinite dimensional control problems. *Math. Control Signals Systems*, **4**, 41–67.

Li, X. and Yong, J. (1995) Optimal Control Theory for Infinite Dimensional Systems. Birkhäuser, Boston.

20
Approximation of nonconvex distributed parameter optimal control problems

I. Chryssoverghi
Department of Mathematics, National Technical University
Zografou Campus, 15773 Athens, Greece.
Phone: 301-7721716. Fax: 301-7721775.
E-mail: ichriso@nisyros.ntua.gr

Abstract
We consider a nonconvex optimal control problem involving nonlinear parabolic systems with several equality and inequality state constraints. Existence and extremality results are obtained using relaxation theory. We then completely discretize the problem and study the limit behaviour of the approximations. Finally, we propose a mixed Frank–Wolfe penalty method using relaxed controls.

Keywords
Optimal control, nonlinear parabolic systems, nonconvexity, relaxed controls, discretization, mixed Frank–Wolfe penalty method

1 INTRODUCTION

It is well known that optimal control problems have no classical solutions in general. In order to prove the existence of optimal controls, some convexity assumptions are usually made on data. In the absence of convexity, the problem is replaced by a generalized, or relaxed, one. Using relaxation theory, one can prove not only the existence of optimal controls for the relaxed problem, but also derive necessary conditions for optimality and develop approximation and optimization methods, for lumped and as well as for distributed parameter systems (see References and bibliography there).

Here we consider an optimal control problem for semilinear parabolic systems with several control and state constraints. We first establish existence and necessary conditions for optimality for the relaxed problem. We then discretize the problem by using a finite element method in space and a semi-implicit finite difference scheme in time, while the controls are independently approximated by piecewise constant relaxed or classical ones and the state constraints slightly perturbed for admissibility reasons. It turns out that the properties of discrete optimality and extremality carry over in the limit to the respective continuous properties. The above results extend those of Chryssoverghi and

Bacopoulos (1993). In addition, we propose a mixed Frank–Wolfe penalty method, which constructs Gamkrelidze relaxed controls, for solving theoretically the continuous relaxed problem and practically the discrete relaxed problem. Finally, using a simple procedure, relaxed controls thus constructed can be approximately simulated by classical controls. A numerical example is given.

2 THE CONTINUOUS OPTIMAL CONTROL PROBLEMS

Let $\Omega \subset I\!R^d$ be a bounded domain, $I := (0,T)$, $T < \infty$, $U \subset I\!R^{d'}$ a compact set, and $A(t)$ a second order linear elliptic differential operator. The continuous classical optimal control (CCP) is the following. The state equation is

$$y_t + A(t)y = f(x,t,y(x,t),w(x,t)) \text{ in } Q := \Omega \times I,$$

$$y = 0 \text{ in } \partial\Omega \times I,\ y(x,0) = y^0(x) \text{ in } \Omega.$$

Defining the functionals

$$J_m(w) := \int_\Omega g_m(x,t,y(x,t),w(x,t))dxdt,\ 0 \leq m \leq q,$$

the constraints are

$$w(x,t) \in U \text{ in } Q,\ J_m(w) = 0,\ 1 \leq m \leq p,\ J_m(w) \leq 0,\ p < m \leq q,$$

and the cost functional to minimize is $J_0(w)$.

The CCP has no classical solutions in general, and to prove the existence of optimal controls some convexity assumptions are usually made on the data. For example, if there are no state constraints, it is assumed that the set

$$S(x,t,y) := \{(\nu,c) |\, \nu = f(x,t,y,u),\ u \in U, g_0(x,t,y,u) \leq c\}$$

is convex, for every $(x,t,y) \in Q \times I\!R$.
In the absence of convexity, we generalize the control problem as follows. Let

$$W := \{w : \overline{Q} \longrightarrow U \text{ measurable}\}$$

be the set of classical controls. The set of relaxed controls is defined by

$$R := \{r : \overline{Q} \longrightarrow M_1(U) \text{ measurable}\}$$

where $M_1(U) \subset C(U)^*$ is the set of probability measures on U with the relative weak star topology. The set R is a convex, metrizable and compact subset of $L^1(\overline{Q}, C(U))^*$ for the relative weak star topology (see Warga, 1972). Moreover, W is embedded in R and W is dense in R. For $r \in span(R)$ and $\phi \in L^1(\overline{Q}, C(U))^*$, we write

$$\phi(x,t,r(x,t)) := \int_U \phi(x,t,u)r(x,t)(du).$$

Now set $V := H_0^1(\Omega)$, and let $a(t,\cdot,\cdot)$ be the bilinear form on V associated to $A(t)$. The continuous relaxed problem (CRP) is the following. For every $r \in R$, the state $y = y_r$ is the unique solution of

$$< y_t, \nu > + a(t,y,\nu) = \int_\Omega f(x,t,y(x,t),r(x,t))\nu(x)dx, \text{ for every } \nu \in V,$$

$$y(0) = y^0 \in L^2(\Omega)(\text{or } V), \;\; y \in L^2(I,V), \;\; y_t \in L^2(I,V^*).$$

The constraints are

$$r \in R, \;\; J_m(r) = 0, \;\; 1 \le m \le p, \;\; J_m(r) \le 0, \;\; p < m \le q,$$

and the cost functional to minimize is $J_0(r)$.

We suppose that the functions f, g_m, $0 \le m \le p$, are defined on $Q \times I\!R \times U$, measurable for fixed y, u, continuous for fixed x, t, and satisfy

$$|f(x,t,y,u)| \le \phi(x,t) + b|y|,$$

$$|f(x,t,y_1,u) - f(x,t,y_2,u)| \le L|y_1 - y_2|,$$

$$|g_m(x,t,y,u)| \le \psi_m(x,t) + c_m y^2,$$

on $Q \times I\!R \times U$, for some $\phi \in L^2(Q)$, $\psi_m \in L^1(Q)$.

Theorem 1 *If there exists an admissible control, then there exists an optimal control for the CRP.*

Next, we suppose in addition that the functions g_{my} are defined on $Q \times I\!R \times U$, measurable for fixed y, u, continuous for fixed x, t, and satisfy

$$|g_{my}(x,t,y,u)| \le \xi_m(x,t) + c_m'|y|, \text{ in } Q \times I\!R \times U, \text{ for some } \xi_m \in L^2(Q).$$

Define the general Hamiltonian (we drop the index m)

$$H(x,t,y,z,u) := zf(x,t,y,u) + g(x,t,y,u),$$

and the general adjoint equation

$$- < z_t, \nu > + a(t,\nu,z) = \int_\Omega [f_y(y,r)z + g_y(y,r)]\nu dx, \text{ for every } \nu \in V,$$

$$z(T) = 0, \;\; z \in L^2(I,V), \;\; z_t \in L^2(I,V^*), \;\; y = y_r.$$

The directional derivative of J is then given by

$$DJ(r, r' - r) = \int_Q H(x,t,y,z,r' - r)dxdt.$$

Theorem 2 *Relaxed Minimum Principle of Optimality*
If r is optimal for either the CRP or the CCP, then r is extremal, i.e. there exist multipliers $\lambda_m \in I\!R$, $0 \le m \le p$, with $\lambda_0 \ge 0$, $\lambda_m \ge 0$, $p < m \le q$, and $\sum_0^q |\lambda_m| = 1$, such that

$$H(x,t,y(x,t),z(x,t),r(x,t)) = \min_{u\in U} H(x,t,y(x,t),z(x,t),u) \text{ a.e. in } Q,$$

where g, g_y are replaced by $\sum_0^q \lambda_m g_m$, $\sum_0^q \lambda_m g_{my}$ *in H, z,*

$$\lambda_m J_m(r) = 0, \; p < m \le q.$$

Further results
Suppose in addition that $f = f_1(y) + f_2(u)$, with f_1 affine, f_2 nonlinear, $g_m = g_{1m}(y) + g_{2m}(u)$, with g_{1m} affine, g_{2m} nonlinear, for $1 \le m \le p$, and g_{1m} convex, g_{2m} nonlinear, for $m = 0$, $p < m \le q$. If $r \in R$ (resp. $r \in W$) is extremal, with some $\lambda_0 > 0$, then r is optimal for the CRP (resp. the CRP and the CCP).

Since $W \subset R$, in general we have

$$\min J_0(r) \le \inf J(w),$$

and it can be proved that the equality holds if (i) there are no state constraints, or (ii) the CCP has a solution but has no admissible abnormal (i.e. with some $\lambda_0 = 0$) extremal controls.

3 DISCRETIZATION

For every n, let $\{S_i^n\}_1^{M(n)}$ be an admissible regular triangulation of $\overline{\Omega}$ into d-simplices, $\{I_j^n\}_0^{N(n)-1}$ a subdivision of $\overline{I} := [0,T]$ into intervals of length $\Delta t^n = T/N(n) \to 0$, as $n \to \infty$, and $\{B_k^n\}_1^{P(n)}$ a partition of $\overline{\Omega}$ into Borel subsets, with $\max_k \mathit{diam}\,(B_k^n) \to 0$ as $n \to \infty$. Let $V^n \subset V$ be the set of continuous functions on Ω which are affine on each S_i^n. Let $R^n \subset R$ be the set of relaxed controls which are constant on each B_k^n (discrete relaxed controls), and $W^n \subset W$ the set of classical controls which are constant on each B_k^n (discrete classical controls). Define the discrete state equations

$$(y_{j+1}^n - y_j^n, v)/\Delta t^n + a(t_{j+1}, y_{j+1}^n, v) = (f_j^n, v), \text{ for every } v \in V^n, \; 0 \le j \le N-1,$$

$$y_0^n \to y^0 \text{ in } V \text{ as } n \to \infty, \; y_j^n \in V^n, \; 0 \le j \le N,$$

$$f_j^n(x) := (1/\Delta t^n) \int_{I_j^n} f(x,t,y_j^n(x),r^n(x,t))dt,$$

the discrete functionals

$$J_m^n(r^n) := \int_Q g_m(x,t,y_-^n(x,t),r^n(x,t))dxdt,$$

where $y_-^n := y_j^n$ in I_j^n, $0 \le j \le N-1$, the constraints

$$r^n \in R^n, \text{ either (1) } |J_m^n(r^n)| \le \varepsilon_m^n,\ 1 \le m \le p, \text{ or (2) } J_m^n(r^n) = \varepsilon_m^n,\ 1 \le m \le p,$$

$$J_m^n(r^n) \le \varepsilon_m^n,\ \varepsilon_m^n \ge 0,\ p < m \le q,$$

where $\varepsilon_m^n \to 0$ as $n \to \infty$, $1 \le m \le q$, and the cost functional to minimize $J_0^n(r^n)$. The discrete relaxed problem DRP_1^n (resp. DRP_2^n) is the above problem, case (1) (resp. case (2)). The discrete classical problems DCP_1^n, DCP_2^n are the problems DRP_1^n, DRP_2^n, where the constraint $r^n \in R^n$ is replaced by $r^n \in W^n$.

Theorem 3 *If there exists an admissible control, then there exists an optimal control for the above discrete problems.*

Now define the discrete adjoint equations

$$(z_j^n - z_{j+1}^n, \nu)/\Delta t^n + a(t_j, \nu, z_j^n) = (f_{yj}^n z_{j+1}^n + g_{yj}^n, \nu), \text{ for every } \nu \in V^n, 0 \le j \le N-1,$$

$$z_N^n = 0,\ z_j^n \in V^n,\ 0 \le j \le N,$$

$$f_{yj}^n(x) := (1/\Delta t^n) \int_{I_j^n} f_y(x,t,y_j^n(x), r^n(x,t))dt,$$

$$g_{yj}^n(x) := (1/\Delta t^n) \int_{I_j^n} g_y(x,t,y_j^n(x), r^n(x,t))dt.$$

Theorem 4 *Discrete Relaxed Minimum Principle*
If r^n is optimal for the DRP_2^n, then r^n is extremal, i.e. there exist multipliers $\lambda_m^n \in \mathbb{R}$, $0 \le m \le q$, with $\lambda_0^n \ge 0$, $\lambda_0^n \ge 0$, $p < m \le q$, and $\sum_0^q |\lambda_m^n| = 1$, such that

$$\int_{B_k^n} H(x,t,y_-^n, z_+^n, r^n)dxdt = \min_{u \in U} \int_{B_k^n} H(x,t,y_-^n, z_+^n, u)dxdt,\ k = 1, \ldots, P,$$

where $z_+^n := z_{j+1}^n$ on I_j^n, $0 \le j \le N-1$, and g, g_y are replaced by $\sum_0^q \lambda_m^n g_m$, $\sum_0^q \lambda_m^n g_{my}$,

$$\lambda_m^n [J_m^n(r) - \varepsilon_m^n] = 0,\ p < m \le q.$$

Note that the minimum principle does not necessarily hold for the DCP_2^n without some additional convexity or similar assumptions.

The next two theorems show the behaviour in the limit of the above discretizations.

Theorem 5 *For each n, let r^n be optimal for the DRP_1^n (resp. DCP_1^n). Under appropriate feasibility assumptions on the ε_m^n (see Chryssoverghi and Kokkinis, 1994), every accumulation point of the sequence $\{r^n\}$ is optimal for the CRP.*

Theorem 6 *For each n, let r^n be an admissible and extremal for the DRP_2^n. Under minimum feasibility assumptions on the ε_m^n, every accumulation point of the sequence $\{r^n\}$ is admissible and extremal for the CRP.*

4 A MIXED FRANK-WOLFE PENALTY METHOD

The method is theoretically applied to the CRP and practically to the DRP_2^n. Let $s_m > 1$, $1 \leq m \leq q$, be chosen numbers, $\{M_m^k\}$, $1 \leq m \leq q$, $\{\delta^k\}$, chosen sequences such that $M_m^k \to \infty$, $1 \leq m \leq q$, $\delta^k \to 0$, as $k \to \infty$, and for each k, define the penalized cost functional (we drop here the index n)

$$G^k(r) := J_0(r) + \sum_1^p M_m^k |J_m(r)|^{s_m} + \sum_{p+1}^q M_m^k [max\,(0, J_m(r))]^{s_m}.$$

Algorithm

Choose $r_1 \in W$.
Step 1. Set $\overline{r_0} := r_1$, $l = 1$, $k = 1$.
Step 2. Find $r_1 \in W$ such that

$$\gamma_l := DG^k(r_l, \overline{r_l} - r_l) = \min_{\overline{r} \in R} DG^k(r_l, \overline{r} - r_l), \ (\gamma_l \leq 0).$$

Step 3. If $\gamma_l < -\delta^k$, go to step 4. Else, set $r^k := r_l$, $\gamma^k := \gamma_l$, $k := k+1$, and go to step 2.
Step 4. Find $\alpha_l \in [0, 1]$ such that

$$G^k(r_l + \alpha_l(\overline{r_l} - r_l)) = \min_{0 \leq \alpha \leq 1} G^k(r_l + \alpha(\overline{r_l} - r_l))$$

Step 5. Set $r_{l+1} := r_l + \alpha_l(\overline{r_l} - r_l)$, $l := l+1$, and go to step 2.

The classical control $\overline{r_l}$ is in fact computed in step 2 by piecewise minimizing a Hamiltonian, and r_l is a convex combination of the classical controls $\overline{r_0}, \ldots, \overline{r_{l-1}}$ (Gamkrelidze relaxed control). Let $\{r^k := r_{l(k)}\}$ be the sequence generated by the algorithm (step 3), and define the sequences of multipliers

$$\lambda_m^k := M_m^k |J_m(r^k)|^{s_m - 1}, \ 1 \leq m \leq p,$$

$$\lambda_m^k := M_m^k [max\,(0, J_m(r^k))]^{s_m - 1}, \ p < m \leq q.$$

Theorem 7 *Let $\{r^k\}$ be a subsequence converging to some $r \in R$ (or R^n).*

(a) If the sequences $\{\lambda_m^k\}$, $1 \leq m \leq q$, are bounded, then r is admissible and extremal.
(b) Suppose that the problem has no admissible abnormal (i.e. with some $\lambda_0 = 0$) extremal controls. If r is admissible, then the sequences $\{\lambda_m^k\}$ are bounded and r is extremal.

5 APPROXIMATION BY CLASSICAL CONTROLS

For simplicity take $B_k^n := S_i^n \times I_j^n$, and let $r_l := \sum_0^{l-1} \beta_\mu \overline{r_\mu}$ be a Gamkrelidze control computed by the algorithm. Subdivide each I_j^n into intervals $I_{l\mu}^n$, $0 \leq \mu \leq l-1$, whose lengths are proportional to the β_l. The control r^k is then approximately simulated by the classical control

$$w_l(x,t) := \overline{r_\mu}(x,t) \text{ in } \Omega^n \times I^n_{l\mu},\ 0 \le \mu \le l-1,\ 0 \le j \le N-1.$$

6 NUMERICAL EXAMPLE

State equation:

$$y_t - y_{xx} = -siny + 6w - 3 \text{ in } Q := (0,\pi) \times (0,1),$$

$$y(0,t) = y(\pi,t) = 0 \text{ in } (0,1),$$

$$y(x,0) = 0 \text{ in } (0,\pi),$$

Constraints:

$$w(x,t) \in \{0,1\} \text{ (two values) in } Q,$$

$$J_1(w) := \int_Q y(x,t)dxdt = 0,$$

$$y(x,t) \le 0.4 \text{ in } Q \Leftrightarrow J_2(w) := \int_Q [max\,(0, y(x,t) - 0.4)]^{1.4} dxdt = 0,$$

Cost functional:

$$J_0(w) := \int_Q [y(x,t) - tsinx]^2 dxdt.$$

After 24 iterations (in l) of the algorithm, and with $\Delta x = \pi/40$, $\Delta t = 1/40$, we found the following approximate functional values

$$J_0 = 0.187101,\quad J_1 = 0.00258348,\quad J_2 = 0.000178508.$$

REFERENCES

Chryssoverghi, I. and Bacopoulos, A. (1993) Approximation of relaxed nonlinear parabolic optimal control problems, *Journal of Optimization Theory and Applications*, 77, 31–50.

Chryssoverghi, I. and Kokkinis, B. (1994) Discretization of nonlinear elliptic optimal control problems, *Systems and Control Letters*, 22, 227–234.

Warga, J. (1972) Optimal control of differential and functional equations, *Academic Press*, New York.

Warga, J. (1977) Steepest descent with relaxed controls, *Journal of Optimization Theory and Applications*, 15, 674–682.

21

A Trotter–type scheme for the generalized gradient of the optimal value function

Cătălin Popa
Facultatea de Matematică, Universitatea "Al.I.Cuza"
Bdul. Copou 11, 6600 Iaşi, România
Fax: 40-32-146330. E-mail: cpopa@uaic.ro

Abstract

An iterative formula for the generalized gradient of a Trotter–type approximation for the optimal value function associated with the control of a certain nonlinear parabolic system is established. This formula is useful in constructing suboptimal feedback controls.

Keywords

Trotter product formula, dynamic programming equation, parabolic variational inequality, (sub)optimal feedback control

1 INTRODUCTION

In this paper we shall reveal an interesting aspect of approximation of dynamic programming Hamilton–Jacobi equations via Trotter product formulas. Also, we shall explain how this aspect is reflected in constructing suboptimal feedback controls.

One of the main objectives of optimal control theory is the construction of optimal feedback controls. Simple heuristic considerations based on dynamic programming lead to feedback laws expressed by means of the optimal value function. In many significant situations, these laws can be rigorously justified (see, for example, Barbu (1984) and Popa (to appear)). However, it remains the question: how to compute (or approximate) the optimal value function in a reasonable manner (so that the feedback law should become effective)? Certainly, this function satisfies a Hamilton–Jacobi equation (called, in this context, the dynamic programming equation), but such an equation is a very complicated mathematical object. In author's opinion, the most promising way to compute solutions of Hamilton–Jacobi equations (satisfying certain initial or final conditions), is offered by a treatment of these equations by Trotter product formulas. These formulas are obtained by breaking Hamilton–Jacobi equations into two or several parts on small intervals (that is, by decomposing the associated Cauchy problem into two or several such problems). This decoupling leads to a better understanding of the making of solutions and, from a numerical viewpoint, to decentralization of calculus. Several convergent Trotter product formulas was proposed by V.Barbu and the author for dynamic programming equations associated with the control of parabolic variational inequalities (see Barbu (1988, 1991) and Popa (1991, 1995)). Numerical tests performed by V.Arnăutu and A.Niemistö (see Arnăutu (1995) and Niemistö (to appear)) show that the Trotter product formula approach is indeed an effective and realistic way in computing solutions of dynamic programming equations, and, implicitly, in constructing suboptimal feedback controls.

The generating idea of this paper consists in two simple observations. First, if we have a Trotter product formula approach to Hamilton–Jacobi equations in view, it is much easier to calculate the gradients of solutions than the solutions themselves. In other words, Trotter approximations for the gradients of solutions are much simpler than those for the solutions. On the other hand, the expression of feedback laws explicitly contains not only the solution of the dynamic programming equation but even its gradient. So, instead of computing solutions of dynamic programming equations by Trotter schemes and then differentiating them, it is preferable to directly compute gradients of solutions by schemes of the same kind.

Our aim here is to give an iterative formula for the generalized gradient of the Trotter approximation for the solution of the dynamic programming equation associated with the control of a certain nonlinear parabolic system. This formula seems to be much more effective in constructing suboptimal feedback controls. Also, it may be interpreted as a Trotter approximation for the vector variant of the conservation law equation obtained by formal differentiation of the dynamic programming equation.

2 AN EXAMPLE

Let us illustrate what we have asserted above by a simple example.

Consider the Hamilton–Jacobi equation of Hamiltonian mechanics corresponding to Hamiltonian $H(p, y) = \frac{1}{2}|p|^2 + U(y)$:

$$\begin{cases} S_t(t, y) + \frac{1}{2}|S_y(t, y)|^2 + U(y) = 0 \text{ in } [0, T] \times \mathcal{H}, \\ S(0, y) = S_0(y), \end{cases} \tag{1}$$

where $\mathcal{H}$ is the state space. Divide the interval $[0, T]$ into N subintervals of the same length $\varepsilon = T/N$. Let us decouple the terms corresponding to the two kinds of energy in (1) on each subinterval by decomposing Cauchy problem (1) into two elementary Cauchy problems. One obtains the following Trotter approximation:

$$\begin{cases} S_t^\varepsilon(t, y) + \frac{1}{2}|S_y^\varepsilon(t, y)|^2 = 0 \text{ in } ((i-1)\varepsilon, i\varepsilon] \times \mathcal{H}, \\ S^\varepsilon((i-1)\varepsilon + 0, y) = S^\varepsilon((i-1)\varepsilon, y) - \varepsilon U(y). \end{cases} \tag{2}$$

We can express the solution of Cauchy problem (2) by using the well-known Lax representation:

$$S^\varepsilon(i\varepsilon, y) = \inf\{\frac{1}{2\varepsilon}|z - y|^2 + S^\varepsilon((i-1)\varepsilon, z) - \varepsilon U(z) : z \in \mathcal{H}\},\ i = 1, 2, ..., N. \tag{3}$$

Now, solving the minimization problem contained in (3), one easily finds the following alternative (but more elaborated) formula:

$$\begin{cases} S^\varepsilon(i\varepsilon, y) & = \frac{\varepsilon}{2}|(\nabla_y S^\varepsilon((i-1)\varepsilon, \cdot) - \varepsilon\nabla U)_\varepsilon y|^2 \\ & +S^\varepsilon((i-1)\varepsilon, (I + \varepsilon(\nabla_y S^\varepsilon((i-1)\varepsilon, \cdot) - \varepsilon\nabla U))^{-1}y) \\ & -\varepsilon U((I + \varepsilon(\nabla_y S^\varepsilon((i-1)\varepsilon, \cdot) - \varepsilon\nabla U))^{-1}y), \; i = 1, 2, ..., N, \\ S^\varepsilon(0, y) & = S_0(y). \end{cases} \tag{4}$$

Next, it is not difficult to show (see Barbu and Precupanu (1986), Thm. 2.3, p.121) that the gradient of S^ε is given by the following iterative formula:

$$\begin{cases} \nabla_y S^\varepsilon(i\varepsilon, y) = (\nabla_y S^\varepsilon((i-1)\varepsilon, \cdot) - \varepsilon\nabla U)_\varepsilon y, \; i = 1, 2, ..., N, \\ \nabla_y S^\varepsilon(0, y) = \nabla S_0(y). \end{cases} \tag{5}$$

In (4), (5) (and throughout in the sequel) the symbol ε subscript after an operator means passing to its Yosida approximation (for instance, $(\nabla S_0 - \varepsilon\nabla U)_\varepsilon = \frac{1}{\varepsilon}(I - (I + \varepsilon(\nabla S_0 - \varepsilon\nabla U))^{-1}))$.

Clearly, the iterative formula (5) for the gradient of S^ε is simpler than that for S^ε given by (4). To make this fact even more striking, apply (4) i times. One obtains

$$\begin{aligned} S^\varepsilon(i\varepsilon, y) & = \frac{\varepsilon}{2}|P_1 Q_2 Q_3 ... Q_i y|^2 + \frac{\varepsilon}{2}|P_2 Q_3 ... Q_i y|^2 + \cdots + \frac{\varepsilon}{2}|P_i y|^2 \\ & -\varepsilon U(Q_1 Q_2 Q_3 ... Q_i y) - \varepsilon U(Q_2 Q_3 ... Q_i y) - \cdots - \varepsilon U(Q_i y) \\ & +S_0(Q_1 Q_2 Q_3 ... Q_i y), \end{aligned} \tag{6}$$

where P_j, Q_j are the nonlinear operators defined by

$$\begin{cases} P_j = (P_{j-1} - \varepsilon\nabla U)_\varepsilon, \; j = 1, 2, ..., N, \\ P_0 = \nabla S_0, \end{cases} \tag{7}$$

$$Q_j = (I + \varepsilon(P_{j-1} - \varepsilon\nabla U))^{-1}, \; j = 1, 2, ..., N. \tag{8}$$

On the other hand, the expression of the gradient of S^ε at $i\varepsilon$ contains only operator P_i :

$$\nabla_y S^\varepsilon(i\varepsilon, y) = P_i y = \underbrace{(...((\nabla S_0 - \varepsilon\nabla U)_\varepsilon - \varepsilon\nabla U)_\varepsilon ... - \varepsilon\nabla U)_\varepsilon}_{i \text{ iterations}} \, y. \tag{9}$$

In conclusion, from a theoretical point of view, the iterative formula (5) for the gradient of S^ε is more attractive than that for S^ε given by (4) at least because the final expression (9) for $\nabla_y S^\varepsilon(i\varepsilon, y)$ is much simpler than that for $S^\varepsilon(i\varepsilon, y)$ given by formulas (6)–(8) (it contains only P_i). Finally, let us point out that $\nabla_y S^\varepsilon(i\varepsilon, y)$ (given by (5)) can be interpreted as a Trotter approximation for the solution $\nabla_y S(t, y)$ of the vector variant of the conservation law equation obtained by formal differentiation of Hamilton–Jacobi equation (1). In other words, $\nabla_y S^\varepsilon(i\varepsilon, y)$ can be formally obtained like $S^\varepsilon(i\varepsilon, y)$, but by treating (this time) the conservation law equation in the same manner as above.

3 THE FRAMEWORK AND THE MAIN RESULT

All the preceding considerations can be repeated with the same effect in the more complex situation of dynamic programming equations associated with the control of nonlinear parabolic systems.

Let $\mathcal{U}$ be a real Hilbert space and set $\mathcal{H} = L^2(\Omega)$, where Ω is an open and bounded subset of $I\!R^n$ having a sufficiently smooth boundary. The control system we deal with is described by the following mixed boundary value problem:

$$\begin{cases} \dfrac{\partial y}{\partial t} + A_0 y + \beta(y) \ni Bu \quad \text{a.e. in } (0,T) \times \Omega, \\ y = 0 \quad \text{on } (0,T) \times \partial\Omega, \\ y(0,x) = y^0(x) \quad \text{in } \Omega. \end{cases} \tag{10}$$

Here A_0 is the elliptic differential operator defined by

$$A_0 y = -\sum_{i,j=1}^{n} \frac{\partial}{\partial x_j}(a_{ij}(x)\frac{\partial y}{\partial x_i}) + a_0(x)y,$$

where $a_{ij} \in C^1(\Omega)$, $a_0 \in L^\infty(\Omega)$, $a_{ij} = a_{ji}$, $a_0(x) \geq 0$ a.e. $x \in \Omega$, and, for some $\omega > 0$,

$$\sum_{i,j=1}^{n} a_{ij}(x)\xi_i\xi_j \geq \omega \sum_{i=1}^{n} |\xi_i|^2 \quad \text{for all } (\xi_1, \xi_2, ..., \xi_n) \in I\!R^n, \text{ a.e. } x \in \Omega.$$

The nonlinear term β is a maximal monotone graph in $I\!R^2$ containing $(0,0)$. Take a convex function $j : I\!R \to (-\infty, +\infty]$ whose subdifferential is β, and define the convex function $\phi : \mathcal{H} \to (-\infty, +\infty]$ as $\phi(y) = \int_\Omega j(y(x))dx$. Operator B from $\mathcal{U}$ to $\mathcal{H}$ is linear and continuous, and $y^0 \in \overline{D(\phi)}$.

A standard existence result (see Barbu (1984), Thm. 4.3, p.131) states that under the above assumptions on A_0, β, B and y^0, problem (10) has a unique solution $y \in C([0,T];\mathcal{H})$ such that $\sqrt{t}\,y' \in L^2(0,T;\mathcal{H})$ and $\sqrt{t}\,y \in L^2(0,T;H_0^1(\Omega) \cap H^2(\Omega))$. Let us also mention that problem (10) can be interpreted as an evolution equation in $\mathcal{H}$, that is, the solution y of (10) satisfies

$$y' + Ay + \beta(y) \ni Bu \quad \text{a.e. in } (0,T),$$

where A is the linear continuous operator from $\mathcal{V} = H_0^1(\Omega)$ to $\mathcal{V}' = H^{-1}(\Omega)$ defined by

$$(Ay, z) = \sum_{i,j=1}^{n} \int_\Omega a_{ij} \frac{\partial y}{\partial x_i} \frac{\partial z}{\partial x_j}\, dx + \int_\Omega a_0 y z\, dx \quad \text{for all } y, z \in H_0^1(\Omega).$$

Consider the following optimal control problem:

(P) Minimize

$$\int_0^T (h(u(t)) + g(y(t)))dt + \ell(y(T))$$

over all $u \in L^2(0,T;\mathcal{U})$, where $y \in C([0,T];\mathcal{H})$ satisfies (10).

Impose on the functions h, g, ℓ the following hypotheses:

(H1) $h : \mathcal{U} \to (-\infty, +\infty]$ is convex, lower semicontinuous, not identically $+\infty$ and, for some $c_1 > 0$ and $c_2 \in \mathbb{R}$, satisfies

$$h(u) \geq c_1|u|^2 - c_2 \quad \text{for all } u \in \mathcal{U}.$$

(H2) $g, \ell : \mathcal{H} \to \mathbb{R}$ are Lipschitz continuous on bounded subsets and bounded from below by affine functions.

We associate with problem (P) the corresponding optimal value function $V : [0,T] \times \overline{D(\phi)} \to \mathbb{R}$:

$$\begin{aligned} V(t,y) = \inf\{ & \int_t^T (h(u(s)) + g(z(s)))ds + \ell(z(T)) : \\ & z' + Az + \beta(z) \ni Bu \ \text{ a.e. in } (t,T), z(t) = y, u \in L^2(t,T;\mathcal{U})\}. \end{aligned}$$

One knows that for every optimal pair of problem (P), the following feedback law holds (see Barbu (1984), Thm. 5.6, p.208, and Popa (to appear), Thm. 2.3):

$$u^*(t) \in \partial h^*(-B^*\partial_y V(t, y^*(t))) \ \text{ a.e. } t \in (0,T). \tag{11}$$

Here h^* is the convex conjugate of h, B^* is the adjoint of B, and $\partial_y V(t,y)$ is the generalized gradient (in Clarke's sense) of $y \mapsto V(t,y)$.

It is also well-known that the optimal value function V satisfies (in a certain generalized sense) the following Hamilton–Jacobi equation:

$$\begin{cases} D_t V(t,y) - h^*(-B^* D_y V(t,y)) - (Ay + \beta(y), D_y V(t,y)) + g(y) = 0 \\ \qquad\qquad\qquad\qquad\qquad\qquad\qquad\qquad \text{in } [0,T] \times \mathcal{H}, \\ V(T,y) = \ell(y), \ y \in \mathcal{H}. \end{cases} \tag{12}$$

One can obtain Trotter–type approximations for V by treating equation (12) in a similar manner as in the preceding section. Let $\varepsilon = T/N$. Decoupling the last three terms in the left–hand side of (12) on each subinterval $[i\varepsilon, (i+1)\varepsilon]$ and then using a Lax–type representation formula, we get (in a heuristic manner) the following Trotter scheme, proposed in Popa (1995) (for simplicity, we shall indicate it only for $t = i\varepsilon$):

$$\begin{cases} V^\varepsilon(i\varepsilon, y) = \inf\{ \ \varepsilon h(u) + \varepsilon g(y + \varepsilon Bu) \\ \qquad\qquad + V^\varepsilon((i+1)\varepsilon, (I + \varepsilon\beta)^{-1}(I + \varepsilon A)^{-1}(y + \varepsilon Bu)) : u \in \mathcal{U}\}, \\ \qquad\qquad\qquad\qquad\qquad y \in \mathcal{H}, \ i = 0, 1, ..., N-1, \\ V^\varepsilon(T,y) = \ell(y), \ y \in \mathcal{H}. \end{cases} \tag{13}$$

The fact that V^ε really approximates V is not trivial in the case of nonlinear and infinite–dimensional control systems (see Popa (1995), Thm. 3.1).

The following question arises at this point: If we take V^ε instead of V in feedback law (11), does the new feedback law provide suboptimal controls (that is, approximately optimal controls) for problem (P)? All that we know at the moment is that the answer is positive in the following sense: Consider the following discrete approximation for (P) (see Popa (1995), Thm. 5.1):

(P^ε) Minimize
$$\sum_{i=1}^{N} \varepsilon(h(u_i) + g(y_i)) + \ell(y_N)$$
over all N–tuples $(u_1, u_2, ..., u_N) \in \mathcal{U}^N$, where $(y_1, y_2, ..., y_N) \in \mathcal{H}^N$ satisfies the scheme

$$\begin{cases} y_i = (I + \varepsilon\beta)^{-1}(I + \varepsilon A)^{-1}(y_{i-1} + \varepsilon B u_i), \; i = 1, 2, ..., N, \\ y_0 = y^0. \end{cases} \tag{14}$$

(Note that scheme (14) can be viewed as a Trotter approximation for the state equation (10).) Then, for every optimal N–tuple $(u_1^\varepsilon, u_2^\varepsilon, ..., u_N^\varepsilon)$ of problem (P^ε), the following discrete version of feedback law (11) holds (see Popa (1995), Thm. 6.1):

$$\begin{cases} u_i^\varepsilon \in \partial h^*(-B^*\partial_y V^\varepsilon((i-1)\varepsilon, y_{i-1}^\varepsilon)), \\ y_i^\varepsilon = (I + \varepsilon\beta)^{-1}(I + \varepsilon A)^{-1}(y_{i-1}^\varepsilon + \varepsilon B u_i^\varepsilon), \; i = 1, 2, ..., N, \\ y_0^\varepsilon = y^0. \end{cases} \tag{15}$$

It is clear (if we regard scheme (15)) that to construct suboptimal feedback controls for problem (P), we must be able to compute $\partial_y V^\varepsilon(i\varepsilon, y)$. Our aim is to give an iterative formula for $\partial_y V^\varepsilon(i\varepsilon, y)$ analogous to that for $\nabla_y S^\varepsilon(i\varepsilon, y)$ established before.

The following additional hypotheses are needed:

(H3) g, ℓ are convex and bounded on bounded subsets.
(H4) If $y, z \in \mathcal{H}$ satisfy $y \leq z$ a.e. in Ω, then $g(y) \leq g(z)$ and $\ell(y) \leq \ell(z)$.

Set $V_-^\varepsilon((i+1)\varepsilon, y) = V^\varepsilon((i+1)\varepsilon, (I + \varepsilon\beta)^{-1}(I + \varepsilon A)^{-1}y) + \varepsilon g(y)$. (We have already met the above expression in the definition (13) of V^ε.)

The following theorem is our main result.

Theorem 1 *Under the above hypotheses on A_0, β, B, y^0 and h, suppose in addition that β is concave and g and ℓ satisfy (H3) and (H4). Then*

$$\partial_y V^\varepsilon(i\varepsilon, y) \subset \bigcup_{u_i} \partial_y V_-^\varepsilon((i+1)\varepsilon, y + \varepsilon B u_i), \tag{16}$$

where u_i runs over all solutions of the inclusion

$$u_i \in \partial h^*(-B^*\partial_y V_-^\varepsilon((i+1)\varepsilon, y + \varepsilon B u_i)), \; i = N-1, ..., 1, 0, \tag{17}$$

and

$$\partial_y V^\varepsilon(N\varepsilon, y) = \partial\ell(y).$$

If $\mathcal{U} = \mathcal{H}$ and $B = I$, then

$$\partial_y V^\varepsilon(i\varepsilon, y) \subset \partial_y V_-^\varepsilon((i+1)\varepsilon, (I - \varepsilon\partial h^*(-\partial_y V_-^\varepsilon((i+1)\varepsilon, \cdot)))^{-1} y). \tag{18}$$

Sketch of the proof. First of all, let us remark that the function $y \mapsto V^\varepsilon(i\varepsilon, y)$ is convex on $\mathcal{H}$. (Consequently, $\partial_y V^\varepsilon(i\varepsilon, y)$ will coincide with the subdifferential of $y \mapsto V^\varepsilon(i\varepsilon, y)$

in the sense of convex analysis.) Indeed, by convexity of ℓ and $r \mapsto (I+\varepsilon\beta)^{-1}(r)$ (do not forget that β is concave), using also (H4), we infer that the function $y \mapsto \ell((I+\varepsilon\beta)^{-1}(I+\varepsilon A)^{-1}y)$ is convex on $\mathcal{H}$. Also, this function is nondecreasing in the sense of (H4). (It suffices to apply the monotonicity of $r \mapsto (I+\varepsilon\beta)^{-1}(r)$ and the maximum principle to the elliptic operator $I+\varepsilon A_0$.) Using both these properties of $y \mapsto \ell((I+\varepsilon\beta)^{-1}(I+\varepsilon A)^{-1}y)$ (and g), it is not difficult to show that $y \mapsto V^\varepsilon((N-1)\varepsilon, y)$ is convex and nondecreasing (see also the proof of Lemma 4.1 in Popa (to appear)). Now we successively argue $N-i$ times as above to obtain that $y \mapsto V^\varepsilon(i\varepsilon, y)$ is convex on $\mathcal{H}$ (and nondecreasing). Consequently, $y \mapsto V_-^\varepsilon((i+1)\varepsilon, y)$ is convex too.

Then, we interpret the minimization problem (13) defining V^ε as a discrete optimal control problem with *convex* performance criterion, where the state variable takes only two values: the initial value y and the final one $y+\varepsilon Bu$. For this problem, it is easy to derive the following discrete Pontryagin–type maximum principle (see also formulas (6.17), (6.18) in Popa (1995)): For any optimal element $u_i \in \mathcal{U}$, there exists $p_i \in \mathcal{H}$ such that

$$\begin{cases} p_i \in -\partial_y V_-^\varepsilon((i+1)\varepsilon, y+\varepsilon Bu_i), \\ u_i \in \partial h^*(B^* p_i). \end{cases} \tag{19}$$

(The characteristic feature of the above optimality conditions is that the discrete costate is constant, that is, the initial costate coincides with the final one.)

Since $y \mapsto V_-^\varepsilon((i+1)\varepsilon, y)$ is convex, we can use the same argument of the proof of Proposition 2.2 from Barbu and Precupanu (1986), p. 317, to prove that

$$\partial_y V^\varepsilon(i\varepsilon, y) = \{-p_i \in \mathcal{H} : \text{there exists } u_i \in \mathcal{U} \text{ such that } (p_i, u_i) \text{ satisfies } (19)\}. \tag{20}$$

But (20) in conjunction with (19) gives (16) and (17). □

A result similar to the above theorem was formulated by Barbu but for convex control problems governed by *linear* parabolic equation (see Barbu (to appear)).

Theorem 1 says that to compute $\partial_y V^\varepsilon(i\varepsilon, y)$, we only need to know $z \mapsto \partial_y V_-^\varepsilon((i+1)\varepsilon, z)$. However, to express the gradient of V_-^ε (with respect to y) in terms of the gradient of V^ε, we need some adequate chain rules. So, one can successively compute $\partial_y V^\varepsilon(i\varepsilon, y)$ $(i = N-1, ..., 1, 0)$ by starting with $\partial_y V^\varepsilon(N\varepsilon, y) = \partial\ell(y)$. Here is an example.

Corollary 1 *If* $\beta \in C^1(\mathbb{R})$, $\mathcal{U} = \mathcal{H}$ *and* $B = I$, *then* (18) *can be written in the form*

$$\partial_y V^\varepsilon(i\varepsilon, y) \subset -\partial h(-(-\partial h^*(-\partial_y V_-^\varepsilon((i+1)\varepsilon, \cdot)))_\varepsilon y), \tag{21}$$

where

$$\begin{aligned} \partial_y V_-^\varepsilon((i+1)\varepsilon, y) \;=\; & (I+\varepsilon A)^{-1}(1+\varepsilon\beta'((I+\varepsilon\beta)^{-1}(I+\varepsilon A)^{-1}y))^{-1} \\ & \partial_y V^\varepsilon((i+1)\varepsilon, (I+\varepsilon\beta)^{-1}(I+\varepsilon A)^{-1}y) + \varepsilon\partial g(y). \end{aligned} \tag{22}$$

(Here, as before, ε subscript means passing to Yosida approximation.)

Proof. One easily derives (22) by using Theorem 2.3.10 from Clarke (1983). □

Let us point out that the iterative formulas (21), (22) for the generalized gradient of V^ε (with respect to y) are a substantial generalization of the iterative formula (5) for the gradient of S^ε. (We obtain a retrograde version of (5) if we take $h(\cdot) = \frac{1}{2}|\cdot|^2$, $g \equiv U$, $A \equiv 0$ and $\beta \equiv 0$ in (21), (22).)

4 REFERENCES

Arnăutu, V. (1995) Numerical results for a product formula approximation of Hamilton–Jacobi equations, *International Journal of Computer Mathematics*, **57**, pp. 75–82.

Barbu, V. (1984) *Optimal control of variational inequalities*, Research Notes in Mathematics, 100, Pitman, London.

Barbu, V. (1988) Approximation of the Hamilton–Jacobi equations via Lie–Trotter product formula, *Control Theory and Advanced Technology,* **4**, pp.189–208.

Barbu, V. (1991) The Fractional Step Method for a Nonlinear Distributed Control Problem, in *Differential Equations and Control Theory* (ed. V.Barbu), Pitman Research Notes in Mathematics Series, 250, Longman Scientific & Technical, Harlow, Essex, pp. 7-16.

Barbu, V. (to appear) Approximation of Hamilton–Jacobi Equations and Suboptimal Feedback Controllers, in *Proceedings of Stefan Banach Institute Conference on Free Boundary Problems*, Warsaw, November 1994.

Barbu, V. and Precupanu, Th. (1986) *Convexity and optimization in Banach spaces*, Second edition, Editura Academiei, Bucureşti and D. Reidel Publ. Co., Dordrecht, Boston, Lancaster.

Clarke, F.H. (1983) *Optimization and nonsmooth analysis,* John Wiley, New York.

Niemistö, A. (to appear) Realization of suboptimal feedback controller for evolution inclusions with numerical tests.

Popa, C. (1991) Trotter product formulae for Hamilton–Jacobi equations in infinite dimensions, *Differential and Integral Equations*, **4**, pp. 1251–1268.

Popa, C. (1995) Feedback laws for nonlinear distributed control problems via Trotter-type product formulae, *SIAM Journal on Control and Optimization*, **33**.

Popa, C. (to appear) The relationship between the maximum principle and dynamic programming for the control of parabolic variational inequalities, submitted to *SIAM Journal on Control and Optimization.*

Optimal control problem for semilinear parabolic equations with pointwise state constraints

J. P. Raymond
Université Paul Sabatier, UMR CNRS MIP
31062 Toulouse Cedex France, Fax: (33) 61 55 83 85.
Phone: (33) 61 55 83 14. E–mail: raymond@cict.fr

Abstract
This paper deals with optimal control problems governed by semilinear parabolic equations in presence of pointwise constraints on the state variable. We obtain necessary optimality conditions in which the adjoint state satisfies a parabolic equation with measures as source terms. We prove new existence and regularity results for solutions of such equations. For simplicity we only consider the case of a boundary control, but the results can be extended to problems with a control in the initial condition and a distributed control.

Keywords
Nonlinear boundary controls, semilinear parabolic equations, pointwise state constraints

1 THE CONTROL PROBLEM

We consider the following state equation

$$\frac{\partial y}{\partial t} + Ay + \Phi(x,t,y) = 0 \quad \text{in} \quad Q = \Omega\times]0,T[, \tag{1a}$$

$$\frac{\partial y}{\partial n_A} + \Psi(y,v) = 0 \quad \text{on} \quad \Sigma = \partial\Omega\times]0,T[, \tag{1b}$$

$$y(x,0) = y_0(x) \quad \text{in} \quad \Omega, \tag{1c}$$

Ω is a bounded regular subset of R^N, A is a second order symmetric uniformly elliptic operator with regular coefficients ($Ay = -\Sigma_{i,j} D_j(a_{ij}(x)D_i y)$), y_0 belongs to $C(\bar{\Omega})$, assumptions on Φ and Ψ are given below, the control v satisfies

$$v \in K_V, \tag{2}$$

K_V is a closed convex subset of $L^{\beta r}(\Sigma)$ (for $r > N+1$ and some $\beta \geq 1$). The pointwise constraints are of the form

$$g(x,t,y(x,t)) \leq 0 \qquad \text{for every } (x,t) \in \bar{Q}. \tag{3}$$

We suppose that y_0 is such that $g(x, 0, y_0(x)) \leq 0$ on $\bar{\Omega}$. These constraints are meaningful since we prove in (Raymond and Zidani, 1995) that the solution of (1) belongs to $C(\bar{Q})$ if assumptions (A1) and (A2) (given below) are satisfied. The cost functional is defined by

$$J(y,v) = \int_Q F(x,t,y)dxdt + \int_\Sigma G(y,v)dsdt + \int_\Omega H(x, y(x,T))dx$$

and the control problem by

$$\inf\{J(y,v) \mid (y,v) \quad \text{satisfies} \quad (1)-(3)\}. \tag{P}$$

In all the sequel, we make the following assumptions.

(A1) For every $y \in R$, $F(.,y)$ and $\Phi(.,y)$ are measurable on Q. For almost every $(x,t) \in Q$, $F(.,y)$ and $\Phi(.,y)$ are of class C^1 and we have the following estimates

$$|\Phi(x,t,y)| + |F(x,t,y)| + |F'_y(x,t,y)| \leq \eta(|y|), \quad c_0 \leq \Phi'_y(x,t,y) \leq \eta(|y|),$$

where η is a nondecreasing function from R^+ into R^+ and $c_0 \in R$. (We have denoted by F'_y and Φ'_y the partial derivatives of F and Φ with respect to y, in all the sequel we adopt the same kind of notation for other functions).

(A2) Ψ and G are continuous on R^2, for every $v \in R$, $\Psi(.,v)$ and $G(.,v)$ are of class C^1 on R and we have the following estimates

$$|G'_y(y,v)| + |G(y,v)| \leq (1+|v|^{\beta r})\eta(|y|), \quad |G'_v(y,v)| \leq (1+|v|^{\beta r-1})\eta(|y|),$$
$$c_0 \leq \Psi'_y(y,v) \leq (1+|v|^{\beta})\eta(|y|), \quad |\Psi(y,v)| \leq (1+|v|^{\beta})\eta(|y|),$$
$$|\Psi'_v(y,v)| \leq (1+|v|^{\beta-1})\eta(|y|),$$

where η is as in (A1).

(A3) For every $x \in \Omega$, $H(x,.)$ is of class C^1 on R, for every $y \in R$, $H(.,y)$ is measurable on Ω,

$$|H'_y(x,y)| + |H(x,y)| \leq \eta(|y|).$$

The function g is continuous on $\bar{Q} \times R$ and, for every (x,t) in $\bar{Q}$, $g(x,t,.)$ is of class C^1 on R.

Under additional assumptions, we have proved that (P) admits solutions (Raymond, 1995). Here, we are mainly interested in optimality conditions.

2 PARABOLIC EQUATIONS WITH MEASURES

Since the adjoint equation in optimality conditions for (P) is a parabolic equation with measures as data, we first consider such equations for which we prove existence and regularity results.

For every $(a,b) \in L^\infty(Q) \times L^r(\Sigma)$, we consider the following terminal boundary value problem:

$$-\frac{\partial p}{\partial t} + Ap + ap = \mu_Q \quad \text{in } Q, \tag{4a}$$

$$\frac{\partial p}{\partial n_A} + bp = \mu_\Sigma \quad \text{on } \Sigma, \tag{4b}$$

$$p(T) = \mu_{\bar{\Omega}_T} \quad \text{in } \bar{\Omega}, \tag{4c}$$

where $\mu = \mu_Q + \mu_\Sigma + \mu_{\bar{\Omega}_T}$ is a bounded Radon measure on $\bar{Q}\backslash\bar{\Omega}\times\{0\}$, μ_Q is the restriction of μ to Q, μ_Σ is the restriction of μ to Σ and $\mu_{\bar{\Omega}_T}$ is the restriction of μ to $\bar{\Omega}\times\{T\}$.

By definition a function $p \in L^1(0,T;W^{1,1}(\Omega))$ is a weak solution of (4) if $bp \in L^1(\Sigma)$ and if

$$\int_Q (p\frac{\partial y}{\partial t} + \sum_{i,j} a_{ij} D_j p D_i y + apy)dxdt + \int_\Sigma bpydsdt = \langle y, \mu\rangle_{C_b(\bar{Q}\backslash\bar{\Omega}_0)\times\mathcal{M}_b(\bar{Q}\backslash\bar{\Omega}_0)} \tag{5}$$

for every $y \in C^1(\bar{Q})$ satisfying $y(x,0) = 0$ in $\bar{\Omega}$.

Thanks to some new regularity results on linear parabolic equations (see Proposition 2) and thanks to duality arguments we can prove the following result (Raymond, 1995).

Theorem 1 *There exists a unique weak solution p of (4) in $L^1(0,T;W^{1,1}(\Omega))$, for every (δ, d) satisfying $\delta > 1$, $d > 1$, $\frac{d}{\delta'} - \frac{d}{2} > \frac{N}{2}$, $\delta \leq r$, $\frac{Nr}{r-2} < d \leq \frac{Nr}{N-1}$, p belongs to $L^{\delta'}(0,T;W^{1,d'}(\Omega))$. The normal trace on ∂Q of the vector field $((\sum_j a_{ij}D_j p)_{1\leq i\leq N}, p)$ belongs to $\mathcal{M}(\partial Q)$ (the space of Radon measure on ∂Q) and p verifies the following Green formula*

$$\int_Q p(\frac{\partial y}{\partial t} + Ay + ay)dxdt + \int_\Sigma p(\frac{\partial y}{\partial n_A} + by)dsdt =$$
$$\langle y, -\frac{\partial p}{\partial t} + Ap + ap\rangle_{C_b(Q)\times\mathcal{M}_b(Q)} + \langle y, \frac{\partial p}{\partial n_A} + bp\rangle_{C_b(\Sigma)\times\mathcal{M}_b(\Sigma)} +$$
$$+\langle y(T), p(T)\rangle_{C(\bar{\Omega}\times\{T\})\times\mathcal{M}(\bar{\Omega}\times\{T\})} - \langle y(0), p(0)\rangle_{C(\bar{\Omega}\times\{0\})\times\mathcal{M}(\bar{\Omega}\times\{0\})},$$

where $p(0)$ is the restriction to $\bar{\Omega}\times\{0\}$ of the normal trace of $(-(\sum_j a_{ij}D_j p)_{1\leq i\leq N}, -p)$.

Remark The conditions $\delta \leq r$ and $\frac{Nr}{r-2} < d \leq \frac{Nr}{N-1}$ can be omitted if $b \in L^\infty(\Sigma)$. Since the divergence in the sense of distributions in Q of the vector field $((\sum_j a_{ij}D_j p)_{1\leq i\leq N}, p)$ is a bounded Radon measure in Q, we can define the normal trace on ∂Q of this vector field in some Sobolev space with a negative exponent which contains the space of Radon measures on ∂Q (see Casas, 1993).

To prove Theorem 1 we proceed as in (Alibert and Raymond, 1994) where the case of elliptic equations is treated. We consider sequences of regular functions $(h_n)_n \subset C_c(Q)$ (the space of continuous functions with compact support in Q), $(k_n)_n \subset C_c(\Sigma)$ and $(\ell_n)_n \subset C(\bar{\Omega})$ such that $(k_n)_n$ converges to μ_Q for the narrow topology of $\mathcal{M}_b(Q)$, $(k_n)_n$ converges to μ_Σ for the narrow topology of $\mathcal{M}_b(\Sigma)$ and $(\ell_n)_n$ converges to $\mu_{\bar{\Omega}_T}$ for the weak star topology of $\mathcal{M}(\bar{\Omega})$. We denote by $(p_n)_n$ the solution of

$$-\frac{\partial p}{\partial t} + Ap + ap = h_n \quad \text{in } Q, \quad \frac{\partial p}{\partial n_A} + bp = k_n \quad \text{on } \Sigma, \quad p(T) = \ell_n \quad \text{in } \Omega.$$

The main point in the proof of Theorem 1 is the following estimate

$$\|p_n\|_{L^{\delta'}(0,T;W^{1,d'}(\Omega))} \leq C(\|h_n\|_{L^1(Q)} + \|k_n\|_{L^1(\Sigma)} + \|\ell_n\|_{L^1(\Omega)}) \tag{6}$$

with C independent of n. To prove such a result, we proceed by duality, we denote by y the solution of

$$\frac{\partial y}{\partial t} + Ay + ay = f - \text{div } \xi \quad \text{in } Q, \quad \frac{\partial y}{\partial n_A} + by = 0 \quad \text{on } \Sigma, \quad y(0) = 0 \quad \text{in } \Omega, \tag{7}$$

where $(f, \xi) \in \mathcal{D}(Q) \times \mathcal{D}(Q; R^N)$. With a Green formula we get

$$|\int_Q (fp_n + \xi \cdot Dp_n) dxdt| = |\int_Q h_n y + \int_\Sigma k_n y + \int_\Omega \ell_n y|. \tag{8}$$

Thanks to (8) and to Proposition 2 (see below), we can finally prove (6) and we can pass to the limit in the weak formulation of the equation satisfied by p_n. The Green formula of Theorem 1 is also obtained by a passage to the limit. The main difficulty is to correctly justify the writing $\langle y(0), p(0)\rangle_{C(\bar{\Omega}\times\{0\})\times\mathcal{M}(\bar{\Omega}\times\{0\})}$.

Proposition 2 *For every $a \in L^\infty(Q)$, every $b \in L^r(\Sigma)$ satisfying $a \geq c_0$, $b \geq c_0$ for some $c_0 \in R$, and for every $(f, \xi) \in \mathcal{D}(Q) \times \mathcal{D}(Q; R^N)$, there exists a unique weak solution y of (7) in $L^2(0, T; H^1(\Omega)) \cap C([0, T]; L^2(\Omega))$. This solution belongs to $C(\bar{Q})$ and satisfies the estimate*

$$\|y\|_{L^2(0,T;H^1(\Omega))} + \|y\|_{C(\bar{Q})} \leq C(\|f\|_{L^\nu(0,T;L^m(\Omega))} + \sum_{i=1}^{N} \|\xi_i\|_{L^\delta(0,T;L^d(\Omega))}),$$

for any (ν, m, δ, d) such that

$$m > 1, \quad \nu > 1, \ m/\nu' > N/2,$$
$$\delta > 1, \quad d > 1, \quad \frac{d}{\delta'} - \frac{d}{2} > \frac{N}{2},$$

and the constant $C = C(c_0, \nu, m, r, \delta, d, \Omega, T)$ does not depend on a, b, f and ξ.

Estimate in Proposition 2 is a classical result for parabolic equations with homogeneous Dirichlet boundary conditions (Ladyženskaya, Solonnikov and Ural'ceva, 1968). Here we cannot use these results because we deal with Robin boundary conditions. Moreover, since b (the coefficient in the boundary condition) can be negative, we cannot use a truncation method to get L^∞-estimates as in (Ladyženskaya, Solonnikov and Ural'ceva, 1968, Chapter 3). To prove Proposition 2, we first consider the solution $\tilde{y}$ of

$$\frac{\partial y}{\partial t} + Ay + c_1 y = e^{-(c_1 - c_0)t}(f - \text{div } \xi) \quad \text{in } Q, \tag{9a}$$

$$\frac{\partial y}{\partial n_A} + c_0 y = 0 \quad \text{on } \Sigma, \quad y(0) = 0 \quad \text{in } \Omega, \tag{9b}$$

where c_1 is a constant sufficiently big. Still by using duality arguments and estimates on analytic semigroups, we can prove the following estimate for $\tilde{y}$

$$\|\tilde{y}\|_{C(\bar{Q})} \leq C(\|f\|_{L^\nu(0,T;L^m(\Omega))} + \sum_{i=1}^{N} \|\xi_i\|_{L^\delta(0,T;L^d(\Omega))}).$$

Thanks to this estimate and to a new comparison principle (Raymond and Zidani, 1995), we obtain

$$0 \leq e^{-(c_1-c_0)t} y^+ \leq \tilde{y}^+$$

where $y^+ = Max(y,0)$ and $\tilde{y}^+ = Max(\tilde{y},0)$. We get a similar estimate for y^- and we can conclude. The comparison principle given in (Raymond and Zidani, 1995) extends a result proved in (Schmidt, 1989) to the case when b is not bounded (only bounded from below) and can be negative.

3 OPTIMALITY CONDITIONS

We now give an existence theorem of Lagrange multipliers for control problems proved in (Alibert and Raymond 1994). We use the following notation

$$Z_0 = Y \times \Pi \quad \text{and} \quad C_0 = Y \times \Pi_{ad},$$

where Y and Π are Banach spaces, Π is separable, Π_{ad} is a nonempty closed convex subset of Π. For $i = 1, 2$, Z_i is a Banach space, Z_i^* its topological dual, G_1 is a mapping from Z_0 into Z_1, G_2 is a mapping from Y into Z_2, J is a functional defined on Z_0 and we denote by C_2 a convex cone in Z_2 with nonempty interior and vertex at the origin. We consider the following abstract control problem.

$$inf\{J(y,\pi) \mid (y,\pi) \in C_0,\ G_1(y,\pi) = 0,\ G_2(y) \in C_2\}. \tag{CP}$$

Theorem 3 *Let $(\bar{y},\bar{\pi})$ be an optimal point for (CP) let us assume that*

(i) J is Fréchet-differentiable at $(\bar{y},\bar{\pi})$ and G_2 is Fréchet-differentiable at $\bar{y}$,
(ii) G_1 is strictly differentiable at $(\bar{y},\bar{\pi})$ and the linear operator $G'_{1y}(\bar{y},\bar{\pi})$ from Y into Z_1 is surjective,
(iii) there exists $y_0 \in Y$ such that

$$G_2(\bar{y}) + G_2'(\bar{y})y_0 \in int\ C_2. \tag{10}$$

There then exists $(p,\mu,\lambda) \in Z_1^* \times Z_2^* \times R$, with $(\mu,\lambda) \neq (0,0)$, satisfying

$$\lambda J_y'(\bar{y},\bar{\pi}) + pG'_{1y}(\bar{y},\bar{\pi}) + \mu G_2'(\bar{y}) = 0, \tag{11}$$

$$\lambda J_\pi'(\bar{y},\bar{\pi})(\pi - \bar{\pi}) + \langle p, G'_{1\pi}(\bar{y},\bar{\pi})(\pi - \bar{\pi})\rangle_{Z_1^* \times Z_1} \geq 0, \tag{12}$$

for every $\pi \in \Pi_{ad}$,

$$\lambda \geq 0,\ \ \mu \in C_2^* \quad \text{and} \quad \langle \mu, G_2(\bar{y})\rangle_{Z_2^* \times Z_2} = 0,$$

(C_2^* is the polar cone of C_2). If moreover there exist $\pi_0 \in \Pi_{ad}$ and $y_0 \in Y$ such that

$$G'_{1y}(\bar{y},\bar{\pi})y_0 + G'_{1\pi}(\bar{y},\bar{\pi})\pi_0 = 0 \quad \textit{and} \quad G_2(\bar{y}) + G'_2(\bar{y})y_0 \in \textit{int } C_2 \tag{13}$$

then we can take $\lambda = 1$ in (11) and (12).

We denote by Y the space

$$Y = \{y \in W(0,T;H^1(\Omega),(H^1(\Omega))') \mid \frac{\partial y}{\partial t} + Ay \in L^\infty(Q),\ \frac{\partial y}{\partial n_A} \in L^r(\Sigma),\ y(0) \in C(\bar{\Omega})\}$$

endowed with the graph norm.

Let $(\bar{y},\bar{v}) \in Y \times K_V$ be a solution of the optimal control problem (P). We can apply Theorem 3 to (P). For this, we set:

$$\begin{array}{l} \Pi = L^{\beta r}(\Sigma), \quad Z_0 = Y \times \Pi, \qquad C_0 = Y \times \Pi_{ad}, \qquad \Pi_{ad} = K_V, \\ Z_1 = L^\infty(Q) \times L^r(\Sigma) \times C(\bar{\Omega}), \qquad Z_2 = C(\bar{Q}), \\ C_2 = \{h \in C(\bar{Q}) \mid h(x,t) \leq 0 \text{ for all } (x,t) \in \bar{Q}\}, \\ G_1(y,v) = (\dfrac{\partial y}{\partial t} + Ay + \Phi(.,y), \dfrac{\partial y}{\partial n_A} + \Psi(y,v), y(0) - y_0) \\ \text{and} \qquad G_2(y) = g(.,y). \end{array}$$

The surjectivity of $G'_{1y}(\bar{y},\bar{v})$ has been proved in (Raymond and Zidani, 1995).

We shall say that $(\bar{y},\bar{v}) \in C(\bar{Q}) \times K_V$ satisfies the *weak qualification condition* for (P) if there exists $z \in C(\bar{Q})$ satisfying

$$g(x,t,\bar{y}(x,t)) + g_y(x,t,\bar{y}(x,t))z(x,t) < 0 \quad \text{on} \quad \bar{Q}.$$

Since Y is dense in $C(\bar{Q})$, this qualification condition corresponds to condition (10) of Theorem 3.

We shall say that $(\bar{y},\bar{v}) \in C(\bar{Q}) \times K_V$ satisfies the *strong qualification condition* for (P) if there exists $(z,v) \in C(\bar{Q}) \times K_V$ satisfying

$$\begin{array}{l} g(.,\bar{y}) + g_y(.,\bar{y})z < 0 \quad \text{on } \bar{Q}, \\ \dfrac{\partial z}{\partial t} + Az + \Phi'_y(.,\bar{y})z = 0 \text{ in } Q, \quad \dfrac{\partial z}{\partial n_A} + \Psi'_y(\bar{y},\bar{v})z + \Psi'_v(\bar{y},\bar{v})v = 0 \text{ on } \Sigma, \quad z(0) = 0 \quad \text{in } \Omega. \end{array}$$

The strong qualification condition corresponds to condition (13) of Theorem 3.

Thanks to Theorem 1 and Theorem 3, we can finally prove the following result.

Theorem 4 *If $(\bar{y},\bar{v})$ is a solution of (P) and if $(\bar{y},\bar{v})$ satisfies the weak qualification condition for (P), there then exist $\lambda \in R^+$, $p \in L^{\delta'}(0,T;W^{1,d'}(\Omega))$ for all (δ,d) satisfying $\delta > 1$, $d > 1$, $\frac{d}{\delta'} - \frac{d}{2} > \frac{N}{2}$, $\delta \leq r$, $\frac{Nr}{r-2} < d \leq \frac{Nr}{N-1}$, and $\mu \in \mathcal{M}_b(\bar{Q} \setminus \bar{\Omega} \times \{0\})$ (the space of bounded Radon measures on $\bar{Q} \setminus \bar{\Omega} \times \{0\}$) such that*

$$-\frac{\partial p}{\partial t} + Ap + \Phi'_y(.,\bar{y})p + \lambda F'_y(.,\bar{y}) + \mu_Q g'_y(.,\bar{y}) = 0 \text{ in } Q, \tag{14a}$$

$$\frac{\partial p}{\partial n_A} + \Psi'_y(\bar{y},\bar{v})p + \lambda G'_y(\bar{y},\bar{v}) + \mu_\Sigma g'_y(.,\bar{y}) = 0 \text{ on } \Sigma, \tag{14b}$$

$$p(T) = -\lambda H'_y(.,\bar{y}(T)) - \mu_{\bar{\Omega}_T} g'_y(.,T,y(T)) \text{ on } \bar{\Omega}, \tag{14c}$$

$$\int_\Sigma (\lambda G'_v(\bar{y},\bar{v}) + p\Psi'_v(\bar{y},\bar{v}))(v-\bar{v})(s,t)dsdt \geq 0 \text{ for all } v \in K_V, \tag{15}$$

$$(\lambda,\mu) \not\equiv 0,\ \mu \geq 0,\ \langle \mu, g(.,\bar{y})\rangle_{\mathcal{M}_b(\bar{Q}\setminus\bar{\Omega}\times\{0\})\times C_b(\bar{Q}\setminus\bar{\Omega}\times\{0\})} = 0, \tag{16}$$

where μ_Q is the restriction of μ to Q, μ_Σ is the restriction of μ to Σ and $\mu_{\bar{\Omega}_T}$ is the restriction of μ to $\bar{\Omega} \times \{T\}$ (Equation (14) is satisfied in the sense of definition given in (5)). If moreover $(\bar{y},\bar{v})$ satisfies the strong qualification condition, we can take $\lambda = 1$ in (14)–(16).

These results extend to problems governed by semilinear parabolic equations previous results obtained when the state equation is elliptic (see Casas (1993), and Alibert and Raymond (1994) where the case of nonlinear boundary conditions and equations with non regular coefficients is considered). The regularity results for equation (1) have been obtained with H. Zidani (Raymond and Zidani (1995)). Existence of a solution in $L^1(Q)$ for equation (4) is obtained in (Mackenroth, 1982) when $a \equiv 0$ and b is a positive constant. Let us stress on that here b can be negative. Optimality conditions are also obtained in (Fattorini and Murthy, 1994) for problems with terminal state constraints.

REFERENCES

Alibert, J. J. and Raymond, J. P. (1994), Optimal control problems governed by semilinear elliptic equations with pointwise state constraints, *submitted to SIAM J. Cont. Optim.*.

Casas, E. (1993) Boundary Control of Semilinear Elliptic Equations with Pointwise State Constraints, *SIAM J. Control Optim.*, **31**, 993-1006.

Fattorini, H. O. and Murthy, T. (1994), Optimal problems for nonlinear parabolic boundary control systems, *SIAM J. Cont. Opt.*, **32**, 1577-1596.

Ladyženskaya, O. A., Solonnikov, V. A. and Ural'ceva, N.N. (1968) Linear and quasilinear equations of parabolic type, *AMS Translations of Mathematical Monographs 23*, Providence.

Mackenroth, U. (1982) Convex Parabolic Boundary Control Problems with Pointwise State Constraints, *J. Math. Anal. Appli.*, **87**, 256-277.

Schmidt, E.J. (1989) Boundary Control for the Heat Equation with Nonlinear Boundary Condition, *J. Diff. Equat.*, **78**, 89-121.

Raymond, J.P. and Zidani, H. (1995) Hamiltonian Pontryagin's Principles for Control Problems Governed by Semilinear Parabolic Equations, *preprint.*

Raymond, J.P. (1995) Nonlinear boundary control of semilinear parabolic equations with pointwise state constraints, *preprint.*

Optimal Control of Plates

23

On the optimal control problem governed by quasistationary von Kárman's equations

Igor Bock
Department of Mathematics, FEI STU
Ilkovičova 3, 812 19 Bratislava, Slovakia.
E–mail: bock@kmat.elf.stuba.sk
and
Ján Lovíšek
Department of Mechanics, Faculty of Constructions STU
Radlinského 11, 813 68 Bratislava, Slovakia

Abstract

Optimal control problems with evolutionary generalized Von Kárman's equations in the role of a state problem are considered. The existence theorems for the state and the control problem are stated. The necessary optimality conditions are investigated.

Keywords

Quasistationary von Kárman's equations, Volterra integral equation, optimal control problem, necessary optimality conditions

1 INTRODUCTION

Optimal control problems for stationary Von Kárman's equations were investigated in the joint papers of (Bock, Hlaváček and Lovíšek, 1984, 1985, 1987) Here we consider in the role of state equations evolutionary Von Kárman's equations describing large deflections of thin viscoelastic plates. The first part of the paper is devoted to the deriving and solving the state problem, whose canonical form is a nonlinear pseudoparabolic equation with an integro-differential part. Optimal control problems with controls in the right-hand sides are solved in the second part.

2 FORMULATION OF THE STATE PROBLEM

Let us assume a thin viscoelastic plate made of a short memory material of the Voigt type (Brilla, 1973), (Christensen, 1971). It occupies the domain

$$Q = \{(x, z) \in R^3;\ x = (x_1, x_2) \in \Omega,\ -h/2 < z < h/2\}, \tag{1}$$

where Ω is a bounded simply connected domain in R^2 with a Lipschitz boundary $\partial\Omega$. After the linearization of the Kirchhoff hypothesis we obtain the strain-displacement relations

$$\varepsilon_{ij} = \frac{1}{2}(\partial_i\omega_j + \partial_j\omega_i + \partial_i y \partial_j y) - z\partial_{ij}y; \quad i,j = 1,2 \tag{2}$$

$$\varepsilon_{13} = \varepsilon_{23} = 0, \quad \varepsilon_{33} = \frac{1}{2}[(\partial_1 y)^2 + (\partial_2 y)^2], \tag{3}$$

where (ω_1, ω_2) is the plane displacement vector, y is the deflection of the middle surface of the plate and

$$\partial_i\omega_j = \frac{\partial\omega_j}{\partial x_i}, \quad \partial_{ij}y = \frac{\partial y}{\partial x_i \partial x_j} \tag{4}$$

The viscoelastic stress-strain relations have the form

$$\sigma^{ij}(t) = A^{(1)}_{ijkl}\partial_t\varepsilon_{kl}(t) + A^{(0)}_{ijkl}\varepsilon_{kl}(t), \quad i,j,k,l \in \{1,2\}; \tag{5}$$

$$\sigma^{33} = 0 \tag{6}$$

The third order tensors $A^{(r)}_{ijkl}$ are symmetric and positively definite

$$A^{(r)}_{ijkl} = A^{(r)}_{jikl} = A^{(r)}_{klij} \tag{7}$$

$$A^{(r)}_{klij}\tau_{ij}\tau_{kl} \geq c_r\tau_{ij}\tau_{ij}, \quad c_r > 0; \tag{8}$$

for all $\{\tau_{ij}\} \in R^4_{sym}$ and $r = 0, 1$.

Let us introduce the matrices

$$\mathbf{A}_r = \begin{pmatrix} A^{(r)}_{1111}, & A^{(r)}_{1112}, & A^{(r)}_{1122} \\ A^{(r)}_{1211}, & A^{(r)}_{1212}, & A^{(r)}_{1222} \\ A^{(r)}_{2211}, & A^{(r)}_{2212}, & A^{(r)}_{2222} \end{pmatrix}, \quad r = 0, 1; \tag{9}$$

and

$$\mathbf{G}(t) = \exp(-\mathbf{A}_1^{-1}\mathbf{A}_0 t)\mathbf{A}_1^{-1} \tag{10}$$

We recall that $\exp \mathbf{A}t$ is the matrix exponential function. The matrix $\mathbf{G}(t)$ is regular and symmetric due to the expansion

$$\mathbf{G}(t)^{-1} = \mathbf{A}_1 \exp(\mathbf{A}_1^{-1}\mathbf{A}_0 t) = \mathbf{A}_0 + \sum_{k=1}^{\infty} \frac{t^k}{k!}\mathbf{A}_1(\mathbf{A}_1^{-1}\mathbf{A}_0)^k \tag{11}$$

We express it in the form

$$\mathbf{G}(t) = \begin{pmatrix} G_{1111}, & G_{1112}, & G_{1122} \\ G_{1211}, & G_{1212}, & G_{1222} \\ G_{2211}, & G_{2212}, & G_{2222} \end{pmatrix} \tag{12}$$

Assuming that the plate is clamped and the forces acting on the parts $\Omega \times \{-h/2\}$ and $\Omega \times \{h/2\}$ are vectors $(0,0,0)$ and $(0,0,v(t,x))$ respectively, we obtain employing the principle of virtual displacements the integro-differential equation connecting the Airy stress function Φ and the perpendicular deflection y:

$$\int_0^t H_{ijkl}(t-s)\partial_{ijkl}\Phi(s)ds = -\frac{1}{2}h[y,y], \tag{13}$$

where

$$H_{ijkl}(t) = H_{jikl}(t) = H_{lkij}(t), \tag{14}$$

$$H_{1111}(t) = G_{2222}(t), \quad H_{1112}(t) = -\frac{1}{2}G_{2212}(t)$$

$$H_{1222}(t) = -\frac{1}{2}G_{1112}(t), \quad H_{1122}(t) = \frac{1}{2}G_{1122}(t) \tag{15}$$

$$H_{1212}(t) = \frac{1}{4}G_{1212}(t), \quad H_{2222}(t) = G_{1111}(t)$$

and

$$[w,y] = \partial_{11}w\partial_{22}y + \partial_{22}w\partial_{11}y - 2\partial_{12}w\partial_{12}y \tag{16}$$

(13) is the Volterra integro-differential equation of the first kind with respect to Φ. After differentiating it with respect to t we obtain the boundary value problem for the Volterra equation of the second kind

$$H_{ijkl}(0)\partial_{ijkl}\Phi(t) + \int_0^t \partial_t H_{ijkl}(t-s)\partial_{ijkl}\Phi(s)ds = -h[\partial_t y, y] \tag{17}$$

$$\Phi(t,\xi) = \frac{\partial \Phi}{\partial \mathrm{n}}(t,\xi) = 0, \quad t>0, \quad \xi \in \partial\Omega \tag{18}$$

In a similar way as in the elastic case the initial-boundary value problem for the pseudoparabolic equation with respect to the deflection $y \equiv y(t,x)$ can be derived :

$$\frac{h^3}{12}(A^{(1)}_{ijkl}\partial_t\partial_{ijkl}y + A^{(0)}_{ijkl}\partial_{ijkl}y) - [\Phi, y] = v, \quad t>0, \quad x \in \Omega \tag{19}$$

$$y(0,x) = y(t,\xi) = \frac{\partial y}{\partial \mathrm{n}}(t,\xi) = 0, \quad t>0, \quad \xi \in \partial\Omega \tag{20}$$

The initial-boundary value problem for the system (17) - (20) represents the strong formulation of the generalized Von Kárman's equations for a viscoelastic plate.

In order to solve it we proceed with a weak formulation of the problem in a Sobolev space $V = H_0^2(\Omega)$. Let us introduce the operators A_r, $H(t): V \to V^*$ (V^* is a dual space of V) by

$$\langle A_r y, w\rangle = \frac{h^3}{12}\iint_\Omega A_{ijkl}^{(r)}\partial_{ij}y\partial_{kl}w dx; \quad y, w \in V; \quad r = 0, 1 \tag{21}$$

$$\langle H(t)y, w\rangle = \iint_\Omega H_{ijkl}(t)\partial_{ij}y\partial_{kl}w dx; \quad y, w \in V \tag{22}$$

The problem (17) - (20) has then a following weak formulation:
For arbitrary $T > 0$ find a couple $\{\Phi, y\} : [0, T] \to V \times V$ fulfilling

$$H(0)\Phi(t) + \int_0^t \partial_t H(t-s)\Phi(s)ds = -h[\partial_t y(t), y(t)] \tag{23}$$

$$A_1\partial_t y(t) + A_0 y(t) - [\Phi(t), y(t)] = v(t) \tag{24}$$

$$y(0) = 0 \tag{25}$$

The operator $H(t)$ is for every $t \in [0, T]$ symmetric and positively definite. Moreover it fulfils the assumptions of Corollary 4.1. from (MacCamy and Wong,1972) and it is nonnegative in the following convolutive sense

$$\int_0^T \langle \int_0^t H(t-s)\Phi(s)ds, \Phi(t)\rangle dt \geq 0 \tag{26}$$

for every $T > 0$ and $\Phi(.) \in C([0, T], V)$,
where $C([0, T], V)$ is the set of all continuous functions defined on $[0, T]$ with values in the space V.

Further, we can define the bilinear and bounded operator $B : V \times V \to V$ by the uniquely solved equation

$$\langle H(0)B(y, w), \phi\rangle = \iint_\Omega [y, w]\phi dx \quad \text{for all} \quad y, w, \phi \in V \tag{27}$$

The equation (23) can then be expressed as the Volterra integral equation of the second kind in the Hilbert space V:

$$\Phi(t) - \int_0^t K(t-s)\Phi(s)ds = -hB(\partial_t y(t), y(t)), \tag{28}$$

where $K(t) : V \to V$ are the operators defined by

$$K(t) = -H(0)^{-1}\partial_t H(t), \quad t > 0 \tag{29}$$

The equation (28) has due to the theory of Volterra integral equations (Balakrishnan,1976) a unique solution $\Phi \in L^2(0,T;V)$, which can be expressed in a form

$$\Phi(t) = -hB(\partial_t y(t), y(t)) - h\int_0^t M(t,s)B(\partial_s y(s), y(s))ds \tag{30}$$

$M(t,s): V \to V$ is the iterated kernel defined as the series of

$$M(t,s) = \sum_{n=1}^{\infty} K_n(t,s) \tag{31}$$

of the iterated kernels

$$K_n(t,s) = \int_s^t K(t-\sigma)K_{n-1}(\sigma,s)d\sigma, \tag{32}$$

$$K_1(t,s) = K(t-s) \tag{33}$$

Inserting the Airy stress function Φ from (30) into (24) we arrive at the canonical initial-boundary value problem for the determining the deflection function $y : [0,T] \to V$:

$$A_1\partial_t y(t) + A_0 y(t) + h[B(\partial_t y(t), y(t)), y(t)] + h[\int_0^t M(t,s)B(\partial_s y(s), y(s))ds, y(t)] = v(t) \tag{34}$$

$$y(0) = 0 \tag{35}$$

Using the method of elliptic regularization (Lions,1969) the problem (34), (35) can be solved in the spaces of Bochner integrable functions

$$\mathcal{V} = L^2(0,T;V), \quad \mathcal{V}^* = L^2(0,T;V^2), \tag{36}$$

$$\mathcal{W} = \{y \in \mathcal{V} : \partial_t y \in \mathcal{V},\ y(0) = 0\}. \tag{37}$$

$\mathcal{V}$ is the Hilbert space with the inner product

$$((u,v)) = \int_0^T (u(t), v(t))_V dt \tag{38}$$

We express the initial value problem (34), (35) in the operator form

$$\mathcal{A}(y) = v, \quad y \in \mathcal{W}, \quad v \in L^2(0,T;L^2(\Omega)) \subset \mathcal{V}^* \tag{39}$$

where the operator $\mathcal{A} : \mathcal{W} \to \mathcal{V}^*$ is defined by

$$\langle\langle \mathcal{A}(y), w\rangle\rangle = \int_0^T \langle A_1\partial_t y(t) + A_0 y(t), w(t)\rangle dt + \tag{40}$$

$$+h\int_0^T ([B(\partial_t y(t), y(t)) + \int_0^t M(t,s)B(\partial_s y(s), y(s))ds, y(t)], w(t))dt, \quad y \in \mathcal{W}, \quad w \in \mathcal{V}.$$

We introduce further the operators $L : \mathcal{W} \to \mathcal{V}$ and $\mathcal{A}_\varepsilon : \mathcal{W} \to \mathcal{W}^*$ by

$$Ly = \partial_t y, \quad y \in \mathcal{W}; \tag{41}$$

$$\langle\langle \mathcal{A}_\varepsilon(y), w \rangle\rangle = \varepsilon((Ly, Lw)) + \langle\langle A(y), w \rangle\rangle, \quad y, w \in \mathcal{W}. \tag{42}$$

$$\mathcal{A}_\varepsilon(y_\varepsilon) = v \tag{43}$$

The operator $\mathcal{A}_\varepsilon : \mathcal{W} \to \mathcal{W}^*$ is coercive and pseudomonotone and then there exists for every $v \in \mathcal{W}^*$ a solution $y_\varepsilon \in \mathcal{W}$ of (43). The convolutive positivity (26) plays the crucial role in the coercivity argument.

After the limit procedure we obtain the existence theorem for the canonical form of the original problem. The theorem has been more detaily verified in (Bock, to appear).

Theorem 1 *For every $v \in L^2(0, T; L^2(\Omega))$ there exists a solution $y \equiv y(v) \in \mathcal{W}$ of the problem (34), (35) or (39).*

3 OPTIMAL CONTROL PROBLEMS

Let $U_{ad} \subset L^2(0, T; L^2(\Omega))$ be an arbitrary convex closed and bounded set of admissible controls. We introduce the cost functional $J : \mathcal{W} \times U_{ad} \to R$ of the form

$$J(y, v) = \mathcal{J}(y) + j(v), \quad y \in \mathcal{W}, \quad v \in U_{ad}. \tag{44}$$

We shall investigate the following

Optimal Control Problem $\mathcal{P}$: To find a couple $(y_0, u) \in \mathcal{W} \times U_{ad}$ such that

$$J(y_0, u) = \min_{(y,v) \in \mathcal{K}} J(y, v), \tag{45}$$

where

$$\mathcal{K} = \{(y, v) \in \mathcal{W} \times U_{ad} : \mathcal{A}(y) = v\}. \tag{46}$$

Theorem 2 *If the functionals $\mathcal{J}, j$ are weakly lower semicontinuous on W and $L^2(0, T; L(\Omega))$ respectively, then there exists a solution $(y_0, u) \in \mathcal{K}$ of the Optimal Control Problem $\mathcal{P}$.*

Proof. A solution is a weak limit of a minimizing sequence $\{y_n, u_n\}$ for a functional J on $\mathcal{K}$. This limit exists due to the boundedness in $\mathcal{V}$ of sequences $\{y_n\}$ and $\{\partial_t y_n\}$ of solutions and their derivatives corresponding to a bounded sequence $\{u_n\} \subset U_{ad}$. □

The operator $\mathcal{A} : \mathcal{W} \to \mathcal{V}^*$ is Fréchet differentiable with the derivative $\mathcal{A}'(y) \in \mathcal{L}(\mathcal{W}, \mathcal{V}^*)$ defined by

$$\langle \mathcal{A}'(y)z, w \rangle_{\mathcal{V}} = \int_0^T \{\langle A_1 \partial_t z + A_0 z, w \rangle +$$

$$+h([B(\partial_t y, y) + \int_0^t M(t,s)B(\partial_s y(s), y(s))ds, z(t)]+ \tag{47}$$

$$+[\partial_t B(y,z) + \int_0^t M(t,s)\partial_s B(y(s), z(s))ds, y(t)], w(t))_0\}dt,$$

where $(y, w) = \iint_\Omega y.wd\Omega$.

Let us define the set $\mathcal{W}_T = \{w \in \mathcal{V} : \partial_t w \in V, w(T) = 0\}$. The adjoint operator $\mathcal{A}'(y)^* : \mathcal{W}_T \to \mathcal{V}^*$ has then the form

$$\langle \mathcal{A}'(y)^* w, z\rangle_\mathcal{V} = \int_0^T \{\langle -A_1\partial_t w + A_0 w, z\rangle +$$

$$+h([B(\partial_t y, y) + \int_0^t M(t,s)B(\partial_s y(s), y(s))ds, w(t)]- \tag{48}$$

$$-[\partial_t B(y,w) + \partial_t \int_t^T M^*(s,t)B(y(s), w(s))ds, y(t)], z(t))_0\}dt.$$

Using the general extremal principle in smoothly convex problems (Joffe and Tichomirov, 1984) we obtain the following necessary optimality conditions:

Theorem 3 *Let the couple $(y_0, u) \in \mathcal{W}\times U_{ad}$ be a solution of the Optimal Control Problem P with a convex functional j and continuously Fréchet differentiable functionals $\mathcal{J}, j$. Then there exists an element $p \in \mathcal{W}_T$ such that*

$$\mathcal{A}(y_0) = u \tag{49}$$

$$\mathcal{A}'(y_0)p = -\mathcal{J}'(y_0) \tag{50}$$

$$\int_0^T (j'(u) - p, v - u)dt \geq 0 \quad \textit{for all} \quad v \in U_{ad} \tag{51}$$

Using the method of penalization an Optimal Control Problem with a cost functional involving all solutions of the state problem can be solved.

Let $\mathcal{F} : \mathcal{W} \times U_{ad} \to \mathcal{V}^*$ be the state operator of the form $\mathcal{F}(y, v) = \mathcal{A}(y) - v$. We introduce the cost functional

$$\hat{J}(v) = \sup_{y\in\mathcal{W}, \mathcal{F}(y,v)=0} [\mathcal{J}(y) + j(v)], \quad v \in U_{ad}. \tag{52}$$

In an analogous way as in the case of stationary Von Kárman's equations (Bock, Hlaváček and Lovíšek, 1987) there can be verified the existence of an optimal control $u \in U_{ad}$ fulfilling $\hat{J}(u) = \min_{v\in U_{ad}} \hat{J}(v)$. The element u is a weak limit of a subsequence $\{u_{\varepsilon_n}\} \subset U_{ad}$ fulfilling $\lim_{n\to\infty} \varepsilon_n = 0$, $\varepsilon_n > 0$ and

$$J_\varepsilon(u_\varepsilon) = \min_{v\in U_{ad}} J_\varepsilon(v), \tag{53}$$

$$J_\varepsilon(v) = \sup_{y \in \mathcal{W}_r} [\mathcal{J}(y) + j(v) - \frac{1}{\varepsilon} \|\mathcal{F}(y,v)\|_{\mathcal{V}^*}], \tag{54}$$

$$\mathcal{W}_r = \{y \in \mathcal{W} : \|y\|_{\mathcal{W}} \leq r\}, \tag{55}$$

where $r > 0$ is chosen such that for every $v \in U_{ad}$ we have $y(v) \in \mathcal{W}_r$.

REFERENCES

Balakrishnan, A.V.(1976) *Applied functional analysis.* Springer Verlag, New York.

Bock, I.(to appear) Von Kárman's equations for viscoelastic plates. *ZAMM.*

Bock, I., Hlaváček,I. and Lovíšek,J. (1984) On the optimal control problems governed by the equations of von Kárman, I. The homogeneous Dirichlet boundary conditions. *Applications of math.*, **29**, 303-314.

Bock, I., Hlaváček, I. and Lovíšek, J. (1985) On the optimal control problems governed by the equations of von Kárman, II. Mixed boundary conditions. *Applications of math.*, **30**, 375-392.

Bock, I., Hlaváček, I. and Lovíšek, J. (1987) On the optimal control problems governed by the equations of von Kárman, III. The case of an arbitrary large perpendicular load. *Applications of math.*, **32**, 315-331.

Brilla, J. (1973) Variational methods in mathematical theory of viscoelasticity. In: Ráb,M.; Vosmanský,J.(eds.): *Proc. Internat. Conf. on Diff. Equations, Equadiff III.* University J.E.Purkyně Press, Brno, 211-216.

Christensen, R.V. (1971) *Theory of viscoelasticity.* Academic Press, New York.

Joffe, A.D. and Tichomirov, V.M. (1984) *The theory of extremal problems* (in Russian). Nauka, Moskva.

Lions, J.L. (1969) *Quelques méthodes de résolution des problèmes aux limites non linéaires.* Dunod, Paris .

MacCamy, R.C. and Wong, J.S.W. (1972) Stability theorems for some functional equations. *Transactions of AMS* 164,**1**, 1-37.

24

Bilinear optimal control of a Kirchhoff plate via internal controllers

M. E. Bradley
Department of Mathematics, University of Louisville
Louisville, KY 40292 U.S.A.
E-mail: mebrad01@homer.louisville.edu
and
S. M. Lenhart
Department of Mathematics, University of Tennessee
Knoxville, TN 37996 U.S.A.
E-mail: lenhart@math.utk.edu

Abstract

We consider the problem of optimal control of a Kirchhoff plate. Bilinear controls are used as forces acting on internal regions, to make the plate close to a desired profile, taking into the account a quadratic cost of control. We prove the existence of an optimal control and characterize it uniquely through the solution of an optimality system.

Keywords

bilinear control, Kirchhoff plate

1 INTRODUCTION

We consider bilinear optimal control of a Kirchhoff plate as modeled below. The controls act on small non-intersecting regions in the interior of the plate. These controls behave as an internal tension or "spring-like" control attached to the plate at specific locations.

In order to define admissible vector controls for our system, we begin by defining an admissible component controller. Let h_i be such that the support of $h_i \subset Q_i \equiv \Omega_i \times [0,T]$ and such that

$$h_i \in U_{M_i} = \{h_i \in L^\infty(Q_i) : \|h\|_{L^\infty(Q_i)} \leq M_i\},$$

where $0 < M_i$. We define our control vector, $\mathbf{h} \equiv (h_1, h_2, ..., h_k)$, where k is the number of controlled regions (and consequently the number of controllers) for the system, requiring that each $h_i \in U_{M_i}$, so that

$$\mathbf{h} \in U \equiv U_{M_1} \times ... \times U_{M_k}.$$

For convenience later, we define $M = \max_{i=1}^{k} M_i$.

Concerning the regions which will be controlled, we require that $\Omega_i \subset\subset \Omega$, $\bar{\Omega}_i \cap \bar{\Omega}_j = \emptyset$ for $i \neq j$ and that $\partial\Omega_i \cap \partial\Omega = \emptyset$.

Under these assumptions, the "displacement" solution $w = w(\mathbf{h})$ of our state equation, satisfies

$$\left.\begin{array}{ll} w_{tt} + \Delta^2 w + w = \sum_{i=1}^{k} h_i(x,y,t)w & \text{on } Q = \Omega \times (0,T) \\ w(x,y,0) = w_0(x,y), w_t(x,y,0) = w_1(x,y) & \text{when } t = 0 \\ \left.\begin{array}{l} \Delta w + (1-\mu)B_1 w = 0 \\ \frac{\partial \Delta w}{\partial \nu} + (1-\mu)B_2 w = 0 \end{array}\right\} \text{ on } \Sigma = \Gamma \times (0,T) & \end{array}\right\} \tag{1.1}$$

where $\Omega \subset \mathbf{R}^2$ with C^2 boundary, $\partial\Omega = \Gamma$, $\vec{\nu} = \langle n_1, n_2 \rangle$ is the outward unit normal vector on $\partial\Omega$, and

$$\begin{aligned} B_1 w &= 2n_1 n_2 w_{xy} - n_1^2 w_{yy} - n_2^2 w_{xx} \\ B_2 w &= \frac{\partial}{\partial \tau}[(n_1^2 - n_2^2)w_{xy} + n_1 n_2 (w_{yy} - w_{xx})]. \end{aligned}$$

The direction τ in $B_2 w$ is the tangential direction along Γ. The plate has free vibrations along Γ. The constant $\mu, 0 < \mu < \frac{1}{2}$, represents Poisson's ratio.

We take as our cost functional

$$J(\mathbf{h}) = \frac{1}{2}\left(\int_Q (w-z)^2 dQ + \sum_{i=1}^{k} \beta_i \int_{Q_i} h_i^2 dQ_i\right), \tag{1.2}$$

where z is the desired evolution for the plate and the quadratic term in h_i represents the cost of implementing the controls. We seek to minimize the cost functional, i.e., find optimal control $\mathbf{h}^* \in U$ such that

$$J(\mathbf{h}^*) = \min_{\mathbf{h} \in U} J(\mathbf{h}).$$

The goal of this paper is to characterize the unique optimal control vector in system consists of the state equation coupled with an adjoint equation. We note that the solution $w = w(h)$ is a nonlinear function of the control, so that uniqueness of the optimal control becomes a delicate issue. We will show that the optimal control is unique, as the unique solution of the optimality system. However, due to the highly nonlinear structure of the optimality system, we obtain this uniqueness only for a small time interval. Consequently, we prove uniqueness of the optimal control for this same small time interval.

For background information on plate equations and control theory, the reader is referred to the classical works of Lagnese (1989), Lagnese and Lions (1988) and Lions (1971).

2 EXISTENCE OF THE OPTIMAL CONTROL

We begin by proving existence, uniqueness, and regularity results for the state equation (1.1). These results will provide the *a priori* estimates needed to prove the existence of an optimal control.

To define our notion of weak solution, we first define the following product Hilbert space: $\mathcal{H} = H^2(\Omega) \times L^2(\Omega)$. We note that the bilinear form

$$a(w,v) = \int_\Omega \{\Delta w \Delta v + (1-\mu)[2w_{xy}v_{xy} - w_{xx}v_{yy} - w_{yy}v_{xx}] + wv\}\, d\Omega \tag{2.3}$$

induces a norm on $H^2(\Omega)$ which is equivalent to the usual norm on $H^2(\Omega)$.

Definition. Given $\mathbf{h} \in U$, $\tilde{w} = \tilde{w}(\mathbf{h}) = (w, w_t)$ is a weak solution of (1.1) if $\tilde{w} \in C([0,T];\mathcal{H})$, $\tilde{w}(0) = (w_0, w_1)$, and $\tilde{w}$ satisfies

$$< w_{tt}, \phi > + a(w,\phi) = \sum_{i=1}^{k} \int_{\Omega_i} h_i w \phi \; d\Omega_i \quad \text{for all} \quad \phi \in H^2(\Omega).$$

Here, we interpret $< \cdot, \cdot >$ as the duality pairing between $H^2(\Omega)$ and $[H^2(\Omega)]'$.

Lemma 1 (Well-posedness and Regularity)

(i) Let $\tilde{w}(0) = (w_0, w_1) \in \mathcal{H}$ and $\mathbf{h} \in U$, then the state equation (1.1) has a unique weak solution $\tilde{w} = \tilde{w}(\mathbf{h}) = (w, w_t)$ with $(w, w_t) \in C([0,T];\mathcal{H})$.

(ii) If in addition, $(w_0, w_1) \in (H^4(\Omega) \cap H^2(\Omega)) \times H^2(\Omega)$ with w_0 satisfying the homogeneous boundary conditions in (1.1), and $\mathbf{h} \in C^2(\bar{Q}_i) \cap U_{M_i}$, then the weak solution $\tilde{w} = \tilde{w}(h)$ satisfies

$$\begin{aligned} \tilde{w} &\in C([0,T]; (H^4(\Omega) \cap H^2(\Omega)) \times H^2(\Omega)) \\ w_{tt} &\in C([0,T]; L^2(\Omega)) \end{aligned}$$

with $\tilde{w}(0) = (w_0, w_1)$. Also $\tilde{w}$ satisfies equation (1.1) in the L^2 sense.

Proof. We refer the reader to techniques used in Bradley and Lenhart (1994) where the authors used semigroup theory combined with a contraction mapping argument to obtain the desired well-posedness and regularity results. □

To prove the existence of an optimal control, we need the following *a priori* estimate.

Lemma 2 *Given $\tilde{w}_0 = (w_0, w_1) \in \mathcal{H}$ and $\mathbf{h} \in U$, the weak solution $\tilde{w} = \tilde{w}(\mathbf{h}) = (w, w_t)$ of (1.1) satisfies*

$$\|\tilde{w}\|_{C([0,T];\mathcal{H})} \le C_1 e^{C_2 k M T} \tag{2.4}$$

where $C_1 = \|\tilde{w}_0\|_{\mathcal{H}}$ and k is the number of control regions.

Proof. The proof is obtained using "multipliers technique" on the smooth solutions guaranteed by Lemma 1 (i) and then passing with a limit for solutions in $\mathcal{H}$. For details, see Bradley and Lenhart (1994). □

We now prove the main result of this section.

Theorem 1 *There exists an optimal control vector* $\mathbf{h}^* \in U$ *which minimizes the cost functional* $J(\mathbf{h})$ *for* $\mathbf{h} \in U$.

Proof. Let $\{\mathbf{h}^n\} \in U$ be a minimizing sequence such that

$$\lim_{n\to\infty} J(\mathbf{h}^n) = \inf_{\mathbf{h}\in U} J(\mathbf{h}).$$

We denote the corresponding solution to (1.1) by $\tilde{w}^n = \tilde{w}(\mathbf{h}^n)$. By Lemma 2,

$$\|\tilde{w}^n\|_{C([0,T],\mathcal{H})} \le C_1 e^{C_2 kMT}.$$

On a subsequence, we have

$$\begin{aligned}
&w^n \rightharpoonup w^* \text{ weakly in } L^2([0,T];H^2(\Omega)),\\
&w^n \to w^* \text{ strongly in } L^2(Q),\\
&w^n_t \rightharpoonup w^*_t \text{ weakly in } L^2(Q),\\
&w^n_{tt} \rightharpoonup w^*_{tt} \text{ weakly in } L^([0,T];[H^2(\Omega)]')\\
&\text{and}\\
&h^n_i \rightharpoonup h^*_i \text{ weakly in } L^2(Q_i).
\end{aligned}$$

We may now pass to the limit on (1.1) as $n \to \infty$, to obtain that $\tilde{w} = \tilde{w}(\mathbf{h}) = (w^*, w^*_t)$ solves the state equation (1.1) with control $\mathbf{h}^*$. Since the cost functional is lower semicontinuous with respect to weak convergence (basically Fatou's Lemma), we obtain $J(\mathbf{h}^*) \le \underline{\lim}_{n\to\infty} J(\mathbf{h}^n) = \inf_{\mathbf{h}\in U} J(\mathbf{h})$. Hence $\mathbf{h}^*$ is an optimal control. □

3 CHARACTERIZATION OF THE OPTIMAL CONTROL

We now derive the optimality system by using the weak partial differentiability of the cost functional $J(\mathbf{h})$ with respect to the controllers h_i. In order to justify that such partial derivatives exist, we first must prove that the mapping $\mathbf{h} \to \tilde{w}(\mathbf{h})$ has the desired weak partial derivatives with respect to controllers h_i.

Lemma 3 *The mapping* $\mathbf{h} \in U \to \tilde{w}(\mathbf{h}) \in \mathcal{H}$ *is has weak partial derivatives in the following sense:*

$$\frac{\tilde{w}(h_1, ..., h_j + \varepsilon l, ..., h_k) - \tilde{w}(\mathbf{h})}{\varepsilon} \rightharpoonup \tilde{\psi}_j \text{ weakly in } L^2(0,T;\mathcal{H})$$

as $\varepsilon \to 0$, *for any* $h_j, h_j + \varepsilon\ell \in U_{M_j}$. *Moreover* $\tilde{\psi}_j = (\psi_j, \psi_{j,t})$ *is a weak solution of the following problem:*

$$\psi_{j,tt} + \Delta^2\psi_j + \psi_j - \sum_{i=1}^{k} h_i\psi_j = \ell w \text{ in } Q \tag{3.5}$$
$$\psi_j(x,0) = \psi_{j,t}(x,0) = 0 \text{ in } \Omega$$

$$\left.\begin{array}{l}\Delta\psi_j + (1-\mu)B_1\psi_j = 0 \\ \frac{\partial}{\partial\nu}\Delta\psi_j + (1-\mu)B_2\psi_j = 0\end{array}\right\} \text{ on } \Sigma$$

where $\tilde{w} = \tilde{w}(\mathbf{h}) = (w, w_t)$.

Proof. Denote $\tilde{w}^\varepsilon = \tilde{w}(h_1, ..., h_j + \varepsilon\ell, ..., h_k) = (w^\varepsilon, w_t^\varepsilon)$ and $\tilde{w} = \tilde{w}(\mathbf{h})$. (We note that w^ε will depend on both j and ε.) Then $\frac{\tilde{w}^\varepsilon - \tilde{w}}{\varepsilon}$ is a weak solution of

$$\left(\tfrac{w^\varepsilon - w}{\varepsilon}\right)_{tt} + \Delta^2\left(\tfrac{w^\varepsilon - w}{\varepsilon}\right) + \left(\tfrac{w^\varepsilon - w}{\varepsilon}\right) = \textstyle\sum_{i=1}^k h_i \left(\tfrac{w^\varepsilon - w}{\varepsilon}\right) + \ell w^\varepsilon \text{ in } Q$$

$$\text{with } \left(\tfrac{w^\varepsilon - w}{\varepsilon}\right)(x,y,0) = \left(\tfrac{w^\varepsilon - w}{\varepsilon}\right)_t (x,y,0) = 0 \text{ in } \Omega$$

and satisfies zero boundary conditions on $\partial\Omega \times (0,T)$. Using a priori estimates like in Lemma 2, we obtain

$$\|\frac{\tilde{w}^\varepsilon - \tilde{w}}{\varepsilon}\|_{C([0,T],\mathcal{H})} \le \|\ell w^\varepsilon\|_{L^2(Q)} e^{CkMT} \le C_3$$

where C_3 depends on the L^∞ bound on ℓ and the number of controlled regions, but is independent of ε, due to a bound on $\|\tilde{w}^\varepsilon\|_{L^2(Q)}$, independent of ε. Hence on a subsequence,

$$\frac{\tilde{w}^\varepsilon - \tilde{w}}{\varepsilon} \rightharpoonup \tilde{\psi}_j \text{ weakly in } L^2(0,T;\mathcal{H}).$$

This convergence and the above *a priori* estimates are sufficient to guarantee that $\tilde{\psi}_j$ is a weak solution of (3.5). □

Finally, we derive our optimality system.

Theorem 2 *Given an optimal control* $\mathbf{h}$ *and corresponding solution* $\tilde{w} = \tilde{w}(\mathbf{h}) = (w, w_t)$, *there exists a weak solution* $\tilde{p} = (p, p_t)$ *in* $\mathcal{H}$ *to the adjoint problem,*

$$\begin{array}{l} p_{tt} + \Delta^2 p + p = \sum_{i=1}^k h_i p + w - z \text{ in } Q \\ \left.\begin{array}{l}\Delta p + (1-\mu)B_1 p = 0 \\ \frac{\partial}{\partial\nu}\Delta p + (1-\mu)B_2 p = 0\end{array}\right\} \text{ on } \Sigma \end{array} \tag{3.6}$$

and transversality conditions $p(x,y,T) = p_t(x,y,T) = 0$ *when* $t = T$. *Furthermore, each control element,* h_i, *satisfies*

$$h_i = \max(-M_i, \min(-\frac{wp}{\beta_i}, M_i)). \tag{3.7}$$

We note that, although we obtain dependence in of ψ_j on the particular partial derivative being taken, we obtain only one adjoint equation, since we have only one state equation.

Proof. Let $\mathbf{h} \in U$ be an optimal control vector and $\tilde{w} = \tilde{w}(\mathbf{h})$ be the corresponding optimal solution. Let $h_j + \varepsilon\ell \in U_{M_j}$ for $\varepsilon > 0$ and $\tilde{w}^\varepsilon = \tilde{w}(h_1, ..., h_j + \varepsilon\ell, ..., h_k)$ be the corresponding weak solution of the state equation (1.1). We compute the partial derivative of the cost functional $J(\mathbf{h})$ with respect to h_j in the direction of ℓ. Since $J(\mathbf{h})$ is a minimum value,

$$
\begin{aligned}
0 &\leq \lim_{\varepsilon\to 0^+} \frac{J(h_1,...,h_j+\varepsilon\ell,...,h_k)-J(\mathbf{h})}{\varepsilon} \\
&= \lim_{\varepsilon\to 0^+} \frac{1}{2\varepsilon}\int_Q ((w^\varepsilon - z)^2 - (w-z)^2)dQ + \frac{\beta_j}{2\varepsilon}\int_{Q_j}((h_j+\varepsilon\ell)^2 - h_j^2)dQ_j \\
&= \lim_{\varepsilon\to 0^+} \int_Q \left(\frac{w^\varepsilon - w}{\varepsilon}\right)\left(\frac{w^\varepsilon + w - 2z}{2}\right) dQ + \frac{\beta_j}{2}\int_Q (2h_j\ell + \varepsilon\ell^2)dQ \qquad (3.8) \\
&= \int_Q \psi_j(w-z)dQ + \beta_j\int_Q h_j\ell\ dQ,
\end{aligned}
$$

where we have used the fact that the support of $h_j \subset\subset Q_i \subset Q$. Also, ψ_j is defined as in Lemma 3.

Let $\tilde{p} = (p, p_t)$ be the weak solution of the adjoint problem (3.6). Existence and uniqueness of $\tilde{p}$ is proved by arguments similar to those in Section 2. Substituting the adjoint solution into (3.8) for $(w-z)$, we obtain

$$0 \leq \int_0^T < p_{tt}, \psi_j > dt + \int_0^T a(p,\psi_j)dt - \alpha\int_Q \psi_j hp\ dQ + \int_Q \beta_j h_j \ell\ dQ.$$

Using the weak form of (3.5), we have

$$0 \leq \int_Q \ell(wp + \beta_j h_j)dQ.$$

By a standard control argument concerning the sign of the variation ℓ depending on the size of h_j, we obtain the desired characterization of $h_j = \max(-M_j, \min(-\frac{wp}{\beta_j}, M_j))$. □

Substituting (3.7) for h_j into the state equation (1.1) and the adjoint equation (3.6), we obtain the optimality system:

$$
\begin{array}{ll}
w_{tt} + \Delta^2 w + w = \sum_{i=1}^k \max(-M_i, \min(-\frac{wp}{\beta}_i, M_i))w & \text{in } Q \\
p_{tt} + \Delta^2 p + p = \sum_{i=1}^k \max(-M_i, \min(-\frac{wp}{\beta}_i, M_i))p + w - z & \text{in } Q \\
\left.\begin{array}{l}\Delta w + (1-\mu)B_1 w = \Delta p + (1-\mu)B_1 p = 0 \\ \frac{\partial}{\partial\nu}\Delta w + (1-\mu)B_2 w = \frac{\partial}{\partial\nu}\Delta p + (1-\mu)B_2 p = 0\end{array}\right\} \text{on } \Sigma & \\
w(x,y,0) = w_0(x,y), \quad w_t(x,y,0) = w_1(x,y) & \text{on } \Omega \\
p(x,y,T) = p_t(x,y,T) = 0. &
\end{array}
\qquad (3.9)
$$

Weak solutions of the optimality system exist by Lemma 1 and Theorems 1 and 2. However, the problem of uniqueness of solutions for this nonlinear optimality system (which implies the uniqueness of the optimal control vector) proves to be more difficult. We will now prove for small time T, that the optimality system (3.9) does, in fact, possess a *unique* solution and thereby show that the optimal control is in fact unique for a small time interval, $[0,T]$. This, then will give a characterization of the unique optimal control in terms of the solution of (3.9).

Theorem 3 *For T sufficiently small, weak solutions of the optimality system (3.9) are unique.*

Proof. Suppose we have two weak solutions,

$$\tilde{w} = (w, w_t),\ \tilde{p} = (p, p_t),\ \hat{w} = (\overline{w}, \overline{w}_t),\ \hat{p} = (\overline{p}, \overline{p}_t).$$

Since $w,\ \overline{w},\ p,\ \overline{p} \in C(0,T; H^2(\Omega))$, we have that $w,\ \overline{w},\ p,\ \overline{p}$ are bounded on $\overline{Q}$.

We change variables

$$w = e^{\lambda t}u, \quad p = e^{-\lambda t}q, \quad \overline{w} = e^{\lambda t}\overline{u}, \quad \overline{p} = e^{-\lambda t}\overline{q}.$$

Then u (and respectively, q) satisfies in a weak sense

$$\begin{aligned} &u_{tt} + 2\lambda u_t + (\lambda^2+1)u + \Delta^2 u + u = \textstyle\sum_{i=1}^k \max(-M_i, \min(-\frac{uq}{\beta_i}, M_i))u \\ &-q_{tt} + 2\lambda q_t - (\lambda^2+1)q - \Delta^2 q - q = \textstyle\sum_{i=1}^k \max(-M_i, \min(-\frac{uq}{\beta_i}, M_i))(-q) \\ &\qquad\qquad - e^{2\lambda t}u + e^{\lambda t}z. \end{aligned}$$

One can check that u, q satisfy similar boundary and initial/terminal conditions as before, so that $u - \overline{u}$ and $q - \overline{q}$ satisfy equations as above (modulo $e^{\lambda t}z$ term in q equation), with homogeneous data.

Using multiplier $(u-\overline{u})_t$ on the $u - \overline{u}$ equation and multiplier $(q-\overline{q})_t$ on the $q - \overline{q}$ equation, and combining, we have the following estimate:

$$\begin{aligned} &\frac{1}{2}\int_\Omega ((u-\overline{u})_t)^2(x,T)d\Omega + \frac{1}{2}\int_\Omega ((q-\overline{q})_t)^2(x,0)d\Omega \\ &\quad + \frac{\lambda^2}{2}\int_\Omega ((u-\overline{u})^2(x,T) + (q-\overline{q})^2(x,0))d\Omega \\ &\quad + a(u-\overline{u}, u-\overline{u})(T) + a(q-\overline{q}, q-\overline{q})(0) \\ &\quad + 2\lambda\int_Q [((q-\overline{q})_t)^2 + ((u-\overline{u})_t)^2]dQ \\ &= \sum_{i=1}^k \int_Q \left[(h_i u - \overline{h}_i\overline{u})(u-\overline{u})_t - (h_i q - \overline{h}_i\overline{q})(q-\overline{q})_t - e^{2\lambda t}(u-\overline{u})(q-\overline{q})_t\right] dQ \end{aligned} \tag{3.10}$$

where $h_i = \max(-M_i, \min(-\frac{uq}{\beta_i}, M_i))$ and $\overline{h}_i = \max(-M_i, \min(-\frac{\overline{uq}}{\beta_i}, M_i))$. It can be shown by direct computation that

$$|h_i - \overline{h}_i| \le \frac{1}{\beta_i}|\overline{uq} - uq| \le \frac{1}{\beta_i}\left(|\overline{u} - u||\overline{q}| + |\overline{q} - q||u|\right),$$

so that we can estimate the right hand side of (3.10) and obtain,

$$\begin{aligned} &2\lambda\int_Q [((q-\overline{q})_t)^2 + ((u-\overline{u})_t)^2]dQ \\ &\quad \le \int_Q [((q-\overline{q})_t)^2 + ((u-\overline{u})_t)^2]dQ \quad + (C_1 e^{C_2(kM+\lambda)T})\int_Q [(u-\overline{u})^2 + (q-\overline{q})^2]dQ, \end{aligned} \tag{3.11}$$

where C_1, C_2 are independent of λ and T but do depend on the number of controlled regions, k and on the L^∞ bounds on u and $\overline{q}$. Noting that

$$\begin{aligned}
\int_Q (u-\overline{u})^2 dQ &= \int_\Omega \int_0^T \left(\int_0^t (u-\overline{u})_t(x,y,s)ds\right)^2 dt d\Omega \\
&\leq \int_\Omega \int_0^T t\left(\int_0^t ((u-\overline{u})_t)^2 ds\right) dt d\Omega \\
&\leq \int_0^T t dt \int_Q ((u-\overline{u})_t)^2 ds d\Omega \\
&\leq \frac{T^2}{2}\int_Q ((u-\overline{u})_t)^2 dQ,
\end{aligned}$$

and using a similar argument for the variable q, we obtain,

$$\begin{aligned}
(2\lambda-1)\int_Q [((q-\overline{q})_t)^2 &+ ((u-\overline{u})_t)^2] dQ \leq \\
&T^2(C_1 e^{C_2(kM+\lambda)T})\int_Q [((q-\overline{q})_t)^2 + ((u-\overline{u})_t)^2] dQ.
\end{aligned}$$

We now fix λ such that $2\lambda - 1 > 0$ and choose T sufficiently small so that

$$2\lambda - 1 > T^2(C_1 e^{C_2(kM+\lambda)T}),$$

and thus $(q-\overline{q})_t = (u-\overline{u})_t \equiv 0$ in Q. Due to agreement of $q, \overline{q}$ and $u, \overline{u}$ at top and bottom of the cylinder Q respectively, we obtain $q = \overline{q}$ and $u = \overline{u}$, as desired. □

4 ACKNOWLEDGMENT

This work was completed while the first author was visiting the Massachusetts Institute of Technology and graciously wishes to acknowledge their support during the 1994 -1995 academic year. The second author was partially supported by grants from the National Science Foundation and the University of Tennessee.

REFERENCES

Bradley, M. E. and Lenhart, S. M. (1994) Bilinear optimal control of a Kirchhoff plate, *Systems and Control Letters*, **22**, 27-38.

Lagnese, J. E. (1989) *Boundary Stabilization of Thin Plates,* Society for Industrial and Applied Mathematics, Philadelphia.

Lagnese, J. E. and Lions, J. L. (1988) *Modelling Analysis and Control of Thin Plates,* Masson, Paris.

Lions, J. L. (1971) *Optimal Control of Systems Governed by Partial Differential Equations,* Springer-Verlag, Berlin.

25

Boundary control problem for a dynamic Kirchhoff plate model with partial observations

Erik Hendrickson
Department of Computer Science and Mathematics
Arkansas State University, University, AR 72401, U.S.A.
and
Irena Lasiecka
Institute of Applied Mathematics
University of Virginia, Charlottesville, Virginia 22901, U.S.A.

Abstract

Boundary control problems of partially observed hyperbolic systems are considered. An algorithm based on FEM leading to a construction of a finite dimensional control is provided. In order to secure the stability of the algorithm, presence of boundary controls, yielding unbounded control operators in the semigroup model, necessitates introducing, prior to the FEM discretization, a specially designed regularization procedure where an additional boundary viscosity term is added. This stability and convergence of the overall regularized and discretized control problem is demonstrated. The theory is applied to a dynamic Kirchhoff plate with controls acting as the moments and boundary observations.

Keywords

Boundary control, Kirchhoff, compensator, regularization, stability

1 ORIENTATION

We consider the problem of the design of a boundary feedback control for a dynamic Kirchhoff plate model where only a partial observation of the state is available, for instance, a boundary observation. Since full knowledge of the state is not available, the primary problem centers around obtaining an estimate of the original state and using this estimate as if it were an exact measurement of the state to solve the original, deterministic control problem. In other words, a feedback control law, called the compensator, is generated using only information from an estimate of the state variable. The generalized linear-quadratic-Gaussian (LQG) theory supporting this method provides a state estimate that reconstructs the full state asymptotically in time and is optimal, in a suitable context, when Gaussian white noise processes pollute the input and output of the dynamical system. Hence, via the separation principle, the original problem is partitioned into solving a deterministic, linear quadratic optimal control problem and an optimal filtering (estimation) problem. The solutions to both the control and estimation problems are based on the solutions to associated Riccati equations with unbounded coefficients (as they arise in the context of boundary control).

Since the original system is infinite dimensional and practicality dictates that both the estimator and compensator must be finite dimensional. A natural way to circumvent this problem is to construct a finite dimensional approximation (eg. via finite elements) of the original control system that will uniformly (in the parameter of discretization) retain the control theoretic properties of the resulting closed loop system, such as the desired performance level and asymptotic stability. To accomplish this, the stability and convergence of solutions to finite dimensional Riccati equations must be provided. The final goal will be to show that the finite dimensional compensator, which is based on a finite dimensional approximation to the estimate and solutions to finite dimensional Riccati equations and when applied to the original system, will produce near optimal performance of the closed loop system. Moreover, the uniform stability of the resulting system should be preserved.

Problems related to finite dimensional approximations to infinite dimensional compensators with boundary controls have received considerable attention in the literature (see Curtain (1986), Ito (1990), Lasiecka (1995), Schumacher (1983) and references therein). However, all these works provide satisfactory convergence and uniform stabilizability results for classes of systems satisfying the so-called *spectrum determined growth condition.* These, in turn, are essentially restricted to delay and parabolic-like dynamics. Moreover, in these cases, stability analysis is much simplified as it amounts to the analysis of the spectrum. The situation is much more complicated in the case of hyperbolic dynamics, where the location of the spectrum does not necessarily determine the stability. In fact, approximations of hyperbolic compensator systems were analyzed in Lasiecka (1992). The results in Lasiecka (1992) provide the optimal theory for bounded control operators, but in the case of unbounded control actions (as in this paper), they require a certain trace-type condition. Indeed, a discrete analogue of (5) is to be satisfied uniformly in the parameter of discretization. As it turns out, this condition may fail for certain popular approximation techniques such as FEM in 2-d problems (Hendrickson, 1993a, 1995). The goal of the present paper is to remove the above restriction. Thus, the novel features of the present paper, with respect to prior literature, are:

(i) hyperbolicity of the dynamics;
(ii) unboundedness of the control actions, such as those which arise in boundary/point control problems;
(iii) no need to impose the discrete trace condition (see Lasiecka, 1992).

In what follows, we shall first treat an *abstract hyperbolic system* and then show how the abstract theory yields the results for a model of a Kirchhoff plate.

2 FORMULATION OF THE PROBLEM

Let H_0 and U be Hilbert spaces. Let $A_0 : H_0 \supset \mathcal{D}(A_0) \to H_0$ be a positive, self-adjoint operator on H_0, and $B_0 : U \to [\mathcal{D}(A_0^{1/2})]'$ be a control operator (i.e. $A_0^{-1/2}B_0 \in \mathcal{L}(U, H_0)$).

Notice that B_0 is unbounded: $U \to H_0$. We consider the following second order dynamics,

$$\begin{aligned} z_{tt}(t) + A_0 z(t) + D z_t(t) &= B_0 u(t) \quad \text{on} \quad [\mathcal{D}(A_0^{1/2})]', \\ z(t=0) = z_0 \in \mathcal{D}(A_0^{1/2}), &\quad z_t(t=0) = z_1 \in H_0. \end{aligned} \tag{1}$$

Here, the operator D is assumed nonnegative, self-adjoint and bounded on H_0. It is well known that (1) can be rewritten as the first order equation

$$\begin{aligned} x_t(t) &= Ax(t) + Bu(t) \quad \text{on} \quad [\mathcal{D}(A)]', \\ x(t=0) &= x_0 = (z_0, z_1) \in H \equiv \mathcal{D}(A_0^{1/2}) \times H_0, \quad \text{where} \end{aligned} \tag{2}$$

$$A \equiv \begin{bmatrix} 0 & I \\ -A_0 & -D \end{bmatrix}; \qquad B \equiv \begin{bmatrix} 0 \\ B_0 \end{bmatrix}; \qquad x \equiv (z, z_t)^T; \tag{3}$$

and A is a generator of a C_0-semigroup e^{At} defined on the Hilbert space H. With the dynamics described by (2) we associate a partial observation given by

$$y(t) = Cx(t) \tag{4}$$

where $C \in \mathcal{L}(H : Y)$, Y is a given Hilbert space representing the output space. We shall be considering the class of problems with the control operator B, generally *unbounded*, satisfying the following *trace* assumption,

$$\int_0^T \|B^* e^{A^* t} x\|_U^2 dt \le C_T \|x\|_H^2, \; x \in \mathcal{D}(A^*), \tag{5}$$

where $(Bu, v)_H = (u, B^* v)_U \; u \in U, \; v \in \mathcal{D}(B^*) \supset \mathcal{D}(A^*)$.

Remark B: continuous from $U \to [\mathcal{D}(\mathcal{A}^*)]'$ implies that $A^{-1}B \in \mathcal{L}(\mathcal{U}, \mathcal{H})$. It is well known (see Lasiecka and Triggiani, 1991a) that assumption (5) is satisfied for a large class of hyperbolic models including boundary control models for waves and plates.

The associated control problem is to develop a feedback control law, based only on the observed state, which minimizes (over $u \in L_2(0, \infty; U)$) the following *cost* functional

$$J(u, x(u)) = \int_0^\infty \|Rx\|_H^2 + \|u\|_U^2 dt, \qquad R \in \mathcal{L}(H). \tag{6}$$

Following ideas developed in finite dimensional estimator theory, we propose the following structure for the estimator. The dynamical estimator for the system (2), (4) is given by

$$\begin{aligned} w_t(t) &= Aw(t) + Bu(t) + K(y(t) - Cw(t)) \quad \text{on} \quad (\mathcal{D}(A))' \\ w(t=0) &= w_0 \in H. \end{aligned} \tag{7}$$

where the operator $K \in \mathcal{L}(Y; H)$.

Remark One can show that, by virtue of hypothesis (5), for a given control $u \in L_2(0,T;U)$, the solution (x,w) of (2), (7) is contained in $C([0,T];H \times H)$ and is continuous with respect to the data, (see Lasiecka and Triggiani, 1991a).

Assuming that the operator K^* is stabilizing for the pair (A^*, C^*) on H, one can prove (see Lasiecka, 1992) that the estimator is uniformly exponentially stable, i.e.,

$$\|x(t) - w(t)\|_H \leq Ce^{-\omega t}\|x_0 - w_0\|_H, \quad for\ some\ \ C, \omega > 0. \tag{8}$$

Thus, the estimator $w(t)$ reconstructs the dynamics of $x(t)$, asymptotically in t. Since the full state $x(t)$ is not available and any reasonable feedback law must be based on the available information, we consider feedbacks of the form,

$$u(t) = Fw(t), \quad where \quad F : H \to U,\ closed\ \ and\ \ densely\ defined. \tag{9}$$

3 OPTIMAL F-D COMPENSATOR AND CLOSED LOOP SYSTEM

Following the LQG theory, the optimal state feedback control law F and Kalman filter gain K are to be sought in the form:

$$F = -B^*P, \qquad K = \hat{P}C^*, \tag{10}$$

where P, $\hat{P} \in \mathcal{L}(H)$ are the positive, semidefinite solutions of the following algebraic Riccati equations (ARE),

$$(A^*Px, y)_H + (PAx, y)_H + (R^*Rx, y)_H = (B^*Px, B^*Py)_U, \quad x, y\ \in \mathcal{D}(A), \tag{11}$$

$$(A\hat{P}x, y)_H + (\hat{P}A^*x, y)_H + (Q^*Qx, y)_H = (C\hat{P}x, C\hat{P}y)_Y, \quad x, y\ \in \mathcal{D}(A^*) \tag{12}$$

and $Q \in \mathcal{L}(H)$.

It is well known that the conditions associated with the unique solvability of the AREs (11), (12) are, together with (5),

$$(A, B),\ (A^*, C^*),\ (A, Q),\ \text{ and } (A^*, R^*) \quad \text{are stabilizable on H.} \tag{13}$$

In fact, the above result is standard if B is *bounded*, see Balakrishnan (1981). In the case considered here, the situation is more complicated due to the *unboundedness* of B. However, due to the *trace* assumption (5), a recent theory of Flandoli, Lasiecka and Triggiani (1988) applies and provides, subject to (13), an existence and uniqueness of a solution $P \in \mathcal{L}(H)$ to (11). Moreover, it is proven that the gain operator is properly defined on the domain of the generator, i.e. $B^*P \in \mathcal{L}(\mathcal{D}(A); H)$. Hence the feedback law, in (9), is unbounded but *densely* defined on H and the nonlinear term in (11) is well-defined on a dense set. If in addition the following *smoothing* assumption on R is satisfied

$$R^*RA \in \mathcal{L}(H), \tag{14}$$

then it is shown in Da Prato, Lasiecka and Triggiani (1986) that $B^*P \in \mathcal{L}(H;U)$, implying that the gain, B^*P, is a *bounded* operator. Using the above described result on the solvability of Riccati Equations, (11), (12), it was shown in Lasiecka (1992) that under the conditions (5), (13), the *optimal compensator*, described by the system

$$\begin{bmatrix} x \\ w \end{bmatrix}_t = \mathcal{A}\begin{bmatrix} x \\ w \end{bmatrix}, \quad \text{where } \mathcal{A} = \begin{bmatrix} A & BF \\ KC & A + BF - KC \end{bmatrix}, \tag{15}$$

with F and K given by (10) is exponentially stable, i.e.

$$\|e^{\mathcal{A}t}\|_{\mathcal{L}(H\times H)} \leq Ce^{-\omega t}, \tag{16}$$

for some $\omega > 0$.

We seek to design a finite dimensional control, $u_h(y) = u_h(Cx)$, where $h \to 0$ (the parameter of discretization), such that the original system (2), with u_h, will provide near optimal performance, i.e. $J(u_h, x(u_h)) \to J(u^0, x(u^0))$, as $h \to 0$ and u^0 corresponds to the optimal, infinite dimensional control of the original system. In addition, we also aim to preserve the stability of the resulting closed-loop system.

The precise formulation of the approximation algorithm and of the main abstract results are given in the following sections.

4 REGULARIZATION

Our first step is to *regularize* the original continuous problem. To accomplish this, we introduce the operator,

$$A_\epsilon \equiv A - \epsilon BB^*, \tag{17}$$

where $\epsilon > 0$ is a parameter of regularization tending to zero. It was shown in Hendrickson and Lasiecka (1993b) that A_ϵ generates a C_0-semigroup such that

$$\|e^{A_\epsilon t}\|_{\mathcal{L}(H)} \leq Me^{\omega_0 t}, \quad \omega_0 > 0 \tag{18}$$

uniformly in $\epsilon > 0$. The following results are also supported in Hendrickson and Lasiecka (1993b):

$$(A_\epsilon, B),\ (A_\epsilon^*, C^*) \quad \textit{are uniformly (in } \epsilon > 0\textit{) stabilizable, and} \tag{19}$$

$$(A_\epsilon^*, Q),\ (A_\epsilon, R) \quad \textit{are uniformly (in } \epsilon > 0\textit{) detectable.} \tag{20}$$

Note that these results do not follow from standard perturbation theory, since ϵBB^* is not a *bounded* perturbation, and are necessary to assert the unique solvability of the algebraic Riccati Equations (11), (12) with A replaced by A_ϵ.

5 APPROXIMATION

We shall approximate the regularized problem. To accomplish this we introduce the following approximation subspaces and operators. Let h denote the discretization parameter which is assumed to tend to zero. $V_h \subset \mathcal{D}(A_0^{1/2})$ be a finite dimensional subspace. Let π_h represent the orthogonal projection of H_0 onto V_h with the properties:

$$\exists y_h = \pi_h y \in V_h \ni \|\pi_h y - y\|_{H_0} \to 0, \quad y \in H_0, \tag{21}$$
$$\|\pi_h y - y\|_{\mathcal{D}(A_0^{1/2})} \to 0, \quad y \in \mathcal{D}(A_0^{1/2}). \tag{22}$$

Assumptions on A_0: Let $A_{0h} : V_h \to V_h$ be a Galerkin approximation of A such that

$$(A_{0h}x_h, \xi_h)_{H_0} = (A_0 x_h, \xi_h)_{H_0}, \quad x_h,\ \xi_h \in V_h. \qquad \text{Hence} \tag{23}$$
$$\|({A_0}^{-1} - {A_{0h}}^{-1}\pi_h)x\|_{\mathcal{D}({A_0}^{1/2})} \to 0, \quad x \in H_0. \tag{24}$$

Assumptions on D: Let $D_h : V_h \to V_h$ be a Galerkin approximation of D such that

$$(D_h x_h, \xi_h)_{H_0} = (D x_h, \xi_h)_{H_0}, \quad x_h,\ \xi_h \in V_h. \tag{25}$$

Assumptions on B_0: Let $B_{0h} : U \to V_h$ be such that

$$\|{B_{0h}}^* \pi_h - {B_0}^*\|_{\mathcal{D}({A_0}^{1/2}) \to U} \to 0; \tag{26}$$

here

$$({B_0}^* v, g)_U = (v, B_0 g)_{H_0} \qquad \text{for} \quad g \in U, v \in \mathcal{D}({B_0}^*) \supset \mathcal{D}({A_0}^{1/2}),$$
$$\|{A_0}^{-1}(B_{0h} - B_0)\|_{U \to \mathcal{D}({A_0}^{1/2})} \to 0; \tag{27}$$
$$\|({A_{0h}}^{-1} - {A_0}^{-1})B_{0h}\|_{U \to \mathcal{D}({A_0}^{1/2})} \to 0. \tag{28}$$

Define $\bar{V_h} \equiv V_h \times V_h$ and $\bar{\pi_h}$ is the orthogonal projection of H onto $\bar{V_h}$ with

$$\bar{\pi_h} = \begin{bmatrix} \pi_h & 0 \\ 0 & \pi_h \end{bmatrix}.$$

Also, $A_h : \bar{V_h} \to \bar{V_h}$ and $B_h : U \to \bar{V_h}$, where

$$A_h \equiv \begin{bmatrix} 0 & \pi_h \\ -A_{0h} & -D_h \end{bmatrix}, \quad B_h \equiv \begin{bmatrix} 0 \\ B_{0h} \end{bmatrix}, \quad A_{h,\epsilon} \equiv A_h - \epsilon B_h^* B_h. \tag{29}$$

Notice that $B_h^* v_h = {B_{0h}}^* v_{2h}$, where $v_h = (v_{1h}, v_{2h}) \subset \bar{V_h}$, and that the above approximation properties on A_0, B_0, D are the usual consistency and stability properties satisfied by most approximation schemes including standard finite element (FE) and finite difference (FD) methods.

The finite dimensional gains associated with (10) are

$$F_{h,\epsilon} = B_h^* P_{h,\epsilon}, \qquad K_{h,\epsilon} = \hat{P}_{h,\epsilon}\pi_h C^*, \tag{30}$$

where $P_{h,\epsilon}, \hat{P_{h,\epsilon}} \in \mathcal{L}(\bar{V_h})$ are positive, self-adjoint, finite dimensional operators which solve the following Riccati equations for all $x_h, y_h \in \bar{V_h}$,

$$(A_{h,\epsilon}{}^* P_{h,\epsilon} x_h, y_h)_H + (P_{h,\epsilon} A_{h,\epsilon} x_h, y_h)_H + (R_h^* R_h x_h, y_h)_H = (B_h^* P_{h,\epsilon} x_h, B_h^* P_{h,\epsilon} y_h)_U, \quad (31)$$

$$(A_{h,\epsilon} \hat{P_{h,\epsilon}} x_h, y_h)_H + (\hat{P_{h,\epsilon}} A_{h,\epsilon}{}^* x_h, y_h)_H + (Q_h^* Q_h x_h, y_h)_H = (C \hat{P_{h,\epsilon}} x_h, C \hat{P_{h,\epsilon}} y_h)_Y. \quad (32)$$

In the above equations, R_h, Q_h are suitable, convergent approximations of R, Q, i.e. $R_h \to R$, $Q_h \to Q$ strongly in H.

The regularized, finite dimensional control, $u_h(t) = F_{h,\epsilon} w_h(t)$, when applied to the original system gives rise to the following compensator problem generated by the operator, $\mathcal{A}_{h,\epsilon} : \mathcal{H} \to \mathcal{H}$, where $\mathcal{H} \equiv H \times H$,

$$\mathcal{A}_{h,\epsilon} = \begin{pmatrix} A & B F_{h,\epsilon} \pi_h \\ K_{h,\epsilon} \pi_h C & A_{h,\epsilon} \pi_h + B_h F_{h,\epsilon} \pi_h - K_{h,\epsilon} \pi_h C \end{pmatrix}. \quad (33)$$

6 ABSTRACT RESULT

The main goal is to show that $e^{\mathcal{A}_{h,\epsilon} t}$ *converges* as $h, \epsilon \to 0$ to the original compensator design described by $e^{\mathcal{A} t}$, where $\mathcal{A}$ is given in (15). This is given in the theorem below, whose proof is given in Hendrickson and Lasiecka (1995).

Theorem 1 *We assume continuous hypotheses (5), (13), (14) and approximation hypotheses (21)-(28). In addition, we shall assume that either Q or C is compact and that*

$$(Dx, x)_{H_0} \geq d\|x\|_{H_0}^2, \quad d > 0, \quad (34)$$

R_h satisfies a discrete counterpart to (14), i.e.

$$\|R_h^* R_h A_h\|_{\mathcal{L}(H)} \leq C, \quad uniformly \quad in \quad h > 0. \quad (35)$$

Then, for all $\bar{x} = (x, w) \in \mathcal{H} \equiv H \times H$,

$$\lim_{\epsilon \to 0} \lim_{h \to 0} \sup_{t \geq 0} \|e^{\mathcal{A}_{h,\epsilon} t} \bar{x} - e^{\mathcal{A} t} \bar{x}\|_{\mathcal{H}} = 0. \quad (36)$$

(Convergence of controls)

$$\lim_{\epsilon \to 0} \lim_{h \to 0} \|u_{h,\epsilon} - u\|_{L_2(0\infty; U)} = 0, \quad (37)$$

(Convergence of Performance Index)

$$\lim_{\epsilon \to 0} \lim_{h \to 0} J(u_{h,\epsilon}, x(u_{h,\epsilon})) - J(u, x(u)) = 0, \quad (38)$$

(Uniform stability)
There exists $\omega_0 > 0$ such that for all $\bar{x} = (x, 0)$,

$$\|e^{\mathcal{A}_{h,\epsilon} t} \bar{x}\|_{\mathcal{H}} \leq C e^{-\omega_0 t} \|\bar{x}\|_{\mathcal{H}}. \quad (39)$$

Remark Under the minimal approximation hypotheses, i.e. the standard stability and consistency hypotheses, the main result of the Main Theorem provides an algorithm for the construction of a finite dimensional control $u_{h,\epsilon}(t)$ which, when inserted into the original system, gives a near optimal performance of the system. The resulting compensator system retains the uniform stability properties (see (39)), provided that the initial condition for the estimator (w) equation, i.e. w_0 is sufficiently regular. Typically $w_0 = 0$, so that this regularity requirement is satisfied automatically. But, the necessity of assuming higher regularity for w_0 results from an interesting new feature of the problem which is the lack of a uniform C_0-semigroup estimate for the operator $e^{A_{h,\epsilon}t}$. We show that $e^{A_{h,\epsilon}t}$ is, in fact, a one time integrated semigroup (Arendt (1983)) while $\|e^{A_{h,\epsilon}t}\|_{\mathcal{L}(\mathcal{H})}$ is of order $(\frac{1}{\sqrt{\epsilon}})$. The main goal of this paper is to show how the abstract theory applies to a model of the Kirchhoff plate.

7 KIRCHHOFF PLATE WITH BOUNDARY CONTROL AND BOUNDARY OBSERVATION

The following Kirchhoff plate model motivates our abstract theory. We shall consider the equation of the Kirchhoff plate with boundary control $u(t)$ acting as a bending moment and the state variable $z(t)$ representing the vertical displacement of the plate.

$$\begin{array}{lll} kz_{tt} - \gamma\Delta z_{tt} + \Delta^2 z + d(z_t - \gamma\Delta z_t) = 0 & \text{on} & \Omega \times (0,\infty), \\ z(x,t) = 0, \quad \Delta z(x,t) = u(t) & \text{on} & \Gamma \times (0,\infty) \equiv \Sigma, \\ z(x,0) = z_0(x), \quad z_t(x,0) = z_1(x) & \text{on} & \Omega, \end{array} \tag{40}$$

where Ω is an open domain in R^2 with boundary Γ of sufficient regularity. k pertains to the flexibility of the plate. $\gamma > 0$ is a parameter that is proportional to the square of the plate thickness. The term $\gamma\Delta z_{tt}$ represents rotational forces (which are neglected in some models). $d \geq 0$ is the damping coefficient. $u(t)$ acts as a boundary control in the form of a bending moment. Notice that the uncontrolled and undamped ($u(t) = 0,\ d = 0$) system is unstable, i.e. there are infinitely many eigenvalues on the imaginary axis. The natural energy associated with the model is

$$E(t) = \int_\Omega |\Delta z(t)|^2 + \gamma|\nabla z_t|^2 + |z_t|^2 d\Omega. \tag{41}$$

With model (40) we associate the boundary observation given by

$$C\begin{pmatrix} z \\ z_t \end{pmatrix} = \frac{\partial}{\partial\nu} z(x,t)|_\Gamma. \tag{42}$$

The cost functional is defined as

$$J(u, z(u)) = \int_0^\infty [\int_\Gamma u^2(x,t)\, d\Gamma + \int_\Omega |\nabla (z(x,t) - z_d(x))|^2\, d\Omega\ ]\, dt, \tag{43}$$

where $z_d(x) \in H^2(\Omega)$ is a given element. Physically, $z_d(x)$ represents a desired trajectory that we wish to track. Hence we seek to minimize the energy of the control and the gradient of the error between the actual trajectory and the desired trajectory. Since only a partial (boundary) observation is available, we need to determine the state estimator and then the desired infinite dimensional control will be of the form

$$u^0(t) = F(w(t), w_t(t)).$$

where w, w_t is an estimator for (40). In order to construct the estimator according to the theory, we rewrite the partial differential equation as an abstract equation within the framework of section 1.

To put problem (40) into an abstract framework, we introduce the following spaces and operators.

$$U = L_2(\Gamma); \quad H_0 = H_0^1(\Omega) \quad \text{with the inner product} \tag{44}$$
$$(u,v)_{H_0} = \int_\Omega uv \, d\Omega + \gamma \int_\Omega \nabla u \nabla v \, d\Omega,$$

$$\begin{aligned} A_D u \equiv -\Delta u \quad &: \quad u \in \mathcal{D}(A_D) \equiv H^2(\Omega) \cap H_0^1(\Omega), \\ A_D \quad &: \quad L_2(\Omega) \supset \mathcal{D}(A_D) \to L_2(\Omega), \\ A_0 \quad &: \quad H^2(\Omega) \cap H_0^1(\Omega) \to L_2(\Omega), \quad \text{where } A_0 = (I + \gamma A_D)^{-1} A_D^2. \end{aligned} \tag{45}$$

$\mathcal{B}_0 g = (I+\gamma A_D)^{-1} A_D D_0 g$, where D_0 is the Dirichlet map defined by $D_0 \; : \; L_2(\Gamma) \to L_2(\Omega)$ and $D_0 g = v$ implies $\{\Delta v = 0 \text{ on } \Omega; \; v|_\Gamma = g\}$.

$$(\mathcal{B}_0 g, h)_{H_0} = (A_D D_0 g, h)_\Omega = (g, D_0^* A_D h)_\Gamma = (g, \frac{\partial}{\partial \nu} h)_\Gamma.$$

Using (44) and (45) we verify that

$$\begin{aligned} \|u\|^2_{\mathcal{D}(A_0^{1/2})} &= (A_0 u, u)_{H_0} = ((I + \gamma A_D)(I + \gamma A_D)^{-1} A_D^2 u, u)_{L_2(\Omega)} \\ &= \|A_D u\|^2_{L_2(\Omega)} = \|\Delta u\|^2_{L_2(\Omega)}. \end{aligned}$$

Hence $E(t) = \|z(t)\|^2_{\mathcal{D}(A_0^{1/2})} + \|z_t(t)\|^2_{H_0}$. With the above notation, (40) is equivalent to the abstract model (1), i.e.,

$$z_{tt} + (I + \gamma A_D)^{-1} A_D^2 z + d z_t = (I + \gamma A_D)^{-1} A_D D_0 u \quad \text{on} \quad [\mathcal{D}(A_D)]'. \tag{46}$$

It is shown (see Lasiecka and Triggiani, 1991a) that

(i) A, defined by (3), is a generator of a C_0-semigroup on $H \equiv \mathcal{D}(A_0^{1/2}) \times H_0(\Omega)$,
(ii) Hypothesis (5), which amounts to the trace regularity (Lasiecka and Triggiani, 1991ab).

$$\int_0^T \int_\Gamma |\frac{\partial}{\partial \nu} \Delta z(t)|^2 \, dx dt \le C_T \; [\; \|z(0)\|^2_{H^2(\Omega)} + \|z_t(0)\|^2_{H^1(\Omega)} \;], \tag{47}$$

which is shown to be satisfied for all $z(t)$, $z_t(t)$ solutions to (40) with $u = 0$. Notice that (47) does not follow *a priori* from the regularity of the solution (see standard trace theory in Kesavan (1989)), rather it is an independent trace regularity result.

(iii) The original problem with pair (A, B) is exponentially stabilizable with the boundary feedback $u(t)$. It has been shown (Horn and Lasiecka (1994), Lasiecka and Triggiani (1991ab)) that the above open loop problem is uniformly stabilizable in the space $H^2(\Omega) \times H^1(\Omega)$ by means of a suitable boundary feedback control, $u(t) = F(z_t(t)) \in L_2(0\infty; L_2(\Gamma))$. (For more stability and controllability results on the Kirchhoff plate with other boundary conditions, see Horn and Lasiecka (1994), Lagnese and Liones (1989), Lasiecka and Triggiani (1991ab) and references therein.) In fact, it was shown in Horn and Lasiecka (1994) that with a feedback control of the form,

$$u(t) = -\frac{\partial}{\partial \nu} z_t \in L_2(0\infty; L_2(\Gamma)). \tag{48}$$

the energy, $E(t)$, of the system given by (41) decays exponentially.

(iv) The observation C given by (42) is a bounded operator, $C: \ H \to Y \equiv L_2(\Gamma)$, by virtue of the Trace Theorem (Pazy (1986)), and is, in fact, compact.

(v) The operator R associated with (43) (**WLOG** assume $z_d = 0$) has form

$$R\begin{pmatrix} z \\ z_t \end{pmatrix} = \begin{pmatrix} A_D^{-1/2} z \\ 0 \end{pmatrix},$$

which is clearly in $\mathcal{L}(H)$ and moreover, since

$$R^* RA \begin{pmatrix} z \\ z_t \end{pmatrix} = \begin{pmatrix} A_D^{-1} z_t \\ 0 \end{pmatrix}, \tag{49}$$

(14) is satisfied. Indeed, from (49)

$$\|R^* RA \begin{pmatrix} z \\ z_t \end{pmatrix} \|_H = \|A_D^{-1} z_t\|_{\mathcal{D}(A_0^{1/2})} = \|(I+\gamma A_D)^{1/2}(I+\gamma A_D)^{-1/2} A_D A_D^{-1} z_t\|_{L_2(\Omega)}$$

$$\leq C\|z_t\|_{L_2(\Omega)} \leq C\| \begin{pmatrix} z \\ z_t \end{pmatrix} \|_H.$$

(vi) If $d > 0$, then the operator D is coercive, i.e.,

$$(Dx, x)_{H_0} = d\int_\Omega (I+\gamma A_D)(I+\gamma A_D)^{-1} A_D x \cdot x \, d\Omega$$

and using Green's formula and Poincare Inequality

$$= d\int_\Omega |\nabla x|^2 \, d\Omega \geq d_1 \|x\|^2_{H_0^1(\Omega)} = d_1\|x\|^2_{H_0}.$$

One can show by the Liapunov function technique (see Horn and Lasiecka, 1994) that the uncontrolled system is exponentially stable, hence for any $Q \in \mathcal{L}(H)$, the pairs (A^*, Q), (A, R) are detectable and (A^*, C^*) is stabilizable.

Therefore, all the assumptions imposed on the continuous model are satisfied and the infinite dimensional estimator takes the form

$$\begin{aligned} &w_{tt} - \gamma\Delta w_{tt} + \Delta^2 w + d\Delta w_t = K(\frac{\partial}{\partial\nu} z_t|_\Gamma - \frac{\partial}{\partial\nu} w_t|_\Gamma), \\ &w = 0, \quad \Delta w = u \quad \text{on } \Sigma, \quad w(t=0) = w_t(t=0) = 0. \end{aligned} \tag{50}$$

$K \in \mathcal{L}(L_2(\Gamma) \to H^2(\Omega) \times H_0^1(\Omega))$ is given by

$$Kg = \hat{P}C^*g = \hat{P}\begin{bmatrix} A_D^{-1}Dg \\ 0 \end{bmatrix}. \tag{51}$$

It can easily be computed that $C^*g = \begin{bmatrix} A_D^{-1} \\ 0 \end{bmatrix}$ and C^* is bounded (and compact): $L_2(\Gamma) \to H$. The operator $\hat{P}$ is the unique, positive definite solution to a Riccati equation (12) with an arbitrary, bounded Q. The control $u(t)$ is given in feedback form (see Hendrickson and Lasiecka, 1993b) as

$$u(t) = -\frac{\partial}{\partial\nu} P_2(w, w_t), \tag{52}$$

where $P = [P_1, P_2]$ satisfies the Riccati equation (11). Theory in (Da Prato, Lasiecka and Triggiani (1986)) together with (49) gives

$$\frac{\partial}{\partial\nu} P_2 \in \mathcal{L}(H^2(\Omega) \times H_0^1(\Omega) \to L_2(\Gamma)). \tag{53}$$

Hence, the gain operator F is *bounded*, despite the *unboundedness* of B.

Since our main task is to construct a finite dimensional control which converges to the original one and is based on a finite dimensional estimator, we introduce approximating spaces and approximating operators and show that all the hypotheses of the Theorem 1 are satisfied.

Let $V_h \subset H^2(\Omega) \cap H_0^1(\Omega)$ be a space of Hermite cubics with orthogonal projection $\pi_h : \; H^2(\Omega) \cap H_0^1(\Omega) \to V_h$. We define $A_{h0} : \; V_h \to V_h$ by the formula

$$(A_{0h}x_h, \xi_h)_{H_0} = \int_\Omega (I + \gamma A_D)A_{0h}u_h v_h \, d\Omega \equiv (A_0 x_h, \xi_h)_{H_0} = \int_\Omega \Delta u_h \Delta v_h \, d\Omega, \tag{54}$$

where (54) results from the self-adjointness of A_D and taking into account zero boundary conditions for u_h and v_h on Γ. We also introduce the boundary operator, $B_{0h}g = \pi_h B_0 g$, defined via duality by

$$(g, B_{0h}^* u_h)_U = (B_{0h}g, u_h)_{H_0} = (B_0 g, u_h)_{H_0}, \quad \textit{and}$$

$$(B_h \begin{pmatrix} x_{1h} \\ x_{2h} \end{pmatrix}, \begin{pmatrix} y_{1h} \\ y_{2h} \end{pmatrix})_H = (B_{0h}x_{2h}, y_{2h})_{H_0} \tag{55}$$

$$= ((I+\gamma A_D)(I+\gamma A_D)^{-1}A_D D_0 x_{2h}, y_{2h})_{L_2(\Omega)}$$
$$= (x_{2h}, \frac{\partial}{\partial \nu} y_{2h})_{L_2(\Gamma)}.$$

It has been verified in (Hendrickson (1993a)) that standard approximation hypotheses (21)-(28) are satisfied by the above operators. As an approximation of operator R^*R we simply take its projection onto $\bar{V}_h$ with respect to the H norm. This way we have

$$\|R_h \begin{pmatrix} z_h \\ z_{th} \end{pmatrix} \|_H^2 = \|R \begin{pmatrix} z_h \\ z_{th} \end{pmatrix} \|_H^2 = \| \nabla z_h \|_{L_2(\Omega)}.$$

It can also be shown that $R_h^* R_h$ satisfies the discrete assumption (35). Moreover, it is straightforward to show (see Hendrickson and Lasiecka, 1993b) that all the approximating assumptions (21)-(28) are satisfied.

Therefore, all the assumptions of Theorem 1 are satisfied and we are in a position to apply its conclusions. We can now construct a finite dimensional compensator that is based only on the available information through the partial observation.

Numerical results illustrating this theory are given in Hendrickson (1995).

REFERENCES

Arendt, W. (1983) Vector Valued Laplace Transforms and Cauchy Problems. *Israel Journal of Mathematical*, **59**, 327-352.

Balakrishnan A.V. (1981). *Applied Functional Analysis*. Springer Verlag.

Curtain, R.F. and Salamon, D. (1986) Finite dimensional compensators for infinite dimensional systems with unbounded input operators. *SIAM Journal of Optimization* **24**, 797-816.

Da Prato, G., Lasiecka, I. and Triggiani, R. (1986) A direct study of riccati equations arising in hyperbolic boundary control problems. *Journal of Differential Equations*, 26-47.

Flandoli, F., Lasiecka, I. and Triggiani, R. (1988) Algebraic riccati equations with nonsmoothing observations arising in hyperbolic and Euler-Bernoulli boundary control problems. *Annali di Matematica Pura et. Applicata* **153**, 307-382.

Hendrickson, E. (1993a) Approximation and regularization methods for the riccati operator of the undamped Kirchhoff plate. *Master's thesis, University of Virginia.*

Hendrickson, E. (1995) Compensator design for the Kirchhoff plate model with boundary control. *Journal of Applied Mathematics and Computer Science* **5**, no. 1.

Hendrickson, E. and Lasiecka, I. (1993b) Numerical approximations and regularizations of Riccati equations arising in hyperbolic dynamics with unbounded control operators. *Computational Optimization and Applications* **2**, 343-390.

Hendrickson, E. and Lasiecka, I. (1995) Finite dimensional approximations of boundary control problems arising in partially observed hyperbolic systems. *Dynamics of Continuous, Discrete and Impulsive Systems* (to appear).

Horn, M.A. and Lasiecka, I. (1994) Asymptotic behavior with respect to thickness of boundary stabilizing feedback for the Kirchhoff plate. *Journal of Differential Equations* **114**, no. 2, 396-433.

Kazufumi, I. (1990) Finite dimensional compensators for infinite dimensional systems via Galerkin-type approximations. *SIAM Journal of Control and Optimization* **28**, 1251-1269.

Kesavan, S. (1989) *Topics in Functional Analysis.* John Wiley and Sons.

Lagnese, J.E. and Liones, J.L. (1989) *Modelling Analysis and Control of thin Plates.* Masson.

Lasiecka, I. (1992) Galerkin approximations of infinite dimensional compensators for flexible structures with unbounded control action. *Acta Applicandae Mathematicae* **28**, 101-133.

Lasiecka, I. (1995) Finite element approximations of compensator design for analytic generators with fully unbounded controls and observations. *SIAM Journal of Control and Optimization* **33**, 67-88.

Lasiecka, I. and Triggiani, R. (1991a) *Differential and algebraic riccati equations with applications to boundary and point control problems: Continuous and approximation theory.* Springer Verlag.

Lasiecka, I. and Triggiani, R. (1991b) Exact controllability and uniform stabilization of Kirchhoff plates with boundary control only on $\Delta w|_\Sigma$ and homogeneous boundary displacement. *Journal of Differential Equations 93*, 62-101.

Pazy, A. (1986) *Semigroups of linear operators and applications to partial differential equations.* Springer Verlag.

Schumacher, J.M. (1983) A direct approach to compensator design for distributed parameter systems. *SIAM Journal of Control and Optimization* **21**, 823-836.

Triggiani, R. (1975) On the stabilizability problem in Banach space. *Journal of Mathematical Analysis and Applications* **54**, 383-403.

26

Exact controllability of anisotropic elastic bodies

Józef Joachim Telega
Institute of Fundamental Technological Research,
Polish Academy of Sciences, Świętokrzyska 21, 00-049 Warsaw, Poland.
E-mail: jtelega@ippt.gov.pl
and
Włodzimierz Robert Bielski
Institute of Geophysics,
Księcia Janusza 64, 01-452 Warsaw, Poland.
E-mail: wbielski@igf.edu.pl

Abstract

The aim of this contribution is to study the exact controllability of linear, anisotropic elastic bodies by applying Lions' Hilbert Uniqueness Method.

Keywords

Unisotropic elasticity, exact controllability, Hilbert uniqueness method (HUM)

INTRODUCTION

It seems that studies of exact controllability of solids and structures like plates and shells have so far been limited to isotropic materials, cf. Lions (1988), Lagnese (1991), Nicaise (1993). A comprehensive review of the relevant literature is out of scope of this limited in space contribution. However, two important aspects of materials properties: anisotropy and inhomogeneity still remain to be included into investigations on exact controllability and stabilization of solids and structures.

Our aim here is to examine exact controllability of linear elastic bodies made of homogeneous, anisotropic materials.

1 BASIC EQUATION AND FORMULATION OF THE PROBLEM OF EXACT CONTROLLABILITY

Let $\Omega \subset R^N$ be a bounded domain with sufficiently regular boundary $\Gamma = \partial\Omega$. Obviously, in physical situations $N = 2$ or 3. The linear elastic body in its undeformed state is identified with $\bar{\Omega}$, the closure of Ω. The elasticity tensor $\mathbf{a} = (a_{ijkl})$ satisfies usual symmetry condition: $a_{ijkl} = a_{jikl} = a_{klij}$. Moreover, we assume that there exists a constant $C > 0$ such that

$$a_{ijkl}E_{ij}E_{kl} \geq CE_{ij}E_{ij}, \qquad \forall \mathbf{E} = (E_{ij}) \in E_s^N \tag{1.1}$$

Here E_s^N is the space of symmetric $N \times N$ matrices. The material of the elastic body is homogeneous, i.e., a_{ijkl} do not depend on $x \in \Omega$. It seems that the general case of exact controllability for $a_{ijkl} \in L^\infty(\Omega)$ still remains an open problem.

By $\mathbf{u} = (u_i), \mathbf{e} = (e_{ij})$ and $\boldsymbol{\sigma} = (\sigma_{ij})$ we denote the displacement vector, the strain tensor and the stress tensor, respectively. The constitutive equation has the form

$$\sigma_{ij} = a_{ijkl}e_{kl}. \tag{1.2}$$

The strain-displacement relation is linear

$$e_{ij}(\mathbf{u}) = u_{(i,j)} = (\frac{\partial u_i}{\partial x_j} + \frac{\partial u_j}{\partial x_i})/2. \tag{1.3}$$

The structure of the elasticity tensor is studied in Chernykh (1988). Particularly, in the case of orthotropy only nine moduli are independent and we have

$$\mathbf{a} = \begin{bmatrix} a_{1111} & a_{1122} & a_{1133} & 0 & 0 & 0 \\ a_{1122} & a_{2222} & a_{2233} & 0 & 0 & 0 \\ a_{1133} & a_{2233} & a_{3333} & 0 & 0 & 0 \\ 0 & 0 & 0 & a_{1212} & 0 & 0 \\ 0 & 0 & 0 & 0 & a_{2323} & 0 \\ 0 & 0 & 0 & 0 & 0 & a_{1313} \end{bmatrix}$$

Here the following change of indices has been used

$$(11) \rightleftharpoons (1), \;\; (22) \rightleftharpoons (2), \;\; (33) \rightleftharpoons (3), \;\; (12) \rightleftharpoons (4), \;\; (23) \rightleftharpoons (5), \;\; (13) \rightleftharpoons (6).$$

In the absence of body forces we shall study the dynamic elasticity problem with Dirichlet control on a part of the boundary:

$$\mathbf{u}'' - div(\mathbf{a}\mathbf{e}(\mathbf{u})) = \mathbf{0} \text{ in } Q = \Omega \times (0,T); \;\; \mathbf{u}(0) = \mathbf{u}^0, \mathbf{u}'(0) = \frac{\partial \mathbf{u}}{\partial t} = \mathbf{u}^1 \text{ in } \Omega, \tag{1.4}$$

$$\mathbf{u} = \begin{cases} \mathbf{v} & \text{on } \Sigma_0 \subset \Sigma = \Gamma \times (0,T), \\ \mathbf{0} & \text{on } \Sigma \setminus \Sigma_0. \end{cases} \tag{1.5}$$

Here $\mathbf{v} = (v_i)$ is a *control* through which the evolution of the solution is influenced. For the sake of simplicity, the density ϱ is assumed to be equal to 1; moreover $\mathbf{u}'' = \frac{\partial^2 \mathbf{u}}{\partial t^2}$ and $[div(\mathbf{a}\mathbf{e}(\mathbf{u}))]_i = (a_{ijkl}e_{kl}(\mathbf{u}))_{,j}$.

Exact Controllability Problem reads: Given $T > 0$ and an "arbitrary" initial state $\{\mathbf{u}^0, \mathbf{u}^1\}$ of the system (1.4), find $\mathbf{v}$ in a suitable function space such that $\mathbf{u}(T) = \mathbf{u}'(T) = 0$.

The part Σ_0 of the boundary has to be suitably chosen. Our problem of exact controllability will be solved by applying Hilbert Uniqueness Method, cf. Lions (1988).

2 BASIC INEQUALITIES

Let $\mathbf{n} = (n_i)$ denote the outward unit normal vector to Γ. Further we set $H_0^1(\Omega)^N = [H_0^1(\Omega)]^N, L^2(\Omega)^N = [L^2(\Omega)]^N$. Essential role in applying HUM plays the system with the homogeneous boundary conditions on Γ:

$$\varphi'' - div(\mathbf{ae}(\varphi)) = 0 \text{ in } Q; \qquad \varphi(0) = \varphi^0, \varphi'(0) = \varphi^1 \text{ in } \Omega, \tag{2.1}$$

$$\varphi = 0 \text{ on } \Sigma, \tag{2.2}$$

$$\varphi^0 \in H_0^1(\Omega)^N,\ \varphi^1 \in L^2(\Omega)^N. \tag{2.3}$$

In the derivation of the so called direct inequality we shall use

Lemma 2. 1 (Lions, 1988) *Let Ω be a bounded domain of R^N with the boundary Γ of class C^2. Then there exists a vector field $\mathbf{h} = (h_i) \in C^1(\bar{\Omega})^N$ such that $\mathbf{h}(x) = \mathbf{n}(x)$ on Γ.*

The total energy $E(t)$ of the system (2.1) - (2.3) is given by

$$E(t) = \frac{1}{2}[\|\varphi'(t)\|_{L^2}^2 + a(\varphi(t), \varphi(t))], \tag{2.4}$$

where $\|\varphi'(t)\|_{L^2}^2 = \int_\Omega |\varphi'(t)|^2 dx, a(\varphi(t), \varphi(t)) = \int_\Omega a_{ijkl} e_{ij}(\varphi(t)) e_{kl}(\varphi(t)) dx$. Since the system is conservative, therefore

$$E_0 := E(0) = E(t), \tag{2.5}$$

where

$$E(0) = \frac{1}{2}\int_\Omega [|\varphi^1|^2 + a_{ijkl} e_{ij}(\varphi^0) e_{kl}(\varphi^0)] dx. \tag{2.6}$$

More precisely to demonstrate (2.6) one has to show that $dE/dt = 0$.

2.1 Direct inequality

The aim of this subsection is to derive the following inequality

$$\int_\Sigma |\frac{\partial \varphi}{\partial \mathbf{n}}|^2 d\Sigma \le C_1 \int_\Sigma a_{ijkl} \frac{\partial \varphi_i}{\partial \mathbf{n}} n_j \frac{\partial \varphi_k}{\partial \mathbf{n}} n_l d\Sigma \le C_2 (T+1) E_0, \tag{2.7}$$

where C_1 and C_2 are positive constants. In the first step we multiply $(2.1)_1$ by $h_m \frac{\partial \varphi_i}{\partial x_m}$ and integrate over Q, where $\mathbf{h}$ is a vector field of class $C^1(\bar{\Omega})^N$:

$$\int_Q [\varphi_i'' - (a_{ijkl} e_{kl}(\varphi))_{,j}] h_k \frac{\partial \varphi_i}{\partial x_k} dx dt = 0. \tag{2.8}$$

Integrating by parts we obtain

$$\int_Q \varphi_i'' h_k \frac{\partial \varphi_i}{\partial x_k} dx dt = (\varphi'(t), h_k \frac{\partial \varphi(t)}{\partial x_k})|_0^T + \frac{1}{2}\int_Q h_{k,k}|\varphi'|^2 dx dt, \tag{2.9}$$

$$\begin{aligned}\int_Q (a_{ijkl}e_{kl}(\varphi))_{,j} h_m \frac{\partial \varphi_i}{\partial x_m} dx dt = -\int_Q a_{ijkl}e_{kl}(\varphi)\frac{\partial h_m}{\partial x_j}\frac{\partial \varphi_i}{\partial x_m} dx dt + \\ +\frac{1}{2}\int_Q h_{m,m} a_{ijkl} e_{ij}(\varphi) e_{kl}(\varphi) dx dt - \frac{1}{2}\int_\Sigma a_{ijkl}e_{ij}(\varphi)e_{kl}(\varphi) h_m n_m d\Sigma + \\ +\int_\Sigma a_{ijkl}e_{kl}(\varphi) n_j h_m \frac{\partial \varphi_i}{\partial x_m} d\Sigma.\end{aligned} \tag{2.10}$$

Here $|\varphi'|^2 = \varphi_i \varphi_i$, $\varphi' = \mathbf{0}$ on Σ (because $\varphi = 0$ on Σ) and $(\varphi,\psi) = \int_\Omega \varphi_i(x)\psi_i(x)dx$ $\forall \varphi,\psi \in L^2(\Omega)^N$. We observe that in order to derive (2.10) we have exploited the fact that a_{ijkl} do not depend on $x \in \Omega$. Substituting (2.9) and (2.10) into (2.8) we get

$$\begin{aligned}\int_\Sigma a_{ijkl}e_{kl}(\varphi) n_j h_m \frac{\partial \varphi_i}{\partial x_m} d\Sigma - \frac{1}{2}\int_\Sigma a_{ijkl}e_{ij}(\varphi)e_{kl}(\varphi) h_m n_m d\Sigma = (\varphi'(t), h_m \frac{\partial \varphi(t)}{\partial x_m})|_0^T + \\ +\frac{1}{2}\int_Q h_{m,m}[|\varphi'|^2 - a_{ijkl}e_{ij}(\varphi)e_{kl}(\varphi)] dx dt + \int_Q a_{ijkl}e_{kl}(\varphi)\frac{\partial h_m}{\partial x_j}\frac{\partial \varphi_i}{\partial x_m} dx dt.\end{aligned} \tag{2.11}$$

Since $\varphi_i = 0$ on Σ therefore we have

$$\frac{\partial \varphi_i}{\partial x_j} = n_j \frac{\partial \varphi_i}{\partial \mathrm{n}}. \tag{2.12}$$

Then the l.h.s of (2.11) takes the form

$$\begin{aligned}\int_\Sigma [a_{ijkl}e_{kl}(\varphi) n_j h_m \frac{\partial \varphi_i}{\partial x_m} - \frac{1}{2} a_{ijkl} e_{ij}(\varphi) e_{kl}(\varphi) h_m n_m] d\Sigma = \\ = \frac{1}{2}\int_\Sigma h_m n_m a_{ijkl} \frac{\partial \varphi_i}{\partial \mathrm{n}} n_j \frac{\partial \varphi_k}{\partial \mathrm{n}} n_l d\Sigma.\end{aligned} \tag{2.13}$$

By virtue of (2.11) we write

$$\begin{aligned}\frac{1}{2}\int_\Sigma h_m n_m a_{ijkl} \frac{\partial \varphi_i}{\partial \mathrm{n}} n_j \frac{\partial \varphi_k}{\partial \mathrm{n}} n_l d\Sigma = (\varphi'(t), h_m \frac{\partial \varphi(t)}{\partial x_m})|_0^T + \\ +\frac{1}{2}\int_Q h_{m,m}\left[|\varphi'|^2 - a_{ijkl}e_{ij}(\varphi)e_{kl}(\varphi)\right] dx dt + \int_Q a_{ijkl}e_{kl}(\varphi)\frac{\partial h_m}{\partial x_j}\frac{\partial \varphi_i}{\partial x_m} dx dt.\end{aligned} \tag{2.14}$$

Inequality (1.1) and Lemma 2.1 yield

$$\frac{1}{2}\int_\Sigma h_m n_m a_{ijkl} \frac{\partial \varphi_i}{\partial \mathrm{n}} n_j \frac{\partial \varphi_k}{\partial \mathrm{n}} n_l d\Sigma \geq \frac{C}{2}\int_\Sigma \frac{\partial \varphi_i}{\partial \mathrm{n}}\frac{\partial \varphi_i}{\partial \mathrm{n}} d\Sigma = \frac{C}{2}\int_\Sigma |\frac{\partial \varphi}{\partial \mathrm{n}}|^2 d\Sigma, \tag{2.15}$$

because $h_m(x)n_m(x) = 1$ for $x \in \Gamma$. Consequently

$$\frac{C}{2}\int_\Sigma |\frac{\partial \varphi}{\partial \mathrm{n}}|^2 d\Sigma \leq \frac{1}{2}\int_\Sigma a_{ijkl}\frac{\partial \varphi_i}{\partial \mathrm{n}}n_j\frac{\partial \varphi_k}{\partial \mathrm{n}}n_l d\Sigma = (\varphi'(t), h_m\frac{\partial \varphi(t)}{\partial x_m})|_0^T +$$
$$+\frac{1}{2}\int_Q h_{m,m}\left[|\varphi'|^2 - a_{ijkl}e_{ij}(\varphi)e_{kl}(\varphi)\right] dxdt + \int_Q a_{ijkl}e_{kl}(\varphi)\frac{\partial h_m}{\partial x_j}\frac{\partial \varphi_i}{\partial x_m}dxdt. \quad (2.16)$$

By using Korn's inequality (Nečas and Hlavaček, 1981) we obtain

$$(\varphi'(t), h_m\frac{\partial \varphi(t)}{\partial x_m})|_0^T = (\varphi'(T), h_m\frac{\partial \varphi(T)}{\partial x_m}) - (\varphi^1, h_m\frac{\partial \varphi^0}{\partial x_m}) \leq$$
$$\leq \max_{x\in\Omega, m=1,\dots,N}|h_m(x)|(\|\varphi'(T)\|_{L^2(\Omega)}\|\nabla\varphi(T)\|_{L^2(\Omega)}+$$
$$+\|\varphi^1\|_{L^2(\Omega)}\|\nabla\varphi^0\|_{L^2(\Omega)}) \leq C_3(E(T)+E(0)). \quad (2.17)$$

Taking account of (2.17) in (2.16), after standard estimations we arrive at

$$\int_\Sigma |\frac{\partial \varphi}{\partial \mathrm{n}}|^2 d\Sigma \leq C_1\int_\Sigma a_{ijkl}\frac{\partial \varphi_i}{\partial \mathrm{n}}n_j\frac{\partial \varphi_k}{\partial \mathrm{n}}n_l d\Sigma \leq C_4(2E_0 + 2TE_0).$$

2.2 Inverse inequality

First, we introduce new notations. Let $x^0 \in R^N$. We define, cf. Lions (1988,chap. I)

$$m_k(x) = x_k - x_k^0, \quad (2.18)$$

$$\Gamma(x^0) = \{x \in \Gamma | \mathrm{m}(x)\cdot \mathrm{n}(x) = m_k(x)n_k(x) > 0\}, \quad (2.19)$$

$$\Gamma^*(x^0) = \Gamma \setminus \Gamma(x^0) = \{x \in \Gamma | \mathrm{m}(x)\cdot \mathrm{n}(x) \leq 0\}, \quad (2.20)$$

$$\Sigma(x^0) = \Gamma(x^0)\times(0,T), \qquad \Sigma^*(x^0) = \Gamma^*(x^0)\times(0,T), \quad (2.21)$$

$$R(x^0) = \max_{x\in\bar\Omega}|\mathrm{m}(x)| = \max_{x\in\bar\Omega}\left[\sum_{i=1}^N (x_i - x_i^0)^2\right]^{1/2}. \quad (2.22)$$

Geometrical interpretation of the set $\Gamma(x^0)$ is given by Lions (1988, pp.79-81).

Now we are in position to derive the inverse inequality. Towards this end we set

$$X = (\varphi'(t), m_k\frac{\partial \varphi(t)}{\partial x_k})|_0^T = (\varphi_i'(t), m_k\frac{\partial \varphi_i(t)}{\partial x_k}), \quad (2.23)$$

$$Y = \int_Q |\varphi'(t)|^2 dxdt - \int_0^T a(\varphi(t),\varphi(t))dt. \quad (2.24)$$

Multiplying $(2.1)_1$ by φ and integrating over Q we infer that

$$Y = (\varphi'(t),\varphi(t))|_0^T. \quad (2.25)$$

Eq. (2.12), (2.14),(2.24) and (2.25) yield the relation

$$\frac{1}{2}\int_{\Sigma} m_k n_k a_{ijpq}\frac{\partial\varphi_i}{\partial \mathbf{n}} n_j \frac{\partial\varphi_p}{\partial \mathbf{n}} n_q d\Sigma = \frac{1}{2}\int_{\Sigma} m_k n_k a_{ijpq} e_{ij}(\varphi) e_{pq}(\varphi) d\Sigma =$$
$$= X + \frac{N-1}{2} Y + \frac{1}{2}\int_Q \left[|\varphi'|^2 + a_{ijkl} e_{ij}(\varphi) e_{kl}(\varphi)\right] dxdt \tag{2.26}$$

provided that $\boldsymbol{h} = \boldsymbol{m}$. We may write

$$X + \frac{N-1}{2} Y + \frac{1}{2}\int_Q \left[|\varphi'|^2 + a_{ijkl} e_{ij}(\varphi) e_{kl}(\varphi)\right] dxdt -$$
$$-\frac{1}{2}\int_{\Sigma\setminus\Sigma(x^0)} m_k n_k a_{ijpq} e_{ij}(\varphi) e_{pq}(\varphi) d\Sigma = \frac{1}{2}\int_{\Sigma(x^0)} m_k n_k a_{ijpq} e_{ij}(\varphi) e_{pq}(\varphi) d\Sigma, \tag{2.27}$$

and

$$|X + \frac{N-1}{2} Y| = |(\varphi'(t), m_k \frac{\partial\varphi(t)}{\partial x_k} + \frac{N-1}{2}\varphi(t))|_0^T \leq$$
$$\leq 2\|(\varphi'(t), m_k \frac{\partial\varphi(t)}{\partial x_k} + \frac{N-1}{2}\varphi(t))\|_{L^\infty(0,T)} \tag{2.28}$$

We shall prove that

$$|X + \frac{N-1}{2}| \leq 2R(x^0) \sup_t \|\varphi'(t)\|_{L^2(\Omega)} \|\nabla\varphi(t)\|_{L^2(\Omega)}. \tag{2.29}$$

To corroborate this statement we find

$$\|(\varphi'(t), m_k \frac{\partial\varphi(t)}{\partial x_k} + \frac{N-1}{2}\varphi(t))\|_{L^\infty(0,T)} \leq$$
$$\leq \sup_t \|\varphi'(t)\|_{L^2(\Omega)} \|m_k \frac{\partial\varphi(t)}{\partial x_k} + \frac{N-1}{2}\varphi(t)\|_{L^2(\Omega)}. \tag{2.30}$$

Next we calculate

$$\|m_k \frac{\partial\varphi(t)}{\partial x_k} + \frac{N-1}{2}\varphi(t)\|^2_{L^2(\Omega)} = \|m_k \frac{\partial\varphi(t)}{\partial x_k}\|^2_{L^2(\Omega)} +$$
$$\left[\frac{(N-1)^2}{4} - \frac{N(N-1)}{2}\right] \|\varphi(t)\|^2_{L^2(\Omega)} \leq \|m_k \frac{\partial\varphi(t)}{\partial x_k}\|^2_{L^2(\Omega)}, \tag{2.31}$$

since

$$(m_k \frac{\partial\varphi(t)}{\partial x_k}, \varphi(t)) = -\frac{N}{2}\|\varphi(t)\|^2_{L^2(\Omega)} \qquad \forall t \in [0,T]$$

and

$$\frac{\partial m_k}{\partial x_k} = N, \qquad \left[\frac{(N-1)^2}{4} - \frac{N(N-1)}{2}\right] \leq 0.$$

We recall that $\varphi = \mathbf{0}$ on Γ. Consequently

$$\|m_k \frac{\partial \varphi(t)}{\partial x_k} + \frac{N-1}{2}\varphi(t)\|_{L^2(\Omega)} \leq \max_{x\in\bar{\Omega}} |\mathrm{m}(x)| \|\nabla \varphi(t)\|_{L^2(\Omega)} = R(x^0)\|\nabla \varphi(t)\|_{L^2(\Omega)}. \quad (2.32)$$

Thus we see that (2.28), (2.30) and (2.32) prove the inequality (2.29). By applying Korn's inequality (Nečas and Hlavaček, 1981), we conclude that there exists a constant $K > 0$ such that $\|\nabla\ \varphi(t)\|^2_{L^2(\Omega)} \leq Ka(\varphi(t), \varphi(t))$. Then (2.29) is estimated as follows

$$|X + \frac{N-1}{2}Y| \leq 2R(x^0)\sqrt{K} \sup_t \|\varphi'(t)\|_{L^2(\Omega)} \left[a(\varphi(t), \varphi(t))\right]^{1/2}. \quad (2.33)$$

Because

$$E_0 = E(t) = \frac{1}{2}\|\varphi'(t)\|^2 + \frac{1}{2}a(\varphi(t), \varphi(t)) \geq \|\varphi(t)\|_{L^2(\Omega)} \left[a(\varphi(t), \varphi(t))\right]^{1/2},$$

therefore

$$|X + \frac{N-1}{2}Y| \leq 2R(x^0)\sqrt{K}E_0. \quad (2.34)$$

Hence

$$-2R(x^0)\sqrt{K}E_0 \leq X + \frac{N-1}{2}Y \leq 2R(x^0)\sqrt{K}E_0. \quad (2.35)$$

We know that $m_k n_k \leq 0$ on $\Sigma \backslash \Sigma(X^0)$. From (2.27), by taking account of (2.35) we obtain

$$-2R(x^0)\sqrt{K}E_0 + TE_0 \leq \max_{x\in\bar{\Omega}} \frac{|\mathrm{m}(x)|}{2} \int_{\Sigma(x^0)} a_{ijkl} e_{ij}(\varphi(t)) e_{kl}(\varphi(t)) d\Sigma.$$

Thus we arrive at the final form of the inverse inequality:

$$\frac{R(x^0)}{2} \int_{\Sigma(x^0)} a_{ijkl} \frac{\partial \varphi_i}{\partial \mathrm{n}} n_j \frac{\partial \varphi_k}{\partial \mathrm{n}} n_l d\Gamma dt \geq \left[T - 2R(x^0)\sqrt{K}\right] E_0. \quad (2.36)$$

Remark 2.1 For isotropic bodies straightforward calculation yield (Lions, 1988)

$$\frac{R(x^0)}{2} \int_{\Sigma(x^0)} \left[\mu(\frac{\partial \varphi}{\partial \mathrm{n}})^2 + (\lambda + \mu)(div\varphi)^2\right] d\Sigma \geq (T - \frac{2R(x^0)}{\sqrt{\mu}})E_0,$$

where λ and μ are the Lamé constants.

3 EXACT CONTROLLABILITY: APPLICATION OF HUM

According to the Hilbert Uniqueness Method an important role is played by the adjoint system. By $\psi = (\psi_i)$ we denote the displacement field of the adjoint system. The system $\{\varphi, \psi\}$ of the HUM is now given by

$$\psi'' - div(\mathrm{ae}(\psi)) = 0 \text{ in } Q; \qquad \psi(T) = \psi'(T) = 0 \text{ in } \Omega, \quad (3.1)$$

$$\psi(t) = \begin{cases} (a_{ijkl}e_{kl}(\varphi(t)n_j) & \text{on} \quad \Sigma(x^0), \\ 0 & \text{on} \quad \Sigma \setminus \Sigma(x^0), \end{cases} \tag{3.2}$$

where φ is the unique solution of the auxiliary system (2.1)-(2.3). In the case of isotropy relation $(3.2)_1$ reduces to, cf. Lions (1988,p.227)

$$\psi(t) = \mu \frac{\partial \varphi(t)}{\partial \mathbf{n}} + (\lambda + \mu)(div\varphi(t))\mathbf{n} \qquad \text{on } \Sigma(x^0),$$

where $div\varphi = \partial\varphi_i/\partial x_i = n_i \partial\varphi_i/\partial\mathbf{n}$, provided that $\varphi = 0$ on $\Gamma(x^0)$. The main result of this paper is formulated as

Theorem 3. 1 *Let $x^0 \in R^N$ and define $\Sigma(x^0)$ and $R(x^0)$ by (2.21) and (2.22), respectively. If $T > 2R(x^0)\sqrt{K}E_0$ then for $\mathbf{u}^0 \in L^2(\Omega)^N, \mathbf{u}^1 \in H^{-1}(\Omega)^N$ there exists $\mathbf{v} \in L^2(\Sigma(x^0))^N$ such that the solution of the system (1.4)-(1.5) satisfies $\mathbf{u}(T) = \mathbf{u}'(T) = 0$, i.e. this system is exactly controllable.*

Proof. According to HUM we define

$$\Lambda\{\varphi^0, \varphi^1\} = \{\psi'(0), -\psi(0)\}, \tag{3.3}$$

a linear and continuous operator. Multiplying Eq. $(3.1)_1$ by φ, the solution of (2.1)-(2.3) and performing integration by parts we obtain

$$-(\psi'(0), \varphi^0) + (\psi(0), \varphi^1) + \int_{\Sigma(x^0)} a_{ijkl} \frac{\partial \varphi_i}{\partial \mathbf{n}} n_j n_k \psi_l d\Sigma = 0,$$

since $\psi = 0$ on $\Sigma \setminus \Sigma(x^0)$. Hence, by virtue of $(3.2)_1$

$$\langle \Lambda\{\varphi^0, \varphi^1\}, \{\varphi^0, \varphi^1\} \rangle = \int_{\Sigma(x^0)} a_{ijkl} n_j \frac{\partial \varphi_k}{\partial \mathbf{n}} n_l \psi_i d\Sigma =$$

$$= \int_{\Sigma(x^0)} a_{ijkl} \frac{\partial \varphi_k}{\partial \mathbf{n}} n_j n_l a_{impq} \frac{\partial \varphi_p}{\partial \mathbf{n}} n_q n_m d\Sigma = \int_0^T \|\psi(t)\|^2_{L^2(\Gamma(x^0))} dt. \tag{3.4}$$

In the Appendix we demonstrate that the matrix $A_{klpq} = a_{ijkl} n_j a_{impq} n_m$ satisfies (λ_0 -a positive constant)

$$A_{klpq} \xi_k n_l \xi_p n_q \geq \lambda_0 a_{ijkl} \xi_i n_j \xi_k n_l \qquad \forall \boldsymbol{\xi} \in R^N. \tag{3.5}$$

Taking account of (2.36) and (3.5) in (3.4) we obtain

$$\langle \Lambda\{\varphi^0, \varphi^1\}, \{\varphi^0, \varphi^1\} \rangle \geq \lambda_0 \int_{\Sigma(x^0)} a_{ijkl} \frac{\partial \varphi_i}{\partial \mathbf{n}} n_j \frac{\partial \varphi_k}{\partial \mathbf{n}} n_l d\Sigma \geq \frac{2\lambda_0}{R(x^0)} \left[T - 2R(x^0)\sqrt{K} \right] E_0. \tag{3.6}$$

Thus Λ is an isomorphism of $H_0^1(\Omega)^N \times L^2(\Omega)^N$ on $H^{-1}(\Omega)^N \times L^2(\Omega)^N$ and may apply HUM. □

Remark 3.1 The r.h.s. of $(3.2)_1$ is the stress vector whilst on the l.h.s. the displacement vector occurs. Therefore we must assume that relevant quantities are non-dimensional.

Acknowledgment. The authors were supported by the State Committee for Scientific Research (Poland) through the grant No 3 P404 013 06.

REFERENCES

Chernykh K.F. (1988) *Introduction to Anisotropic Elasticity*, Nauka, Moskva, in Russian.

Lagnese J.F. (1991) Uniform asymptotic energy estimates for solutions of the equations of dynamic plane elasticity with nonlinear dissipation at the boundary. *Nonlinear Anal., TMA*, **16**, 35-54.

Lions J.-L. (1988) *Controlabilité Exacte, Perturbations et Stabilisation de Systèmes Distribués, t.1.Controlabilité Exacte*, Masson, Paris.

Nečas J., Hlavaček I. (1981) *Mathematical Theory of Elastic and Elastic-Plastic Bodies: An Introduction.* Elsevier, Amsterdam.

Nicaise S. (1993) About the Lamé system in a polygonal or a polyhedral domain and coupled problem between the Lamé system and the plate equation. II. Exact controllability, *Annali della Scuola Norm. Sup. di Pisa*, **20**, 327-361

APPENDIX

Our aim now is to show that there exists a positive constant λ_0 depending on (a_{ijkl}) such that (3.5) is satisfied. We are considering the case $N = 3$ only; the case $N = 2$ is obviously simpler. We set $B_{ik} = a_{ijkl}n_j n_l$. The matrix $\mathbf{B}$ is symmetric and positive definite (p.d.), what follows from the properties of the elasticity tensor $\mathbf{a} = (a_{ijkl})$. Then (3.5) takes the form $B_{ij}B_{jk}\xi_i\xi_k \geq \lambda_0 B_{ij}\xi_i\xi_j$. Hence $B_{ij}(B_{jk} - \lambda_0\delta_{jk})\xi_i\xi_j \geq 0$. It is thus sufficient to show that the matrix $\mathbf{C} = (C_{ij}), C_{ik} = B_{ij}(B_{jk} - \lambda_0\delta_{jk})$ is p.d. for some $\lambda_0 > 0$. $\mathbf{C}$ is p.d. iff all its principal minors are p.d. The principal minor of $\mathbf{C}$ of rank 1 is $C_{11} = B_{1j}(B_{j1} - \lambda_0\delta_{j1}) = B_{11}^2 + B_{12}^2 + B_{13}^2 - \lambda_0 B_{11}$. Since $B_{11} > 0$, therefore $C_{11} > 0$ if $0 < \lambda_0 < B_{11}$, say $\lambda_0 = 1/2B_{11}$. Consider now the principal minor of $\mathbf{C}$ of rank 2. We have $C_{\alpha\beta} = B_{\alpha j}(B_{j\beta} - \lambda_0\delta_{j\beta})$, $j = 1,2,3$; $\alpha, \beta = 1,2$, and $\det(C_{\alpha\beta}) = a\lambda_0^2 - b\lambda_0 + c \equiv g(\lambda_0)$, where $a = B_{11}B_{22} - B_{12}^2 > 0$, $b = (B_{11} + B_{22} + \frac{B_{23}^2}{B_{22}})a + B_{22}(B_{13} - \frac{B_{12}B_{23}}{B_{22}})^2 > 0$, $c = a^2 + (B_{11}B_{23} - B_{13}B_{12})^2 + (B_{12}B_{23} - B_{13}B_{22})^2 > 0$, because (B_{ij}) is p.d. The following cases are possible: (i) The determinant of the quadratic function $g(\lambda_0)$ is negative $(b^2 - 4ac < 0)$ and then $g(\lambda_0) > 0$ for all λ_0, (ii) The determinant is nonnegative and then there exists two positive roots $(\lambda_0^{(1)} \leq \lambda_0^{(2)})$ of the equation $g(\lambda) = 0$. Thus $g(\lambda_0) > 0$ for $0 < \lambda_0 < \lambda_0^{(1)}$, for instance $\lambda_0 = 1/2\lambda_0^{(1)}$. Consider now the determinant of the matrix $\mathbf{C} : \det \mathbf{C} = \det(\mathbf{B}(\mathbf{B} - \lambda_0\mathbf{I})) = (\det \mathbf{B}) \det(\mathbf{B} - \lambda_0\mathbf{I})$. Since $\det \mathbf{B} > 0$ therefore $\det \mathbf{C} > 0$ iff $\det(\mathbf{B} - \lambda_0\mathbf{I}) > 0$. We have $f(\lambda_0) = \det(\mathbf{B} - \lambda_0\mathbf{I}) = -\lambda_0^3 + \lambda_0^2 I_B - \lambda_0 II_B + III_B$, where I_B, II_B, III_B are principal invariants of $\mathbf{B}$. We see that $f(0) = III_B = \det \mathbf{B} > 0$ and $f(\lambda) < 0$ for large λ. Thus there exists at least one positive root λ_1, i.e. $f(\lambda_1) = 0$. Let $\lambda_1 > 0$ be the smallest of such roots. Then $f(\lambda) > 0$ for $0 < \lambda < \lambda_1$. Finally a good candidate for λ_0 is $\lambda_0 = \frac{1}{2}\min\{B_{11}, \lambda_1\}$ in case (i) or $\lambda_0 = \frac{1}{2}\min\{B_{11}, \lambda_0^{(1)}, \lambda_1\}$ in case (ii).

Abstract and Stochastic Problems

27
Viability for differential inclusions in Banach spaces

Ovidiu Cârjă
Facultatea de Matematică, Universitatea "Al.I.Cuza"
Bdul. Copou 11, 6600 Iaşi, România
Fax: 40-32-146330. E-mail: carja@uaic.ro

Abstract
The aim of this paper is to give a sufficient condition in order that a subset of a Banach space be a viability domain for a semilinear differential inclusion.

Keywords
Viability, semilinear differential inclusion, tangency condition

1 INTRODUCTION AND MAIN RESULT

Let X be a Banach space, D a nonempty subset of X, $F : D \to X$ a multifunction with nonempty values and A the infinitesimal generator of a c_0 semigroup $S(t)$ on X. Consider the differential inclusion

$$x'(t) \in Ax(t) + F(x(t)). \tag{1}$$

By a viable solution to (1) we mean a continuous function $x : [0, \sigma] \to X$ which satisfies $x(t) \in D$ for all $t \in [0, \sigma]$ and there exists a strongly measurable function f from $[0, \sigma]$ into X with $f(s) \in F(x(s))$ a.e., such that

$$x(t) = S(t)x(0) + \int_0^t S(t-s)f(s)\,ds, \ \forall t \in [0, \sigma].$$

The set D is called a viability domain for the differential inclusion (1) if for every $x_0 \in D$ there exists a viable solution to (1) with $x(0) = x_0$. The viability problem is to give conditions in order that the set D be a viability domain for (1).

In the finite dimensional case (where $A = 0$, hence $S(t) = I$), the main viability result is due to Gautier (1973) and Haddad (1981). See also Aubin and Cellina (1984), Cârjă and Ursescu (1993). It asserts that if F is upper semicontinuous with compact convex values and D is locally closed, then D is a viability domain for (1) if and only if the tangency condition

$$F(x) \cap T_D(x) \neq 0, \quad \forall x \in D, \tag{2}$$

is satisfied. Here the tangency concept T_D is that of Bouligand and Severi: $u \in T_D(x) \Longleftrightarrow \liminf_{t\downarrow 0} \frac{1}{t} d(x+ty, D) = 0$. where $d(a, B) = \inf_{b\in B} \|a-b\|$.

The viability result mentioned above has served as a main tool in a general study of the characteristics method for a first order partial differential equation (see Cârjă and Ursescu (1993)). In particular, this leads to a study of the Hamilton–Jacobi–Bellman equations from the point of view of contingent solutions (see Cârjă (1996)). In infinite dimensions, for differential inclusion of type (1) we mention two important results. The first one is due to Pavel and Vrabie (1979) (see also Pavel (1984), Chapter 5), where $S(t)$ is compact, F is locally bounded, demiclosed with closed convex values, and D is locally closed. They got viable solutions to (1) for every $x_0 \in D$ from the tangency condition

$$\lim_{t\downarrow 0} \frac{1}{t} d(S(t)x + ty, D) = 0, \quad \forall x \in D, \ \forall y \in F(x). \tag{3}$$

The second result is due to ShiShuzhong (1989), where $S(t)$ is a compact differentiable semigroup, D is compact and F is upper semicontinuous with compact convex values. He proved that under these conditions D is a viability domain for (1) if and only if the tangency condition

$$F(x) \cap T_D^S(x) \neq 0, \quad \forall x \in D, \tag{4}$$

is satisfied, where the tangency concept T_D^S is defined by

$$u \in T_D^S(x) \Longleftrightarrow \liminf_{t\downarrow 0} \frac{1}{t} d(S(t)x + tu, D) = 0. \tag{5}$$

In many situations, especially in the study of the Bellman equation associated with control problems, the hypotheses in ShiShuzhong (1989) that D is compact and F has compact values are too strong. A very simple example is the linear control system $x'(t) \in Ax(t) + BU$, where U is a bounded closed convex set and B is a linear continuous operator.

On the other hand, the tangency condition in Pavel and Vrabie (1979) is also too strong in comparison to that of the finite dimensional case.

The aim of this note is to show that the tangency condition (3) in Pavel and Vrabie (1979) can be replaced by the weaker condition (4). The price to be paid is to strengthen the conditions on F. Namely, in addition to the other hypotheses we assume that F is "strongly-weakly" lower semicontinuous. Precisely, we have

Theorem 1 *Let X be a reflexive Banach space and D a locally closed subset of X. Assume: $S(t)$ is compact for $t > 0$; F is locally bounded, demiclosed and lower semicontinuous with closed convex values; the tangency condition (4) holds true. Then for every $x_0 \in D$ there exists a viable solution to the differential inclusion (1).*

Let us define the notions *lower semicontinuous* and *demiclosed.*

We say that the multifunction $F : D \to X$ is lower semicontinuous in $x \in D$ if $\forall y \in F(x)$, $\forall (x_n) \in D, x_n \to x$, there exists $y_n \in F(x_n)$ such that $y_n \rightharpoonup y$.

We say that the multifunction F is demiclosed if the hypotheses: $y_n \in F(x_n)$, with $x_n \in D, x_n \to x$ and $y_n \rightharpoonup y$ imply $x \in D$ and $y \in F(x)$.

Mention that we use the same idea of proof as in Pavel and Vrabie (1979) and Pavel (1984), therefore some steps will only be sketched. Finally, note that in Pavel and Vrabie (1979) and Pavel (1984) both D and F depend on t but we consider the autonomous case here for simplicity only.

2 PROOF OF THE MAIN RESULT

As we remarked earlier, we follow the same line as in Pavel and Vrabie (1979) and Pavel (1984). By the nonexpansivity of the distance function it follows that the tangency concept T_D^S defined in (5) is equivalent to the following one:

$$u \in T_D^S(x) \Longleftrightarrow \liminf_{h\downarrow 0}(1/h)d(S(h)x + \int_0^h S(h-s)u\,ds, D) = 0. \tag{6}$$

Since F is locally bounded and D is locally closed, for $x_0 \in D$ there are constants $M > 0$, $r > 0$ such that $B(x_0, r) \cap D$ is closed and

$$\|f(y)\| \leq M, \ \forall y \in B(x_0, r) \cap D, \ \forall f(y) \in F(y). \tag{7}$$

Here $B(x_0, r)$ is the closed ball of center x_0 and radius r.

Let T be sufficiently small such that

$$\max_{0\leq t\leq T} \|S(t)x_0 - x_0\| + T(M+1)K \leq r, \tag{8}$$

where $K = C\exp(\omega T)$ and $C \geq 1$, $\omega \geq 0$ are such that $\|S(t)\| \leq C\exp(\omega t)$ for $t \geq 0$.

The following lemma concerns the construction of an approximate solution to (1) on $[0, T)$. This is the main step of the proof of Theorem 1.

Lemma 1 *Suppose that the hypotheses of Theorem 1 hold. Let $x_0 \in D$ and choose r, T, M as in (7) and (8). Then there is an $(1/n)-$approximate solution x_n to (1) on $[0,T)$ in the following sense: For each positive integer n, there is an infinite partition $\{t_i^n\}_{i\geq 0}$ of $[0,T]$ with the following properties:*

(P1) $t_0^n = 0, \ \ t_{i+1}^n - t_i^n := d_i^n \in (0, \frac{1}{n}], \ \ \lim_{i\to\infty} t_i^n = T$;

(P2) $x_n(0) = x_0, \ \ x_n(t_i^n) := x_i^n \in B(x_0, r) \cap D$;

(P3) $x_n(t) = S(t-t_i^n)x_i^n + \int_{t_i^n}^t S(t-s)f(x_i^n)\,ds + (t-t_i^n)p_i^n \in B(x_0,r)$ *for* $t \in [t_i^n, t_{i+1}^n]$ *with* $\|p_i^n\| \leq \frac{1}{n}$, *where* $f(x_i^n) \in F(x_i^n)$.

Proof. To simplify notation, suppress n as a superscript for t_i, d_i, x_i and p_i. The construction is by induction. Set $t_0 = 0, x_n(0) = x_0$, and assume that x_n is constructed on $[0, t_i]$. If $t_i = T$, set $t_{i+1} = t_i$. Consider now the case $t_i < T$. Taking into account that $x_i \in D$, by (4) and (6) there exists $f(x_i) \in F(x_i)$ such that

$$\liminf_{h\downarrow 0}(1/h)d(S(h)x_i + \int_0^h S(h-s)f(x_i)\,ds, D) = 0.$$

This implies the existence of $h_i \in (0, 1/n]$ with $t_i + h_i \leq T$ such that

$$d(S(h_i)x_i + \int_0^{h_i} S(h_i - s)f(x_i)\,ds, D) \leq \frac{h_i}{2n}.$$

We can thus define

$$\begin{aligned}\delta_i = \sup\{h \in (0, \tfrac{1}{n}]; t_i + h \le T, \exists f(x_i) \in F(x_i),\\ d(S(h)x_i + \textstyle\int_0^h S(h-s)f(x_i)\,ds, D) \le \tfrac{h}{2n}\}.\end{aligned} \tag{9}$$

Therefore, there exist $d_i \in (\frac{\delta_i}{2}, \delta_i], 0 < d_i \le \frac{1}{n}, t_i + d_i \le T$ and $f(x_i) \in F(x_i)$ such that

$$d(S(d_i)x_i + \int_0^{d_i} S(d_i - s)f(x_i)\,ds, D) < \frac{d_i}{2n}. \tag{10}$$

Set $t_{i+1} := t_i + d_i$. By (10), there exists $x_{i+1} \in D$ such that

$$\|S(d_i)x_i + \int_0^{d_i} S(d_i - s)f(x_i)\,ds - x_{i+1}\| \le \frac{d_i}{n},$$

i.e.,

$$x_{i+1} = S(t_{i+1} - t_i)x_i + \int_{t_i}^{t_{i+1}} S(t_{i+1} - s)f(x_i)\,ds + (t_{i+1} - t_i)p_i, \quad with\|p_i\| \le \tfrac{1}{n}.$$

Define x_n on $[t_i, t_{i+1}]$ as indicated in (P3). It is easy to see that

$$x_n(t) = S(t)x_0 + \int_0^t S(t-s)f_n(s)\,ds + g_n(t), \qquad t \in [0, t_{i+1}], \tag{11}$$

where

$$f_n(s) = f(x_i), \qquad s \in [t_j, t_{j+1}], \; j = 0, \cdots, i \tag{12}$$

and

$$\begin{aligned}g_n(t) = \textstyle\sum_{k=0}^{j-1}(t_{k+1} - t_k)S(t - t_{k+1})p_k + (t - t_j)p_j, \quad t \in [t_j, t_{j+1}], \; j = 1, \cdots, i,\\ g_n(t) = tp_0, \quad t \in [0, t_1].\end{aligned} \tag{13}$$

By induction hypothesis, $x_n(t) \in B(x_0, r)$ for $t \in [0, t_i]$. We have to prove that $x_n(t) \in B(x_0, r)$ for $t \in [t_i, t_{i+1}]$. Indeed, for $t \in [t_i, t_{i+1}]$,

$$\|x_n(t) - x_0\| \le \|S(t)x_0 - x_o\| + TMK + \frac{TK}{n} \le r$$

because of (7), (8) and the fact that

$$\|g_n(t)\| \le \frac{Kt}{n}.$$

Therefore, if $t^\star := \lim_{i\to\infty} t_i$, we can define $x_n(t)$ by (11) for every $t \in [0, t^\star)$ where f_n is given by (12) for every $j \in \mathbb{N}$ and g_n is given by (13) for every $j \in \mathbb{N}$. As results from Pavel (1984), (x_i) is a Cauchy sequence, and let $x^\star$ be its limit. Clearly, $x^\star \in B(x_0, r) \cap D$.

We prove now that $t^\star = T$. It is this step where we use the fact that F is lower semicontinuous instead of the condition (3). suppose to the contrary that $t^\star < T$. By (2), there exist $h^\star \in (0, \frac{1}{n}]$ with $h^\star + t^\star < T$ and $f^\star \in F(x^\star)$ such that

$$d(S(h^\star)x^\star + \int_0^{h^\star} S(h^\star - s)f^\star\,ds, D) \le \frac{h^\star}{4n}. \tag{14}$$

Since $d_i \to 0$, $\delta_i \to 0$ hence there exists i_0 such that for $i \geq i_0$, $h^\star > \delta_i$. Because F is lower semicontinuous, there exists $y_i \in F(x_i)$ such that $y_i \rightharpoonup f^\star$. Since $t_i \leq t^\star$, we have $h^\star + t_i < T$. Along with $h^\star > \delta_i$, the very definition of δ_i implies

$$d(S(h^\star)x_i + \int_0^{h^\star} S(h^\star - s)y_i\,ds, D) > \frac{h^\star}{2n}, \quad i \geq i_0. \tag{15}$$

On the other hand by a standard argument,

$$\int_0^{h^\star} S(h^\star - s)y_i\,ds \to \int_0^{h^\star} S(h^\star - s)f^\star\,ds.$$

Passing to limit in (15) we get a contradiction with (14). Therefore $t^\star = T$ as claimed. This ends the proof of Lemma 1.

Proof of Theorem 1. It follows as in Pavel (1984), p.190. One proves that the sequence (x_n) constructed in Lemma 1 has a convergent subsequence in $C([0,T];X)$ to a solution of (1). The Ascoli-Arzela theorem and a result of Kato are the main tools.

3 REFERENCES

Aubin, J. P. and Cellina, A. (1984) *Differential Inclusions,* Springer-Verlag, Berlin.

Cârjă, O. (1996) Lower semicontinuous solutions for a class of Hamilton–Jacobi–Bellman equations, *Journal of Optimization Theory and Applications,* **89**.

Cârjă, O. and Ursescu, C. (1993) The characteristics method for a first order partial differential equation, *Analele Ştiinţifice ale Universităţii Al.I.Cuza Iaşi,* **39**, pp. 367–396.

Cârjă, O. and Ursescu, C. (1994) Viscosity solutions and partial differential inequations, in *Evolutions Equations, Control Theory and Biomathematics* (eds. P. Clement and G. Lumer) , Marcel Dekker, New York, pp. 39–44.

Gautier, S. (1973) Equations différentielles multivoque sur un fermè , *Publications de l'Université de Pau,* pp. 1–5.

Haddad, G. (1981) Monotone trajectories of differential inclusions and functional differential inclusions with memory, *Israel Journal of Mathematics,* **39**, pp. 83–100.

Pavel, N. H. and Vrabie, I. I. (1979) Semi-linear evolution equations with multivalued right hand side in Banach spaces, *Analele Ştiinţifice ale Universităţii Al.I.Cuza Iaşi,* **25**, pp. 137–157.

Pavel, N. H. (1984) *Differential Equations, Flow Invariance and Applications,* Research Notes in Mathematics, 113, Pitman, London.

ShiShuzhong, (1989) Viability theorems for a class of differential-operator inclusions, *Journal of Differential Equations*, **79**, pp. 232–257.

28

Relaxation of optimal control problems coercive in L^p-spaces

T. Roubíček
Mathematical Institute, Charles University,
Sokolovská 83, CZ-186 00 Praha 8, Czech Republic
and
Institute of Information Theory and Automation, Academy of Sciences,
Pod vodárenskou věží 4, CZ-182 08 Praha 8, Czech Republic.
E–mail: `roubicek@karlin.mff.cuni.cz`

Abstract

This contribution surveys a relaxation method applicable for optimal control problems whose admissible controls are not a priori bounded so that the coercivity can be expected only in L^p- but not L^∞-spaces. Correct relaxed problem is constructed by a continuous extension of the original one and proper optimality conditions are derived and analyzed. For this, the relative L^1-weak compactness of energy of minimizing sequences is essential. The method is illustrated on an elliptic optimal control problem.

Keywords

Nonconvex optimal-control problems, unbounded control, oscillations, relaxation, nonconcentration, Young measures, Pontryagin maximum principle, elliptic problems

1 THE ORIGINAL PROBLEM

Optimization problems involving controls from Lebesgue spaces usually impose control constraints ensuring the set of admissible controls to be bounded in an L^∞-space. If the control constraints have a more general structure, one can expect only a boundedness of the set of admissible controls in an L^p-space provided a suitable coercivity of the problem is supposed. The aim of this contribution is to show that, in this more general case, basically the standard results can be obtained but more advanced techniques must be used. Though such phenomena can be pursued also in nonconvex variational calculus or game theory, we want to illustrate them on a rather simpler situation.

Let us consider a model optimal control problem in a general form

$$
\text{(P)} \quad \begin{cases} \text{Minimize} & \int_\Omega \varphi_0(x,y(x),u(x))\mathrm{d}x & & \text{(cost functional)} \\ \text{subject to} & Ay = \varphi(x,y(x),u(x)) & \text{on } \Omega, & \text{(state equation)} \\ & u(x) \in S(x) & \text{on } \Omega, & \text{(control constraints)} \\ & y \in L^q(\Omega;\mathbb{R}^n), \quad u \in L^p(\Omega;\mathbb{R}^m), & & \end{cases}
$$

where u is a control distributed on a sufficiently regular domain $\Omega \subset \mathbb{R}^{n_1}$, y is a state, $\varphi_0 : \Omega \times (\mathbb{R}^n \times \mathbb{R}^m) \to \mathbb{R}$ and $\varphi : \Omega \times (\mathbb{R}^n \times \mathbb{R}^m) \to \mathbb{R}^{m_1}$ are Carathéodory mappings, the later one satisfying the growth condition

$$|\varphi(x,r,s)| \le a(x) + b|r|^{q/p_1} + c|s|^{p/p_1} \tag{1}$$

with some $a \in L^{p_1}(\Omega)$ and $b, c \in \mathbb{R}$ so that $\varphi(y,u)$ lives in $L^{p_1}(\Omega;\mathbb{R}^{m_1})$, $p_1 > 1$. Moreover, we suppose that $S : \Omega \rightrightarrows \mathbb{R}^m$ is a measurable multivalued mapping, $A : \mathcal{D}_A \to L^{p_1}(\Omega;\mathbb{R}^{m_1})$ is a (for simplicity linear) mapping whose domain $\mathcal{D}_A$ is a subspace of $L^q(\Omega;\mathbb{R}^n)$, and the state equation $Ay = \varphi(x,y(x),u(x))$ has, for any $u \in L^p(\Omega;\mathbb{R}^m)$, precisely one solution $y = \pi(u) \in \mathcal{D}_A$. A natural growth condition for φ_0 is

$$|\varphi_0(x,r,s)| \le a_0 + b_0|r|^q + c_0|s|^p \tag{2}$$

with some $a_0 \in L^1(\Omega)$ and $b_0, c_0 \in \mathbb{R}$ so that $\varphi_0(y,u)$ lives in $L^1(\Omega)$.

Usually, the control-constraint mapping S is assumed bounded to ensure coercivity of the problem in $L^\infty(\Omega;\mathbb{R}^m)$ but here we want to put off such assumption. Instead of this, we suppose the coercivity of φ_0 in the sense

$$\varphi_0(x,r,s) \ \ge \ c_1|s|^p \tag{3}$$

for some $c_1 > 0$. This ensures that every sequence of controls $\{u_k\}_{k\in\mathbb{N}}$ minimizing for (P) is inevitably bounded in $L^p(\Omega;\mathbb{R}^m)$ (but not in $L^\infty(\Omega;\mathbb{R}^m)$, however!).

2 THE RELAXED PROBLEM

As neither convexity of $\varphi_0(x,r,\cdot)$ and of $S(x)$ nor linearity of $\varphi(x,r,\cdot)$ is supposed, the original optimal control problem (P) need not have any solution, and therefore its relaxation is urgent. We will make it by a continuous extension of (P). For this reason, let us consider a suitable separable normed linear space H of Carathéodory integrands $\Omega \times \mathbb{R}^m \to \mathbb{R}$ with the growth at most p (i.e. each $h \in H$ fulfils $|h(x,s)| \le a_h(x) + c_h|s|^p$ with some $a_h \in L^1(\Omega)$ and $c_h \in \mathbb{R}$) which is sufficiently rich, namely $\varphi_0 \circ y \in H$ and $\varphi \circ y \in H^{m_1}$ for any $y \in L^q(\Omega;\mathbb{R}^n)$, where $[\varphi \circ y](x,s) = \varphi(x,y(x),s)$. In concrete cases, there is usually enough freedom in the choice of H.

A natural imbedding of $L^p(\Omega;\mathbb{R}^m)$ into H^*, denoted by i_H, is defined by $\langle i_H(u), h\rangle = \int_\Omega h(x,u(x))\mathrm{d}x$. It is known [13] that the weak* closure of $i_H(L^p(\Omega;\mathbb{R}^m))$, denoted by $Y_H^p(\Omega;\mathbb{R}^m)$, is a convex, locally compact, locally sequentially compact, and σ-compact subset of H^*. These properties of the full of $Y_H^p(\Omega;\mathbb{R}^m)$ enables us to employ techniques like if the relaxed problems were defined on a finite-dimensional space.

Let us still impose a slight requirement on H, namely that $gh \in H$ and $\|gh\|_H \leq C\|g\|_{C(\bar{\Omega})}\|h\|_H$ for any $g \in C(\bar{\Omega})$ and any $h \in H$, where $[gh](x,s) = g(x)h(x,s)$. Then we can define the bilinear mapping $(h,\eta) \mapsto h \bullet \eta$ from $H \times H^*$ to the space of Radon measures $\mathcal{M}(\bar{\Omega}) \cong C(\bar{\Omega})^*$ by $\langle h \bullet \eta, g\rangle = \langle \eta, gh\rangle$ for any $g \in C(\bar{\Omega})$. Note that $\eta \mapsto h \bullet \eta : Y_H^p(\Omega;\mathbb{R}^m) \to \mathcal{M}(\bar{\Omega})$ represents the affine (weak*,weak*)-continuous extension of the Nemytskii mapping $u \mapsto h \circ u : L^p(\Omega;\mathbb{R}^m) \to L^1(\Omega)$. Also note that, as φ satisfies the growth condition (1), $(\varphi \circ y) \bullet \eta$ lives not only in $\mathcal{M}(\bar{\Omega};\mathbb{R}^{m_1})$ but even in $L^{p_1}(\Omega;\mathbb{R}^{m_1})$.

Supposing still that, for some $a_2 \in L^{q/(q-1)}(\Omega)$ and $b_2, c_2 \geq 0$,

$$|\varphi(x,r_1,s) - \varphi(x,r_2,s)| \leq (a_2(x) + b_2|r_1|^{q-1} + b_2|r_2|^{q-1} + c_2|s|^{p(q-1)/q})|r_1 - r_2| \tag{4}$$

$$|\varphi_0(x,r_1,s) - \varphi_0(x,r_2,s)| \leq (a_2(x) + b_2|r_1|^{q-1} + b_2|r_2|^{q-1} + c_2|s|^{p(q-1)/q})|r_1 - r_2| \tag{5}$$

the two-argument Nemytskii mappings $(y,u) \mapsto \varphi(\cdot, y(\cdot), u(\cdot)) : L^q(\Omega;\mathbb{R}^n) \times L^p(\Omega;\mathbb{R}^m) \to L^{p_1}(\Omega;\mathbb{R}^{m_1})$ and $(y,u) \mapsto \varphi_0(\cdot, y(\cdot), u(\cdot)) : L^q(\Omega;\mathbb{R}^n) \times L^p(\Omega;\mathbb{R}^m) \to L^1(\Omega)$ admit jointly (norm$\times$weak*,weak*)-continuous extensions $(y,\eta) \mapsto (\varphi \circ y) \bullet \eta : L^q(\Omega;\mathbb{R}^n) \times Y_H^p(\Omega;\mathbb{R}^m) \to L^{p_1}(\Omega;\mathbb{R}^{m_1})$ and $(y,\eta) \mapsto (\varphi_0 \circ y) \bullet \eta : L^q(\Omega;\mathbb{R}^n) \times Y_H^p(\Omega;\mathbb{R}^m) \to \mathcal{M}(\bar{\Omega})$, respectively.

Then the relaxed problem, created by a continuous extension of (P), takes the form

$$(\mathsf{RP}) \quad \begin{cases} \text{Minimize} & \int_\Omega (\varphi_0 \circ y) \bullet \eta \, dx \\ \text{subject to} & Ay = (\varphi \circ y) \bullet \eta \ \text{ on } \Omega, \\ & y \in L^q(\Omega;\mathbb{R}^n), \quad \eta \in \bar{U}_{\text{ad}} = \text{w*-cl}\, i_H(U_{\text{ad}}) \subset Y_H^p(\Omega;\mathbb{R}^m)\,, \end{cases}$$

where $U_{\text{ad}} = \{u \in L^p(\Omega;\mathbb{R}^m);\ u(x) \in S(x) \text{ for a.a. } x \in \Omega\}$ denotes the set of admissible controls for (P). We will suppose that the extended state equation $Ay = (\varphi \circ y) \bullet \eta$ has, for any $\eta \in Y_H^p(\Omega;\mathbb{R}^m)$, precisely one solution $y = \bar{\pi}(\eta) \in \mathcal{D}_A$ and

$$\bar{\pi} : Y_H^p(\Omega;\mathbb{R}^m) \to L^q(\Omega;\mathbb{R}^n) \ \text{ is continuous}, \quad \bar{\pi} \circ i_H = \pi. \tag{6}$$

Then the coercivity (3) together with the local compactness of $Y_H^p(\Omega;\mathbb{R}^m)$, density of U_{ad} (imbedded via i_H) in $\bar{U}_{\text{ad}}$ and the continuity of all data involved in (RP) enable us to build in a standard way an existence, a stability, and an approximation theories for the relaxed problem, as well as to show that (RP) is actually a correct relaxation (P). In particular:

Proposition 1. *Let (1)–(6) be fulfilled. Then:*

1. *inf(P) = min(RP).*
2. *If $\{u_k\}_{k\in\mathbb{N}}$ is a minimizing sequence for (P), then every weak* cluster point of $\{i_H(u_k)\}_{k\in\mathbb{N}}$ is an optimal relaxed control for (RP).*
3. *For every optimal relaxed control η for (RP) there is a minimizing sequence $\{u_k\}_{k\in\mathbb{N}}$ for (P) such that $\eta = \text{w*-lim}_{k\to\infty} i_H(u_k)$.*

An important property of elements of $Y_H^p(\Omega;\mathbb{R}^m)$ is their possible "nonconcentration". More precisely, we say that $\eta \in Y_H^p(\Omega;\mathbb{R}^m)$ is p-nonconcentrating if it is attainable by a sequence $\{u_k\}_{k\in\mathbb{N}}$ (i.e. $\eta = \text{w*-}\lim_{k\to\infty} i_H(u_k)$) such that the set $\{|u_k|^p;\ k \in \mathbb{N}\}$ is not only bounded in $L^1(\Omega)$ but even relatively weakly compact in $L^1(\Omega)$. Using Ball's theorem [2], it is possible to show [13] that every p-nonconcentrating η admits an L^p-Young-measure representation [14] in the sense that there is a weakly measurable mapping $x \mapsto \nu_x$ from Ω to the set of all probability Radon measures on $\mathbb{R}^m$ such that $x \mapsto \int_{\mathbb{R}^m} |s|^p \nu_x(\mathrm{d}s)$ belongs to $L^1(\Omega)$ and

$$\forall h \in H: \qquad \langle \eta, h\rangle = \int_\Omega \int_{\mathbb{R}^m} h(x,s)\nu_x(\mathrm{d}s)\mathrm{d}x. \tag{7}$$

Then also $[h \bullet \eta](x) = \int_{\mathbb{R}^m} h(x,s)\nu_x(\mathrm{d}s)$ for a.a. $x \in \Omega$.

We say that $\mathring{\eta} \in Y_H^p(\Omega;\mathbb{R}^m)$ is a p-nonconcentrating modification of a given $\eta \in Y_H^p(\Omega;\mathbb{R}^m)$ if $\mathring{\eta}$ is p-nonconcentrating and $\langle \mathring{\eta}, h\rangle = \langle \eta, h\rangle$ for any $h \in H$ such that $|h(x,s)| \le a(x) + o(|s|^p)$ with some $a \in L^1(\Omega)$ and $o: \mathbb{R}^+ \to \mathbb{R}$ satisfying $\lim_{r\to\infty} o(r)/r = 0$. It is a nontrivial result, based on Chacon's biting lemma [7] and the Dunford-Pettis theorem, that every $\eta \in Y_H^p(\Omega;\mathbb{R}^m)$ possesses precisely one p-nonconcentrating modification. Also one can prove that $\langle \eta - \mathring{\eta}, h\rangle > 0$ provided $\eta \ne \mathring{\eta}$ and $h \in H$ is coercive in the sense $h(x,s) \ge a_0(x) + b|s|^p$ with some $a_0 \in L^1(\Omega)$ and $b > 0$.

Let us assume, for a moment, that an optimal relaxed control η for (RP) is not p-nonconcentrating. This means $\eta \ne \mathring{\eta}$. Due to our special form of the set of the admissible controls U_{ad}, one can show that $\eta \in \bar{U}_{\mathrm{ad}}$ implies $\mathring{\eta} \in \bar{U}_{\mathrm{ad}}$. Thanks to (1), $|[\varphi \circ y](x,\cdot)|$ has the growth lesser than p, namely p/p_1. Therefore $(\varphi \circ y) \bullet \eta = (\varphi \circ y) \bullet \mathring{\eta}$, which yields $\bar{\pi}(\eta) = \bar{\pi}(\mathring{\eta})$. Due to the coercivity (3), we have $\int_{\bar{\Omega}} (\varphi_0 \circ y) \bullet \eta \,\mathrm{d}x = \langle \eta, \varphi_0 \circ y\rangle > \langle \mathring{\eta}, \varphi_0 \circ y\rangle = \int_{\bar{\Omega}} (\varphi_0 \circ y) \bullet \mathring{\eta} \,\mathrm{d}x$. Therefore, $\mathring{\eta}$ is an admissible relaxed control that achieves a strictly lower cost than η, which demonstrates that η cannot be optimal, a contradiction. This shows:

Proposition 2. *Let (1)–(6) be valid and η be an optimal relaxed control for* (RP). *Then η is p-nonconcentrating.*

This assertion also implies that the minimizing sequences for the original problems (P) do not concentrate "energy":

Proposition 3. *Let (1)–(6) be valid and $\{u_k\}_{k\in\mathbb{N}} \subset U_{\mathrm{ad}}$ be a minimizing sequence for* (P), *i.e.* $\lim_{k\to\infty} \int_\Omega \varphi_0(x, \pi(u_k), u_k)\mathrm{d}x = \inf(\mathrm{P})$. *Then the set $\{|u_k|^p;\ k \in \mathbb{N}\}$ is relatively weakly compact in $L^1(\Omega)$.*

3 FIRST-ORDER OPTIMALITY CONDITIONS

The relaxed problem consists, in fact, in minimization of the cost functional $J(\eta) = \int_{\bar{\Omega}} (\varphi_0 \circ \bar{\pi}(\eta)) \bullet \eta \,\mathrm{d}x$ over the set $\bar{U}_{\mathrm{ad}}$ assumed now to be convex, which is basically a slight requirement of H and S. If J is Gâteaux differentiable, the first-order necessary optimality condition for $\eta \in \bar{U}_{\mathrm{ad}}$ to be an optimal relaxed control for (RP) is

$$\nabla J(\eta) \in -N_{\bar{U}_{\mathrm{ad}}}(\eta), \tag{8}$$

where ∇J denotes the Gâteaux differential of J and $N_{\bar{U}_{\mathrm{ad}}}(\eta)$ the normal cone to $\bar{U}_{\mathrm{ad}}$ at the point η. To ensure the smoothness of the extended problem, we have to suppose that both

$\varphi(x,\cdot,s) : \mathbb{R}^n \to \mathbb{R}^{m_1}$ and $\varphi_0(x,\cdot,s) : \mathbb{R}^n \to \mathbb{R}$ are differentiable and the derivatives, denoted by $\varphi' = (\frac{\partial \varphi_k}{\partial r_l})_{l=1,\ldots,n}^{k=1,\ldots,m_1}$ and $\varphi_0' = (\frac{\partial \varphi_0}{\partial r_l})_{l=1,\ldots,n}$, satisfy $g \cdot (\varphi' \circ y) \cdot \tilde{y} \in H$ and $g_0 \cdot (\varphi_0' \circ y) \cdot \tilde{y} \in H$ for any $g \in L^{p_1/(p_1-1)}(\Omega;\mathbb{R}^{m_1})$, $g_0 \in C(\bar{\Omega})$ and $\tilde{y} \in L^q(\Omega;\mathbb{R}^n)$. Moreover, for some $a \in L^{qp_1/(q-p_1)}(\Omega)$, $a_0 \in L^{q/(q-1)}(\Omega)$, $a' \in L^{qp_1/(q-2p_1)}(\Omega)$, $a_0' \in L^{q/(q-2)}(\Omega)$ and $b, c < +\infty$, the derivatives φ' and φ_0' satisfy

$$|\varphi'(x,r,s)| \le a(x) + b|r|^{(q-p_1)/p_1} + c|s|^{p(q-p_1)/qp_1}, \tag{9}$$

$$|\varphi_0'(x,r,s)| \le a_0(x) + b|r|^{q-1} + c|s|^{p(q-1)/q}, \tag{10}$$

$$|\varphi'(x,r_1,s) - \varphi'(x,r_2,s)| \le (a'(x) + b|r_1|^{(q-2p_1)/p_1} + b|r_2|^{(q-2p_1)/p_1} + c|s|^{p(q-2p_1)/qp_1})|r_1 - r_2| \tag{11}$$

$$|\varphi_0'(x,r_1,s) - \varphi_0'(x,r_2,s)| \le (a_0'(x) + b|r_1|^{q-2} + b|r_2|^{q-2} + c|s|^{p(q-2)/q})|r_1 - r_2|. \tag{12}$$

If $q \ge 2p_1$ and (9)–(12) is valid, the extended Nemytskii mappings $(y,\eta) \mapsto (\varphi \circ y) \bullet \eta$ and $(y,\eta) \mapsto (\varphi_0 \circ y) \bullet \eta$ possess Gâteaux differentials which are, if H is normed suitably, locally Lipschitz continuous. Then J is Gâteaux differentiable and, evaluating (8) by the adjoint equation technique, one obtains the following maximum principle:

Proposition 4. *Let $q \ge 2p_1$, (1)–(6) and (9)–(12) be valid, let (y,η) solve* (RP), *and let there is $\lambda \in L^{p_1/(p_1-1)}(\Omega;\mathbb{R}^{m_1})$ solving the adjoint equation*

$$A^*\lambda = [(\varphi' \circ y) \bullet \eta]^{\mathrm{T}} \lambda - (\varphi_0' \circ y) \bullet \eta \,. \tag{13}$$

Then the integral maximum principle

$$\int_{\bar{\Omega}} [h_{y,\lambda} \bullet \eta](\mathrm{d}x) \;=\; \sup_{u \in U_{\mathrm{ad}}} \int_\Omega h_{y,\lambda}(x,u(x))\mathrm{d}x \tag{14}$$

holds, where the "Hamiltonian" $h_{y,\lambda} \in H$ is defined by $h_{y,\lambda} = -\varphi_0 \circ y + \lambda \cdot (\varphi \circ y)$. Moreover, if also the coercivity (3) holds, then (14) can be localized to the point-wise (also called Pontryagin) maximum principle

$$h_{y,\lambda} \bullet \eta \;=\; \max_{s \in \mathbb{R}^m} h_{y,\lambda}(\cdot,s) \qquad \textit{in the sense of } L^1(\Omega). \tag{15}$$

We saw in Proposition 2 that (RP) possesses typically a p-nonconcentrating solutions which have, due to (7), a Young-measure representation. We say that η is k-atomic if it has the Young-measure representation ν which is k-atomic, i.e. for a.a. $x \in \Omega$ the probability measure ν_x is a convex combination of at most k Dirac masses. The nonconcentration principle enables us to exploit in detail the maximum principle (15):

Proposition 5. *Let $q \ge 2p_1$, (1)–(6) and (9)–(12) be valid. Then* (RP) *always possesses some solution such that η is (m_1+1)-atomic. Moreover, if the Hamiltonian $h_{y,\lambda}(x,\cdot) : \mathbb{R}^m \to \mathbb{R}$ attains its maximum at no more that k-points, then every solution to* (RP) *is k-atomic.*

Let us remark that information about a finite number of atoms can be exploited in numerical implementation of (RP).

4 ILLUSTRATION: AN ELLIPTIC OPTIMAL CONTROL PROBLEM

The general problem (P) covers, in particular, the following optimal control problem for a system of elliptic equations with homogeneous Neumann boundary conditions:

$$(\mathsf{P}_1)\begin{cases} \text{Minimize} & \int_\Omega \varphi_0(x,y(x),u(x))\mathrm{d}x \\ \text{subject to} & \sum_{j,l=1}^{n_1}\sum_{i=1}^{n} \frac{\partial}{\partial x_l}(c_{ijkl}(x)\frac{\partial y_i}{\partial x_j}) = \varphi_k(x,y(x),u(x)) \quad \text{on } \Omega, \quad k=1,...,n \\ & \sum_{j,l=1}^{n_1}\sum_{i=1}^{n} \nu_l c_{ijkl}(x)\frac{\partial y_i}{\partial x_j} = 0 \quad \text{on } \partial\Omega, \ k=1,...,n \\ & u(x)\in S(x) \quad \text{on } \Omega, \\ & y\in W^{1,2}(\Omega;\mathbb{R}^n), \quad u\in L^p(\Omega;\mathbb{R}^m), \end{cases}$$

where $\nu = (\nu_1,...,\nu_{n_1})$ denotes the unit outward normal to the boundary $\partial\Omega$. We will suppose the following symmetry condition:

$$c_{ijkl}(x) = c_{klij}(x)\ , \qquad \frac{\partial\varphi_k(x,r,s)}{\partial r_i} = \frac{\partial\varphi_i(x,r,s)}{\partial r_k}\ . \tag{16}$$

Under this condition, y solves the state problem in (P_1) (in the weak sense) if it minimizes the functional Φ_u defined by

$$\Phi_u(y) = \int_\Omega \Big(\frac{1}{2}\sum_{j,l=1}^{n_1}\sum_{i,k=1}^{n} c_{ijkl}(x)\frac{\partial y_i}{\partial x_j}\frac{\partial y_k}{\partial x_l} + \hat{\varphi}(x,y(x),u(x))\Big)\,\mathrm{d}x$$

over the Sobolev space $W^{1,2}(\Omega;\mathbb{R}^n)$, where $\hat{\varphi}:\Omega\times\mathbb{R}^n\times\mathbb{R}^m\to\mathbb{R}$ is given by the formula $\hat{\varphi}(x,r,s) = \int_0^1\sum_{k=1}^n r_k\varphi_k(x,\tau r,s)\mathrm{d}\tau$. Let us still suppose that, for some ε positive and every matrix $\xi = [\xi_{ij}]\in\mathbb{R}^{n\times n_1}$, the coefficients $c_{ijkl}\in L^\infty(\Omega)$ satisfy

$$\sum_{j,l=1}^{n_1}\sum_{i,k=1}^{n} c_{ijkl}\xi_{ij}\xi_{kl} \geq \varepsilon||\xi||^2 \tag{17}$$

and that φ satisfies

$$\sum_{k=1}^{n} r_k\varphi_k(x,r,s)\geq\varepsilon|r|^2, \quad \sum_{k=1}^{n}(\varphi_k(x,r,s)-\varphi_k(x,\tilde{r},s))(r_k-\tilde{r}_k)\geq\varepsilon|r-\tilde{r}|^2. \tag{18}$$

Then Φ is coercive and uniformly convex, which ensures uniqueness and continuous dependence on u of the solution $y=\pi(u)$ to the state problem in (P_1) provided $p_1 > 2n/(n+2)$ which just ensures the imbedding $L^{p_1}(\Omega;\mathbb{R}^n)\subset W^{-1,2}(\Omega;\mathbb{R}^n)$.

It is clear that (P_1) takes the form (P) with $n = m_1$, $q < 2n_1/(n_1-2)$, $\mathcal{D}_A = W^{1,q}(\Omega;\mathbb{R}^n)$, and $p_1 > 2n_1/(n_1+2)$ so that we can immediately apply the preceding results to (P_1). Then, under the assumptions (1)–(2) and (4)–(5), the relaxed problem (RP_1) looks as

$$(\mathsf{RP}_1)\quad\left\{\begin{array}{lll}\text{Minimize} & \int_{\bar\Omega}(\varphi_0\circ y)\bullet\eta\,\mathrm{d}x & \\ \text{subject to} & \sum_{j,l=1}^{n_1}\sum_{i=1}^{n}\frac{\partial}{\partial x_l}(c_{ijkl}\frac{\partial y_i}{\partial x_j}) = (\varphi_k\circ y)\bullet\eta & \text{on } \Omega,\quad k=1,\dots,n\\ & \sum_{j,l=1}^{n_1}\sum_{i=1}^{n}\nu_l c_{ijkl}\frac{\partial y_i}{\partial x_j} = 0 & \text{on } \partial\Omega,\ k=1,\dots,n\\ & u(x)\in S(x) & \text{on } \Omega,\\ & y\in W^{1,2}(\Omega;\mathbb{R}^n),\quad \eta\in\bar U_{\mathrm{ad}} = \mathrm{w}^*\text{-cl}\; i_H(U_{\mathrm{ad}})\subset Y_H^p(\Omega;\mathbb{R}^m)\ , & \end{array}\right.$$

We can apply Proposition 1 to see that (RP_1) is actually a correct relaxation scheme for (P_1). The nonconcentration of every optimal relaxed control for (RP_1) can be obtained by Proposition 2, while Proposition 3 yields nonconcentration of energy of every minimizing sequence for (P_1). We will additionally suppose $2p_1 \le q$; note that this restriction basically enables us to treat at most five-dimensional problems, i.e. $n_1 \le 5$ (if boundary controls were considered, even $n_1 \le 3$ would have to be supposed). Then Proposition 4 justifies the Pontryagin maximum principle (15) involving the Hamiltonian $h_{y,\lambda}\in H$ given by

$$h_{y,\lambda}(x,s) = -\varphi_0(x,y(x),s) + \sum_{k=1}^{n}\lambda_k(x)\varphi_k(x,y(x),s)$$

with $\lambda\in W^{1,2}(\Omega;\mathbb{R}^n)$ solving the adjoint elliptic system

$$\left.\begin{array}{ll}\sum_{j,l=1}^{n_1}\sum_{i=1}^{n}\frac{\partial}{\partial x_l}(c_{ijkl}\frac{\partial\lambda_i}{\partial x_j}) = \sum_{i=1}^{n}[(\frac{\partial\varphi_k}{\partial r_i}\circ y)\bullet\eta]\lambda_i - (\frac{\partial\varphi_0}{\partial r_i}\circ y)\bullet\eta & \text{on } \Omega,\\ \sum_{j,l=1}^{n_1}\sum_{i=1}^{n}\nu_l c_{ijkl}\frac{\partial\lambda_i}{\partial x_j} = 0 & \text{on } \partial\Omega.\end{array}\right\}\qquad(19)$$

with $k = 1,\dots,n$. Finally, by Proposition 5 we can claim that at least one optimal relaxed control for (RP_1) is $(n+1)$-atomic. Let us emphasize that our assumptions here, namely (16)–(18), ensures automatically (6) as well as the existence of the adjoint state λ which had to be supposed in Propositions 1–5.

The above outlined example reproduces (and sometimes generalizes) standard results, e.g. by Alibert and Raymond (preprint), Bonnans and Casas (1991, 1992, 1995), Bonnans and Tiba (1991), Casas (1993,to appear), Casas and Fernández (1991, 1995), Mackenroth (1986), Raitums (1986)], or Zolezzi (1972) where only scalar case (i.e. $n = 1$) has been considered without any relaxation, i.e. the maximum principle has been obtained only for ordinary optimal controls whose existence is, however, guaranteed only if the problem has a convex/linear structure. The presented example can be also modified for a boundary control (like in Alibert and Raymond(preprint), Bonnans and Casas (1992), Casas and Fernandez (1995) or nonlinear operator A (like in Casas and Fernandez (1995)).

REFERENCES

Alibert, J.J. and Raymond, J.P. Optimal control problems governed by semilinear elliptic equations with pointwise state constraints. (preprint)

Ball, J.M. (1989) A version of the fundamental theorem for Young measures. In: *PDEs and Continuum Models of Phase Transition.* (Eds. M.Rascle, D.Serre, M.Slemrod.) Lecture Notes in Physics **344**, Springer, Berlin, 207–215.

Bonnans, J.F. and Casas, E. (1991) Un principe de Pontryagine pour le contrôle des systèmes semilinéaires elliptiques. *J. Diff. Equations* **90**, 288-303.

Bonnans, J.F. and Casas, E. (1992) A boundary Pontryagin's principle for the optimal control of state-constrained elliptic systems. In: *Optimization, Optimal Control, and P.D.E.* (V.Barbu, J.F.Bonnans, D.Tiba, Eds.) ISNM **107**, Birhäuser, Basel, 241–249.

Bonnans, J.F. and Casas, E. (1995) An extension of Pontyagin's principle for state-constrained optimal control of semilinear elliptic equations and variational inequalities. *SIAM J. Control Optim.* **33**, 274–298.

Bonnans, J.F. and Tiba, D. (1991) Pontryagin's principle in the control of semilinear elliptic variational equations. *Appl. Math. Optim.* **23**, 299-312.

Brooks, J.K. and Chacon, R.V. (1980) Continuity and compactness of measures. *Adv. in Math.* **37**, 16–26.

Casas, E. (1993, to appear) Pontryagin Principle for optimal control problems governed by semilinear elliptic equations. In: *Proc. Int. Conf. on Control and Estimation of Distributed Parameter Systems.* Vorau.

Casas, E. and Fernández, L.A. (1991) State-constrained control problems of quasilinear elliptic equations. In: Optimal Control of Partial Differential Equations. (Eds. K.-H.Hoffmann, W.Krabs.) *Lecture Notes Control Inf. Sci.* **149**, Springer, Berlin, 11–25.

Casas, E. and Fernández, L.A. (1995) Dealing with integral state constraints in boundary control problems of quasilinear elliptic equations. *SIAM J. Control Optim.* **33**, 568–589.

Mackenroth, U. (1986) On some elliptic optimal control problems with state constraints. *Optimization* **17**, 595–607.

Raitums, U.E. (1986) Maximum principle in optimal control problems for elliptic equations. *A. Anal. Anwendungen* **5**, 291–306. (In Russian.)

Roubíček, T. *Relaxation in Optimization Theory and Variational Calculus.* W. de Gruyter, Berlin, 1996 (in preparation)

Young, L.C. (1937) Generalized curves and the existence of an attained absolute minimum in the calculus of variations. *Comptes Rendus de la Société des Sciences et des Lettres de Varsovie*, Classe III **30**, 212–234.

Zolezzi, T. (1972) Necessary conditions for optimal controls of elliptic or parabolic problems. *SIAM J. Control* **10**, 594–602.

29

Adaptive control of semilinear stochastic evolution equations

L. Stettner
Institute of Mathematics Polish Academy of Sciences
Sniadeckich 8, 00-950 Warsaw, Poland

Abstract
In the paper a nearly self optimal strategy is constructed for the control problem of stochastic semilinear evolution equation depending on an unknown parameter with path wise average per unit time cost functional.

Keywords
Stochastic evolution equations, adaptive control, self-optimality

1 INTRODUCTION

Let $(H, |\cdot|_H)$ be a separable Hilbert space and (Ω, F, F_t, P^0) be a probability space. Consider the following semilinear stochastic evolution equation on H

$$dX = (AX + F(X))dt + B(X)dW \quad X(0) = x \in H \tag{1}$$

where A is a generator of a C_0 semigroup $S(t)$ on H and $W(t)$ is a cylindrical Wiener process on H adapted to F_t.

Assume

(A1) $\exists L > 0$ such that $|F(z) - F(y)|_H \leq |z - y|_H$ and $||B(z) - B(y)||_{L(H)} \leq |z - y|_H$, for $z, y \in H$
(A2) $\sup_{z \in H} ||B^{-1}(z)||_{L(H)} < \infty$
(A3) $\exists \beta \in (0, 1)$ such that for each $T > 0$, $\int_0^T (1 + t^{-\beta})||S(t)||_2^2 dt < \infty$

It follows from Da Prato and Zabczyk (1992) that under (A1)-(A3) there exists a unique mild solution $X(t)$ to the equation (1) i.e. an H valued process $X(t)$ such that for $t \geq 0$

$$X(t) = S(t)x + \int_0^t S(t-s)F(X(s))ds + \int_0^t S(t-s)B(X(s))dW(s) \tag{2}$$

Moreover $X(t)$ has a version with continuous trajectories that is a Markov process on H with transition operator P_t^0 and (see Peszat and Zabczyk (1994)) (a1) $\exists c_t$ that is bounded for t from compact subsets of $(0,\infty)$ such that $|P_t^0\psi(x) - P_t^0\psi(y)| \leq c_t|x-y|_H \sup_{z\in H}|\psi(z)|$, for $x,y \in H$ and bounded Borel measurable functions ψ on H,

(a2) X is irreducible i.e. $P_t^0(x,\mathcal{O}) > 0$ for $x \in H$, $t > 0$, and open set $\mathcal{O}$

Let U and A be given compact metric spaces of control and unknown parameters resp., and let $D : H \times U \times A \mapsto H$ and $c : H \times U \mapsto R$. Furthermore assume

(A4) the mappings D and c are bounded and continuous in $v \in U$, and D is moreover Lipschitz in α, uniformly with respect to other parameters,

Denote by $\mathcal{A}$ the class of all Borel measurable functions from H in U. Given $\alpha^0 \in A$ and $u \in \mathcal{A}$, by an infinite dimensional version of Girsanov theorem (see Duncan Pasik-Duncan and Stettner (1995)) we can define a new measure $P^{\alpha^0,u}$ on Ω, such that the restrictions $P^{\alpha^0,u}_{|t}$ and $P^0_{|t}$ of $P^{\alpha^0,u}$ and P^0 resp., to σ-field F_t satisfy $P^{\alpha^0,u}_{|t}(d\omega) = exp[\zeta^{\alpha^0,u}(x,t)]P^0_{|t}(d\omega)$ with

$$\zeta^{\alpha^0,u}(x,t) = \int_0^t < B^{-1}(X(s))D(X(s),u(X(s)),\alpha^0), dW(s) > \\ -\frac{1}{2}\int_0^t |B^{-1}(X(s))D(X(s),u(X(s)),\alpha^0)|_H^2 ds \tag{3}$$

and where $X(0) = x$. Under the measure $P^{\alpha^0,u}$, the process $X(t)$ becomes a solution to the following integral equation

$$X(t) = S(t)x + \int_0^t S(t-s)(F(X(s)) + D(X(s),u(X(s)),\alpha^0))ds \\ + \int_0^t S(t-s)B(X(s))dW(s) \tag{4}$$

and the solution to (4) in the above weak sense is unique. In what follows we shall study the case when the state process $X(t)$ is a solution to (4) and α^0 is unknown. Our aim is to construct a control $v_t = u_t(X(t))$, where $u_t \in \mathcal{A}$ such that under measure $P^{\alpha^0,u}$ the path wise cost functional

$$J((v_t)) = \limsup_{t\to\infty} t^{-1}\int_0^t c(X(s),v_s)ds \tag{5}$$

is minimal. The study of the cost functional (5) requires certain ergodic properties of the controlled process $X(t)$. For this purpose we shall impose the following assumption

(A5) there exists a Markov time τ such that for $\tau_{n+1} = \tau_n + \tau \circ \theta_{\tau_n}$, ($\theta_t$ is a Markov shift operator corresponding to $X(t)$), the process $X(\tau_n)$ is Markov on a compact set Γ and uniformly ergodic which means that $\forall \alpha \in A, u \in \mathcal{A}$ $\exists 0 < \gamma < 1$ $\exists \eta_u^\alpha \in \mathcal{P}(\Gamma)$ such that

$$\sup_{u\in\mathcal{A}}\sup_{\alpha\in A}\sup_{x\in\Gamma}\sup_{B\in\mathcal{B}(\Gamma)} |P_x^{\alpha,u}\{X(\tau_n) \in B\} - \eta_u^\alpha(B)| \leq \gamma^n \tag{6}$$

Moreover we have $\sup_{u\in\mathcal{A}}\sup_{\alpha\in A}\sup_{x\in\Gamma}E_x^{\alpha,u}\{\tau^2\} < \infty$, $\sup_{x\in\Gamma}E_x^0\{\tau^2\} < \infty$, $\inf_{u\in\mathcal{A}}\inf_{\alpha\in A}\inf_{x\in\Gamma}E_x^{\alpha,u}\{\tau^2\} > 0$ and $E_x^{\alpha,u}\{T_\Gamma\} < \infty$, for $x \in H$, $u \in \mathcal{A}$, $\alpha \in A$, with $T_B = inf\{s \geq 0, X(s) \in B\}$ for $B \in \mathcal{B}(H)$.

For the model we study the assumption (A5) is not very restrictive. In section 3 a particular form of Markov time τ and set Γ will be shown. It is known (see e.g. Duncan Pasik-Duncan and Stettner (1994b)) that under (A5) the optimal value of cost functional (5) is $P^{\alpha^0,u}$ a.e. equal to that of

$$\bar{J}_x^{\alpha^0}((v_t)) = \limsup_{t\to\infty} t^{-1}E_x^{\alpha^0}\{\int_0^t c(X(s),v_s)ds\} \tag{7}$$

We shall also assume

(A6) (Roxin condition) for each $x \in H$ the set $\left\{\binom{D(x,v)}{c(x,v)}, v \in U\right\}$ is convex in the Hilbert space $H \times R$.

which is a technical assumption that can be removed by considering a larger class of so called relaxed controls (see e.g. Gatarek and Sobczyk (1994)).

Adaptive control of LQ infinite dimensional models has been studied in Duncan Goldys and Pasik-Duncan (1991), Duncan Maslowski Pasik-Duncan (1994), (1995), and Aihara (1995). In the paper adaptive control of semilinear stochastic evolution equations is considered. The adaptive procedure is an infinite dimensional adaptation of Duncan Pasik-Duncan and Stettner (1994b). Results of the paper can in particular be applied to a system of controlled reaction diffusion equations (see e.g. Duncan Pasik-Duncan and Stettner (1995)).

2 CONTINUITY PROPERTIES

In this section we list a family of consequences of the assumptions (A1)-(A6), which will be used in the proof of the main Theorem in section 6.

Notice first that by (a1) and (a2) the measures $P_t^0(x,\cdot)$, for $x \in H$ and $t > 0$ are equivalent. Fix $\bar{x} \in H$ and $\bar{t} > 0$ and let $\mu = P_{\bar{t}}^0(\bar{x},\cdot)$. By Theorem 1 of Duncan Pasik-Duncan and Stettner (1995) the measures μ and $P_t^{\alpha,u}(x,\cdot)$, for $x \in H$ and $u \in \mathcal{A}$ are equivalent. Moreover we have

Lemma 1 *Under (A1)-(A6) the measures*

$$\pi_u^\alpha(B) = \frac{\int_\Gamma E_x^{\alpha,u}\{\int_0^\tau \chi_B(X(s))ds\}\eta_u^\alpha(dx)}{\int_\Gamma E_x^{\alpha,u}\{\tau\}\eta_u^\alpha(dx)} \tag{8}$$

for $B \in \mathcal{B}(H)$, *are unique invariant for the transition semigroup* $P_t^{\alpha,u}$ *and the family* $\{\pi_u^\alpha, \alpha \in A, u \in \mathcal{A}\}$ *is tight.*

Proof. One can easily check that the measures π_u^α are invariant for the semigroup $P_t^{\alpha,u}$. Since transition operators $P_t^{\alpha,u}(x,\cdot)$ are equivalent for $x \in H$ and $t > 0$, the invariant measures π_u^α are unique. It remains to show the tightness of the family $\{\pi_u^\alpha, \alpha \in A, u \in \mathcal{A}\}$. For this purpose we shall prove that the family of measures $\Psi_x^{\alpha,u}(B) = \frac{E_x^{\alpha,u}\{\int_0^\tau \chi_B(X(s))ds\}}{E_x^{\alpha,u}\{\tau\}}$, $x \in \Gamma$, $\alpha \in A$, $u \in \mathcal{A}$ is tight. In fact, for given sequences $x_n \in \Gamma$, $\alpha_n \in A$, $u_n \in \mathcal{A}$, by compactness of Γ, A and Lemma 2 of Duncan Pasik-Duncan and Stettner (1995) one can find subsequences, for simplicity denoted by n and $x \in \Gamma$, $\alpha \in A$, $u \in \mathcal{A}$ such that $x_n \to x$, $\alpha_n \to \alpha$ and $D(x, u_n(x), \alpha_n) \to D(x, u(x), \alpha)$ in the weak $*$ topology of $L_\infty(H, \eta, H)$. Mimicking the proof of the first part of the Theorem 2 of Duncan Pasik-Duncan and Stettner (1995) for each $t > 0$ and $\phi \in C(H, R)$ we have

$$\int_\Omega \int_0^{\tau \wedge t} \phi(X(s))ds \quad exp[\zeta^{\alpha_n,u_n}(x_n,t)]dP^0 \to \int_\Omega \int_0^{\tau \wedge t} \phi(X(s))ds \quad exp[\zeta^{\alpha,u}(x_n,t)]dP^0 \tag{9}$$

Since by (A5), $\sup_{u\in\mathcal{A}} \sup_{\alpha\in A} \sup_{x\in\Gamma} E_x^{\alpha,u}\{\tau^2\} < \infty$, for any $\phi \in C(H, R)$

$$E_{x_n}^{\alpha_n,u_n}\{\int_0^\tau \phi(X(s))ds\} \to E_x^{\alpha,u}\{\int_0^\tau \phi(X(s))ds\} \tag{10}$$

and consequently $\Psi_{x_n}^{\alpha_n,u_n}(\phi) \to \Psi_x^{\alpha,u}(\phi)$, which completes the proof of the tightness of π_u^α with $\alpha \in A$, $u \in \mathcal{A}$. $\square$

By Remark 6 of Duncan Pasik-Duncan and Stettner (1995) and Lemma 1 we immediately obtain

Corollary 1 *Under (A1)-(A6) for each $\alpha \in A$ there exists an optimal control function $u^\alpha \in \mathcal{A}$ for the cost functional $\bar{J}_x^\alpha((u(X(t))))$.*

Let ρ_A denote a metric on A. From an infinite dimensional version of Proposition 2.2 and (19) of Duncan Pasik-Duncan and Stettner (1994a), using Lipschitian of D in α we obtain

Lemma 2 *Under (A1)-(A5) for each $\epsilon > 0$ there exists $\delta > 0$ such that if $\rho_A(\alpha, \alpha') < \delta$ we have*

$$\sup_{u\in\mathcal{A}} \sup_{x\in\Gamma} \sup_{B\in\mathcal{B}(\Gamma)} |P_x^{\alpha,u}\{X(\tau) \in B\} - P_x^{\alpha',u}\{X(\tau) \in B\}| < \epsilon \tag{11}$$

and

$$\sup_{u\in\mathcal{A}} \sup_{x\in\Gamma} \sup_{B\in\mathcal{B}(H)} |E_x^{\alpha,u}\{\int_0^\tau \chi_B(X(s))ds\} - E_x^{\alpha',u}\{\int_0^\tau \chi_B(X(s))ds\}| < \epsilon \tag{12}$$

Denote by $||\eta||_{var}$ the variation norm of the measure η. By the proof of Proposition 1 (in particular (18)) of Stettner (1993) we have

Lemma 3 *Under (A5) there exist constants K_1 and K_2 such that for α, $\alpha' \in A$*

$$\sup_{u\in\mathcal{A}} ||\eta_u^\alpha - \eta_u^{\alpha'}||_{var} \le K_1 K_2 \sup_{u\in\mathcal{A}} \sup_{x\in\Gamma} \sup_{B\in\mathcal{B}(\Gamma)} |P_x^{\alpha,u}\{X(\tau) \in B\} - P_x^{\alpha',u}\{X(\tau) \in B\}| \tag{13}$$

Combining (8) and (11), (12), (13) we obtain

Corollary 2 *Under (A1)-(A5) the mapping $A \ni \alpha \mapsto \pi_u^\alpha$ is uniformly in $u \in \mathcal{A}$ continuous in variation norm of $\mathcal{P}(H)$.*

Consequently by Corollary 3 and Lemma 2 of Stettner (1993) we obtain

Corollary 3 *Under (A1)-(A5) for every $\epsilon > 0$, there exists a finite class $\mathcal{A}(\epsilon) = \{u_1, \dots, u_{r(\epsilon)}\}$ of ϵ - optimal control functions for the cost functionals $\bar{J}_x^\alpha$ with $\alpha \in A$ i.e.*

$$\forall \alpha \in A \quad \exists i \in \{1, \dots, r(\epsilon)\} \quad \bar{J}_x^\alpha((u_i(X(t)))) \leq \inf_{u \in \mathcal{A}} \bar{J}_x^\alpha((u(X(t)))) + \epsilon \tag{14}$$

For α, $\alpha' \in A$ and $u \in \mathcal{A}$ define an information measure $K_u(\alpha, \alpha')$ as follows

$$K_u(\alpha, \alpha') = \int_H |B^{-1}(z)(D(z, u(z), \alpha) - D(z, u(z), \alpha'))|_H^2 \pi_u^{\alpha'}(dz) \tag{15}$$

It has the following important properties

Lemma 4 *Under (A1)-(A5) for every $u \in \mathcal{A}$ the mapping $A \times A \ni (\alpha, \alpha') \mapsto K_u(\alpha, \alpha')$ is continuous. Moreover if $K_u(\alpha, \alpha') = 0$ we have $\pi_u^\alpha = \pi_u^{\alpha'}$.*

Proof. The continuity follows from Corollary 2 and Lipschitian in α of D. If $K_u(\alpha, \alpha') = 0$, by equivalence of measures π_u^α and μ, $D(x, u(x), \alpha) = D(x, u(x), \alpha')$ for μ almost all $x \in H$. Therefore by the first part of the proof of Theorem 2 of Duncan Pasik-Duncan and Stettner (1995), $P_x^{\alpha,u}(x, \cdot) = P_x^{\alpha',u}(x, \cdot)$ for $x \in H$, and consequently $\pi_u^\alpha = \pi_u^{\alpha'}$. □

Let p_m be projection operators on $\mathcal{D}(A^*)$, the domain of the adjoint operator to A such that for $x \in H$, $|x - p_m x|_H \to 0$ as $m \to \infty$. By analogy to (15) define

$$K_u^m(\alpha, \alpha') := \int_H |p_m B^{-1}(z)(D(z, u(z), \alpha) - D(z, u(z), \alpha'))|_H^2 \pi_u^{\alpha'}(dz) \tag{16}$$

Lemma 5 *Under (A1)-(A5) for every $u \in \mathcal{A}$ we have*

$$\sup_{\alpha, \alpha' \in A} |K_u^m(\alpha, \alpha') - K_u(\alpha, \alpha')| \to 0 \tag{17}$$

as $m \to \infty$.

Proof. The proof is based on the Lipschitzianity in α of D, boundedness of $B^{-1}(x)$ (by (A2)) and Corollary 2. □

3 CONSTRUCTION OF THE MARKOV TIME τ

This section is devoted to the construction of a particular Markov time τ, for which assumption (A5) is satisfied. Recall first Lemma 5 of Da Prato Gatarek Zabczyk (1992)

Lemma 6 *Under (A1)-(A4) for $p > \frac{2}{\beta}$ $\exists C > 0$ such that $\forall R > 0$*

$$P_1^{\alpha,u}(x, K(R,\beta)) \geq 1 - CR^{-1}(1 + |x|_H^p) \qquad P_1^0(x, K(R,\beta)) \geq 1 - CR^{-1}(1 + |x|_H^p) \quad (18)$$

provided $|x|_H^p \leq R$, where $K(R,\beta)$ is a compact set in H given by the formula

$$K(R,\beta) = \{x \in H : x = S(1)z + G_1 g(1) + G_{2^{-1}\beta} h(1), |z|_H^p \leq R, |g|_{L_p}^p \leq R, |h|_{L_p}^p \leq R\} \quad (19)$$

with $G_{\beta'} f(1) = \int_0^1 (1-s)^{\beta'-1} S(1-s) f(s) ds$ for $\beta' \in (0,1]$ and any bounded Borel measurable $f : [0,1] \mapsto H$.

Let $B(r) = \{x \in H, |x|_H < r\}$, $\quad \bar{B}(r) = \{x \in H, |x|_H \leq r\}$. Assume

(B1) $\exists r_0$ such that for $r_1 > r_0$ we have $\sup_{|x|_H = r_1} \sup_{u \in \mathcal{A}} \sup_{\alpha \in A} E_x^{\alpha,u}\{T_{\bar{B}(r_0)}^4\} < \infty$ and $\sup_{|x|_H = r_1} E_x^0\{T_{\bar{B}(r_0)}^4\} < \infty$.

For given r_0 such that (B1) holds we choose R such that $CR^{-1}(1 + r_0^p) \leq \theta < 1$ with $p > \frac{2}{\beta}$. Then we find $r_1 > 2r_0$ such that $K(R,\beta) \subset B(2^{-1}r_1)$ and for some $\kappa > 0$ $\sup_{|x|_H \leq r_0} \sup_{u \in \mathcal{A}} \sup_{\alpha \in A} P_x^{\alpha,u}\{\exists t \in [0,1], X(t) \in H \setminus B(r_1)\} \leq 1 - \kappa$ and $\sup_{|x|_H \leq r_0} P_x^0\{\exists t \in [0,1], X(t) \in H \setminus B(r_1)\} \leq 1 - \kappa$. Let now $\Gamma = K(R,\beta)$ and

$$\tau = T_{H \setminus B(r_1)} + \sigma \circ \theta_{T_{H \setminus B(r_1)}} \qquad \sigma = inf\{s + 1 : s \geq 0, X(s) \in \bar{B}(r_0), X(s+1) \in \Gamma\} \quad (20)$$

We have

Lemma 7 *Under (A1)-(A4) and (B1) we have*

$$\sup_{u \in \mathcal{A}} \sup_{\alpha \in A} \sup_{x \in \Gamma} E_x^{\alpha,u}\{\tau^2\} < \infty \qquad \sup_{x \in \Gamma} E_x^0\{\tau^2\} < \infty \quad (21)$$

Proof. Let $\bar{\tau} = T_{H \setminus B(r_1)} + T_{\bar{B}(r_0)} \circ \theta_{T_{H \setminus B(r_1)}}$, and $\bar{\tau}_1 = \tau$, $\bar{\tau}_{n+1} = \bar{\tau}_n + \bar{\tau} \circ \theta_{\bar{\tau}_n}$, for $n = 1, 2, \ldots$. Define $s(0) = inf\{j \geq 1 : \bar{\tau}_j \geq 1\}$ and for $i \geq 1$, $s(i) = inf\{j \geq i : \bar{\tau}_j \geq \bar{\tau}_i + 1\}$. Let $\tilde{\tau} = inf\{\bar{\tau}_{s^n(0)} + 1, X(\bar{\tau}_{s^n(0)} + 1) \in \Gamma, n = 0, 1, \ldots\}$ with $s^{n+1}(0) = s(s^n(0))$ and $\tau_{s^0(0)} := 0$. If $|X(0)|_H \leq r_0$, we have $\tau \leq \tilde{\tau}$. Since by the proof of Lemma 2.3 of Maslowski and Seidler (1993), for any $r > 0$, $\sup_{|x|_H \leq r} \sup_{u \in \mathcal{A}} \sup_{\alpha \in A} E_x^{\alpha,u}\{T_{H \setminus B(r)}^4\} < \infty$ and $\sup_{|x|_H \leq r} E_x^0\{T_{H \setminus B(r)}^4\} < \infty$, and $\theta, \kappa < 1$, it can be shown (the details are left to the reader) that

$$\sup_{u \in \mathcal{A}} \sup_{\alpha \in A} \sup_{x \in \Gamma} E_x^{\alpha,u}\{\tilde{\tau}^2\} < \infty, \qquad \sup_{x \in \Gamma} E_x^0\{\tilde{\tau}^2\} < \infty \quad (22)$$

from which the assertion of lemma follows. □

Let τ_n be defined as in (A5) with τ given in (20). To complete the proof of (A5) we need to show (6). We have

Proposition 1 *Under (A1)-(A4) and (B1), the process $X(\tau_n)$ is uniformly ergodic i.e. (6) of (A5) holds.*

Proof. By the proof of Theorem 2.1 of Duncan Pasik-Duncan and Stettner (1994a) it suffices to show that for $\delta > 0$

$$\sup_{x,y\in\Gamma}\sup_{u\in\mathcal{A}}\sup_{\alpha\in A}\sup_{B\in\mathcal{B}(H)} |P_x^{\alpha,u}\{X(\tau)\in B\} - P_y^{\alpha,u}\{X(\tau)\in B\}| \leq 1-\delta \tag{23}$$

Suppose (23) does not hold i.e. for some $x_n, y_n \in \Gamma$, $u_n \in \mathcal{A}$, $\alpha_n \in A$ and $B_n \in \mathcal{B}(H)$ we have $P_{x_n}^{\alpha_n,u_n}\{X(\tau)\in B_n\} \to 1$ and $P_{y_n}^{\alpha_n,u_n}\{X(\tau)\in B_n\} \to 0$ as $n\to\infty$. Without loss of generality we can assume that $x_n \to \bar{x}$ and $y_n \to \bar{y}$. By Girsanov theorem and finiteness of the second moment of τ, we obtain $P_{y_n}^0\{X(\tau)\in B_n\} \to 0$ and consequently from (a1), $P_{\bar{y}}^0\{X(\tau)\in B_n\}\to 0$. Using the equivalence of transition probabilities of $X(\tau_n)$ we have $P_{\bar{x}}^0\{X(\tau)\in B_n\}\to 0$ and then by (a1) again, $P_{x_n}^0\{X(\tau)\in B_n\}\to 0$. The last convergence by Girsanov theorem implies $P_{x_n}^{\alpha_n,u_n}\{X(\tau)\in B_n\}\to 0$, a contradiction. □

4 ESTIMATION AND CONTROL

To make MLE procedure feasible we add to the ML estimator the projection operator p_m and additionally assume

(B2) for each $x \in H$, $B^{-1}(x) : \mathcal{D}(A^*) \mapsto \mathcal{D}(A^*)$

Then we estimate the unknown parameter α^0 maximizing over $\alpha \in A$ value of

$$L_m^t(\alpha) := \int_0^t < p_m B^{-1}(X(s))D(X(s),u(X(s)),\alpha), p_m B^{-1}(X(s))[dX(s) \tag{24}$$
$$-(AX(s)+F(X(s)))ds] > -2^{-1}\int_0^t |p_m B^{-1}(X(s))D(X(s),u(X(s)),\alpha)|_H^2 ds$$

which is equivalent to the maximization of

$$\int_0^t < p_m B^{-1}(X(s))(D(X(s),u(X(s)),\alpha) - D(X(s),u(X(s)),\alpha^0), p_m dW(s) > \tag{25}$$
$$-2^{-1}\int_0^t |p_m B^{-1}(X(s))(D(X(s),u(X(s)),\alpha) - D(X(s),u(X(s)),\alpha^0))|_H^2 ds$$

To construct a nearly optimal adaptive strategy we fix $\epsilon > 0$. By Corollary 3 we can find a finite class of ϵ optimal control functions $\mathcal{A}(\epsilon) = \{u_1,\ldots,u_{r(\epsilon)}\}$. Let

$T(\epsilon) := \{a_i \in R^+, a_i + r \leq a_{i+1}, a_0 = 0$ such that $\frac{\epsilon}{2r||c||} \leq \liminf_{n\to\infty} n_1 \sum_{i=0}^{n-1}\chi_{T(\epsilon)}(i)$ and $\limsup_{n\to\infty} n_1 \sum_{i=0}^{n-1}\chi_{T(\epsilon)}(i) \leq \frac{\epsilon}{r||c||}\}$,

with $||c||$ standing for supremum norm, and $r := r(\epsilon)$. We have the following

Lemma 8 *Under (A1)-(A5) for any $\epsilon' > 0$ there exists a $\delta > 0$ and a positive integer m such that $\sup_{\alpha,\alpha'\in A}|K_u^m(\alpha,\alpha') - K(\alpha,\alpha')| \leq \frac{\delta}{6}$, and for any $\{0,1\}$ valued sequence $\beta^i(k)$,*

$i = 1, 2, \ldots$, $k \in \{1, \ldots, r\}$ which for each i it takes value 1 only at one $k \in \{1, \ldots, r\}$, and $\beta^i(k) = 1$ for $i = a_j + k - 1$, $a_j \in T(\epsilon)$, we have for $\alpha, \alpha' \in A$
if $\liminf_{n\to\infty} n^{-1} \sum_{i=0}^{n-1} \sum_{k=1}^{r} \beta^i(k) K_{u_k}^m(\alpha, \alpha') < \delta$ then $sup_{u\in\mathcal{A}(\epsilon)} ||\pi_u^\alpha - \pi_u^{\alpha'}||_{var} < \epsilon'$.

Proof. It follows from Lemma 5, Lemma 4 and the definition of the set $T(\epsilon)$. □

We continue our construction. Take δ and m such that Lemma 8 holds with $\epsilon' \leq \frac{\epsilon}{\|c\|}$. By Corollary 2 and there exists a finite set $A(\delta) \subset A$ such that for any $\alpha \in A$ one can find $\bar{\alpha} \in A(\delta)$ for which $\sup_{u\in\mathcal{A}} ||\pi_u^\alpha - \pi_u^{\bar{\alpha}}||_{var} \leq \frac{\epsilon}{\|c\|}$ and $\sup_{u\in\mathcal{A}(\epsilon)} K_u^m(\bar{\alpha}, \alpha) \leq \frac{\delta}{3}$.

Then we choose a positive integer N such that (its existence follows from the uniform ergodicity of $X(\tau_n)$) for all $x \in \Gamma$, $u \in \mathcal{A}(\epsilon)$, $\alpha \in A(\delta)$, $\alpha' \in A$ we have

$$|E_x^{\alpha',u}\{\int_0^{\tau_N} |p_m B^{-1}(X(s))(D(X(s), u(X(s)), \alpha) - D(X(s), u(X(s)), \alpha'))|_H^2 ds\} \\ (E_x^{\alpha',u}\{\tau_N\})^{-1} - K_u^m(\alpha, \alpha')| \leq \frac{\delta}{3} \tag{26}$$

and

$$|E_x^{\alpha',u}\{\int_0^{\tau_N} c(X(s), u(X(s)))ds\}(E_x^{\alpha',u}\{\tau_N\})^{-1} - \int_E c(z, u(z))\pi_u^\alpha(dz)| \leq \epsilon. \tag{27}$$

For a given $\epsilon > 0$ we have just constructed $\mathcal{A}(\epsilon)$, $T(\epsilon)$, δ, m, N, $A(\delta)$. We are now in position to define an adaptive control strategy $(\hat{v}_t)$, which can be described as follows.

We wait until the process $X(s)$ enters the set Γ. Then we start our procedure. At moments τ_{Ni}, $i = 1, \ldots$ we choose control functions u from the set $\mathcal{A}(\epsilon)$ accordingly to the following algorithm

- if $i = a_r + k - 1$, with $a_r \in T(\epsilon)$ and $k \in \{1, \ldots, r\}$ we use the function u_k for the next τ_N units of time,

- otherwise we maximize $L_m^{\tau_{Ni}}(\alpha)$ over $\alpha \in A(\delta)$ i.e. determine the value $\hat{\alpha}_{Ni} := argmax_{\alpha\in A(\delta)}\{L_m^{\tau_{Ni}}(\alpha)\}$ and use for the next τ_N units of time the control function u from $\mathcal{A}(\epsilon)$ that is ϵ optimal for the parameter $\hat{\alpha}_{Ni}$

In other words the strategy is to force all control functions from $\mathcal{A}(\epsilon)$ in rare moments of time (determined by $T(\epsilon)$) while in the other moments use control functions that are ϵ optimal for the current value of the estimation over the finite set $A(\delta)$. An alternative to the forced use all control functions from $\mathcal{A}(\epsilon)$ is a randomization as in Duncan Pasik-Duncan and Stettner (1994b).

Theorem 1 *Under (A1)-(A6) and (B2) or (A1)-(A4), (A6), (B1), (B2) we have $J((\hat{v}_t) \leq \inf_{u\in\mathcal{A}} \bar{J}_x^{\alpha^0}((u(X(t))) + 5\epsilon \quad P_x^{\alpha^0} a.e.$*

Proof. We follow the consideration of the proof of Theorem 1 of Duncan Pasik-Duncan and Stettner (1994b). Notice that if $\alpha \in A(\delta)$ is a frequent point of the estimation, we have $\liminf_{n\to\infty} n^{-1} \sum_{i=0}^{n-1} \sum_{k=1}^{r} \beta^i(k) K_{u_k}^m(\alpha, \alpha^0) < \delta$ and consequently by Lemma 8 we have $sup_{u\in\mathcal{A}(\epsilon)} ||\pi_u^\alpha - \pi_u^{\alpha^0}||_{var} < \frac{\epsilon}{\|c\|}$. The details are left to the reader. □

REFERENCES

Aihara, S.I. (1995) On adaptive boundary control for stochastic parabolic systems with unknown potential coefficient. *IEEE Trans. Aut. Cont.*, submitted.

Da Prato, G., Gatarek, D. and Zabczyk, J. (1992) Invariant measures for semilinear stochastic equations. *Stoch. Anal. Appl.*, **10**, 387–408.

Da Prato, G. and Zabczyk, J. (1992) Stochastic equations in infinite dimensions. *Cambridge University Press.*

Duncan, T.E., Goldys, B. and Pasik-Duncan, B. (1991) Adaptive Control of linear stochastic evolution systems. *Stochastics & Stochastics Rep.*, **36**, 71–90.

Duncan, T.E., Maslowski, B. and Pasik-Duncan, B. (1994) Adaptive boundary and point control of linear stochastic distributed parameter systems. *SIAM J. Control Optimiz.*, **32**, 648–672.

Duncan, T.E., Maslowski, B. and Pasik-Duncan, B. (1995) Adaptive boundary control of linear stochastic distributed parameter systems described by analytic semigroup. Preprint.

Duncan, T.E., Pasik-Duncan, B. and Stettner, L. (1994a) Almost self-optimizing strategies for the adaptive control of diffusion processes. *JOTA*, **81**, 479–507.

Duncan, T.E., Pasik-Duncan, B. and Stettner, L. (1994b) On discretized MLE in adaptive control of ergodic Markov models. Preprint.

Duncan, T.E., Pasik-Duncan, B. and Stettner, L. (1995) On ergodic control of stochastic evolution equations. *Stoch. Anal. Appl.*, submitted.

Gatarek, D. and Sobczyk, J. (1994) On the existence of optimal controls of Hilbert-valued diffusions. *SIAM J. Control Optimiz.*, **32**, 170–175.

Maslowski, B. and Seidler, J. (1993) Ergodic properties of recurrent solutions of stochastic evolution equations. Preprint AVCR MU 81.

Peszat, S. and Zabczyk, J. (1994) Strong Feller property and irreducibility for diffusion on Hilbert spaces. *Annals of Prob.*, to appear.

Stettner, L. (1993) On nearly selfoptimizing strategies for a discrete-time uniformly ergodic adaptive model. *JAMO*, **27**, 161–177.

30

Active control of mechanical distributed systems with stochastic parametric excitations

*Andrzej Tylikowski**
Warsaw University of Technology
Institute of Machine Design Fundamentals, Narbutta 84,
02-524 Warsaw, Poland. Phone: 0482-6608244. Fax: 04822-490306.
E–mail: aty@syriusz.simr.pw.edu.pl

Abstract

The study is based on the application of piezoelectric distributed sensors, actuators, and an appropriate feedback and is adopted for stability problems of system consisting of plate with control part governed by partial differential equations with stochastic coefficients. The application of Liapunov method to the Itô equation leads to the effective analytical estimation of a gain velocity feedback implying nonincreasing of the functional along an arbitrary plate motion and in consequence to balance the supplied energy by the parametric excitation and the dissipated energy by the inner and control damping.

Keywords

stochastic systems, distributed control, Liapunov method, stabilization, elastic plates, piezoelectric transducers

1 INTRODUCTION

Piezoelectric materials show great advantages as sensors and actuators in intelligent structures i.e. structures with highly distributed actuators, sensors, and processor networks. Piezoelectric sensors and actuators have been applied successfully in the closed-loop control (cf. Bailey and Hubbard, 1985). Crawley and de Luis (1987) presented a comprehensive static model for a piezoelectric actuator glued to a beam. This one-dimensional theory was extended by Dimitriadis, Fuller and Rogers (1991) on thin plates with two-dimensional piezoelectric patches. The general dynamic coupling model of the beam with bonding sensors and actuators was used to derive the Liapunov control strategy, which

*This research was supported by the grant from the Polish State Committee for Scientific Research (KBN Nr 3P4 04 009 07)

is especially useful in the collocated sensor-actuator systems (Tylikowski, 1993). The dynamic extensional strain on the beam surface was calculated by considering the dynamic coupling between the actuator and the beam, and by taking into account a finite bonding layer with the finite stiffness (Tylikowski, 1994). Lee and Moon (1990) introduced novel distributed sensors/actuators, which sense and actuate the particular modal coordinate without extensive signal processing. Dimitriades, Fuller and Rogers (1991) showed that two dimensional patch type actuators show large potential for controlling vibration in distributed systems. Lee, Chiang and O'Sulivan (1991) demonstrated advantages of combining distributed modal sensor/actuator pairs into flexible structures. Tzou and Fu (1992) analysed models of a plate with segmented distributed piezoelectric sensors and actuators, and showed that segmenting improves the observability and the controllability of the system. Recently designed by Kumar, Bhalla, and Cross (1994) the piezoelectric actuators with constant properties over a wide-band frequencies seems to be appropriate for applying in systems subjected to wide-band stochastic excitations. Chow and Maestrello (1993) examined the exponential stabilization of panels with time-dependent parametric excitations and concluded that a stronger mode of control, such as a distributed control, should be used to stabilize time-dependent systems. The direct Liapunov method was applied to the stabilization problem of the beam subjected to a wide-band parametric excitation (Tylikowski, 1995).

The purpose of the present paper is to solve an active control problem of parametric vibrations excited by the biaxial in-plane wide-band Gaussian forces. The problem is solved using the concept of distributed piezoelectric sensors and actuators with a sufficiently large value of velocity feedback. Real mechanical systems are subjected not only to nontrivial initial conditions but also to permanently acting excitations and the active vibration control should be modify in order to balance the supplied energy by external parametric excitation. The applicability of active vibration control is extended to distributed systems with stochastic parametric excitation. The effective estimation of the feedback constant stabilizing the plate parametric vibration is derived analytically. The minimal value of the feedback constant is effectively expressed by the constant component of in-plane forces, intensities of stochastic components, geometry, mechanical and piezoelectric properties of actuators and sensors.

2 DYNAMICS EQUATION

Consider a Kirchhoff plate of a bending stiffness D biaxially loaded in the plate middle plane by time-dependent forces F_x, and F_y uniformly distributed over the simply supported edges. The rectangular plate domain is denoted by $\mathcal{P} \equiv (0,a) \times (0,b)$. The dynamics equation of plate motion includes both an internal passive damping with the proportionality coefficient α and an active damping. Thin piezoelectric patches are perfectly mounted on opposite sides of the plate. It is assumed that the transverse motion dominates the in-plane plate vibrations. The thickness of the plate, the actuator and the sensor is denoted by t_p, t_a and t_s, respectively. The sensing and actuating effects of piezoelectric layers are used to extract the mechanical energy and in a final result to stabilize both the free vibration due to initial disturbances and the parametric vibration excited by the axial force. Assume a negligible stiffness of the sensor in comparison with that of the plate and reducing the influence of the piezoelectric actuator on the plate to bending moments M_x, M_y distributed over the plate surface.

2.1 Sensor equation

Sensor electric displacement

$$D_3 = -e_{s31}\epsilon_{s_1} - e_{s32}\epsilon_{s_2} \tag{1}$$

Substituting Hook's law and the geometric formulas relating deformations on the plate surface with the curvatures of the middle plane $\epsilon_{s_1} = (t_p+t_s)w_{,xx}/2$, $\epsilon_{s_2} = (t_p+t_s)w_{,yy}/2$ we have

$$D_3 = -\frac{t_p + t_s)E_s}{2(1-\nu_s^2)}[(d_{s31} + \nu d_{s32})w_{,xx} + (d_{s32} + \nu d_{s31})w_{,yy}]. \tag{2}$$

Integrating the electric displacement over the sensor area $\mathcal{S}$ and dividing by the capacity we obtain the sensor voltage of the open circuit

$$\mathcal{V}_s = -\frac{t_s(t_p + t_s)E_s}{2A_s\varepsilon_{33}(1-\nu_s^2)}\int_S[(d_{s31} + \nu_s d_{s32})w_{,xx} + (d_{s32} + \nu_s d_{s31})w_{,yy}]dS. \tag{3}$$

2.2 Actuator equation

Normal stresses in the actuator due to the direct piezoelectric effect are as follows

$$\sigma_{a_1} = \frac{d_{a31}\mathcal{V}_a}{t_a}. \tag{4}$$

$$\sigma_{a_2} = \frac{d_{a32}\mathcal{V}_a}{t_a}. \tag{5}$$

Assuming the uniform stress distribution over the actuator thickness the control bending moments are given by the following formulas

$$\begin{bmatrix} M_x \\ M_y \end{bmatrix} = \frac{E_a\mathcal{V}_a(t_p + t_a)}{2(1-\nu_s^2)t_a}\chi_a(x,y)\begin{bmatrix} d_{a31} + \nu_a d_{a32})w_{,xx} \\ (d_{a32} + \nu_a d_{a31})w_{,yy} \end{bmatrix}. \tag{6}$$

The shape of the piezoelectric actuator $\mathcal{A}$ is described by the characteristic function $\chi_a(x,y)$.

2.3 Plate equation of motion

Assuming the velocity feedback control with the gain factor K_a the voltage applied to the actuator is given by

$$\mathcal{V}_a = K_a\frac{d\mathcal{V}_s}{dt}. \tag{7}$$

The plate differential equation of motion with closed loop control has the form

$$\rho t_p w_{,tt} + \alpha w_{,t} + F_x(t) w_{,xx} + F_y(t) w_{,yy} + D\Delta^2 w + M_{x,xx} + M_{y,yy} = 0, \quad (x,y) \in \mathcal{P}. \qquad (8)$$

Substituting the bending moments and introducing dimensionless coordinates yield

$$w_{,tt} + 2\beta w_{,t} + (f_{ox} + f_x(t)) w_{,xx} + (f_{oy} + f_y(t)) w_{,yy} + \Delta^2 w +$$

$$+2(\beta_x \chi_{a,xx} + \beta_y \chi_{a,yy}) S(w_{,t}) = 0 \qquad (9)$$

where the sensor linear functional has the form

$$S(w) = \int_S ((d_{31} + \nu d_{32}) w_{,xx} + (d_{32} + \nu d_{,31}) w_{,yy}) dS. \qquad (10)$$

The actuator shape is described by the characteristic function $\chi_a(x,y)$. Equation (9) with zero initial conditions $w(x,0) = \frac{\partial w}{\partial t}(x,0) = 0$ possess a trivial solution $w(x,t) = 0$, which corresponds to an undeflected plate middle plane.

From the mathematical point of view, the feature common to all parametric vibrations is that they are described by differential equations with coefficients depending explicitly on time. In deterministic parametric vibrations it is well known that the stability properties are determined from the Mathieu equation together with the corresponding Ince-Strutt diagram. If the parametric excitation becomes random, the stability criteria depend on the statistical characteristics of the excitation and the system parameters. Specifically, if the excitation is sufficiently narrow-banded or it has one latent periodicity, a series of wedges on the amplitude-frequency plane can be expected, analogously to the deterministic parametric resonance. In the present analysis the direct Liapunov method is proposed so as to solve stochastic parametric vibration and to establish a stability criterion for the equilibrium state of plate with closed-loop control in a stochastic sense.

3 ENERGY EXTRACTION

Vibration damping of the plate with parametric excitation can be examined by means of the total energy considerations. The mechanical energy consists of the kinetic energy, the bending energy, and the elastic energy of compression due to the constant components of in-plane forces f_{ox}, and f_{oy}

$$V = \frac{1}{2} \int_{\mathcal{P}} (w_{,t}^2 + (\Delta w)^2 - f_{ox} w_{,x}^2 - f_{oy} w_{,y}^2) d\mathcal{P} \qquad (11)$$

The energy is positive-definite if the classic buckling condition is fulfilled by the constant components of in-plane forces. By differentiating equation (11) with respect to time, the rate of energy extraction is

$$\frac{dV}{dt} = \int_{\mathcal{P}} (w_{,t} w_{,tt} + \Delta w \Delta w_{,t} - f_{ox} w_{,x} w_{,xt} - f_{oy} w_{,y} w_{,yt}) d\mathcal{P}. \qquad (12)$$

Eliminating the acceleration by means of dynamics equation (9), and integrating by parts (12) gives

$$\frac{dV}{dt} = -\int_{\mathcal{P}} w_{,t}(f_x(t)w_{,xx} + f_y(t)w_{,yy})d\mathcal{P} - 2\beta\int_{\mathcal{P}} w_{,t}^2 d\mathcal{P} +$$

$$-2S(w_{,t})\int_{\mathcal{A}}(\beta_x\chi_{a,xx} + \beta_y\chi_{a,yy})w_{,t}d\mathcal{P} \tag{13}$$

The first component with an undefined sign in equation (13) is the power flow due to the parametric excitation. The second negative component represents the rate at which the energy is extracted from the plate by the passive damping. Assuming the same shape of sensor and actuator the third component in equation (13) also describes the energy extraction as it can be expressed by the square of integral. Therefore, for the sufficiently large gain factor it is possible to stabilize parametric vibrations excited by the time-dependent axial force.

The analogous analyses of active damping vibration caused due to initial nonzero disturbances were presented by a number of researchers (eg. Bailey, and Hubbard, (1985), and recently Pourki, (1993)). However, inequality (13) does not provide an effective quantitive estimation of the minimal active damping coefficients β_x, β_y stabilizing the parametric vibration. In order to derive an analytical relation involving characteristics of the parametric excitation, and parameters of passive and active damping it is necessary to precisely define the class of parametric excitations $f_x(t)$, $f_y(t)$ randomly fluctuating over a wide-band of frequencies and reformulate the stabilization problem of plate parametric vibration as a qualitative analysis of stochastic partial differential equations (Curtain, and Falb, 1965). More advanced methods of stability analysis introduced for distributed parameter systems with wide-band Gaussian coefficients (Tylikowski, 1991) have to be applied to derive the formula describing the stability domains as a function of the passive and active damping coefficients, and statistical characteristics of parametric excitation.

4 STOCHASTIC STABILITY PROBLEM

If the parametric excitation is a wide-band Gaussian stochastic process equation (9) should be understood as the Itô stochastic partial differential equation (Curtain, and Falb, 1965), and rewritten in the form

$$dw = w_{,t}dt \tag{14}$$

$$dw_{,t} = -\Big(2\beta w_{,t} + \Delta^2 w + f_{ox}w_{,xx} + f_{oy}w_{,yy} + 2(\beta_x\chi_{\mathcal{A},xx} + \beta_y\chi_{\mathcal{A},yy})S(w)\Big)dt +$$

$$-\sigma_x w_{,xx}d\mathcal{W}_x - \sigma_y w_{,yy}d\mathcal{W}_y, \qquad (x,y)\in\mathcal{P}. \tag{15}$$

where $\mathcal{W}_x$, $\mathcal{W}_y$ are independent Wiener processes , and σ_x σ_y the intensities of the compressive forces equal to the square roots of their maximal spectral densities. As there is no randomness in the first equation we notice that the Stratonovich-Wong-Zakai correction term is equal to zero.

We assume that the solutions for equations (14) and (15) exist and belong to the appropriate Hilbert space. The purpose of the present paper is to derive criteria for solving the following problem: will the deviations of the plate middle plane from the unperturbed state (trivial solution) be sufficiently small in some mathematical sense in the case when the in-plane forces are the wide-band Gaussian processes. The plate dynamically buckles when the in-plane forces get so large that the plate with closed-loop control does not oscillate and a new increasing mode of oscillations occurs. To estimate a perturbed solution of equations (14) and (15) it is necessary to introduce a measure of distance $\|.\|$ of the solution of equations (14) and (15) with nontrivial initial conditions from the trivial one. The equilibrium state of equation (14) and (15) is said to be uniformly stochastically stable, if the following logic sentence is true

$$\bigwedge_{\epsilon\geq 0}\bigwedge_{\delta\geq 0}\bigvee_{r\geq 0}\|w(.,0)\|\leq r\Rightarrow P(\sup_{t\geq 0}\|w(.,t)\|\geq\epsilon)\leq\delta \tag{16}$$

In the present paper the direct Liapunov method is proposed to establish criteria for the uniform stochastic stability of the unperturbed (trivial) solution of the plate with closed-loop control.

5 STABILITY ANALYSIS

We construct the Liapunov functional as a sum of the modified kinetic energy and the elastic energy of the plate (Tylikowski, 1991)

$$V=\frac{1}{2}\int_{\mathcal{P}}\Big[(\Delta^2 w)^2-f_{ox}w_{,xx}^2-f_{oy}w_{,yy}^2+\big(w_{,t}+2\beta w+2(\beta_x\chi_{a,xx}+\beta_y\chi_{a,yy})S(w)\big)^2\Big]d\mathcal{P}. \tag{17}$$

If the classical condition for the static buckling is fulfilled, functional (17) satisfies the positive-definiteness condition, and the measure of distance between the perturbed solution and the trivial one can be chosen as the square root of the functional $\|.\|=V^{1/2}$. As realizations of the Wiener process are not differentiable the Itô calculus has to be applied to calculate the differential of functional (17)

$$dV=\mathcal{F}(w)dt+\mathcal{G}_x d\mathcal{W}_x+\mathcal{G}_x d\mathcal{W}_y, \tag{18}$$

where

$$\mathcal{F}=\int_{\mathcal{P}}\Big[2\big(\beta+(\beta_x\chi_{a,xx}+\beta_y\chi_{a,yy}S(w))\big)(\Delta w)^2-\frac{1}{2}(\sigma_x^2 w_{,xx}^2+\sigma_y^2 w_{,yy}^2)+$$
$$-2\beta(f_{ox}w_{,x}^2+f_{oy}w_{,y}^2)-2S(w)(\beta_x\chi_{a,xx}+\beta_y\chi_{a,yy})(f_{ox}w_{,xx}+f_{oy}w_{,yy})\Big]d\mathcal{P}. \tag{19}$$

Integrating the differential (18) with respect to time from s to $\tau_\delta(t)$, where $\tau_\delta(t)=\min(t,\tau_\delta)$, and τ_δ is a random time of the first exit from the domain $\|.\|=\delta$, conditionally averaging (**E**) with respect to a σ-algebra $\mathcal{N}_t\subset$ ß generated by the Wiener processes

$\{\xi(s')|s' \subset [s,t]\}$ and taking into account the fact that the conditional average of the second and third terms in equation (18) are equal to zero, we have

$$\mathbf{E}V(\tau_\delta(t)) = V(s) - \mathbf{E}\int_s^{\tau_\delta(t)} \mathcal{F} dt \tag{20}$$

Assuming that

$$\mathcal{F} \geq 0 \tag{21}$$

we can write the following inequality

$$\mathbf{E}V(\tau_\delta(t)) \leq V(s) \tag{22}$$

Proceeding similarly to the proof of Chebyshev's inequality, we find that the trivial solution (corresponding to the equilibrium state) of the stochastic equations (14) and (15) governing the dynamics of the plate with closed-loop control is uniformly stochastically stable with respect to the measure $\|.\| = V^{1/2}$, if the inequality (21) holds.

6 STABILIZATION CONDITIONS

Using the derived formula (21) it is easy to obtain particular results relating to basic problems of active and passive damping of vibration. Neglecting the constant components of in-plane forces and the passive damping, and introducing the ratio of piezoelectric properties $r = (d_{31}+\nu d_{32})/(d_{32}+\nu d_{31})$ the minimum active damping coefficient stabilizing the parametric vibration due to wide-band Gaussian force with the known maximum spectral densities is defined in the following way

$$\frac{\sigma_x^2}{4\beta_x} \leq \min_{m,n=1,2,..} \frac{\int_{\mathcal{A}} \Delta^2(w_{,xx} + \frac{\beta_y}{\beta_x} w_{,yy}) d\mathcal{A} \int_{\mathcal{S}} (w_{,xx} + r w_{,yy}) d\mathcal{S}}{\int_{\mathcal{P}} \left(w_{,xx}^2 + (\frac{\sigma_y}{\sigma_x})^2 w_{,yy}^2\right) d\mathcal{P}}. \tag{23}$$

We look for the minimum of functional in the class of functions satisfying simply supported boundary conditions $\sin(n\pi x)\sin(n\pi x)$ for $m, n = 1, 2, ...$ For the same shape of piezoelectric sensor and actuator described by Pourki's function $y = x^2(x-1)$ (1993), and assuming the same properties of sensor and actuator (r=1), as well as $\sigma_y/\sigma_x = 1$ the stabilization conditions of the square plate $(a = b = 1)$ has the form

$$\frac{\sigma_x^2}{4\beta_x} \leq 0.0135. \tag{24}$$

7 CONCLUSIONS

By means of the direct Liapunov method the active stabilization of a vibrating plate with distributed piezoelectric sensor, actuator, and the velocity feedback has been studied. The elastic plate is simply supported and loaded biaxially by in-plane compressive forces

randomly fluctuating over a wide band of frequencies. Without any passive damping and control, the plate motion is unstable due to the parametric excitation.
The stabilization of stochastic parametric vibrations needs sufficiently large active damping coefficient proportional to the gain factor. The minimum gain factor stabilizing the system depends on the shape of sensor and actuator.

For no in-plane compression, this is the case of free vibration due to the nontrivial initial conditions. As long as the active or passive damping is present, the system is stable and oscillations decay.

REFERENCES

Bailey, T, and Hubbard, J. E. Jr, (1985) Distributed piezoelectric-polymer active vibration control of a cantilever beam, *J. Guidance, Control, and Dynamics*, **8**, 605-11.

Chow, P. L, and Maestrello (1993), Stabilization of the nonlinear vibration of an elastic panel by boundary control, [in:] R. A. Burdiso, (ed.) *Second Conference on Recent Advances in Active Control of Sound and Vibration*, Technomic Publishing, Lancaster–Basel, 660-7.

Crawley, E. F, and de Luis, J, (1987) Use of piezoelectric actuators as elements of intelligent structures, it AIAA J, **25**, 1373-85.

Curtain, R. F, and Falb, P. L, (1965) Stochastic differential equations in Hilbert space, *J. Diff. Equations*, **10**, 412-30.

Dimitriadis, E, Fuller, C. E, and Rogers, C. A, (1991) Piezoelectric actuators for distributed vibration excitation of thin plates, *J. Appl. Mech.*, **113**, 100-7.

Kumar, S, Bhalla, A. S, and Cross, L. E, (1994) Smart ceramics for broadband vibration control *J. Intelligent Material Systems and Structures*, **5**, 673-7.

Lee, C. K, Chiang, W. W, and O'Sulivan, T. C, (1991) Piezoelectric modal sensor/actuator pairs for critical active damping vibration control, *J. Acoust. Soc. Am.*, **90**, 374-84.

Pourki, F, (1993), Active distributed damping of flexible structures using piezoelectric actuator/sensors, **20**, 279-85.

Tylikowski, A, (1991) *Stochastic stability of continuous systems*, Polish Scientific Publishers, Warszawa, (in Polish).

Tylikowski, A, (1993) Stabilization of beam parametric vibrations, *Journal of Theoretical and Applied Mechanics*, **32**, 657-70.

Tylikowski, A, (1995) Active stabilization of beam vibrations parametrically excited by wide-band Gaussian force, [in:] S. D. Sommerfelt, and H. Hamada (eds.), *Proceedings of The International Symposium on Active Control of Sound and Vibration*, Newport Beach, CA, 91-102.

Tzou, H. S, and Fu, H. Q, (1992) A study on segmentation of distributed piezoelectric sensors and actuators; Part 1 - Theoretical analysis, *Active Control of Noise and Vibration*, ASME, DSC - **38**, 239-46.

PART FOUR

Numerical Modelling

31
Mixed finite element for stationary flow of a mixture with barodiffusion

Piotr Krzyżanowski
Warsaw University
Institute of Applied Mathematics, ul. Banacha 2, 02–097 Warszawa, Poland
E–mail: Piotr.Krzyzanowski@appli.mimuw.edu.pl

Abstract
We analyse a mixed finite element discretization of a system of equations describing the stationary, isothermic flow of a mixture with nonlinear barodiffusion. Using Taylor–Hood elements for velocity and pressure, and linear elements for concentration, we provide results on existence, uniqueness and approximation for derived discrete problem.

Keywords
Mixture flow, Stokes equations, barodiffusion, mixed finite element

1 INTRODUCTION

The stationary isothermic flow of a mixture of two incompressible viscous fluids with diffusion is described by the boundary value problem (Petrosyan, 1984), (Łukaszewicz, 1991):

$$-\nu\Delta u + (u\cdot\nabla)u + \nabla p = f + cg \text{ in } \Omega, \tag{1}$$
$$\operatorname{div} u = 0 \text{ in } \Omega, \tag{2}$$
$$-\operatorname{div}(D(c)\nabla c) + u\cdot\nabla c = \operatorname{div}(K(c)\nabla p) \text{ in } \Omega, \tag{3}$$
$$u = u_0 \text{ on } \partial\Omega, \tag{4}$$
$$c = c_0 \text{ on } \partial\Omega. \tag{5}$$

Here, Ω denotes a bounded open subset in R^3 or R^2. The unknowns are: the mean mass velocity vector u, the pressure p in the mixture and the first component concentration c. The vectors f, g denote external forces acting on the components of the mixture, D and K denote diffusion and barodiffusion coefficients, dependent on the concentration c. We assume that the viscosity ν is positive and constant. The boundary conditions on u and c are imposed by the trace of given functions u_0, c_0, defined on entire Ω.

The above system of PDEs appears, for example, in mathematical models of the flow of some suspensions, such as blood (Popel et al., 1974). Existence and uniqueness of weak solutions to (1) – (5) has been investigated in (Łukaszewicz, 1991). This system is a basis for analysis of more advanced models of suspensions (e.g. (Krzyżanowski, 1994), where the micropolar fluid mixture is considered).

In case when the velocity is relatively small, it is physically reasonable to skip the nonlinear term in the Naver – Stokes equations, obtaining the following system:

$$-\nu\Delta u + \nabla p = f + cg \text{ in } \Omega, \tag{6}$$

$$\operatorname{div} u = 0 \text{ in } \Omega, \tag{7}$$

$$-\operatorname{div}(D(c)\nabla c) + u \cdot \nabla c = \operatorname{div}(K(c)\nabla p) \text{ in } \Omega, \tag{8}$$

$$u = u_0 \text{ on } \partial\Omega, \tag{9}$$

$$c = c_0 \text{ on } \partial\Omega. \tag{10}$$

In this paper we present an analysis of finite element approximation of the solution of system (6) – (10), with additional assumption that the diffusion coefficient D is a positive constant. For simplicity, we also assume homogeneous boundary condition on u.

Throughout the paper we assume, unless otherwise stated, that Ω is a bounded polygon in R^2 or a bounded polyhedron in R^3.

1.1 Notation

We shall use several function spaces, which properties are described, for example, in (Adams, 1975). By $W^{k,p}(\Omega)$ we shall denote the usual Sobolev spaces, identifying $W^{0,p}(\Omega)$ with the $L^p(\Omega)$ space of measurable functions with their p-th power Lebesgue integrable. The standard norm in $W^{k,p}$ shall be denoted by $||\cdot||_{k,p}$, while the seminorm – by $|\cdot|_{k,p}$. For the space $W^{k,2}(\Omega)$ we shall use a symbol $H^k(\Omega)$, and the norm in that space we shall abbreviate as $||\cdot||_k$.

By $H_0^1(\Omega)$ we shall understand the subspace of $H^1(\Omega)$ of functions with their trace on $\partial\Omega$ equal to zero. By $L_0^2(\Omega)$ we denote the subspace of $L^2(\Omega)$, defined as

$$L_0^2(\Omega) = \{w \in L^2(\Omega) : \int_\Omega w = 0\}.$$

We denote the inner product in $L^2(\Omega)$ by brackets:

$$(u,v) := \int_\Omega u\, v\, dx$$

for any $u, v \in L^2(\Omega)$. Following (Temam, 1979), we also introduce a trilinear form

$$b(u,v,w) := \frac{1}{2}\Big((u\cdot\nabla v, w) - (u\cdot\nabla w, v)\Big)$$

for any $u, v, w \in H^1(\Omega)$. This continuous form is by definition antisymmetric with respect to the last two arguments (which reflects the antisymmetry of $(u\cdot\nabla v, w)$ on the solution u of (6)–(10)). In particular, we have

$$b(u,v,v) = 0. \tag{11}$$

By symbol "Const" we denote generic constant, independent of h, which, where necessary, we shall distinguish by subscripts.

Where there is no risk of confusion, we shall write $W^{k,p}$, H^k, H_0^1, L_0^2 instead of $W^{k,p}(\Omega)$, $H^k(\Omega)$, $H_0^1(\Omega)$, $L_0^2(\Omega)$.

1.2 General assumptions.

We shall assume there holds the following regularity condition:

(R1) For every $f \in L^2(\Omega)$ the weak solution (u,p) of Stokes equation

$$-\Delta u + \nabla p = f \quad \text{in} \quad \Omega, \tag{12}$$
$$\operatorname{div} u = 0 \quad \text{in} \quad \Omega, \tag{13}$$

with homogeneous Dirichlet boundary condition on u, and with $\int_\Omega p = 0$, belongs to $(H^2(\Omega) \cap H_0^1(\Omega)) \times H^1(\Omega)$ and

$$||u||_2 + ||p||_1 \leq \text{Const}\, ||f||_0.$$

for some Const independent of f.

For example, if $\Omega \subset R^2$ is a convex polygon, then assumption (R1) holds (Kellogg and Osborn, 1976).

We shall make the following assumptions on the data (see also (Łukaszewicz, 1991)):

(A1) $f, g \in L^3$;
(A2) $c_0 \in H^2$ and $0 \leq c_0(x) \leq 1$ for $x \in \partial\Omega$;
(A3) $K : R \to R$ is Lipschitz continuous function:

$$|K(s) - K(t)| \leq L_K|s-t|, \quad \forall t, s \in R$$

and such that $K(s) \equiv 0$ for $s \notin (0,1)$.

1.3 Discrete problem

We shall work with the following function spaces:

$V := H_0^1(\Omega)^d$, where $d = 2,3$ is the dimension of $\Omega \subset R^d$,
$W := L_0^2(\Omega)$,
$X := H_0^1(\Omega)$,
$X(c_0) := X + c_0$.

For homogeneous Dirichlet boundary condition on u, the variational formulation of the simplified problem (6) – (10) is as follows:

Problem 1 *Find* $(u,p,c) \in V \times (W \cap H^1) \times X(c_0)$, *such that*

$$\nu(\nabla u, \nabla v) - (p, div\, v) = (f + cg, v) \quad \forall v \in V, \tag{14}$$
$$(div\, u, w) = 0 \quad \forall w \in W, \tag{15}$$
$$D(\nabla c, \nabla \xi) + b(u, c, \xi) = -(K(c)\nabla p, \nabla \xi) \quad \forall \xi \in X. \tag{16}$$

This problem has a unique solution for sufficiently small and regular data in regular domains, and the proof is similar to that in (Lukaszewicz, 1991), where full problem (1) –(5) is considered.

We approximate Problem 1 in finite dimensional subspaces $V_h \subset V$, $W_h \subset W$, $X_h \subset X$, $X_h(\tilde{c_0}) := X_h + \tilde{c_0}$, where $\tilde{c_0}$ is a (finite element) approximation of the boundary condition c_0, using mixed method:

Problem 2 *Find* $(u_h, p_h, c_h) \in V_h \times W_h \times X_h(\tilde{c_0})$, *such that*

$$\nu(\nabla u_h, \nabla v_h) - (p_h, div\, v_h) = (f + c_h g, v_h) \quad \forall v_h \in V_h, \tag{17}$$
$$(div\, u_h, w_h) = 0 \quad \forall w_h \in W_h, \tag{18}$$
$$D(\nabla c_h, \nabla \xi_h) + b(u_h, c_h, \xi_h) = -(K(c_h)\nabla p_h, \nabla \xi_h) \quad \forall \xi_h \in X_h. \tag{19}$$

1.4 Finite element assumptions.

In our analysis we shall assume that the finite dimensional spaces V_h, W_h, X_h are specific finite element spaces. We cover $\bar{\Omega}$ with a quasi-uniform, shape regular triangulation (Ciarlet, 1991) $\mathcal{T}_h$, dividing $\bar{\Omega}$ into triangles K (or tetrahedra in three dimensional case)

$$\bar{\Omega} = \bigcup_{K \in \mathcal{T}_h} K,$$

so that any $K \in \mathcal{T}_h$ has at least one vertex not on $\partial\Omega$ (Bercovier and Pironneau, 1979). The mesh parameter h is defined as

$$h = \max_{K \in \mathcal{T}_h} \operatorname{diam} K.$$

Let $P_j(K)$ denote the space of polynomials of degree not greater than j on single triangle $K \in \mathcal{T}_h$. We define the finite element spaces V_h, W_h, X_h as follows.

For approximation of the velocity and pressure we use the Taylor – Hood finite elements (Brezzi and Fortin, 1991),

$$V_h = \{v \in V \cap C(\bar{\Omega}) : v|_K \in P_2(K) \quad \forall K \in \mathcal{T}_h\},$$

and

$$W_h = \{w \in W \cap C(\bar{\Omega}) : w|_K \in P_1(K) \quad \forall K \in \mathcal{T}_h\}.$$

Continuous finite element approximation of the pressure is necessary in our case, due to the ∇p_h term in (17).

As concerns the space in which we approximate the concentration c, we shall consider X_h consisting of linear elements, i.e.

$$X_h = \{\xi \in X \cap C(\bar{\Omega}) : \xi|_K \in P_1(K) \quad \forall K \in \mathcal{T}_h\}.$$

Properties of the above finite element spaces may be found in, for example, (Ciarlet, 1991).

We choose the finite element approximation $\tilde{c_0} \in C(\bar{\Omega})$ of the boundary condition c_0 so that $\tilde{c_0}|_K \in P_1(K)$ for all $K \in \mathcal{T}_h$ and such that in the nodal points x of triangulation $\mathcal{T}_h$ we have

$$\tilde{c_0}(x) = \begin{cases} c_0(x) & \text{if } x \in \partial\Omega, \\ 0 & \text{otherwise.} \end{cases}$$

1.5 Main result

Theorem 1 *Suppose Problem 1 admits a solution* $(u, p, c) \in (H^3 \cap V) \times (H^2 \cap W) \times (H^2 \cap X(\tilde{c_0}))$ *satisfying* $\max\{||u||_3, ||p||_2, ||c||_2\} \leq Const_1$. *Under assumptions from paragraphs 1.2 and 1.4, there exist* $H > 0$ *and some polynomial* $P : R^6 \to R$, *with positive coefficients and with property* $P(0,0,0,0,0,0) = 0$, *such that if the data of the problem satisfy condition*

$$P(D^{-1}, \nu^{-1}, |K|_\infty, ||f||_{0,3}, ||g||_{0,3}, ||c_0||_1) \leq 1, \tag{20}$$

then for any $0 < h < H$ *there exists a unique solution* (u_h, p_h, c_h) *of Problem 2, which satisfies the following error estimate:*

$$||u - u_h||_1 + ||p - p_h||_1 + ||c - c_h||_1 = O(h). \tag{21}$$

Let us comment on this theorem. The coefficients of polynomial P depend only on Ω, Const_1 and H. Since P vanishes at $(0,0,0,0,0,0)$, then, by continuity, condition (20) is satisfied for sufficiently small values of $D^{-1}, \nu^{-1}, |K|_\infty, ||f||_{0,3}, ||g||_{0,3}, ||c_0||_1$. Hence, (20) is kind of small data requirement. This condition, evidently, might be replaced by simpler expression, like $D^{-1} + \nu^{-1} + |K|_\infty + ||f||_{0,3} + ||g||_{0,3} + ||c_0||_1 \leq \text{Const}_2$, but polynomial condition is more descriptive, as it contains information that "smallness" of the data is relative one to another. Similar polynomial condition appears within uniqueness theorem for Problem 1 (Lukaszewicz, 1991), but, apparently, is weaker than derived here.

In our case we explicitly restrict ourselves only to the class of solutions which are uniformly bounded by Const_1. However, if the domain boundary was smoother, then the assumption $\max\{||u||_3, ||p||_2, ||c||_2\} \leq \text{Const}_1$ could have been replaced by "small data" requirement again. This follows from continuous dependence on data in above norms (Lukaszewicz, 1991).

The remaining of the paper is devoted to the proof of Theorem 1, and is organized as follows. In the next section, we briefly discuss a linearization of Problem 2, which decouples the system into two independent ones. Then in Section 3 we outline the proof of the existence statement of Theorem 1, based on Brouwer's fixed point theorem. Section 4 contains a scheme of the proof of the approximation part of Theorem 1, while in Section 5 we sketch the proof of the uniqueness result.

2 LINEARIZED DISCRETE PROBLEM

Let us introduce an auxiliary linear problem:

Problem 3 *Given $c_h^* \in X_h(\tilde{c_0})$, find $(u_h, p_h, c_h) \in V_h \times W_h \times X_h(\tilde{c_0})$, such that*

$$\nu(\nabla u_h, \nabla v_h) - (p_h, div\, v_h) = (f + c_h^* g, v_h) \quad \forall v_h \in V_h, \tag{22}$$

$$(div\, u_h, w_h) = 0 \quad \forall w_h \in W_h, \tag{23}$$

$$D(\nabla c_h, \nabla \xi_h) + b(u_h, c_h, \xi_h) = -(K(c_h^*)\nabla p_h, \nabla \xi_h) \quad \forall \xi_h \in X_h. \tag{24}$$

This problem may be seen as a linearization scheme for Problem 2, and may be used as a basis for an algorithm for iterative solution of Problem 2, with Stokes and diffusion equations decoupled. We shall use this linearization for the proof of existence and uniqueness of Problem 2.

Lemma 2 *There exists exactly one solution $(u_h, p_h, c_h) \in V_h \times W_h \times X_h(\tilde{c_0})$ of Problem 3. Moreover,*

$$D||c_h||_1 \leq 2D||\tilde{c_0}||_1 + |K|_\infty ||\nabla p_h||_0 + Const\, ||u_h||_1 ||\tilde{c_0}||_1. \tag{25}$$

Proof. The existence follows from (Girault and Raviart, 1986) and (Ciarlet, 1991). Taking $\xi = c - c_0$ in (24), we easily get (25).

3 EXISTENCE OF SOLUTIONS TO DISCRETE PROBLEM

Let us introduce a mapping $\Phi : X_h \to X_h$, defined as

$$\Phi(c_h^* - \tilde{c_0}) = c_h - \tilde{c_0},$$

where $c_h \in X_h(\tilde{c_0})$ is the solution of Problem 3 for given $c_h^* \in X_h(\tilde{c_0})$. According to Lemma 2, this mapping is well defined.

First, we show that Φ is continuous. To this end, we consider the difference $\Phi(c_1^* - \tilde{c_0}) - \Phi(c_2^* - \tilde{c_0})$ for arbitrary $c_1^*, c_2^* \in X_h(\tilde{c_0})$. By definition, this difference is equal to $c_1 - c_2$, where c_1, c_2 are the solutions of Problem 3 with given c_1^*, c_2^*, respectively. Subtracting equations (24) we observe that $\bar{c} = c_1 - c_2$ satisfies

$$D(\nabla \bar{c}, \nabla \xi) + b(u_1, c_1, \xi) - b(u_2, c_2, \xi) = (K(c_1^*)\nabla p_1 - K(c_2^*)\nabla p_2, \nabla \xi)$$

for every $\xi \in X_h$. Taking $\xi = \bar{c}$ we obtain, due to (11) and imbedding $H^1 \hookrightarrow L^6$,

$$D|\bar{c}|_1^2 \leq |b(u_1 - u_2, c_1, \bar{c})| + |\bar{c}|_1 \text{Const}\, (L_K ||\nabla p_2||_{0,3} ||c_1^* - c_2^*||_1 + |K|_\infty ||p_1 - p_2||_1) \tag{26}$$

Using inverse inequalities (Ciarlet, 1991) and estimates on solutions of discrete Stokes equations (Girault and Raviart, 1986), we obtain

$$D||\bar{c}|| \leq M||c_1^* - c_2^*||_1,$$

with constant $M = M(h, f, g, \tilde{c_0}, \nu, K, c_1^*)$, whence the continuity of Φ. Next, we show that, for sufficiently small data (in the sense of Theorem 1), there exists a ball $\mathcal{M} \subset X_h$ such that $\Phi(\mathcal{M}) \subset \mathcal{M}$ and the diameter of $\mathcal{M}$ is *independent* of h.

We estimate $||\Phi(c_h^* - \tilde{c_0})||_1$ in terms of $||c_h^* - \tilde{c_0}||_1$, obtaining

$$||\Phi(c_h^* - \tilde{c_0})||_1 \leq A \cdot ||c_h^* - \tilde{c_0}||_1 + B,$$

where

$$A = \frac{\text{Const}_3}{D}\left(\frac{||\tilde{c_0}||_1}{\nu} + |K|_\infty\right)||g||_{0,3},$$

$$B = \frac{\text{Const}}{D}\left(\frac{||\tilde{c_0}||_1}{\nu} + |K|_\infty\right)||f||_{0,3} + (A+1)||\tilde{c_0}||_1.$$

Now, we take any positive real A_0 such that $A_0 < 1$. We define polynomial P_1 as

$$P_1(x_1, x_2, \ldots, x_6) = \frac{\text{Const}_3}{A_0} x_1 x_5 (x_2 x_6 + x_3).$$

Suppose $P_1(D^{-1}, \nu^{-1}, |K|_\infty, ||f||_{0,3}, ||g||_{0,3}, ||c_0||_1) \leq 1$, that is, the data are small enough in the sense of Theorem 1. Then there exists $r > 0$, such that Φ maps the ball $\mathcal{M} = \{\xi \in X_h : ||\xi||_1 \leq r\}$ into itself. Let us stress that this invariant ball is independent of the mesh parameter h, so Brouwer's fixed point theorem yields the existence of solutions of Problem 2 for *any* value of h.

4 ERROR ESTIMATE

In this section we outline the derivation of estimate (21). We shall assume that Problem 1 admits solution $(u, p, c) \in (H^3 \cap V) \times (H^2 \cap W) \times (H^2 \cap X(\tilde{c_0}))$. Without loss of generality we can assume that the boundary condition on c is represented exactly by $\tilde{c_0}$, since the solutions of Problem 1 depend continuously (Lukaszewicz, 1991) on $\tilde{c_0}$, and $\tilde{c_0}$ approximates c_0 on the boundary in $H^{1/2}$-norm with order $O(h)$.

Suppose there exists a solution (u_h, p_h, c_h) of Problem 2. Using standard procedures and approximation results on discrete solutions of Stokes equations (Girault and Raviart, 1986), (Bercovier and Pironneau, 1979) we obtain

$$||u - u_h||_1 \leq O(h^2) + \frac{\text{Const}}{\nu}||g||_{0,3}||c - c_h||_{0,6} \tag{27}$$

$$||p - p_h||_1 \leq O(h) + \text{Const}\, ||g||_{0,3}||c - c_h||_{0,6} \tag{28}$$

and

$$\begin{aligned} D||c_h - c||_1 \;\; \leq \;\; & (2D + \text{Const}\, ||u_h||_1)h||c||_2 + L_K \text{Const}\,_4 |p|_{1,4}||c - c_h||_1 \\ & + \text{Const}\,_5 ||c||_1 ||u - u_h||_1 + \text{Const}\,_6 |K|_\infty ||p - p_h||_1. \end{aligned}$$

Substituting (27) and (28) and using imbedding $H^k \hookrightarrow W^{k-1,6}$, $k = 1, 2$, we conclude that if the data are small in the sense of Theorem 1, namely, if

$$D > (L_K \text{Const}_4 \text{Const}_1 + \frac{\text{Const}_5}{\nu} \text{Const}_1 ||g||_{0,3} + |K|_\infty \text{Const}_6 ||g||_{0,3}),$$

then $||c - c_h||_1 = O(h)$, whence, using (27) and (28) again, we obtain (21).

Remark 1 *In view of approximation properties of V_h, it seems that the approximation error in u_h is not optimal in h (contrary to $||p - p_h||_1$ and $||c - c_h||_1$, which are of optimal order). However, there are chances that the actual order of approximation of the velocity may be improved, or, in case of more regular data, even restored to optimal level, since by (27) this error depends on a c_h error in a norm which doesn't involve derivatives.*

5 UNIQUENESS OF APPROXIMATE SOLUTIONS

We estimate the difference between two possibly different solutions $c_1, c_2 \in \mathcal{M}$ for the same data. The ball $\mathcal{M}$ is defined as in Section 3.

Using (26), (21) and estimates of discrete solutions to Stokes equations, we conclude that, for data and h sufficiently small in the sense of Theorem 1, the difference $\bar{c} = c_1 - c_2$ satisfies $||\bar{c}||_1 \leq 0$, so the solution must be unique.

ACKNOWLEDGMENT

I would like to thank Prof. M. Dryja for his valuable remarks and comments.

REFERENCES

Adams, R.A. (1975) *Sobolev Spaces*, Academic Press, New York.

Brezzi F., Fortin M. (1991) *Mixed and Hybrid Finite Element Methods*, Springer-Verlag, New York.

Bercovier, M. and Pironneau, O. (1979) Error estimates for Finite Element Method solution of the Stokes problem in the primitive variables, *Num. Math.*, **33**, 211-224.

Ciarlet, P.G. (1991) Basic Error Estimates for Elliptic Problems, *Handbook of Numerical Analysis, vol II, Finite Element Methods (Part I)*, Elsevier Science Publishers B.V. (North-Holland).

Gilbarg, D. and Trudinger, N.S. (1983) *Elliptic Partial Differential Equations of Second Order*, Springer-Verlag, Berlin.

Girault, V. and Raviart, P.A. (1986) *Finite Element Method for Navier–Stokes Equations. Theory and Algorithms*, Springer-Verlag, Berlin Heidelberg New York.

Hackbush, W. (1992) *Elliptic Differential Equations. Theory and Numerical Treatment*, Springer-Verlag, Berlin Heidelberg.

Kellogg, R.B. and Osborn, J.E. (1976) A regularity result for the Stokes problem in a convex polygon, *J. Funct. Anal.*, **21**, 397-431.

Krzyżanowski, P. (1994) On stationary flow of asymmetric fluids with diffusion, *Math. Meth. Appli. Sci.*, **17** 837-854.

Łukaszewicz, G. (1991) On diffusion in viscous fluids. Mixed boundary conditions, *Proc. 'First European Conference on Elliptic and Parabolic Problems', Pont-a-Mousson, June 1991.*

Petrosyan, L.G. (1984) *Some Problems of Mechanics of Fluids with Antisymmetric Stress Tensor*, Izd. Erev. Univ., Erevan (in Russian).

Popel, A.S., Regirer, S.A. and Usick, P.I. (1974) A continuum model of blood flow, *Biorheology*, **XI**, 427-437.

Temam, R. (1979) *Navier – Stokes Equations. Theory and Numerical Analysis*, North-Holland, Amsterdam New York Oxford.

32

Additive Schwarz method with strip substructures

Monika Mróz
Institute of Applied Mathematics and Mechanics, Faculty of Mathematics, Computer Science and Mechanics, Warsaw University, Banacha 2, 02-097 Warszawa, Poland, tel/fax +48 2 6583236.
E-mail: monika@appli.mimuw.edu.pl

Abstract

Domain Decomposition Method for finite element discretization of elliptic problems with discontinuous coefficients is analyzed. The domain is divided into overlapping strip shaped subdomains for 2-D problems and slice or pillar shape subdomains in 3-D case. The coefficients of the problem are discontinuous and constant on these subdomains. The approximate solution is obtained iteratively by solving local problems associated with each subdomain and the global problem associated with the coarse triangulation.

The framework for Schwarz methods, as developed by Dryja and Widlund (1990), is used. The estimates of convergence are established for strips in 2-D and slices or pillars in 3-D, which are independent of the jumps of coefficients and discretization parameters.

Keywords

Elliptic problems, domain decomposition, iterative methods, Schwarz method

1 INTRODUCTION

Domain Decomposition Methods are very powerful iterative technique for solving linear systems arising from discretization of partial differential equations. We apply this technique in order to solve elliptic equations with discontinuous and piecewise constant coefficients. A two level nested triangulation is defined. The original problem is discretized with respect to the fine triangulation. The domain is divided into a number of subdomains with boundaries matching the fine triangulation. The arising linear system is solved by an iterative method. Each iteration step consists of solving a number of linear systems, which correspond to restrictions of the original problem to subdomains and a small global problem defined on the coarse triangulation.

There are two classes of Domain Decomposition Methods: Schwarz methods (for overlapping subdomains) and Substructuring methods (for non overlapping subdomains). In this paper we construct and analyze the Additive Schwarz Algorithms in the case when the domain is divided into strip shape subdomains (i.e. strips in 2–D and slices or pillars in 3–D).

The partition of the domain into strips has several advantages. The bandwidth of local matrices is narrow, which minimizes computations and memory requirements. Also the structure of local problems is useful for vectorization of an algorithm. We also have better estimates on convergence than for partition into boxes, since less number of subdomains intersect with each other.

The special attention is paid to the problems with coefficients constant on these subdomains, thus these algorithms may be used for solving real problems arising in mechanics, fluid dynamics and similar areas.

For construction of Additive Schwarz Algorithms we use the framework for Schwarz methods, as developed by Dryja and Widlund (1990). The convergence of algorithms is independent of jumps of the coefficients.

Optimal estimates on convergence with respect to parameters of discretization is established in the case of strips in 2–D and slices in 3–D. The analysis of these cases is very similar to analysis for boxes cf. Dryja, Sarkis and Widlund (1994). In the case of pillars in 3–D the convergence is bounded by $(1 + \ln(H/h))$. This result is better then known results for boxes, cf. Dryja, Sarkis and Widlund (1994).

Results of the related numerical experiments can be found in Mróz (1995).

2 MODEL PROBLEM

We consider the problem of finding an approximate solution of the following elliptic boundary value problem.

For given a bilinear form $a(\cdot,\cdot)$ and a linear functional $l(\cdot)$ on $H_0^1(\Omega)$ we want to find $u \in H_0^1(\Omega)$ such that

$$a(u,v) = l(v) \quad \forall v \in H_0^1(\Omega)\ , \tag{1}$$

where Ω is a Lipschitz bounded domain in R^2 or R^3. For simplicity of presentation we assume that Ω is a polygon in R^2 or polyhedron in R^3.

The bilinear form $a(\cdot,\cdot)$ is defined as follows

$$a(u,v) = \int_\Omega \rho(x) \sum_{i=1}^{2} \frac{\partial u}{\partial x_i}\frac{\partial v}{\partial x_i} dx\ . \tag{2}$$

The function $\rho(x)$ is piecewise constant i.e.

$$\rho(x) = \rho_i > 0\ , \quad x \in \Omega_i\ , \tag{3}$$

where Ω_i denote the strip shape subdomain consisting of elements of coarse triangulation (see Section 4). The jumps of coefficients between subdomains may be large. This model problem can be applied to the case when the function $\rho(x)$ varies moderately on each subdomain and is discontinuous between subdomains. If $\rho \equiv 1$ we have the case of Poisson's equation.

Let $l(v)$ denote the linear form defined by

$$l(v) = (f,v)_{L^2(\Omega)} = \int_\Omega fv dx\ .$$

The bilinear forms $a(\cdot,\cdot)$ define scalar product on $H_0^1(\Omega)$.

3 FINITE ELEMENT APPROXIMATION

A two level triangulation is defined on the domain Ω. First, we construct a coarse triangulation Ω_H that consists of shape regular, quasi uniform (cf. Ciarlet (1978)), non overlapping simplices Ω^j of diameter of order H. In second step, we further divide each element of the triangulation Ω_H into smaller, shape regular, quasi uniform simplices of diameter $O(h)$. They form the fine triangulation Ω_h.

Spaces of piecewise linear, continuous functions on Ω_H and Ω_h are denoted by $V^H(\Omega)$ and $V^h(\Omega)$. The restriction to subspaces of functions vanishing on $\partial\Omega$ is denoted by $V_0^h(\Omega)$ and $V_0^H(\Omega)$ respectively.

A finite element formulation is obtained by projecting the space $H_0^1(\Omega)$ onto its finite dimensional subspace $V_0^h(\Omega)$. The corresponding approximate problem for (1) is then:

Find $u^* \in V_0^h(\Omega)$ such that

$$a(u^*, v) = l(v) \quad \forall v \in V_0^h(\Omega) \; . \tag{4}$$

Let $\{\phi_j^h\}$ be the set of standard, piecewise linear, nodal basis functions, thus $V_0^h(\Omega) = span\{\phi_j^h\}$. In this basis, the discrete variational problem (4) can be rewritten as a system of linear equations

$$\mathrm{Au} = \mathrm{f} \; , \tag{5}$$

where coefficients of the matrix A and the vector f are given by

$$\mathrm{A}_{jl} = a(\phi_j^h, \phi_l^h) \; , \quad \mathrm{f}_j = l(\phi_j^h) \; .$$

The matrix A is positive definite and symmetric. The condition number of A is proportional to $\frac{\max_j \rho^j}{\min_j \rho^j} h^{-2}$.

4 ADDITIVE SCHWARZ METHOD

In this section we present a construction of the Additive Schwarz Method with overlaps for model problem (1). We use the abstract framework of Schwarz method developed by Dryja and Widlund (cf. Dryja and Widlund (1990) or Dryja and Widlund (1994)).

The domain Ω is divided into n subdomains Ω_i, $i = 1, \ldots, n$. In 2–D case the subdomains are called strips. We assume that the boundary of each strip consists only of sides of elements from the coarse triangulation Ω_H. The strip Ω_i has common boundary with at most two neighboring strips.

In similar way we define slices in 3–D case. The boundary of each slice consists only of faces and edges of elements from the coarse triangulation Ω_H. Each slice has a common boundary with at most two neighboring slices.

The partition into 3–D pillars is defined in the following way. We first divide the domain Ω into slices. Each slice is further divided into pillars according to the same rules as for division into slices. We assume that the interface between two pillars is empty or consists of an edge of pillar (a number of adjacent edges of the elements of Ω_H) or face of a pillar (a number of adjacent sides of elements of Ω_H). The common interface between more then two pillars consists of edge or is empty. We assume that the boundary points of each edge lies on $\partial\Omega$.

Each subdomain Ω_i is extended to Ω_i' such that the distance between $\partial\Omega_i$ and $\partial\Omega_i'$ is positive and equal to $\delta_i < H$. We now restrict each Ω_i' to Ω. The boundaries of these subdomains match with the fine triangulation Ω_h. The parameter δ denotes the minimal overlap,

$$\delta = \min_i \delta_i \,.$$

The Additive Schwarz Method is described in the terms of subspaces, variational forms and projections. The space $V_0^h(\Omega)$ is decomposed into a sum of $N+1$ subspaces V_i.

$$V_0^h(\Omega) = \sum_{i=0}^{N} V_i \,.$$

The subspace $V_0 = V_0^H(\Omega)$ is defined in Section 3. Each of the remaining subspaces V_i is related to the subdomain Ω_i',

$$V_i = V_0^h(\Omega) \cap H_0^1(\Omega_i') \,,$$

and extended by zero off Ω_i'.

Let P_i, $i = 1, \ldots, N$, denote the orthogonal global projections from $V_0^h(\Omega)$ to V_i with respect to bilinear form $a(\cdot,\cdot)$, i.e.

$$a(P_i u, v) = a(u, v) \quad \forall v \in V_i \,.$$

The mapping P from $V_0^h(\Omega)$ into $V_0^h(\Omega)$ is defined as the sum of projections P_i,

$$P = \sum_{i=0}^{N} P_i \,. \tag{6}$$

The operator P is invertible (see Theorems 1 and 2), thus (4) is equivalent to the auxiliary problem of finding $u^* \in V_0^h(\Omega)$ which satisfies

$$Pu^* = g \,. \tag{7}$$

if $g = \sum_{i=0}^{N} g_i$, where $g_i = P_i w$ is the solution of the equation

$$a(g_i, v) = l(v) \quad \forall v \in V_i$$

since

$$a(g_i, v) \equiv a(P_i u^*, v) = a(u^*, v) \equiv l(v) \quad \forall v \in V_i \,.$$

It should be pointed out that g can be found without knowing the solution of (1).

The operator P is well conditioned (see Theorems 1 and 2) and the right hand side is known, thus one can use the preconditioned conjugate gradient (PCG) method (cf. Concus, Golub, O'Leary (1976)) to solve (7).

5 IMPLEMENTATION

In each step of the PCG method a vector is multiplied by the matrix P corresponding to the operator P from (6). We describe the algorithm that determines the function $v \in V_0^h(\Omega)$

$$v = Pu\,,$$

where $u \in V_0^h(\Omega)$ is a given function.

Algorithm

1. *Find projections $P_i u$, $i = 1, \cdots, N$ by solving*

$$a(P_i u, \phi_j^h) = a(u, \phi_j^h) \quad \forall \phi_j^h \in V_i = V_0^h(\Omega_i')\,.$$

These systems correspond to solving the subproblems individually for each extended subdomain Ω_i' with homogeneous Dirichlet boundary conditions on $\partial\Omega_i'$.

2. *Find the projection P_0 by solving the global system*

$$a(P_0 u, \phi_j^H) = a(u, \phi_j^H) \quad \forall \phi_j^H \in V_0 = V_0^H(\Omega)\,.$$

3. *Compute the vector v from*

$$v = \sum_{i=0}^{N} P_i u\,.$$

This algorithm admits easy parallelization since the systems form Step 1 and 2 are independent. If we assign $N+1$ processors to the algorithm , then we achieve the maximal degree of parallelism. If the number of unknowns in the global system (Step 2) is of the same order as in the strip, then we have balanced loads on each processor.

Let us now describe the matrix representation of the operator P defined by (6). Computing the projection $P_i u$, $i = 1, \ldots, N$ of an arbitrary function u involves the solution of linear system defined only for nodes of interiors of Ω_i'. Let us denote by A_i the matrix that corresponds the variational problem (4) defined on the subdomain Ω_i'. The matrix representation of local projection P_i, $i = 1, \ldots, N$ is given by

$$\mathrm{P}_i = \mathrm{R}_i^T \mathrm{A}_i^{-1} \mathrm{R}_i \mathrm{A} = \mathrm{B}_i \mathrm{A}$$

where R_i is the restriction operator which returns only those unknowns which are associated with the interior nodes of Ω_i'. The matrix representation of projection P_0 involves solving the global system defined on the coarse triangulation Ω_H. Let us assume that A_0 denotes the stiffness matrix for the original problem constructed by the basis functions of $V_0^H(\Omega)$. The matrix R_0^T represents the linear interpolation from $V^H(\Omega)$ to $V^h(\Omega)$, thus the projection matrix P_0 is of the form

$$\mathrm{P}_0 = \mathrm{R}_0^T \mathrm{A}_0^{-1} \mathrm{R}_0 \mathrm{A} = \mathrm{B}_0 \mathrm{A}\,.$$

The matrix representation of the problem (7) is the following system

$$\mathrm{BAu} = \mathrm{Bf} \tag{8}$$

where

$$\mathrm{B} = \sum_{i=0}^{N} \mathrm{B}_i\,.$$

Thus the matrix B can be interpreted as a inverse of a preconditioner for matrix A.

6 CONVERGENCE ESTIMATES

In this section we formulate theorems, that gives us estimates on condition number of the operator P, see (6). The proof of these theorems can be found in Mróz (1995). We first establish estimates in the case of 2–D strips or 3–D slices.

Theorem 1 *The operator $P : V_0^h(\Omega) \to V_0^h(\Omega)$ defined by (6) (in the case of 2–D strips or 3–D slices) is symmetric and positive definite in the scalar product $a(\cdot,\cdot)$, and hold*

$$m\left(1+\frac{H}{\delta}\right)^{-1} a(u,u) \le a(Pu,u) \le M a(u,u) \quad \forall u \in V_0^h(\Omega)\,, \tag{9}$$

where positive constants m and M are independent of H, h, overlap δ, and the jumps of coefficient.

As a consequence of this theorem, the condition number of BA, see (8), is proportional to

$$cond(\mathrm{BA}) \sim \left(1+\frac{H}{\delta}\right)\,.$$

Note that if δ is bounded from below by fixed fraction of H, then the algorithm is optimal.

In the case of partitioning into 3–D pillars we have the following estimate

Theorem 2 *The operator $P : V_0^h(\Omega) \to V_0^h(\Omega)$ defined by (6) (in the case of 3–D pillars) is symmetric and positive definite in scalar product $a(\cdot,\cdot)$, and hold*

$$m\left(1+\frac{H}{\delta}\right)^{-1}\left(1+\ln\frac{H}{h}\right)^{-1} a(u,u) \le a(Pu,u) \le M\, a(u,u) \quad \forall u \in V_0^h(\Omega)\,, \tag{10}$$

where constants m and M are independent of H, h, the overlap δ and the jumps of coefficient.

The condition number of BA defined by (8) in the case of pillars is proportional to

$$cond(\mathrm{BA}) \sim \left(1 + \frac{H}{\delta}\right)\left(1 + \ln\frac{H}{h}\right).$$

7 REMARKS

The algorithm presented in this paper has one disadvantage. The coefficients of local matrices A_i, see Section 5 are not constant. In order to omit such disadvantage we can define the extended subdomain Ω_i' as a union of two subdomains Ω_i and Ω_{i+1}. Now the local problem is corresponds to two adjacent subdomains. Now we can use any substructuring methods for two subdomains, cf. Mróz (1995). Similar algorithms are described in Nepomnyaschikh (1992).

The algorithm presented in this paper can be easily extended for parabolic problems.

REFERENCES

Ciarlet, P. G. (1978) *The Finite Element Method for Elliptic Problems.* North–Holland.

Concus, P., Golub, G. H. and O'Leary,D. P. (1976) A generalized conjugate gradient method for the numerical solution of elliptic PDE, in *Sparse Matrix Computations* (ed. J. R. Bunch and D. J. Rose). Academic Press, N.J.

Dryja, M., Widlund, O. B. (1990) Towards a Unified Theory of Domain Decomposition Algorithms for Elliptic Problems, in *Third International Symposium on Domain Decomposition Methods for Partial Differential Equations, held in Houston, Texas, March 20-22, 1989* (ed. T. Chan and others) SIAM, Philadelphia, PA.

Dryja, M., Widlund, O. B. (1994) Domain Decomposition Algorithms with Small Overlap. *SIAM J. Sci.Comput.,* **15**.

Dryja, M, Sarkis, M., Widlund, O. B.(1994) Multilevel Schwarz Methods for Elliptic Problems with Discontinuous Coefficients in Three Dimensions. *Numer. Math.* to appear.

Mróz, M.(1995) Domain Decomposition Methods with Strip Substructures for Finite Element Elliptic and Parabolic Problems, PHD Thesis, Institute of Applied Mathematics and Mechanics, Warsaw Unversity.

Nepomnaschikh, S. V.(1992) Domain Decomposition Method for the Elliptic Problem with Jumps in the Coefficients in Thin Strips, *Siberian Journal of Comp. Math.,* **1**.

33
Modelling in numerical simulation of electromagnetic heating

*J. Rappaz and M. Swierkosz**
Department of Mathematics, Swiss Federal Institute of Technology
CH–1015 Lausanne, Switzerland. Phone: +41 21 693 25 55.
Fax: +41 21 693 43 03. E–mail: marek@dma.epfl.ch

Abstract
This paper deals with numerical simulation of induction heating for tri-dimensional time-varying axisymmetric geometries. The modelling used for the eddy current solver is presented in detail, from physical equations to a numerical scheme.

Keywords
Numerical simulation, electromagnetic heating, coupled problems, finite element method, boundary element method

1 INTRODUCTION

Induction heating involves both electromagnetic and thermal phenomena, described by coupled nonlinear partial differential equations. An induction heating setup usually consists of one or several inductors and one or several workpieces to be heated (Figure 1). The inductors may move with respect to the workpieces.

The alternating current flowing through the inductor generates a rapidly oscillating magnetic field. This in turn induces eddy currents inside the workpiece, which results in temperature increase due to the Joule effect and to the hysteresis effect.

In this research, we consider axisymmetric induction heating setups. The aim is to obtain a model which leads to an efficient numerical simulation of the induction heating phenomenon. It is assumed that the coils are supplied with a sinusoidal alternating current, so as to obtain a steady-state electromagnetic problem. The total voltage in each coil v_k and the angular frequency ω are considered to be given.

*This research was supported by the Swiss "Nationaler Energie–Forschungs–Fonds" and performed in collaboration with the company Amysa Yverdon SA, Switzerland.

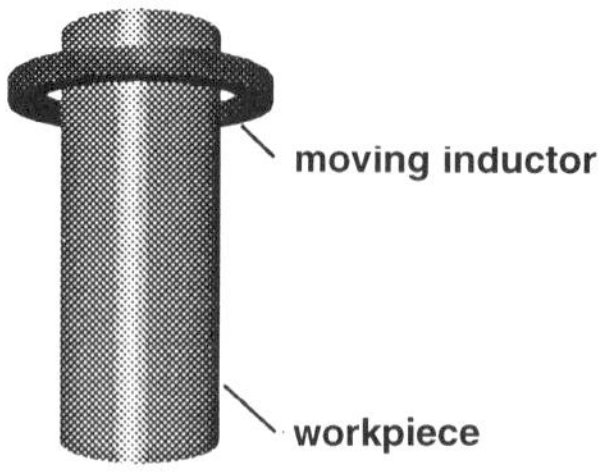

Figure 1 A sample induction heating setup

2 MATHEMATICAL MODEL

2.1 Scalar magnetic potential formulation

Let us consider an axisymmetric induction heating setup consisting of N conductors (inductors and workpieces). Let $(\Delta_i)_{i=1}^N$ be N bounded open sets of $\mathbf{R}^3$ corresponding to the areas in space occupied by the conductors. These sets are obtained by revolution of N open simply connected sets $\Omega_i \subset \mathbf{R}^2$, $i = 1, \ldots, N$, around a straight line that we consider to be the Oz axis of a Cartesian coordinate system (x, y, z). Let us denote by Δ the union of the sets Δ_i: $\Delta = \cup_{i=0}^N \Delta_i$, by Ω the union of the sets Ω_i: $\Omega = \cup_{i=0}^N \Omega_i$ and by Δ' the complementary of $\overline{\Delta}$ in $\mathbf{R}^3$ ($\overline{\Delta}$ denotes the adherence of Δ).

The starting point of our modelling are Maxwell equations with displacement currents neglected, and Ohm law. The following equations hold in whole $\mathbf{R}^3$:

$$\begin{aligned} \mathrm{div}\mathbf{B} &= 0, && (1) \\ \mathbf{curl}\,\mathbf{E} &= -\frac{\partial \mathbf{B}}{\partial t}, && (2) \\ \mathbf{curl}\,\mathbf{H} &= \mathbf{j}, && (3) \\ \nu \mathbf{B} &= \mathbf{H}, && (4) \end{aligned}$$

Moreover, in Δ, we have:

$$\mathbf{j} = \sigma \mathbf{E}. \qquad (5)$$

Here t denotes the time, $\mathbf{E}$ the electric field, $\mathbf{H}$ the magnetic field, $\mathbf{B}$ the magnetic induction, $\mathbf{j}$ the electric current density, σ the electric conductivity, and ν the inverse of the magnetic permeability μ (called magnetic reluctivity). For the moment, we assume that the values of σ, μ and ν do not depend on the time. In the reality, they will usually vary with the temperature and to a certain extent with the magnetic field.

Let us now consider a cylindrical coordinate system (r, θ, z) with its associated natural tangent reference system $(\mathbf{e}_r, \mathbf{e}_\theta, \mathbf{e}_z)$, The following assumptions are made:

1. The fields $\mathbf{B}$, $\mathbf{H}$, $\mathbf{E}$ are invariant along θ, i.e. their components in the reference system $(\mathbf{e}_r, \mathbf{e}_\theta, \mathbf{e}_z)$ do not depend on θ.
2. The electric current density is of the form $\mathbf{j} = \mathbf{j}(\mathbf{r}, \mathbf{z})\mathbf{e}^{\mathbf{i}\omega \mathbf{t}}\mathbf{e}_\theta$, where $j : (r, z) \in \mathbf{R}^+ \times \mathbf{R} \mapsto \mathbf{j}(\mathbf{r}, \mathbf{z}) \in \mathbf{C}$ is some complex-valued function.

It will also be assumed that there are no surface currents i.e. no Dirac δ-like current "concentration" on the surface of the conductors.

Suppose that a periodic voltage of the form $v_k e^{i\omega t}$ is imposed in each of those sets Δ_k which are not simply connected. Those sets are necessarily toroidal and correspond to the inductor coils. In those sets Δ_k that are simply connected, we shall set $v_k \equiv 0$ by convention. Due to the linearity of the problem (with constant coefficients σ, μ and ν), we can look for the fields $\mathbf{B}$, $\mathbf{H}$, $\mathbf{E}$ in the form:

$$\mathbf{B} = \mathbf{B}(\mathbf{r}, \mathbf{z})\mathbf{e}^{\mathbf{i}\omega\mathbf{t}}, \qquad \mathbf{H} = \mathbf{H}(\mathbf{r}, \mathbf{z})\mathbf{e}^{\mathbf{i}\omega\mathbf{t}}, \qquad \mathbf{E} = \mathbf{E}(\mathbf{r}, \mathbf{z})\mathbf{e}^{\mathbf{i}\omega\mathbf{t}}. \tag{6}$$

Here $\mathbf{B} : (\mathbf{r}, \mathbf{z}) \in \mathbf{R}^+ \times \mathbf{R} \mapsto \mathbf{B}(\mathbf{r}, \mathbf{z}) \in \mathbf{C}^3$, $\mathbf{H} : (\mathbf{r}, \mathbf{z}) \in \mathbf{R}^+ \times \mathbf{R} \mapsto \mathbf{H}(\mathbf{r}, \mathbf{z}) \in \mathbf{C}^3$, $\mathbf{E} : (\mathbf{r}, \mathbf{z}) \in \mathbf{R}^+ \times \mathbf{R} \mapsto \mathbf{E}(\mathbf{r}, \mathbf{z}) \in \mathbf{C}^3$ are complex vector fields to be found. In the sequel, we shall omit the term $e^{i\omega t}$.

Our aim is now to show that under the above assumptions, the magnetic induction $\mathbf{B}$ can be expressed in terms of a scalar potential $\phi : (r, z) \in \mathbf{R}^+ \times \mathbf{R} \mapsto \phi(\mathbf{r}, \mathbf{z}) \in \mathbf{C}$.

Let us consider equation (3) expressed in cylindrical coordinates. According to Assumption 1, the components of $\mathbf{H}$ in the natural tangent reference system are $\mathbf{H} = (\mathbf{H}_\mathbf{r}(\mathbf{r}, \mathbf{z}), \mathbf{H}_\theta(\mathbf{r}, \mathbf{z}), \mathbf{H}_\mathbf{z}(\mathbf{r}, \mathbf{z}))$.

Therefore, equation (3) can be rewritten in the form

$$\left(-\frac{\partial H_\theta}{\partial z}\right)\mathbf{e}_r - \left(\frac{\partial H_z}{\partial r} - \frac{\partial H_r}{\partial z}\right)\mathbf{e}_\theta + \left(\frac{1}{r}\frac{\partial (rH_\theta)}{\partial r}\right)\mathbf{e}_z = \mathbf{j}. \tag{7}$$

Using assumption 2, we get the following system:

$$-\frac{\partial H_\theta}{\partial z} = 0, \tag{8}$$

$$-\frac{\partial H_z}{\partial r} + \frac{\partial H_r}{\partial z} = j(r, z), \tag{9}$$

$$\frac{1}{r}\frac{\partial (rH_\theta)}{r} = 0. \tag{10}$$

Equation (8) implies that $H_\theta = H_\theta(r)$, and from (10) we get $H_\theta = \dfrac{c}{r}$, where c is a constant. If c were nonzero, we would have $\lim_{r \to 0} H_\theta = \infty$, which is absurd. Therefore, the magnetic field $\mathbf{H}$ has the form

$$\mathbf{H} = \mathbf{H}_\mathbf{r}(\mathbf{r}, \mathbf{z})\mathbf{e}_\mathbf{r} + \mathbf{H}_\mathbf{z}(\mathbf{r}, \mathbf{z})\mathbf{e}_\mathbf{z}. \tag{11}$$

This result and equation (4) together imply that the magnetic induction $\mathbf{B}$ also has the form

$$\mathbf{B} = \mathbf{B_r}(\mathbf{r},\mathbf{z})\mathbf{e_r} + \mathbf{B_z}(\mathbf{r},\mathbf{z})\mathbf{e_z}. \tag{12}$$

Equation (1) yields thus

$$\frac{\partial}{\partial r}(rB_r) + \frac{\partial}{\partial z}(rB_z) = 0. \tag{13}$$

Equation (13) states the fact that the field $(r,z) \in \mathbf{R}^+ \times \mathbf{R} \mapsto (\mathbf{rB_r}, \mathbf{rB_z}) \in \mathbf{C}^2$ is divergence–free when we consider (r,z) as Cartesian coordinates. A well-known result (see e.g. Dautray and Lions, 1988) allows us to conclude that this field can be expressed in terms of the curl of a scalar function. In other words, there exists a function $\psi : (r,z) \in \mathbf{R}^+ \times \mathbf{R} \mapsto \psi(\mathbf{r},\mathbf{z}) \in \mathbf{C}$ such that:

$$rB_r = -\frac{\partial\psi}{\partial z}, \qquad rB_z = \frac{\partial\psi}{\partial r}. \tag{14}$$

Let $\phi : (r,z) \in \mathbf{R}^+ \times \mathbf{R} \mapsto \phi(\mathbf{r},\mathbf{z}) \in \mathbf{C}$ be defined by $\phi(r,z) = \frac{1}{r}\psi(r,z)$. Equations (14) can then be rewritten in the form:

$$B_r = -\frac{\partial\phi}{\partial z}, \qquad B_z = \frac{1}{r}\frac{\partial(r\phi)}{\partial r}. \tag{15}$$

We conclude that there exists a vector magnetic potential $\mathbf{A}$ of the form

$$\mathbf{A}(\mathbf{r},\mathbf{z}) = \phi(\mathbf{r},\mathbf{z})\mathbf{e}_\theta, \tag{16}$$

such that

$$\mathbf{B} = \mathbf{curl}\,\mathbf{A}. \tag{17}$$

Moreover, we clearly have

$$\mathrm{div}\mathbf{A} = \mathbf{0}, \tag{18}$$

since $\mathrm{div}\mathbf{A} = \frac{\mathbf{1}}{\mathbf{r}}\frac{\partial\phi}{\partial\theta}$. Using Biot–Savart law, one can show that $\mathbf{B} \sim \dfrac{\mathbf{1}}{(\mathbf{r}^2+\mathbf{z}^2)^{\frac{3}{2}}}$ when $(r^2+z^2)^{\frac{1}{2}} \to \infty$. In (14), ψ is defined up to a constant. Consequently, ϕ can be chosen so that $\phi \sim \dfrac{1}{r^2+z^2}$ when $(r^2+z^2)^{\frac{1}{2}}$ tends to infinity, or, in other words,

$$\phi = O\left(\frac{1}{r^2+z^2}\right) \text{when} |\mathrm{r}| + |\mathrm{z}| \to \infty. \tag{19}$$

Taking into account (17), (4) and Assumption 2, equation (3) yields

$$\mathbf{curl}\,(\nu\mathbf{curl}\,\mathbf{A}) = \mathbf{j}\mathbf{e}_\theta, \tag{20}$$

which can be expanded to the form

$$-\left(\frac{\partial}{\partial r}\left(\frac{\nu}{r}\frac{\partial (r\phi)}{\partial r}\right)+\frac{\partial}{\partial z}\left(\nu\frac{\partial \phi}{\partial z}\right)\right)\mathbf{e}_\theta=\mathbf{j}. \tag{21}$$

This result holds both inside the conductors and outside them.

Outside the conductors, j is zero and ν is constant, so that (21) yields

$$\frac{\partial}{\partial r}\left(\frac{1}{r}\frac{\partial (r\phi)}{\partial r}\right)+\frac{\partial^2 \phi}{\partial z^2}=0. \tag{22}$$

This equation does not imply that $\Delta\phi=0$. However, multiplying it by $\sin\theta$, we get $\Delta\left(\phi\sin\theta\right)=0.$ (23)

2.2 Eddy current model

So far, we have expressed the magnetic field and the magnetic induction in terms of a scalar magnetic potential ϕ. Our aim is now to find a relationship between the current density j, the potential ϕ, and the voltage v_k imposed in the conductor.

From Ohm law (5) and Assumption 2, we get that $\mathbf{E}=\mathbf{E(r,z)}\mathbf{e}_\theta$ in each open set Δ_k (corresponding to the location of a conductor). By rewriting the equation (2) in cylindrical coordinates, taking into account (6) and (12), we get that

$$i\omega(B_r\mathbf{e}_r+B_z\mathbf{e}_z)+\left(-\frac{\partial E}{\partial z}\right)\mathbf{e}_r+\left(\frac{1}{r}\frac{\partial (rE)}{\partial r}\right)\mathbf{e}_z=0, \tag{24}$$

i.e. $i\omega B_r=\dfrac{\partial E}{\partial z}$, $i\omega B_z=-\dfrac{1}{r}\dfrac{\partial (rE)}{\partial r}$. Substituting (15) into these equations, we get

$$\frac{\partial}{\partial z}\left(rE+i\omega r\phi\right)=0,\qquad \frac{\partial}{\partial r}\left(rE+i\omega r\phi\right)=0. \tag{25}$$

Consequently, there exist constants $c_k\in\mathbf{C},\mathbf{k}=\mathbf{0},..,\mathbf{N}$ such that

$$rE+i\omega r\phi=c_k\mathrm{in}\Delta_\mathrm{k},\qquad \mathrm{k}=0,..,\mathrm{N}. \tag{26}$$

Ohm law (5) yields

$$j=\sigma\left(-i\omega\phi+\frac{c_k}{r}\right),\qquad k=1,..,N. \tag{27}$$

Let Δ_k be a toroidal conductor to which a voltage v_k is applied. Then, there exists a disk Σ_k, centered on the Oz axis of the cylindrical coordinates system and orthogonal to this axis, such that the $\Delta_k\cup\Sigma_k$ is simply connected (Figure 2). The voltage v_k is then defined as a line integral of $\mathbf{E}+\mathbf{i}\omega\mathbf{A}$ along $\partial\Sigma_k$. Since $\mathbf{A}$ takes the form (16), we get: $v_k=\int_{\partial\Sigma_k}\left(E+i\omega\phi\right)\mathbf{e}_\theta\mathbf{d}\tau$. Taking into account (26) and the fact that $\mathbf{d}\tau=\mathbf{e}_\theta\mathbf{r}\mathbf{d}\theta$ in this case, we obtain: $v_k=\int_0^{2\pi}c_k\,d\theta=2\pi c_k$. Therefore, in any non simply connected set Δ_k, we have

$$j=\sigma\left(-i\omega\phi+\frac{v_k}{2\pi r}\right). \tag{28}$$

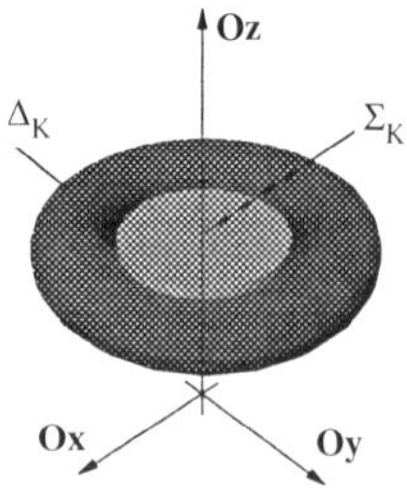

Figure 2 Application of a voltage to a toroidal conductor

Any simply connected axisymmetric set must have a non-empty intersection with the Oz axis, where r is zero. Since the current density cannot be infinite, equation (27) implies that c_k is zero in those of the sets Δ_k that are simply connected.

Let us remark that an inductor cannot be simply connected. In all workpieces, the voltage v_k is zero, which implies that c_k is zero also in those workpieces that are not simply connected.

Equations (28), (21) and Assumption 2 yield the following condition, valid in any conductor Δ_k:

$$-\left(\frac{\partial}{\partial r}\left(\frac{\nu}{r}\frac{\partial\,(r\phi)}{\partial r}\right)+\frac{\partial}{\partial z}\left(\frac{\nu}{r}\frac{\partial\,(r\phi)}{\partial z}\right)\right)+i\omega\sigma\phi=\frac{\sigma v_k}{2\pi r}. \tag{29}$$

2.3 Interface conditions

We shall now attempt to find conditions holding on the boundary of the conductors. Let $[f]$ denote the jump of a function f on $\partial\Delta$, and let $\mathbf{n}=(\mathbf{n_r},\mathbf{n_z})$ be the normal vector to $\partial\Delta$. Equations (16) and (17) imply that ϕ must be continuous. Since we assumed that there are no surface currents, we have $[\mathbf{H}\times\mathbf{n}]=\mathbf{0}$ on $\partial\Delta$. Since $\mathbf{H}\times\mathbf{n}=(\mathbf{H_z n_r}-\mathbf{H_r n_z})\mathbf{e}_\theta$, we get $[H_r n_z - H_z n_r]=0$. Equations (15) and (4) yield $\left[\frac{\nu}{r}\left(\frac{\partial(r\phi)}{\partial z}n_z+\frac{\partial(r\phi)}{\partial r}n_r\right)\right]=0$, or shortly $\left[\frac{\nu}{r}\frac{\partial(r\phi)}{\partial n}\right]=0$. We have thus the following interface conditions:

$$[\phi]=\left[\frac{\nu}{r}\frac{\partial(r\phi)}{\partial n}\right]=0. \tag{30}$$

In conclusion, the model consists of equation (22) (or 23) in Δ' (outside the conductors), equation (29) in Δ (inside the conductors), interface condition (30) and condition (19) at infinity.

3 NUMERICAL MODEL

Our aim is now to formulate a set of equations that could be used in a numerical simulation code. The difficulty lies in the fact that equation (22) is defined over an unbounded domain. Therefore a straightforward solution by finite element method is not possible.

Equation (22) could be solved over the whole unbounded domain using so-called "infinite elements". Another commonly found way of dealing with such problems is to approach the infinite domain by a "sufficiently large" finite domain, with suitable conditions at the boundary, and to solve the whole problem by standard finite element method.

In our case, these two solutions must be ruled out. In fact, the assumption is that inductors may move with respect to the workpieces. Therefore, the mesh outside the conductors would have to vary with the inductor movement. Apart from other considerations, the generation of such a mesh would lead to efficiency problems.

We opted therefore for a mixed boundary element – finite element formulation. The behaviour of the magnetic potential outside the conductors will be expressed in terms of integrals over their boundary, while inside the conductors a classical finite element formulation was opted for.

The key of the boundary element part is the so-called simple–double layer formulation. Let $G(y,x) = \dfrac{1}{4\pi|x-y|}$ denote the Green kernel in $\mathbf{R}^3$. Let f denote any function of $C^2(\mathbf{R}^3)$, harmonic in Δ' and such that $f = O\left(\frac{1}{|x|}\right)$ when $|x| \to \infty$. Then, the value of f at any point $y \in \partial\Delta$ is given by the equation (Nédélec, 1977):

$$\frac{f(y)}{2} = \int_{\partial\Delta} \frac{\partial f(x)}{\partial n_x} G(y,x) ds_x - \int_{\partial\Delta} \frac{\partial G(y,x)}{\partial n_x} f(x) ds_x, \tag{31}$$

where $\mathbf{n_x}$ denotes the normal vector to $\partial\Delta$ at the point x, oriented outside Δ'.

We apply the above formulation to the function $\phi \sin\theta$ which satisfies equation (23). Let the points $x, y \in \mathbf{R}^3$ be denoted by $x = (r,\theta,z)$, $y = (r_b,\theta_b,z_b)$ in cylindrical coordinates. Then the Green kernel is

$$G(r_b,\theta_b,z_b,r,\theta,z) = \frac{1}{4\pi} \frac{1}{[(r_b\cos\theta_b - r\cos\theta)^2 + (r_b\sin\theta_b - r\sin\theta)^2 + (z_b - z)^2]^{1/2}}. \tag{32}$$

Since the normal vector to $\partial\Delta$ has no component along θ, we have $\frac{\partial(\sin\theta)}{\partial n_x} = 0$. Therefore, equation (31) takes the form

$$\frac{\phi(r_b,z_b)\sin\theta_b}{2} = \int_{\partial\Delta} \frac{\partial(\phi(r,z)\sin\theta)}{\partial n_x} G(y,x)\, ds_x - \int_{\partial\Delta} \frac{\partial G(y,x)}{\partial n_x} \phi(r,z)\sin\theta\, ds_x. \tag{33}$$

Since the problem is axisymmetric, we can choose an arbitrary value for θ_b in the above equation. As $\Delta = \Omega \times [0, 2\pi]$, we get:

$$\phi(r_b,z_b) = \frac{1}{2\pi} \int_{\partial\Omega} \frac{\partial\phi(r,z)}{\partial n_x} g(r,z,r_b,z_b) - \phi(r,z)\tilde{g}(r,z,r_b,z_b)\, ds_{r,z}, \text{ where} \tag{34}$$

$$g(r,z,r_b,z_b) = 4\pi \int_0^{2\pi} G(r,\theta,z,r_b,\frac{\pi}{2},z_b)\sin\theta\, d\theta, \tag{35}$$

$$\tilde{g}(r,z,r_b,z_b) = 4\pi \int_0^{2\pi} \frac{\partial G(r,\theta,z,r_b,\frac{\pi}{2},z_b))}{\partial n_x} \sin\theta \, d\theta. \tag{36}$$

The integrands in (35) and (36) have an elementary expression in terms of r, r_b, z, z_b, θ and $\mathbf{n}$. However, computation of integrals involving these functions gives rise to challenging numerical problems which are beyond the scope of this paper.

Equations (34) and (29) give rise to a finite element approximation. A triangular mesh τ_h is built over Ω. A mesh on the boundary $\partial\Omega$ is induced by τ_h. Since the normal derivative of ϕ is not continuous on $\partial\Omega$, we introduce a new variable $\lambda = \frac{\nu}{r}\frac{\partial(r\phi)}{\partial n}$ which is continuous on $\partial\Omega$ according to (30). To approximate ϕ, we use standard $\mathbf{P}_1$ elements inside Ω, while $\mathbf{P}_0$ elements are used to approximate λ on $\partial\Omega$. We obtain thus a linear system where the unknowns are the values of ϕ at the mesh nodes and the values of λ on the boundary edges. We obtain thus a boundary element–finite element model which was implemented into an eddy current computation code.

4 SIMULATION SOFTWARE

The eddy current computation code presented above was coupled with a module computing the temperature evolution. The heat equation, together with suitable radiation and convection conditions on the boundary is solved by standard finite elements, using an enthalpic formulation and taking into account the possible phase transformations. An additional module computing solid state phase transformations in steels was added to the whole software. This allows to simulate stream quenching of steels.

The simulation algorithm can be summarized as follows: we solve first the electromagnetic problem for a given time t_0 and a given initial temperature field. We suppose that this solution is valid for a short time step τ, during which the temperature variation did not influence too much the physical characteristics of the conductors. We solve then the thermal problem over the interval $[t_0, t_0+\tau]$. We update the physical characteristics of the conductors according to the new temperature field, and we also update the coordinates of the inductor in case of inductor movement. We proceed then to another solution of the electromagnetic problem. The whole procedure is repeated as many times as necessary.

An account of this work, as well as comparisons of numerical results with experiments, can be found in Jacot, Swierkosz et al.(1995). The software is now being used in industrial practice.

REFERENCES

Jacot, A., Swierkosz, M., Rappaz, J., Rappaz, M. and Mari, D. (1995) *Modelling of Electromagnetic Heating, Cooling and Phase Transformations During Surface Hardening of Steels*, submitted to: Proceedings of Mecamat'95, La Bresse, France, May 16–19.

R. Dautray, J.-L. Lions, *Analyse mathématique et calcul numérique pour les sciences et les techniques*, Masson, Paris.

Nédélec, J.C. (1977) *Approximation des équations intégrales en mécanique et en physique*, cours du Centre de Mathématiques Appliquées, Ecole Polytechnique, Paris.

34

Effective coefficients of thermoconductivity on some symmetric periodically perforated plane structures

J. Vucans
Department of Mathematical Analysis, University of Latvia
Rainis boulevard 19, Riga, LV-1586, Latvia.
Phone: +371-2-615646. Fax: +371-7-820113. E-mail: `fmmak@cclu.lv`

Abstract

In this article we discuss an auxiliary problem which arises in the homogenization theory for the Laplacian on the plane with periodic array of square holes and homogeneous Neumann boundary conditions on those. Independently, this problem describes the process of thermoconductivity. We find the explicit formulas for effective coefficients of thermoconductivity (homogenized modula). We make also the asymptotic analysis of these formulas in the cases of big and small holes.

Keywords

Homogenization, Laplacian, periodic structure, holes, asymptotic analysis

1 INTRODUCTION

First let's introduce for the numerical parameters γ and θ, $0 < \gamma < 1$, $0 \leq \theta \leq 1$, the sets

$$H_{\gamma,\theta} = \Big\{(x_1, x_2) \in R^2 \mid x_1 = \frac{1}{2} + r\cos\left(\varphi - \theta\frac{\pi}{2}\right), \ x_2 = \frac{1}{2} + r\sin\left(\varphi - \theta\frac{\pi}{2}\right), \\ 0 \leq \varphi < 2\pi, \ 0 \leq r \leq \frac{\gamma}{2\max\{|\cos\varphi|, |\sin\varphi|\}}\Big\}.$$

For fixed γ and θ the set $H_{\gamma,\theta}$ is a square with the center (1/2, 1/2) and the length of side equal to γ; the parameter θ determines its orientation in the plane. We shall concentrate our attention to the two cases $\theta = 0$ and $\theta = 1$, which correspond to the situations when the sides (for $\theta = 0$) or the diagonals (for $\theta = 1$) of the squares $H_{\gamma,\theta}$ are parallel to the coordinate axis.

Let's introduce also the periodic perforated plane structures

$$B_{\gamma,\theta} = R^2 \setminus \bigcup_{k=-\infty}^{\infty} \bigcup_{j=-\infty}^{\infty} \Big((k,j) + H_{\gamma,\theta} \Big),$$

their unit cells of periodicity $C_{\gamma,\theta}$ and the quarters $Q_{\gamma,\theta}$ of these unit cells:

$$C_{\gamma,\theta} = B_{\gamma,\theta} \cap (-1/2,\ 1/2)^2, \qquad Q_{\gamma,\theta} = B_{\gamma,\theta} \cap (0,\ 1/2)^2.$$

Let's for an arbitrary set $S \subset R^2$ denote by ∂S the boundary of this set.

This article refers to the homogenization theory for the Dirichlet problem of the Laplacian on the perforated plane structures $B_{\gamma,0}$ and $B_{\gamma,1}$ with homogeneous Neumann conditions on the boundaries of holes.

In order to find for such problems (with $\theta = 0$ *or* 1) the homogenized matrix, which is due to symmetry equal to $(K_{\gamma,\theta}\ \delta_{ij})$, we have to solve the auxiliary problem on the unit cell $C_{\gamma,\theta}$:

$$\begin{array}{lll} \Delta U_{\gamma,\theta} = 0 & in & C_{\gamma,\theta}\,, \\ \dfrac{\partial U_{\gamma,\theta}}{\partial n} = 0 & in & \partial C_{\gamma,\theta} \setminus \partial(-1/2,\ 1/2)^2, \\ (U_{\gamma,\theta} - x_1) & is & 1 - periodic\ in\ both\ variables\ x_1\ and\ x_2\,, \end{array} \tag{1}$$

where n is the unit outward normal of the set $C_{\gamma,\theta}$. The homogenized modula, which in this case is the effective coefficient of thermoconductivity, is equal to

$$K_{\gamma,\theta} = \frac{1}{mes\ C_{\gamma,\theta}} \iint_{C_{\gamma,\theta}} \frac{\partial U_{\gamma,\theta}(x_1,x_2)}{\partial x_1}\, dx_1\, dx_2. \tag{2}$$

We note that such problems for more general periodic structures were discussed in the books of Bakhvalov and Panasenko (1989) and Bensoussan, Lions and Papanicolaou (1978). Some results concerning the asymptotic analysis of $K_{\gamma,0}$ and $K_{\gamma,1}$ as the functions of the parameter γ are obtained in the papers of Kozlov (1989), Kozlov and Vucans (1992) and Vucans (1992). In the present paper, using the method similar as in the article of Kozlov and Vucans (1992), we continue these studies.

2 FORMULAS FOR EFFECTIVE COEFFICIENTS $K_{\gamma,0}$ AND $K_{\gamma,1}$

In order to formulate the main results of this article we shall use the following functions:

$$\varphi(t,\alpha) = \frac{1}{\sqrt{(1+t)(\alpha^2-t^2)}}\,,$$

$$\psi_0(t,\alpha) = \frac{t}{(1-t^2)(\alpha^2-t^2)}\,, \qquad \psi_1(t,\beta) = \frac{1}{(1-t^2)(\beta^2-t^2)^2}\,,$$

$$I_{10}(\alpha) = \int_1^{\alpha} \sqrt{-\psi_0(t,\alpha)}\, dt\,, \qquad I_{11}(\beta) = \int_1^{\beta} \sqrt[4]{-\psi_1(t,\beta)}\, dt\,,$$

$$I_{20}(\alpha) = \int_0^1 \sqrt{\psi_0(t,\alpha)}\, dt\,, \qquad I_{21}(\beta) = \int_0^1 \sqrt[4]{\psi_1(t,\beta)}\, dt\,,$$

$$I_{30}(\alpha) = \int_0^1 \sqrt{\psi_0(z,\alpha)} \int_z^\alpha \varphi(t,\alpha)\, dt\, dz\,, \qquad I_{31}(\beta) = \int_{-1}^1 \sqrt[4]{\psi_1(z,\beta)} \int_z^\beta \varphi(t,\beta)\, dt\, dz,$$

$$I_{40}(\alpha) = \int_0^1 \sqrt{\psi_0(z,\alpha)} \int_{-1}^z \varphi(t,\alpha)\, dt\, dz\,, \qquad I_{41}(\beta) = \int_{-1}^1 \sqrt[4]{\psi_1(z,\beta)} \int_{-1}^z \varphi(t,\beta)\, dt\, dz,$$

$$I_5(\alpha) = \int_{-1}^\alpha \varphi(t,\alpha) dt = \sqrt{\frac{2}{\alpha}} K\left(\sqrt{\frac{\alpha+1}{2\alpha}}\right),$$

where $K(\cdot)$ is the complete Legendre elliptic integral of the first kind.
In these notations from the paper of Kozlov and Vucans (1992) immediately follows:

Theorem 1 *The effective coefficient of thermoconductivity $K_{\gamma,0}$ of the structure $B_{\gamma,0}$ for a fixed parameter $\gamma \in (0,1)$ is given by the two equivalent expressions*

$$K_{\gamma,0} = \frac{1}{1+\gamma}\left[1 + \frac{I_{30}(\alpha)}{I_{10}(\alpha)\, I_5(\alpha)}\right] = \frac{1}{1-\gamma^2}\left[1 - \gamma\frac{I_{40}(\alpha)}{I_{20}(\alpha)\, I_5(\alpha)}\right],$$

where $\alpha = \alpha(\gamma)$ is the unique solution from the interval $(1, +\infty)$ of the equation

$$\gamma\,[I_{10}(\alpha) + I_{20}(\alpha)] = I_{20}(\alpha)\,.$$

Moreover, from the papers of Kozlov and Vucans (1992) and Vucans (1992) follows:

Theorem 2

$$K_{\gamma,0} = 1 + \left(1 - \frac{\Gamma^4(1/4)}{8\pi^2}\right)\gamma^2 + o\left(\gamma^2\right) \approx 1 - 1.18844\gamma^2 + o\left(\gamma^2\right), \quad \text{when } \gamma \to 0_+\,;$$

$$K_{\gamma,0} = \frac{1}{2} + \frac{1}{2}\left(\frac{1}{2} + c_1\right)(1-\gamma) + \frac{1}{2}\left[\frac{1}{2}\left(\frac{1}{2} + c_1\right) + c_2\right](1-\gamma)^2 + o\left((1-\gamma)^2\right) \approx$$
$$\approx 0.5 + 0.389682\,(1-\gamma) + 0.230222\,(1-\gamma)^2 + o\left((1-\gamma)^2\right), \qquad \text{when } \gamma \to 1_-\,,$$

with

$$c_1 = \frac{4}{\pi^2}\int_0^{arcsh\, 1} t\sqrt{csch^2\, t - 1}\, dt \approx 0.279364$$

and

$$c_2 = -c_1\left\{\frac{1}{2\pi}\left[5\ln 2 + 4\arctan\left(2^{-\frac{1}{4}}\right)\right] - \frac{5}{4}\right\} \approx 0.253296\, c_1 \approx 0.070762\,.$$

We must note here that these asymptotic equalities coincide with results of Kozlov (1989) and Kozlov and Panasenko (1991).

Now using the method, similar as in the paper of Kozlov and Vucans (1992), we are going to prove the corresponding theorem for the structure $B_{\gamma,1}$.

Theorem 3 *The effective coefficient of thermoconductivity $K_{\gamma,1}$ of the structure $B_{\gamma,1}$ for a fixed parameter $\gamma \in (0, 1/\sqrt{2})$ is given by the two equivalent expressions*

$$K_{\gamma,1} = \frac{1-\sqrt{2}\,\gamma}{1-\gamma^2}\left[1 + \frac{I_{31}(\beta)}{\sqrt{2}\,I_{11}(\beta)\,I_5(\beta)}\right] = \frac{1}{1-\gamma^2}\left[1 - \frac{\gamma}{\sqrt{2}}\frac{I_{41}(\beta)}{I_{21}(\beta)\,I_5(\beta)}\right], \tag{3}$$

where $\beta = \beta(\gamma)$ is the unique solution from the interval $(1, +\infty)$ of the equation

$$\gamma\, I_{11}(\beta) = \left(1 - \sqrt{2}\,\gamma\right) I_{21}(\beta)\,. \tag{4}$$

Proof. First let's note that in our case $\theta = 1$ the quarter $Q_{\gamma,1}$ of the periodic structures unit cell $C_{\gamma,1}$ is equal to the pentagon

$$Q_{\gamma,1} = \left\{(x_1, x_2) \in (0,\, 1/2)^2 \mid x_1 + x_2 < 1 - \gamma/\sqrt{2}\right\}$$

with sides

$$\begin{aligned}
l_1 &= \left\{(x_1, x_2) \in R^2 \mid x_1 = 0,\ 0 < x_2 < 1/2\right\}, \\
l_2 &= \left\{(x_1, x_2) \in R^2 \mid x_2 = 0\ ,\ 0 < x_1 < 1/2\right\}, \\
l_{3,\gamma} &= \left\{(x_1, x_2) \in R^2 \mid x_1 = 1/2,\ 0 < x_2 < 1/2 - \gamma/\sqrt{2}\right\}, \\
l_{4,\gamma} &= \left\{(x_1, x_2) \in R^2 \mid x_1 + x_2 = 1 - \gamma/\sqrt{2},\ 1/2 - \gamma/\sqrt{2} < x_2 < 1/2\right\}, \\
l_{5,\gamma} &= \left\{(x_1, x_2) \in R^2 \mid x_2 = 1/2,\ 0 < x_1 < 1/2 - \gamma/\sqrt{2}\right\}.
\end{aligned}$$

Latter, without special references, we shall identify these sides of the boundary $\partial Q_{\gamma,1}$ in R^2 with corresponding sets of points in the complex plane C. Due to the properties of periodicity of $(U_{\gamma,1} - x_1)$ and the properties of symmetry of the set $C_{\gamma,1}$ we can pass from the periodic boundary problem (1) on the unit cell $C_{\gamma,1}$ to the "pure" third type boundary problem on its quarter $Q_{\gamma,1}$ (for more details of this pass see, for example, the chapter 6 of the book of Bakhvalov and Panasenko (1989)):

$$\begin{aligned}
\Delta U_{\gamma,1} &= 0 \quad in \quad Q_{\gamma,1}\,, & U_{\gamma,1} &= 0 \quad in \quad l_1\,, \\
U_{\gamma,1} &= 1/2 \quad in \quad l_{3,\gamma}\,, & \frac{\partial U_{\gamma,1}}{\partial n_Q} &= 0 \quad in \quad l_2 \cup l_{4,\gamma} \cup l_{5,\gamma}\,,
\end{aligned} \tag{5}$$

where n_Q is the unit outward normal of the set $Q_{\gamma,1}$. Moreover, from these properties of symmetry and from (2) and (5) by integration with respect to x_1 we get:

$$\begin{aligned}
K_{\gamma,1} &= \frac{1}{mes\, Q_{\gamma,1}} \iint_{Q_{\gamma,1}} \frac{\partial U_{\gamma,1}(x_1, x_2)}{\partial x_1}\, dx_1\, dx_2 = \\
&= \frac{4}{1-\gamma^2}\left[\frac{1}{4}\left(1 - \sqrt{2}\gamma\right) + \int_{1/2-\gamma/\sqrt{2}}^{1/2} U_{\gamma,1}\left(1 - \gamma/\sqrt{2} - x_2\,,\, x_2\right) dx_2\right] = \\
&= \frac{1}{1-\gamma^2}\left[1 - \sqrt{2}\,\gamma + \frac{4}{i-1}\int_{l_{4,\gamma}} U_{\gamma,1}(\xi)\, d\xi\right],
\end{aligned} \tag{6}$$

where the last integral is presented in terms of the complex variable ξ along the side $l_{4,\gamma}$ of the boundary $\partial Q_{\gamma,1}$ from point $(1/2 + i\,(1/2 - \gamma/\sqrt{2}))$ to point $(1/2 - \gamma/\sqrt{2} + i/2)$.

In order to obtain the solution $U_{\gamma,1}$ of the problem (5) we construct the conformal representation of the upper half plane to $Q_{\gamma,1}$. According to Schwarz-Christoffel theorem such one-to-one mapping $\{z : Im\, z > 0\} \stackrel{\omega}{\longmapsto} \{\xi : \xi \in Q_{\gamma,1}\}$ is given by

$$\xi = \omega(z) = (i-1)\, c_0 \int_0^z \frac{dt}{\sqrt[4]{1-t^2}\,\sqrt{\beta^2 - t^2}} + \frac{1+i}{2\sqrt{2}}\left(\sqrt{2} - \gamma\right), \tag{7}$$

where β satisfies (4) and for the real constant c_0 we have equivalent expressions:

$$c_0 = \frac{\gamma}{2\sqrt{2}\, I_{21}(\beta)} = \frac{1 - \sqrt{2}\gamma}{2\sqrt{2}\, I_{11}(\beta)}. \tag{8}$$

The mapping ω transforms the real axis into the boundary $\partial Q_{\gamma,1}$ in the following way:

$$\begin{array}{lll} \omega((-\infty, -\beta)) = l_2, & \omega((-\beta, -1)) = l_{3,\gamma}, & \omega((-1,1)) = l_{4,\gamma}, \\ \omega((1,\beta)) = l_{5,\gamma}, & \omega((\beta, +\infty)) = l_1. & \end{array} \tag{9}$$

We remark here that equality (4) defines for $\beta > 1$ the strictly decreasing function $\gamma = \gamma(\beta)$. This function has the following limiting values: $\gamma(1_+) = 1/\sqrt{2}$, $\gamma(+\infty) = 0$. So, in the interval $(0, 1/\sqrt{2})$ there exists a strictly decreasing inverse function $\beta = \beta(\gamma)$.

Now let's introduce the complex function

$$\Phi(z) = c_1 \left[\int_0^z \frac{dt}{\sqrt{(1+t)\,(\beta^2 - t^2)}} + c_2 \right] \tag{10}$$

of the complex variable $z = z_1 + i\, z_2$ $(z_1, z_2 \in R)$ where c_1 and c_2 are real constants:

$$c_1 = \frac{-1}{2\, I_5(\beta)}, \qquad c_2 = -\int_0^\beta \varphi(t,\beta)\, dt. \tag{11}$$

For points z in the upper half plane and on the real axis the integrals in (7) and (10) are defined along any path from 0 to z which completely lies in the upper half plane.

It's easy to verify the following properties of the function Φ:

(i) $Re\, \Phi$ is harmonic in the upper half plane;

(ii) $Re\, \Phi$ has the following limiting values on the real axis:

$$\begin{array}{lll} Re\, \Phi(z) = -c_1 \int_{-\infty}^z \varphi(t,\beta)\, dt & \text{if} & z \in (-\infty, -\beta], \\ Re\, \Phi(z) = 1/2 & \text{if} & z \in [-\beta, -1], \\ Re\, \Phi(z) = -c_1 \int_z^\beta \varphi(t,\beta)\, dt & \text{if} & z \in [-1, \beta], \\ Re\, \Phi(z) = 0 & \text{if} & z \in [\beta, +\infty); \end{array}$$

(iii) in the intervals $(-\infty, -\beta)$ and $(-1, \beta)$ on the real axis the derivative $\frac{\partial \Phi(z_1 + i\, z_2)}{\partial z_2}$ is purely imaginary, therefore $\frac{\partial\, Re\, \Phi(z_1 + i\, z_2)}{\partial z_2} = 0$ there.

From the properties i), ii), iii) and from (7) follows that the solution of the problem (5) is equal to

$$U_{\gamma,1}(\xi) = Re\,\Phi\left(\omega^{-1}(\xi)\right) .$$

From the property ii) changing the variable and taking into account (7), (8), (9) and (11) we get:

$$\begin{aligned}\int_{l_{4,\gamma}} U_{\gamma,1}(\xi)\,d\xi &= \int_{l_{4,\gamma}} Re\,\Phi\left(\omega^{-1}(\xi)\right)\,d\xi = \int_{-1}^{1} Re\,\Phi(z)\,\omega'(z)\,dz = \\ &= -\,c_0\,c_1\,(i-1)\int_{-1}^{1}\frac{1}{\sqrt[4]{1-z^2}\,\sqrt{\beta^2-t^2}}\int_z^{\beta}\frac{dt}{\sqrt{(1+t)\,(\beta^2-t^2)}} = \\ &= \frac{(i-1)\,\left(1-\sqrt{2}\,\gamma\right)\,I_{31}(\beta)}{4\,\sqrt{2}\,I_{11}(\beta)\,I_5(\beta)}\,.\end{aligned}$$

This formula jointly with (6), (8) and (11) gives the result of Theorem 3. □

The obtained formulas (3) allow us to make the asymptotic analysis of the $K_{\gamma,1}$ in the cases when in our periodic structure the squareic holes are small ($\gamma \to 0_+$) or when they are in the position near to touch one another ($\gamma \to 1/\sqrt{2}_-$).

Theorem 4

$$K_{\gamma,1} = 1 + \left(1 - \frac{\Gamma^4(1/4)}{8\,\pi^2}\right)\gamma^2 + o\left(\gamma^2\right) \approx 1 - 1.18844\gamma^2 + o\left(\gamma^2\right), \quad when\ \gamma \to 0_+; \qquad (12)$$

$$K_{\gamma,1} = -\frac{\pi}{2}\,\frac{1}{\ln(1/\sqrt{2}-\gamma)} + o\left(\frac{1}{\ln(1/\sqrt{2}-\gamma)}\right), \qquad when\ \gamma \to 1/\sqrt{2}_-\,. \qquad (13)$$

Proof. a) Let $\gamma \to 0_+$; in this case from (4) and from our definitions we have:

$$\lim_{\beta\to+\infty}\sqrt{\beta}\,I_{11}(\beta) = \frac{\Gamma^2(1/4)}{2\sqrt{2\pi}}, \qquad \lim_{\beta\to+\infty}\beta\,I_{21}(\beta) = \frac{(2\pi)^{3/2}}{\Gamma^2(1/4)},$$

$$\lim_{\beta\to+\infty}\sqrt{\beta}\,\gamma(\beta) = \frac{8\pi^2}{\Gamma^4(1/4)}, \qquad \lim_{\gamma\to 0_+}\beta(\gamma) = +\infty,$$

$$\lim_{\beta\to+\infty}\beta^2\,I_{41}(\beta) = \sqrt{2}\,\pi, \qquad \lim_{\beta\to+\infty}\sqrt{\beta}\,I_5(\beta) = \frac{\Gamma^2(1/4)}{2\sqrt{2\pi}},$$

The asymptotic equality (12) follows from these properties.

b) Let $\gamma \to 1/\sqrt{2}_-$; in this case from (4) and from our definitions we have:

$$\lim_{\beta\to 1_+}\frac{I_{11}(\beta)}{\sqrt[4]{\beta-1}} = \frac{2^{7/4}\,\pi^{3/2}}{\Gamma^2(1/4)}, \qquad \lim_{\beta\to 1_+} I_{21}(\beta) = \frac{\Gamma^2(1/4)}{2\sqrt{2\pi}},$$

$$\lim_{\beta\to 1_+}\frac{1/\sqrt{2}-\gamma}{\sqrt[4]{b-1}} = \frac{2^{9/4}\,\pi^2}{\Gamma^4(1/4)}, \qquad \lim_{\gamma\to 1/\sqrt{2}_-}\beta(\gamma) = 1_+,$$

$$\lim_{\beta\to 1_+} I_{31}(\beta) = \frac{\sqrt{\pi}}{2}\Gamma^2(1/4), \qquad \lim_{\beta\to 1_+} \frac{I_5(\beta)}{\ln(\beta-1)} = -\frac{1}{\sqrt{2}}.$$

The asymptotic equality (13) follows from these properties.

□

We note that in the case of small holes ($\gamma \to 0_+$) the asymptotic formulas for $K_{\gamma,0}$ and $K_{\gamma,1}$ are equal up to the terms with γ^2. This agrees with the result of Kozlov (1989).

REFERENCES

Bakhvalov, N. and Panasenko, G.P. (1989) Homogenization: averaging processes in periodic media. Dordrecht/Boston/London: Kluwer Academic Publishers.

Bensoussan, A., Lions, J.-L. and Papanicolaou, G. (1978) Asymptotic analysis for periodic structures. Amsterdam, North-Holland Publ. Comp.

Kozlov, S.M. (1989) Geometric aspects of homogenization. *Russian Mathematical Surveys*, **44, 2**, 91–144.

Kozlov, S.M. and Panasenko, G.P. (1991) Corrections to the strength materials theory for the lattice structure. *Publication of the Laboratory of Numerical Analysis of the University Paris-VI*, **R91020**, 9p.

Kozlov, S.M. and Vucans, J. (1992) Explicit formulæ for effective thermoconductivity on the squareic lattice structure. *Comptes Rendus de l'Académie des Sciences, I*, **314**, 281–6.

Vucans, J. (1992) Approximation of the third order to effective thermoconductivity of the squareic lattice structure. *Acta Universitatis Latviensis*, **575**, 125–32.

35

An approach to infinite domains, singularities and superelements in FEM computations

Antoni Żochowski
Systems Research Institute of the Polish Academy of Sciences
Newelska 6, 01–447 Warszawa, Poland

Abstract

In the paper the representation method for treating the infinite parts of computational domains is derived. It is based on a discrete formulation of the problems and uses the formal series technique. The method applies to wedge–like domains and static problems in general. For finite domains, it makes possible a uniform treatment of problems with corner singularities and improvement of accuracy in ordinary FEM computations.

Keywords

Finite elements, elliptic equations, infinite domains, singular solutions

1 INTRODUCTION

The method presented in this paper is quite general and may be used in finite element computations for partial differential equations or systems of any order (2–nd, 4–th or higher if it makes sense), but we shall concentrate here on the second order examples and plane problems.

Let us imagine a bounded star–shaped domain Ω_s, having the center at $0 \in \mathrm{R}^2$, and define an exterior domain $\Omega_e = \mathrm{R}^2 - \Omega_s$ with the boundary $\Gamma^0 = \partial\Omega_e$. Next we approximate Γ^0 by Γ_h^0 using linear segments and extend a set of radii originating from 0 and going through nodes of Γ_h^0. Now we may define consecutive crossections of Ω_e by similarity transformation with center in 0, taking $\Gamma_h^i = r^i \cdot \Gamma_h^0$, where $r > 1$ and $i = 0, 1, 2, \ldots$. Between Γ_h^i and Γ_h^{i+1} lay the ring–like parts of Ω_{eh}, denoted by Ω_{eh}^i. These rings are also similar, since $\Omega_{eh}^{i+1} = r \cdot \Omega_{eh}^i$, and they sum up to the whole Ω_{eh}, giving in this way the discretization of the exterior of the "hole' in R^2. It is obvious, that we may as well consider a part of Ω_{eh} limited by two radii, on which homogeneous Neumann or Dirichlet boundary conditions are imposed, while the loading of Γ_h^0 is arbitrary.

Our goal is to solve elasticity or Laplace equations in such a domain using finite element method. Let the parts between crossections $k, k+1$ be discretized with some kind of elements. We shall denote by u_k the vector of all nodal values of solution corresponding to the k–th crosssection, discounting those, which may be eliminated immediately due to the homogeneous Dirichlet conditions on the bounding radii. Now the elastic energy of the whole body after discretization can be written as

$$E = \sum_{k=0}^{\infty} E_k(u_k, u_{k+1}), \tag{1}$$

where E_k denotes the energy of the k–th ring. Let us concentrate on E_0. By eliminating internal nodes between sections Γ^0 and Γ^1 we get

$$E_0(u_0, u_1) = \frac{1}{2}[u_0^T, u_1^T] \cdot M \cdot \begin{bmatrix} u_0 \\ u_1 \end{bmatrix}, \tag{2}$$

where $2n \times 2n$ symmetric stiffness matrix M ($n = dim\ u_k$) has the form

$$M = \begin{bmatrix} A_1 & , & B \\ B^T & , & A_2 \end{bmatrix}. \tag{3}$$

Observe, that both $n \times n$ matrices A_1, A_2 are symmetric and positive definite.

Now comes the crucial observation. Assume, that Ω^i_{eh} has been triangulated and the linear finite elements have been used. Then the matrix M is proportional to the area of the ring, that is r^2, and inversely proportional to the squares of lengths of the sides of triangles, since the gradient of u is inversely proportional to r. As a result, M is the same for all rings, and E_k have the same form for all k.

Using the above notation we may write the energy of the whole Ω_{eh} as

$$E = \frac{1}{2} \begin{bmatrix} u_0 \\ u_1 \\ u_2 \\ \vdots \\ \vdots \end{bmatrix}^T \cdot \begin{bmatrix} A_1 & B & & & \\ B^T & A & B & & \\ & B^T & A & B & \\ & & B^T & \ddots & \\ & & & & \ddots \end{bmatrix} \cdot \begin{bmatrix} u_0 \\ u_1 \\ u_2 \\ \vdots \\ \vdots \end{bmatrix}, \tag{4}$$

where $A = A_1 + A_2$. If we could express E as a function of u_0 only, $E = E(u_0)$, then it would be possible to write down the whole energy in terms of finite number of nodal values.

The problem of representing the infinite body has received some attention in FEM literature. In (Givoli,1992),(Grote,Keller,1994) the use is made of the exact analytic solutions in certain kinds of infinite domains in order to get substitute boundary conditions. Here we treat the problem from the beginning in its discrete formulation and use energetic approach. In (Sharau,1994) a completely different method is used, depending on treating the infinite body as a certain kind of elastic support. The method presented here may be proved to be correct (convergent), but of course has also limitations. It applies to periodic or "radially" periodic 2–D and 3–D bodies with finite arbitrary part. Nevertheless, it may be generalized as well, as it will be shown later.

2 PROBLEM FORMULATION

Let us return to the expression for E, (4), and solve the elasticity equations in the whole right part imposing the boundary conditions on u_0. The necessary condition for the minimum of energy takes on the form

$$M_\infty \cdot u_\infty = \begin{bmatrix} A & B & & & \\ B^T & A & B & & \\ & B^T & A & B & \\ & & B^T & \ddots & \\ & & & & \ddots \end{bmatrix} \cdot \begin{bmatrix} u_1 \\ u_2 \\ u_3 \\ \vdots \\ \vdots \end{bmatrix} = \begin{bmatrix} -B^T \\ 0 \\ 0 \\ \vdots \\ \vdots \end{bmatrix} \cdot u_0. \tag{5}$$

As a result, we must solve the matrix equation of the infinite order

$$M_\infty \cdot Q_\infty = \begin{bmatrix} A & B & & & \\ B^T & A & B & & \\ & B^T & A & B & \\ & & B^T & \ddots & \\ & & & & \ddots \end{bmatrix} \cdot \begin{bmatrix} Q_1 \\ Q_2 \\ Q_3 \\ \vdots \\ \vdots \end{bmatrix} = \begin{bmatrix} -B^T \\ 0 \\ 0 \\ \vdots \\ \vdots \end{bmatrix} \tag{6}$$

where $Q_1, Q_2, \ldots$ have dimensions $n \times n$.

It is well known (Cooke,1950), that such systems may have infinitely many solutions. Therefore we impose the physical condition, that the consecutive energy terms (2) diminish, or the requirement, that the necessary condition gives minimum, not the saddle point or maximum of the elastic energy. As it will turn out, this makes the solution unique.

If we could obtain the expression for Q_i in the multiplicative form, $Q_i = Q^i$, then taking into account, that $u_k = Q_{k-1}u_0 = Q^{k-1}u_0$, the energy on the whole domain takes on the form

$$\begin{aligned} E(u_k, u_{k+1}) &= \frac{1}{2}(u_k^T A_1 u_k + u_{k+1}^T B^T u_k + u_k^T B u_{k+1} + u_{k+1}^T A_2 u_{k+1}) \\ &= \frac{1}{2} u_1^T \cdot (Q^T)^k \cdot [A_1 + Q^T B^T + BQ + Q^T A_2 Q] \cdot Q^k \cdot u_1. \end{aligned} \tag{7}$$

Hence, assuming the convergence of the infinite sum, the whole E may be computed as $E = \frac{1}{2} u_0^T \cdot S \cdot u_0$ where $S = \sum_{k=0}^{\infty} (Q^T)^k \cdot R \cdot Q^k$, $R = A_1 + Q^T B^T + BQ + Q^T A_2 Q$. The series for S can be, as we shall see later, computed exactly in a closed form.

3 FORMAL SERIES APPROACH

In this section, we shall solve the equation (6) by embedding the problem into the framework of operations on infinite series (Stanley,1986). Let us establish the correspondence between the infinite vector $f_\infty = [f_1, f_2, \ldots]^T$ and the formal power series:

$$f(x) = \sum_{i=1}^{\infty} f_i \frac{x^{i-1}}{(i-1)!}. \tag{8}$$

Differentiating this series gives

$$Df(x) = \sum_{i=2}^{\infty} f_i \frac{x^{i-2}}{(i-2)!},$$

or, in vector representation,

$$f_\infty = [f_1, f_2, \ldots]^T, \quad Df_\infty = [f_2, f_3, \ldots]^T. \tag{9}$$

This shows, that the differentiation may be represented as multiplication by the matrix:

$$Df_\infty = \begin{bmatrix} 0 & 1 & & & \\ & 0 & 1 & & \\ & & 0 & 1 & \\ & & & \ddots & \ddots \end{bmatrix} \cdot f_\infty. \tag{10}$$

and similarly for integration. Let us notice, that M_∞ has a block structure, i.e. every n–th row repeats itself after shifting n places to the right. Therefore we must introduce a whole vector of functions

$$w^k(x) = \sum_{j=1}^{\infty} w_j^k \frac{x^{j-1}}{(j-1)!}, \qquad k = 1, \ldots, n, \tag{11}$$

and denote $u_i = [w_i^1, \ldots, w_i^n]^T$, so that

$$u(x) = \begin{bmatrix} w^1(x) \\ \vdots \\ w^n(x) \end{bmatrix} = \sum_{i=1}^{\infty} u_i \frac{x^{i-1}}{(i-1)!}. \tag{12}$$

If we neglect first n rows, the system (6) is equivalent to $B^T \cdot \int u + A \cdot u + B \cdot Du = 0$, and putting $\bar{u} = \int u$ gives finally the differential equation $B \cdot \bar{u}'' + A \cdot \bar{u}' + B^T \cdot \bar{u} = 0$. The solution must have the form $\bar{u} = r_\lambda \cdot e^{\lambda x}$, where $dim\ r_\lambda = n$. Furthermore, λ is the root of the 2n–th order characteristic polynomial

$$det(B \cdot \lambda^2 + A \cdot \lambda + B^T) = 0, \tag{13}$$

and r_λ should be the right eigenvector: $(B \cdot \lambda^2 + A \cdot \lambda + B^T) \cdot r_\lambda = 0$. In general, (13) has 2n roots. However, from the particular form of (13) it follows, that these roots occur in pairs, $(\lambda_i, 1/\lambda_i)$, $i = 1, \ldots n$. Let us eliminate at this point the roots with absolute values bigger than 1, and consider (after rearranging) only $\lambda_1, \ldots \lambda_n$. The corresponding solutions have the form:

$$\bar{u} = c_{1,1} r_{\lambda_1} \exp(\lambda_1 x) + c_{2,1} r_{\lambda_2} \exp(\lambda_2 x) + \ldots + c_{n,1} r_{\lambda_n} \exp(\lambda_n x). \tag{14}$$

The constants $c_{1,1}, \ldots, c_{n,1}$ are chosen in such a way, that the first n rows of (6) are satisfied. Let us notice, however, that equation (6) has n right–hand sides. Thats why we have double subscript here: $c_{p,q}$ denotes p–th constant corresponding to the q–th column on the right.

Let us now introduce the following notation: $R_\lambda = [r_{\lambda_1}, \ldots, r_{\lambda_n}]$, $C = [c_{j,k}]_{j,k=1,..n}$, $\Lambda = diag[\lambda_1, \ldots, \lambda_n]$. It may be proved, that the solution of (6) takes on the form $Q_1 = R_\lambda \cdot \Lambda \cdot C$, $Q_2 = R_\lambda \cdot \Lambda^2 \cdot C, \ldots, Q_n = R_\lambda \cdot \Lambda^n \cdot C$. The choice of the matrix of constants C must ensure satisfaction of the first block of equations, what leads to the formula $C = R_\lambda^{-1}$. In consequence we get, as was required: $Q_1 = Q = R_\lambda \cdot \Lambda \cdot R_\lambda^{-1}$, $Q_i = Q^i = R_\lambda \cdot \Lambda^i \cdot R_\lambda^{-1}$, $i = 1, 2, \ldots$ In general there may appear $\lambda = 1$, what corresponds to the constant solution. For scalar Laplace equation it must have multiplicity 2, with only one eigenvector. For elasticity system, there exist pairs consisting of eigenvalue $\lambda = 1$ and eigenvector responsible for constant displacements in 2 or 3 independent directions (rotations are excluded, since they imply $\lambda > 1$). However, they do not contribute to the energy, so the convergence of the energy series is assured by the next eigenvalue strictly smaller than 1.

We may also prove, that after regrouping the terms, the matrix S takes on the simple form: $S = A_1 + BQ$.

4 GENERALIZATIONS

Three–dimensional case. Let us consider the three–dimensional domains, where the layers $\Omega_1, \Omega_2, \ldots$ are cut out from the 3–D space by the sides of the cone with origin in 0. The derivatives of discretized functions still contain terms proportional to $1/r$, but the volumes of elements are proportional to r^3, so as a result

$$E_i = \frac{1}{2}[(u_i^h)^T, (u_{i+1}^h)^T] \cdot (r^{i-1}M) \cdot \begin{bmatrix} u_i^h \\ u_{i+1}^h \end{bmatrix}, \quad i = 1, 2, \ldots \tag{15}$$

with M having the form as in (3). Hence the system $M_\infty \cdot Q_\infty = B_\infty$ becomes

$$\begin{bmatrix} A_2 + rA_1 & , & rB & , & 0 & & \cdots \\ rB^T & , & rA_2 + r^2A_1 & , & r^2B & , & 0 & \cdots \\ 0 & , & r^2B^T & , & r^2A_2 + r^3A_1 & , & r^3B & \cdots \\ & & & & \ddots & & \ddots \\ & & & & & & \ddots \end{bmatrix} Q_\infty = \begin{bmatrix} -B^T \\ 0 \\ 0 \\ \vdots \\ \vdots \end{bmatrix}. \tag{16}$$

The rows do not repeat here exactly, so the solution requires some scaling. Let I be an $n \times n$ identity matrix, and define $P = \text{diag}[r^{-1/2}I, r^{-1}I, \ldots, r^{-i/2}I, \ldots]$. Then (16) may be rewritten (diagonal infinite matrix is invertible) as $(P^{-1} \cdot \tilde{M}_\infty \cdot P^{-1}) \cdot Q_\infty = B_\infty$, where $\tilde{M}_\infty = P \cdot M_\infty \cdot P$. Moreover, the system $\tilde{M}_\infty \cdot \tilde{Q}_\infty = \tilde{B}_\infty$, where $\tilde{Q}_\infty = P^{-1}Q_\infty$, $\tilde{B}_\infty = PB_\infty$, has the form

$$\begin{bmatrix} \frac{1}{r}A_2 + A_1 & , & \frac{1}{\sqrt{r}}B & , & 0 & & \cdots \\ \frac{1}{\sqrt{r}}B^T & , & \frac{1}{r}A_2 + A_1 & , & \frac{1}{\sqrt{r}}B & , & 0 & \cdots \\ 0 & , & \frac{1}{\sqrt{r}}B^T & , & \frac{1}{r}A_2 + A_1 & , & \frac{1}{\sqrt{r}}B & \cdots \\ & & & & \ddots & & \ddots \\ & & & & & & \ddots \end{bmatrix} \cdot \tilde{Q}_\infty = \begin{bmatrix} -\frac{1}{\sqrt{r}}B^T \\ 0 \\ 0 \\ \vdots \\ \vdots \end{bmatrix}, \tag{17}$$

which falls into the framework of our method, with the eigenvalue problem (13)

$$\det[\lambda^2 \frac{1}{\sqrt{r}} + \lambda(\frac{1}{r}A_2 + A_1) + \frac{1}{\sqrt{r}}B^T] = 0. \tag{18}$$

The same procedure applies obviously to 1–D and $2\frac{1}{2}$–D domains, only some directions would require different scaling and we would have to define matrix P in a more complicated way, but the idea would remain the same.

Finite domain – superelements. Consider the finite, star–shaped body. We may discretize the boundary and then use radii originating from the star center for the construction of similar layers of domain discretizations with the decreasing size, so now $r < 1$. Again, the layers form infinite series filling the body, and all reasoning may be repeated. Thus we get a method for calculating the stiffness matrix for the arbitrary star–shaped super–element. The situation reminds somehow a discrete version of the boundary integral method, since we consider only a thin boundary layer and use the fact, that the unknown function satisfies the state equation inside the domain.

Singular elements. Consider a discretized problem on some domain Ω_h (finite or infinite). Let p_0 be a point on $\partial\Omega_h$ where a geometrical singularity of the solution may occur (reentrant corner, change of the type of boundary conditions), belonging also to the nodes of triangulation. The standard way of dealing with such problems is to refine locally the discretization at the cost of increasing dimensionality (Grisvard,1985). Our method suggests another approach. Create a star–shaped domain consisting of all triangles having p_0 as a vertex. Then we may treat this star as a finite body and construct for it a stiffness matrix using p_0 as a similarity origin. In this way we have mesh refinement without dimensionality increase. Moreover, the rate of convergence is the same as in the smooth case.

Improving accuracy. Let us consider the discretization consisting of convex quadrilaterals. Taking centers of these quadrilaterals as similarity centers we may construct stiffness matrices in the same way as for superelements. Such an approach does not increase dimensionality, but, as shown by experiments, improves accuracy in comparison to ordinary linear elements on triangles. This phenomenon occurs in all cases, singular, homogeneous as well as nonsingular and nonhomogeneous.

5 NUMERICAL EXPERIMENTS

Infinite domains. In order to test the method, we have conducted several computational experiments involving the known exact solutions to the 2–D Laplace equation. The domain constituted a lower half–plane $y < 0$ with the unit circle $x^2 + y^2 < 1$ cut out. The boundary of this domain can be divided into two parts: $\Gamma_0 = \{(x, y) \mid (x < -1 \text{ or } x > 1) \text{ and } (y = 0)\}$, $\Gamma_1 = \{(x, y) \mid (-1 \leq x \leq 1) \text{ and } (y = -\sqrt{(1 - x^2)})\}$, As test functions we have chosen: [A] $u = x/r^2$; [B] $u = (y^2 - x^2)/r^4$. Both of these functions satisfy $\frac{\partial u}{\partial n} = 0$ on Γ_0 and their total flux over Γ_1 vanishes, so they also vanish at infinity. The test consisted in imposing a computed (exact) flux along Γ_1 and getting a numerical approximation of u . The domain has been discretized into consecutive ring–like layers of growing thickness, the layers themself being cut into $n + 1$

Table 1 Maximal error along Γ_1 for the case A

	n=16	n=32
2 layers	0.660	0.815
4 layers	0.396	0.643
8 layers	0.128	0.369
16 layers	0.031	0.100
Series app.	0.025	0.006

Table 2 Maximal error along Γ_1 for the case B

	n=16	n=32
2 layers	0.416	0.646
4 layers	0.158	0.374
8 layers	0.063	0.109
16 layers	0.057	0.021
Series app.	0.057	0.016

quadrilaterals by equally spaced radii. The ratio r has been chosen in such a way, that the quadrilaterals resembled squares as well as possible (so r and the thicknesses of layers depended on n).

Two methods of the solution have been compared. First, it has been computed by the method presented in this paper and therefore using only the first layer of discretization in order to compute the stiffness matrix S. Second, the ordinary FEM solutions have been found over the growing number of layers, with null boundary condition at the outermost boundary of the last layer. The approximation was in both cases linear, i.e. the quadrilaterals have been divided into four triangles on which the linear elements were used, and the internal node was eliminated. The results for the function (A) are shown in Table.1. The fact, that the error of ordinary FEM for $n = 32$ approaches its counterpart for series method slower than for $n = 16$, is connected with thickness of layers: for $n = 32$ they are thinner and therefore 16 layers (n=32) contain smaller area than 16 layers (n=16). The results for function (B) look qualitatively the same. Observe, however, that now the convergence of the ordinary FEM model to the corresponding solutions obtained using our method is faster. The explanation is simple: the function decays faster and more far away layers may be disregarded.

Corner singularity. In this case the test domain constituted a unit circle with one

Table 3 Maximal error and convergence rate for corner singularity

	n=6	n=12	n=24	$L_2(\Gamma)$–conv. rate
FEM	0.083	0.023	0.008	1.27
Series app.	0.057	0.012	0.003	1.60

Table 4 Maximal error and convergence rate for homogeneous case

	n=6	n=12	n=24	L_2–conv. rate
FEM	9.79	3.34	1.03	2.25
Series app.	4.76	1.49	0.45	2.22

Table 5 Maximal error and convergence rate for nonhomogeneous case

	n=6	n=12	n=24	L_2–conv. rate
FEM	0.246	0.072	0.021	2.18
Series app.	0.197	0.049	0.013	2.17

quarter cut out. On the radii bounding the cut out part the homogeneous Neumann boundary conditions have been imposed. A well known singular solution $u = r^{2/3} \cdot \cos(\frac{2}{3}\phi)$ has been used as a test function (Grisvard,1985). The discretization consisted of n evenly spaced radii and $[n/3]$ rings, similarly as in former case. We have computed two error indicators: maximal error along the line $\Gamma = r = 0.5$ and $L_2(\Gamma)$ convergence rate. The star mentioned in the last section consisted of all triangles having a vertex in 0. The results are summarized in Table 3. Since convergence rate concerns a projection on a line, theoretically it should be equal to 1.166... for ordinary FEM and 1.5 for FEM using local singular elements (Grisvard,1985). As we see, our approach is as good as the second case.

Improving accuracy. Here the computational domain consisted of the square $[0,4] \times [0,4]$ divided into $n \times n$ subsquares. We have compared the performance of the ordinary linear finite element with our modified one for two cases: [A] Test function $u = \exp(x)\sin y$ satisfying homogeneous Dirichlet equation. The results are in Table 4. [B] Test function $u = y^2\exp(x)$ satisfying nonhomogeneous Dirichlet equation. The results are in Table 5. In both cases we have obtained the same rate of convergence as for ordinary FEM, with twice better accuracy.

REFERENCES

Cooke, R.G. (1950) *Infinite Matrices and Sequence Spaces.* MacMillan and Co., London.

Givoli, D. (1992) *Numerical methods for problems in infinite domains.* Elsevier, Amsterdam.

Grisvard, P. (1985) *Elliptic problems in nonsmooth domains*, Pitman, London. 1985.

M. Grote, M. and Keller, J.B. (1994) Nonreflecting boundary conditions. Danish Center for Applied Mathematics and Mechanics Anniversary volume, Technical University of Denmark, (submitted to J. Comput. Phys.).

Sharau, S.K. (1994) Finite element analysis of infinite solids using elastic supports. *Computers and Structures*, vol. 13, No. 5, 1145–1152.

Stanley, R.P. (1986) *Enumerative Combinatorics.* Wadsworth & Brooks/Cole, Monterey, California.

PART FIVE

Mechanical Applications

36

Evolution law for shock strength in simple elastic structures

Sławomir Kosiński
Dept. of Mechanics of Materials K-61,
Łódź University of Technology, Al. Politechniki 6, 93-590 Łódź, Poland.
Phone/Fax: (048) (042) 313551.
E-mail: `slawek@kmm-sun.p.lod.edu.pl`

Abstract
The longitudinal shock wave propagates in the semi-infinite thin elastic layer and the material region in front of it is unstrained and at rest. Rods and layers are simple elastic structures and they are usually modeled as one-dimensional objects. In the paper a one-dimensional continuum theory with one internal scalar variable is used to describe the thin layer with propagating shock. In such a way the effect of small but finite transversal dimensions can be considered. From the singular surface theory and the equations of motion results a set of coupled evolution equations for the shock amplitude and the amplitudes of higher order discontinuities which accompany the shock. Using the perturbation method with a small perturbation parameter - the initial amplitude of the shock, we obtain the approximate solution of the problem; the longitudinal (or quasilongitudinal) shock propagates accompanied by the discontinuity of third order connected with the transverse motion.

Keywords
shock waves, singular surface theory, nonlinear elastic material, evolution laws, perturbation methods, unidimensional theory

1 INTRODUCTION

In the simple nonlinear theories for wave propagation a slender layer can be treated as one-dimensional object endowed with structure i.e. additionally one internal scalar variable is introduced. This variable represent the axisymmetric motion and includes the effects connected with the transverse dimension (Wright, 1981). In this paper we wish to apply this approximation method to the studies of shock wave propagation in slender elastic layer. For above problem we are interested in the differences in quality between the averaged unidimensional theory for 3-D slender elastic layer (Wright, 1981) and the simplest 1-D theory for nonlinear wave propagation in thin uniform layer (Fu and Scott, 1989). The simplest perturbation solution of the problem gives the uncoupled with the transverse motion system of equations for functions connected with the longitudinal motion only, identical as in Fu and Scott (1989). The influence of transversal shear and inertia gives additional effects for shock propagation. According to the boundary conditions, the longitudinal shock wave corresponding with the main motion, propagates accompanied by either third order discontinuity (see Figure 1) or third order discontinuity with transversal shock (see Figure 2). We shall work with the isentropic approximation and with nonlinear Murnaghan material. Numerical analysis (Kosiński, 1995) shows that only shocks of a relatively small (of order up to 10^{-3}) strength can propagate in Murnaghan material. Thus it is reasonable then to consider perturbation methods as a means of finding an approximate solution of the evolution problem.

2 BASIC EQUATIONS

Suppose that a plane shock wave propagates through a thin elastic layer in the direction of the X_1 axis in region $X_1>0$. See Figure 1. The axial symmetric motion (according to X_2 axis) of the semi-infinite layer is assumed in the form :

$$x_1 = X_1 + u_1(X_1,t), \quad x_2 = X_2 + f(X_2)u_2(X_1,t), \quad x_3 = X_3, \tag{1}$$

where: $f(X_2)$ is an arbitrary *odd* function $f \in C^1(\langle -h,h\rangle \to \mathrm{R})$, and functions $u_k \in C^3(\langle 0,\infty)\times\langle 0,\infty) \to R)$ $k=1,2$ (see (16)) x_i (i=1,2,3) is the position at time t of a particle which is at the position X_α (α=1,2,3) in the undeformed configuration B_R and the principal assumption (Wright, 1981)) is made that this motion occurs without application of perceptible contact forces to the lateral boundaries $X_2=\pm h$ (see Figure 1). Here and throughout this paper we assume the simplest form of $f(X_2)=X_2$. The displacements U_1, U_2 the strains e_1, e_2 and the particle velocities v_1, v_2 are given by

$$U_1 = x_1 - X_1 = u_1(X_1,t), \quad U_2 = x_2 - X_2 = X_2 u_2(X_1,t)$$
$$\dot{x}_1 = v_1 = \dot{u}_1(X_1,t), \quad \dot{x}_2 = v_2 = X_2\dot{u}_2(X_1,t) \qquad e_1 = u_{1,1}, \; e_2 = u_{2,1} \tag{2}$$

For the assumed motion (1) the deformation gradient is given by

$$[x_{i\alpha}] = \begin{bmatrix} 1+u_{1,1}(X_1,t) & 0 & 0 \\ X_2 u_{2,1}(X_1,t) & 1+u_2(X_1,t) & 0 \\ 0 & 0 & 1 \end{bmatrix}, \tag{3}$$

The material region F in front of the propagating shock $\Sigma(X_1,t)$ is initially unstrained and at rest, the region B behind of it (comp. (1)) is in a state of plane strain deformation. The jumps of both displacements U_1 and U_2 on the discontinuity surface $\Sigma(X_1,t)$ are equal zero, for this reason because of continuity we obtain that the jumps of the functions $u_1(X_1,t), u_2(X_1,t)$ are also equal zero

$$[U_1] = 0 \Rightarrow [u_1] = 0, \quad [U_2] = 0 \Rightarrow [u_2] = 0, \tag{4}$$

but we assume that the both derivatives $u_{1,1}$, $u_{2,1}$ can suffer a jump on the discontinuity surface $\Sigma(X_1,t)$. The deformation gradient on both sides of $\Sigma(X_1,t)$ has the form

$$[x_{i\alpha}]^B = \begin{bmatrix} 1+u_{1,1}(X_1,t) & 0 & 0 \\ X_2 u_{2,1}(X_1,t) & 1 & 0 \\ 0 & 0 & 1 \end{bmatrix}, \qquad [x_{i\alpha}]^F = \delta_{i\alpha} \tag{5}$$

3 AVERAGED EQUATION OF MOTION

The equations of motion for the deformation gradient (3) are reduced to the system of equations for plane deformation:

$$\begin{aligned} &T_{R11,1} + T_{R12,2} = \rho_R \ddot{u}_1 \\ &T_{R21,1} + T_{R22,2} = \rho_R X_2 \ddot{u}_2 \\ &T_{R33,3} = 0, \ \textit{(identity)} \end{aligned} \tag{6}$$

where the stress component $T_{Ri\alpha}$ used here are the unsymmetric Piola-Kirchhoff stress components and ρ_R is the density in B_R. The boundary conditions on the lateral surfaces of the layer require that two stress components vanish.

$$\left.\begin{aligned} &t_i = T_{Ri\alpha} K_\alpha = 0, \\ &K_\alpha = (0,1,0) \end{aligned}\right\} \Rightarrow T_{R12} = T_{R22} = 0 \text{ for } X_2 = \pm h \tag{7}$$

The equation $(6)_2$ is multiplied by X_2 and we average now the both equations of motion (6) over the unit length cross section A (A =1 x $2h$ has unit length in the direction of the X_3 axis and is perpendicular to the X_1 axis). After integration of (6) and for the stress free lateral surfaces $T_{R22}\big|_{-h}^{h} = T_{R12}\big|_{-h}^{h} = 0$ we obtain equations

$$\begin{aligned} &\frac{\partial}{\partial X_1}\left(\frac{1}{2h}\int_{-h}^{h} T_{R11} dX_2\right) = \rho_R \ddot{u}_1, \\ &\frac{\partial}{\partial X_1}\left(\frac{1}{2h}\int_{-h}^{h} X_2 T_{R21} dX_2\right) - \frac{1}{2h}\int_{-h}^{h} T_{R22} = \rho_R \ddot{u}_2 \frac{h^2}{3}, \end{aligned} \tag{8}$$

The above system of averaged equations is consistent with the system of Euler-Lagrange equations which for such two independent functions have the form

$$u_1(X_1,t) \Rightarrow \quad \frac{\partial}{\partial t}\left(\frac{\partial L}{\partial \dot{u}_1}\right)+\frac{\partial}{\partial X_1}\left(\frac{\partial L}{\partial u_{1,1}}\right)-\frac{\partial L}{\partial u_1}=0,$$
$$u_2(X_1,t) \Rightarrow \quad \frac{\partial}{\partial t}\left(\frac{\partial L}{\partial \dot{u}_2}\right)+\frac{\partial}{\partial X_1}\left(\frac{\partial L}{\partial u_{2,1}}\right)-\frac{\partial L}{\partial u_2}=0 \tag{9}$$

where $L = K - W(e_1, e_2, u_2)$ is the Lagrangian function, K- kinetic energy and W- internal elastic energy is function of both strains e_1, e_2 and displacement u_2. After some calculations we finally obtain the form. (The prime indicates differentiation with respect to X_1).

$$S' = \rho_1 \ddot{u}_1, \quad Q' - P = \rho_2 \ddot{u}_2, \quad \text{with} \quad S = \frac{\partial W}{\partial u_{1,1}}, Q = \frac{\partial W}{\partial u_{2,1}}, P = \frac{\partial W}{\partial u_2} \tag{10}$$

The comparison of the two equivalent equations forms gives the following relationships between the stresses and averaged values and densities

$$S = \frac{1}{2h}\int_{-h}^{h} T_{R11} dX_2\,, \quad Q = \frac{1}{2h}\int_{-h}^{h} X_2 T_{R21} dX_2\,, \quad P = \frac{1}{2h}\int_{-h}^{h} T_{R22} dX_2\,,$$
$$\rho_1 = \rho_R, \quad \rho_2 = \rho_R \frac{h^2}{3} \tag{11}$$

The averaging of the stresses over the cross section, gives the system of three forces P, Q, S and every such quantity is function of two variables; variable X_1, and internal scalar variable h - half thickness of the layer.

4 APPLICATION FOR SHOCK ANALYSIS

Very important for the study of shock propagation in 1-D are the kinematical conditions of compatibility (Chen (1971), Fu and Scott (1989)) whichever of them can be derived from:

$$\frac{d}{dt}[f] = [\dot{f}] + U_N [f_{X_1}], \qquad \frac{d}{dt} = U_N \frac{d}{dX_1} \tag{12}$$

The space derivative d/dX_1 following the wave front $\Sigma(X_1,t)$ is related to the displacement derivative d/dt by $(12)_2$, U_N is the normal shock speed. Replacing $f(X_1$,t) in (12) in turn by: u_1, $v_p = \dot{u}_p$ and $v_{p,X_1} = \dot{u}'_p$ we obtain through continual use of $(12)_1$ relation which are very useful for transformation of the equation of motion.

From the equation expressing the balance of momentum on the discontinuity surface and assuming that only two components of the deformation gradient $[x_{11}]\neq 0$ and $[x_{21}]\neq 0$ suffer a jump on $\Sigma(X_1,t)$ we obtain (see (5)) the expression for the shock wave speed .

$$U_N^2 = \frac{[T_{R11}][x_{11}]+[T_{R21}][x_{21}]}{[x_{11}]^2+[x_{21}]^2} \tag{13}$$

The entropy jump $[\eta]$ ($[\eta]>0$ for stable shocks) can be calculated from the balance laws on the discontinuity surface (Wesołowski (1978)), because $\left(T_{Ri\alpha}\right)^F = 0$

$$2\rho_R U_N[\sigma]+[T_{Ri\alpha}][\dot{x}_i]N_\alpha = -2\left(T_{Ri\alpha}\right)^F N_\alpha[\dot{x}_i] \Rightarrow \quad [\sigma]=\frac{[T_{Ri\alpha}]}{2\rho_R}[x_{i\alpha}] \tag{14}$$

The averaged equations of motion are satisfied on both sides of the discontinuity surface $\Sigma(X_1,t)$. Differentiating these equations with respect to X_1 we obtain

$$S' = \rho_1\ddot{u}_1, \quad Q'-P = \rho_2\ddot{u}_2, \quad \text{and} \quad S'' = \rho_1\ddot{u}_1', \quad Q''-P' = \rho_2\ddot{u}_2', \tag{15}$$

On taking the jumps of last two above equations across $\Sigma(X_1,t)$ we finally obtain the basic system of equations for the analysis.

$$[S'] = \rho_1[\ddot{u}_1], \quad [Q']-[P] = \rho_2[\ddot{u}_2], \qquad 16)$$

Using the compatibility conditions for the jumps of the right hand sides of them we can obtain system of equations for shock amplitudes and the amplitudes of higher order discontinuities which accompany the shock.

5 SELECTED PARTICULAR CASES

Analysis of the problem is restricted to a special kind of second order elastic material, called Murnaghan material

$$\begin{aligned} W(e_1,e_2,u_2) = \rho_R\sigma &= \frac{1+2m}{24}(I_1-3)^3+\frac{\lambda+2\mu+4m}{8}(I_1-3)^2+\frac{8\mu+n}{8}(I_1-3) \\ &-\frac{m}{4}(I_1-3)(I_2-3)-\frac{4\mu+n}{8}(I_2-3)+\frac{n}{8}(I_3-1) \end{aligned} \tag{17}$$

where $I_1=B_{ii}$, $I_2=(B_{ii}B_{jj}-B_{ij}B_{ij})/2$, $I_3=\det(B_{ij})$ are the invariants of the left Cauchy-Green strain tensor B_{ij} , λ and μ are Lamé coefficients, and l , m , n are the elastic constants of second order.

On $\Sigma(X_1,t)$ all functions depend on one single variable X_1, the deformation gradient has the form (5) and we seek perturbation solutions in the form:

$$[e_1]=\varepsilon Y_1+\varepsilon^2 Y_2+.., \quad [e_{1,1}]=Z_0+\varepsilon Z_1+.., \quad [e_{1,11}]=W_0+\varepsilon W_1+..,$$
$$[e_2]=\varepsilon\hat{Y}_1+\varepsilon^2\hat{Y}_2+.., \quad [e_{2,1}]=\hat{Z}_0+\varepsilon\hat{Z}_1+.., \quad [e_{2,11}]=\hat{W}_0+\varepsilon\hat{W}_1+.., \tag{18}$$

where ε is a small dimensionless perturbation parameter which can be taken to be initial shock strength and we assume the boundary conditions at $X_1=0$

$$[e_1]\big|_{\hat{X}_1=0}=\hat{h}\Rightarrow\varepsilon Y_1(0)\approx\hat{h}, \quad [e_{1,1}]\big|_{\hat{X}_1=0}=k \Rightarrow Z_0(0)\approx k,$$
$$[e_2]\big|_{X_1=0}=0\Rightarrow\varepsilon\hat{Y}_1(0)\approx 0, \quad [e_{2,1}]\big|_{X_1=0}=0 \Rightarrow \hat{Z}_0(0)\approx 0 \tag{19}$$

Substituting these expansions into the basic system of equations we obtain the *uncoupled* (with the transverse motion) system of equations for functions connected with the longitudinal motion only. Equating the coefficients of like power ε^0 ε^1 we have:

$$\varepsilon^1\Rightarrow \quad \frac{1}{4}c_1Y_1(X_1)Z_0(X_1)+\frac{dY(X_1)}{dX_1}=0, \quad \varepsilon^0\Rightarrow \quad \frac{1}{2}c_1Z_0^2(X_1)+\frac{dZ_0(X_1)}{dX_1}=0 \tag{20}$$

where $c_1=\dfrac{(3\lambda+6\mu+2l+4m)}{(\lambda+2\mu)}$

There are exactly the equations and solutions obtained (Fu and Scott, (1989)) after the application of the simplest theory for nonlinear wave propagation in a straight uniform rod.

$$[e_{1,1}]=\frac{k}{1+\dfrac{c_1kX_0}{2}X_1}, \quad [e_1]=\frac{\hat{h}}{\sqrt{1+\dfrac{c_1kX_0}{2}X_1}} \tag{21}$$

For the transverse motion the equations are partly coupled with the longitudinal motion and the first two terms in expansions have the form

$$[e_{2,1}]\approx\hat{Z}_0=0, \; [e_{2,11}]\approx\hat{W}_0(X_1)=-\frac{\lambda}{\lambda+\mu}\frac{3}{h^2}Z_0(X_1)=-\frac{\lambda}{\lambda+\mu}\frac{3}{h^2}\frac{k}{1+\dfrac{c_1kX_0}{2}X_1} \tag{22}$$

For the isentropic approximation used here the entropy condition is satisfied and the shock is stable. According to our perturbation procedure, from the fact that $e_2^F=e_{2,1}^F=0\Rightarrow[e_{2,1}]=e_{2,1}^B$ and that the jump $[e_{2,1}]\approx\hat{Z}_0=0$, it follows that on the discontinuity surface (where all the functions depend on one single co-ordinate X_1 only) the function $[e_2]$ remains unchanged $[e_2]\approx\varepsilon\hat{Y}_1(0)=\varepsilon\hat{Y}_1(X_1)=const$,

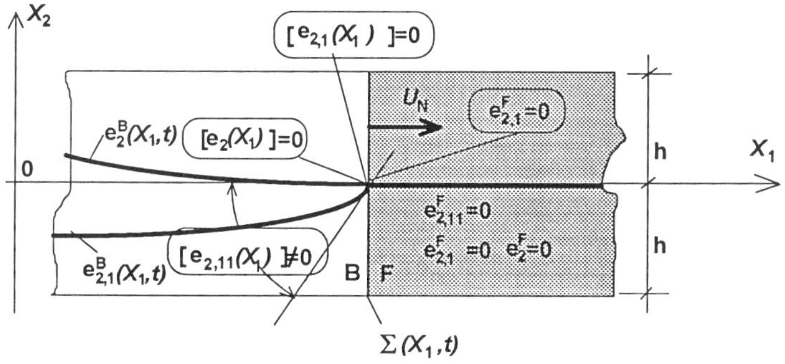

Figure 1 Third order discontinuity $[e_{2,11}]$ corresponding with the transverse motion which accompany the main longitudinal discontinuity of first order

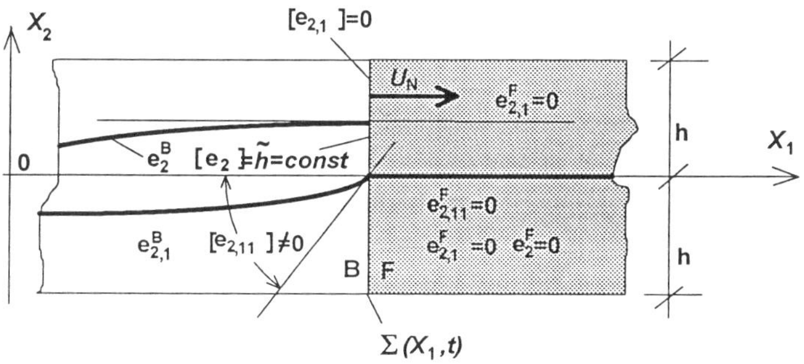

Figure 2 Constant discontinuity of first order $[e_2]=\tilde{h}=const$ and the third order discontinuity $[e_{2,11}]$ caused by transverse motion coupled with the main longitudinal motion.

Assuming now (see Figure 2) that the initial longitudinal shock $\hat{h}$ is accompanied by the transverse one $\tilde{h}$ i.e.

$$[e_2]\Big|_{X_1=0}=\tilde{h}\Rightarrow \varepsilon\hat{Y}_1(0)\approx\tilde{h}\,,\quad [e_{2,1}]\Big|_{X_1=0}=0\quad\Rightarrow \hat{Z}_0(0)\approx 0 \tag{23}$$

When a shock is initiated by longitudinal loading only the relationship between the initial shock values in the boundary conditions can be given as $T_{R21}(\hat{h},\tilde{h})=0$.

REFERENCES

Chen, P.J. (1971) One dimensional shock waves in elastic non-conductors. *ARMA*, **43**, 350-362.

Fu, Y.B.and Scott, N.H. (1989) The evolution law of one-dimensional weak nonlinear shock waves in elastic non-conductors. *Q.Jl Mech. Appl. Math.*, **42**, 23-39.

Kosiński, S. (1995) Transport equation for shock strength in hyperelastic rods. *Engineering Transactions* **43**, 205-224

Wesołowski, Z. (1978) Strong discontinuity wave in initially strained elastic medium. *Archives of Mechanics*, **30**, 309-322.

Wright, T.W. (1981) Nonlinear waves in rods, in *Proceedings of the IUTAM Symposium on Finite Elasticity, Lehigh University,August 10-15,1980* (ed. D.E.Carlsohn and R.T.Shield) 423-443, Martinus Nijhoff Publishers, The Hague/Boston/London

37

On a class of composite plates of maximal compliance

T.Lewiński and A.M.Othman
Warsaw University of Technology
Civil Engineering Faculty, Institute of Structural Mechanics,
al. Armii Ludowej 16, 00-637 Warsaw, Poland

Abstract

Within the class of thin, transversely loaded plates of composite microstructure undergoing bending as well as transverse shearing an effective elastic potential of the weakest plate is found and its nonlinear characterization is explicitly given. The optimal design of the plate of maximal compliance turns out to coincide with the optimal design of the weakest plate of microstructure upon which Kirchhoff' s constraints are imposed.

Keywords

Optimum plate design, composite plates, homogenization

1 INTRODUCTION

A correct designing of thin two-phase elastic plates of extremal compliance requires a relaxation, cf Kohn and Vogelius(1986). In the relaxed problem one admits mixing the phases at the microstructural level. The plate is then characterized by effective stiffnesses given by Duvaut-Metellus formulae for Kirchhoff plate homogenization. Such optimal designs can be found in Lur'e and Cherkaev(1986) .This approach goes along similar lines as that developed for plane elasticity (cf Allaire and Kohn(1993)). However, its application to plate problems is controvertible, since the three-dimensional analysis of accuracy of Duvaut-Metellus formulae shows that they assess the stiffnesses incorrectly, in particular those responsible for torsion. An essential improvement in evaluating stiffnesses can be achieved if one admits transverse shear deformation of the microstructure, see Lewiński (1992). Therefore it is proposed in this paper to endow the optimized plate with a Hencky-Reissner (not Kirchhoff) microstructure by using homogenization formulae derived by the conformal (or refined) scaling, cf Lewiński (1992), Telega (1992). The local subproblem is solved by the translation method, cf Lur'e and Cherkaev (1986).

2 FINDING THE WEAKEST TWO-PHASE PLATE

Consider a thin plate whose middle plane occupies a plane domain Ω parametrized by a Cartesian coordinate system x_α, e_α being its versors, α and other small Greek letters run over 1,2. Tensors (a_i) defined by

$$a_1 = a_{11} + a_{22}\ , \quad a_2 = a_{11} - a_{22}\ , \quad a_3 = a_{12} + a_{21}\ , \quad a_4 = a_{12} - a_{21} \tag{1}$$

where $a_{\alpha\beta} = 2^{-\frac{1}{2}} e_\alpha \otimes e_\beta$ constitute an orthonormal basis in the space of second order tensors. The stiffness tensor of a two-phase thin plate is given by

$$D(x) = D_1\chi_1(x) + D_2\chi_2(x) \tag{2}$$

χ_α being indicator functions of the plate phases characterized by the isotropic stiffnesses

$$D_\alpha = 2k_\alpha a_1 \otimes a_1 + 2\mu_\alpha(a_2 \otimes a_2 + a_3 \otimes a_3)\ ; \tag{3}$$

k_α and μ_α represent elastic moduli. The ordered case is dealt with: $k_2 > k_1 > 0$ and $\mu_2 > \mu_1 > 0$. The constitutive relationships are of Kirchhoff' s type

$$M^{\alpha\beta} = D^{\alpha\beta\lambda\mu}\kappa_{\lambda\mu}(w)\ , \qquad \kappa_{\lambda\mu}(w) = -w_{,\lambda\mu}\ , \tag{4}$$

$D^{\alpha\beta\lambda\mu}$ being components of D in the basis $e_\alpha \otimes e_\beta \otimes e_\lambda \otimes e_\mu$ and $()_{,\lambda} = \partial/\partial x_\lambda$. A scalar field v defined on Ω is said to be kinematically admissible (kin.adm.) if $v \in H_0^2(\Omega)$, i.e. the plate is clamped along its boundary. The plate is subject to transverse loading $p = p(x)\ ,\ x = (x_\alpha)$. The compliance of the plate

$$J(\chi_1) = \int_\Omega pw dx \tag{5}$$

is viewed as a functional over all possible $\chi_1(\chi_2 = 1 - \chi_1)$, i.e. over all layouts of the phases; the deflection function w represents the solution to the Kirchhoff plate problem

$$-\frac{1}{2}J(\chi_1) = \inf\left\{\frac{1}{2}\int_\Omega [\kappa_{\lambda\mu}(v)D^{\lambda\mu\alpha\beta}\kappa_{\alpha\beta}(v) - 2pv]dx \,\middle|\, v \text{ kin.adm.}\right\} \tag{6}$$

with D depending upon χ_1 according to Eq.(2).

Assume that the amounts of both plate materials are given, i.e.

$$\int_\Omega \chi_1 dx = A_1 \tag{7}$$

A_1 being a constant. Define

$$J_\lambda(\chi_1) = \frac{1}{2}J(\chi_1) - \lambda\int_\Omega \chi_1 dx \tag{8}$$

λ being a Lagrangian multiplier. Since the weakest plate is characterized by a maximal compliance $J(\chi_1)$ it is sufficient to consider the auxiliary problem $\sup\{J_\lambda(\chi_1)|\chi_1\}$ that by (6) can be cast in the form

$$\inf_{\chi_1}\ \inf_{v\ kin.adm.}\left\{\frac{1}{2}\int_\Omega [\kappa_{\alpha\beta}(v)D^{\alpha\beta\lambda\mu}(x)\kappa_{\lambda\mu}(v) - 2pv + 2\lambda\chi_1]dx\right\} \tag{9}$$

This problem is ill-posed, cf Kohn and Strang (1986), Lipton (1994). To relax and thus make the problem well-posed one should endow the plate with a two-phase microstructure of thin plate (Kirchhoff) type, cf Lipton(1994). This formulation will not be recalled. In the present paper we shall consider a seemingly slight modification of the relaxed formulation (9). In place of the Kirchhoff's microstructure we assume Hencky-Reissner's microstructure undergoing additionally a transverse shear deformation. Let us introduce indispensable notions. The mid-plane of the periodicity cell is denoted by $Y = (0, l_1) \times (0, l_2)$. The layout of the bending stiffnesses within Y is determined by indicator functions $\chi_\alpha^Y(x, \cdot)$

$$D(x,y) = D_1\chi_1^Y(x,y) + D_2\chi_2^Y(x,y) \ , \quad y \in Y \ , \tag{10}$$

where D_α are given by (3). The transverse shear stiffness tensor H is assumed to be common for both phases: $H = 2\frac{1}{2} g a_1$, where g is the value of the transverse shear stiffness and tensor a_1 is given by (1) . Such a mismatch between the layouts of D and H is possible if the microstructure is transversely isotropic . The averages over Y

$$\langle \chi_\alpha^Y(x,y) \rangle = \theta_\alpha(x) \ , \quad \langle \cdot \rangle = \frac{1}{l_1 l_2} \int\limits_Y (\cdot) dy \tag{11}$$

are functions : $\theta_\alpha : \Omega \to [0,1]$ interrelated with χ_α by the relations

$$\int\limits_\Omega \theta_\alpha dx = \int\limits_\Omega \chi_\alpha dx \ . \tag{12}$$

Let us define bending and shearing deformations as functions of the fields $\phi = (\phi_\alpha)$, w defined on Y:

$$e_{\alpha\beta}(\phi) = \frac{1}{2}(\phi_{\alpha|\beta} + \phi_{\beta|\alpha}) \ , \ \gamma_\alpha(\phi, w) = \phi_\alpha + w_{|\alpha} \ , \tag{13}$$

where $()_{|\alpha} = \partial / \partial y_\alpha$. Problem (9) is replaced with

$$\inf_{\theta_1 : \Omega \to [0,1]} \ \inf_{v \ kin.adm.} \left\{ \frac{1}{2} \int\limits_\Omega [W_H(\kappa(v); \theta_1) - 2pv + 2\lambda\theta_1] dx \right\} \ , \tag{14}$$

where

$$W_H(\kappa; \theta_1) = \inf \left\{ W(\chi_1^Y; \kappa, \theta_1) \,\middle|\, \langle \chi_1^Y \rangle = \theta_1 \right\} \tag{15}$$

and

$$\begin{aligned} W(\chi_1^Y; \kappa, \theta_1) = \inf \Big\{ & \underline{\langle e_{\alpha\beta}(\phi) D^{\alpha\beta\lambda\mu}(x,y) e_{\lambda\mu}(\phi) + \gamma_\alpha(\phi, w) H^{\alpha\beta} \gamma_\beta(\phi, w) \rangle} \ \Big| \\ & e = (e_{\alpha\beta}) \ , \ \gamma = (\gamma_\alpha) \text{ such that they are of the form (13), are } Y-\text{periodic and} \\ & \langle e_{\alpha\beta}(\phi) \rangle = \kappa_{\alpha\beta} \Big\} \ , \end{aligned} \tag{16}$$

where $H = H^{\alpha\beta}e_\alpha \otimes e_\beta$. On can show that $\langle\gamma_\alpha\rangle = 0$ due to assumption of H being constant. The expression underscored represents a sum of bending and transverse shearing energies. Definition of the effective potential W follows from the formula for effective stiffnesses of moderately thick plates, cf Lewiński (1992), Telega (1992). Formulation (14) is correct similarly as the relaxed problem within Kirchhoff's theory (then $\phi_\alpha = -w_{|\alpha}$ in (16)). Obviously, formulation (14) is neither full nor partial relaxation of problem (9) (notions introduced in Kohn and Vogelius (1986), see also Bonnetier and Conca (1994)), since it exceeds the framework of the Kirchhoff' s theory. This problem is an approximation of Hencky-Reissner type of the optimization problem of the two-phase transversely homogeneous plate within the three-dimensional description.

3 BOUNDING THE EFFECTIVE POTENTIAL $W_H(\chi_1^Y;\kappa,\theta_1)$

To take into account differential constraints (13) involved in (16) it is helpful to apply the translation method, see Lur'e and Cherkaev (1986), Allaire and Kohn (1993), Cherkaev and Gibianski (1993). For technical aims we define

$$\begin{aligned} &E = [\phi_{1|1}\ ,\ \phi_{1|2}\ ,\ \phi_{2|1}\ ,\ \phi_{2|2}\ ,\ w_{|1} + \phi_1\ ,\ w_{|2} + \phi_2]^T \\ &\mathcal{D} = diag[D, \tilde{H}]\ , \quad \tilde{H} = diag[H, H] \end{aligned} \tag{17}$$

Then one can write

$$e_{\alpha\beta}D^{\alpha\beta\lambda\mu}e_{\lambda\mu} + \gamma_\alpha H^{\alpha\beta}\gamma_\beta = E^T\mathcal{D}E \tag{18}$$

We need the estimate

$$\langle E^TTE\rangle \geq \langle E^T\rangle T\langle E\rangle \quad \forall E \text{ of the form (17)}\ , \tag{19}$$

where T is a constant 6×6 matrix. A good candidate for T is

$$T = \begin{bmatrix} c & & & d & b & b \\ & c & -d & & -b & b \\ & -d & c & & b & -b \\ d & & & c & b & b \\ b & -b & b & b & a & \\ b & b & -b & b & & a \end{bmatrix} \tag{20}$$

where if

(a) $c = 0$, then $b = 0$, $a \geq 0$

(b) $c = 0$, then $a \geq 0$, $c > 0$, $2b^2 \leq ac$

If $c = a = b = 0$, then (19) becomes an equality for all d and E . The proof of (19) can be performed by Fourier analysis. It is omitted to save space.

Now let us change the basis

$$\phi_{\beta|\alpha}e_\alpha \otimes e_\beta = \delta_i a_i\ ,\ \kappa_{\alpha\beta}e_\alpha \otimes e_\beta = \kappa^i a_i\ ,\ i = 1, \ldots, 4 \tag{21}$$

hence

$$E^T T E = \tilde{E}^T \tilde{T} \tilde{E}\ ,\ \ \tilde{E} = [\delta, \gamma]\ ,\ \ \delta = [\delta_1, \delta_2, \delta_3, \delta_4]\ , \gamma = [\gamma_1, \gamma_2] \tag{22}$$

where

$$\tilde{T} = \begin{bmatrix} 2r & & & & t & t \\ & 2s & & & & \\ & & 2s & & & \\ & & & 2r & -t & t \\ t & & & -t & a & \\ t & & & t & & a \end{bmatrix}$$

and

$$t = \sqrt{2}b\ ,\quad 2r = c + d\ ,\quad 2s = c - d \tag{23}$$

Parameter t couples bending and transverse shearing deformations. The constitutive relationships assume a diagonal form

$$\sigma = \tilde{D}\tilde{E}\ ,\quad \sigma = [M^1, M^2, M^3, M^4, Q^1, Q^2] \tag{24}$$

$$\tilde{D} = diag[2k, 2\mu, 2\mu, 0, g, g] \tag{25}$$

and

$$M = M^{\alpha\beta} e_\alpha \otimes e_\beta = M^i a_i\ ,\quad Q = Q^\alpha e_\alpha \tag{26}$$

with k, μ defined similarly as D in Eq. (10). Now let us express W in a new form

$$\begin{aligned} &W(\chi_1^Y; \kappa, \theta_1) = \inf\left\{\langle \delta^T \tilde{D}\delta + \gamma^T \tilde{H}\gamma\rangle \,\middle|\, \langle\delta_i\rangle = \kappa^i \text{ for } i = 1, 2, 3\ ;\right. \\ &\left.\delta, \gamma \text{ are } Y-\text{periodic and defined by(22); } \langle\delta_4\rangle = 0\ ,\ \langle\gamma_\alpha\rangle = 0\right\} \end{aligned} \tag{27}$$

or

$$\begin{aligned} &W(\chi_1^Y; \kappa, \theta_1) = \inf\left\{\langle \tilde{E}^T \tilde{\mathcal{D}} \tilde{E}\rangle \,\middle|\, \tilde{E} \text{ given by (22)with conditions given above}\right\} \\ &\tilde{\mathcal{D}} = diag[\tilde{D}, \tilde{H}] \end{aligned} \tag{28}$$

The idea of the translation method is to decompose $\langle \tilde{E}^T \tilde{\mathcal{D}} \tilde{E}\rangle$

$$\langle \tilde{E}^T (\tilde{\mathcal{D}} - \tilde{T})\tilde{E}\rangle + \langle \tilde{E}^T \tilde{T} \tilde{E}\rangle\ , \tag{29}$$

omit differential constraints while estimating the first term and apply the estimate (19) in which these differential constraints are fully used. Hence

$$W(\chi_1^Y; \kappa, \theta_1) \geq E_0^T[\langle(\tilde{\mathcal{D}} - \tilde{T})^{-1}\rangle^{-1} + \tilde{T}]E_0 \qquad E_0 = \langle\tilde{E}\rangle = [\kappa^1, \kappa^2, \kappa^3, 0, 0, 0] \tag{30}$$

and parameters involved in $\tilde{T}$ should satisfy condition $(\tilde{\mathcal{D}} - \tilde{T}) \geq 0$ (in particular: $-2r \geq 0$, etc.). Note that the right hand-side is independent of χ_1^Y .

By virtue of a specific form of matrices $\tilde{\mathcal{D}}$ and $\tilde{T}$ one can evaluate the right hand side of (30) explicitly

$$\frac{1}{2}W(\chi_1^Y;\kappa,\theta_1) \geq \eta^2 \max\{(K^0+\xi^2G^0) \,\Big|\, \text{admissible } r,s,t,a\} \tag{31}$$

where $\eta = \kappa^1$, $\xi = [(\kappa^2)^2+(\kappa^3)^2]^{\frac{1}{2}}//\kappa^1$

$$K^0 = \frac{(g-a)(k_1k_2-r\bar{k})-\bar{k}t^2}{(g-a)(\tilde{k}-r)-t^2} , \quad G^0 = \frac{\mu_1\mu_2-s\bar{\mu}}{\tilde{\mu}-s} \tag{32}$$

and

$$\tilde{f} = f_1\theta_2+f_2\theta_1 , \quad \bar{f} = f_1\theta_1+f_2\theta_2 , \quad \Delta f = f_2-f_1 \tag{33}$$

for $f\in\{k,\mu\}$.

Note that η and ξ are invariants of κ ; η represents the spherical part of κ , ξ being the ratio of the deviatoric part of κ to η . One can show that maximum in (31) is attained for $t=0$ and then $c=0, r=-s$; condition $-2r\geq 0$ becomes irrelevant. Function $f(r) = K^0(r)+\xi^2G^0(r)$ attains maximum at

$$r_0 = (\tilde{k}\Delta\mu\cdot\xi-\Delta k\tilde{\mu})//(\Delta\mu\cdot\xi+\Delta k) \tag{34}$$

Since $K^0>0$ and $G^0>0$, we have

$$-r\leq\mu_1 \Rightarrow \xi\geq\xi_2 = \frac{\tilde{\mu}+k_1}{\theta_1\Delta\mu} \tag{35}$$

$$-r\geq -k_1 \Rightarrow \xi\leq\xi_1 = \frac{\theta_1\Delta k}{\tilde{k}+\mu_1} \tag{36}$$

Thus $W\geq \underline{W}(\kappa,\theta_1)$ and

$$\underline{W}(\kappa,\theta_1) = \begin{cases} W_2(\theta_1\ ,\eta,\xi) & \text{for} \quad \xi\leq\xi_2 \\ W_{int}(\theta_1\ ,\eta,\xi) & \text{for} \quad \xi_2\leq\xi\leq\xi_1 \\ W_1(\theta_1\ ,\eta,\xi) & \text{for} \quad \xi\geq\xi_1 \end{cases} \tag{37}$$

where $W_J(\theta_1\ ,\eta,\xi) = 2\eta^2(a_J+c_J\xi^2)$, $J=1,2$;

$$W_{int}(\theta_1\ ,\eta,\xi) = 2\eta^2[a_J+c_J\xi^2-A_J(\xi-\xi_J)^2] , \tag{38}$$

the last equation being true for both $J=1,2$; coefficients a_J, c_J, A_J depend on the data and will not be reported here. We see that $\underline{W}$ is composed of three parabolas ; the intermediate one is stitched smoothly with W_2 at $\xi=\xi_2$ and with W_1 at $\xi=\xi_1$. Hence the effective constitutive relationships are continuous.

4 CONCLUSIONS

The final result (37) coincides with that concerning entirely Kirchhoff relaxation (substitution $\phi_\alpha = -w_{|\alpha}$, in (16)), cf Lur'e and Cherkaev (1986). Thus the lower bound $\underline{W}$ can be attained by second rank laminates (regimes: $\xi \leq \xi_2$ and $\xi \geq \xi_1$) and by first rank laminates $(\xi_2 \leq \xi \leq \xi_1)$. In all cases eigendirections of κ coincide with directions of layering. Consequently torsion is absent along these directions, hence formulae for $D^{\alpha\alpha\beta\beta}_{hom}$ suffice for proving attainability of the bound. Since these effective stiffnesses are common for both conformal (refined) scaling- based approach and plane scaling-based Duvaut-Metellus' approach (see Lewiński (1991)) one concludes that the design of the weakest Kirchhoff plate with Kirchhoff's microstructure coincides with the weakest Kirchhoff's plate with Hencky-Reissner's microstructure.

ACKNOWLEDGMENTS

The work of the first author was supported by the Polish State Committee for Scientific Research through the grant No 3 P404 013 06; the work of the second author was supported by Statutory Project No 504//072//239//1 coordinated by the Faculty of Civil Engineering, Warsaw University of Technology.

REFERENCES

Allaire, G. and Kohn, R.V. (1993) Explicit optimal bounds on the elastic energy of a two-phase composite in two space dimensions, *Q.Appl.Math.*, **LI**, 675-699.

Bonnetier, E. and Conca, C. (1994) Approximation of Young measures by functions and application to a problem of optimal design for plates with variable thickness, *Proc.Roy.Soc.Edinb.*, **124** A, 399-422.

Cherkaev, A. and Gibianski, L.V. (1993) Coupled estimates for the bulk and shear moduli of a two-dimensional isotropic elastic composite, *J.Mech.Phys.Solids*, **41**, 937-980.

Kohn, R.V. and Strang, G. (1986) Optimal design and relaxation of variational problems -I,II,III. *Comm. Pure Appl. Math.*, **39**, 113-138; 139-182; 353-377.

Kohn, R.V. and Vogelius, M. (1986) Thin plates with varying thickness and their relation to structural optimization. In: *Homogenization and Effective Moduli of Materials and Media.* IMA Vol. Math. Appl. **1** .Eds.: J.L. Ericksen et al. pp 126-149, Berlin. Springer.

Lewiński, T. (1991) Effective models of composite periodic plates -III. Two-dimensional approaches. *Int.J.Solids Structures*, **27**, 1185-1203.

Lewiński, T. (1992) Homogenizing stiffnesses of plates with periodic structure. *Int.J.Solids Structures*, **29**, 309-326.

Lipton, R. (1994) Optimal design and relaxation for reinforced plates subject to random transverse loads. *Probab.Eng.Mech.*, **9**, 167-177.

Lur'e, K. and Cherkaev, A. (1986) Effective characteristics of composites and optimum structural design (in Russian) *Adv.Mech.(Uspekhi Mekhaniki)*, **9**, 3-81.

Telega, J.J. (1992) Justification of a refined scaling of stiffnesses of Reissner plates with fine periodic structure. *Math.Models Meth.Appl.Sci.*, **2**, 375-406.

38

An inaccuracy in semi-analytical analysis for Timoshenko beam

Niels Olhoff
University of Aalborg, Department of Mechanical Engineering
Pontoppidanstraede 101, DK-9220 Aalborg East, Denmark
Phone: +45 98 15 85 22 ext.3513. Fax: +45 98 15 14 11
and
Lucyna Bogdan
Systems Research Institute, Polish Academy of Sciences
Newelska 6, 01-447 Warsaw, Poland. Phone: 36 41 03. Fax: 37 27 72
E-mail: `bogdan@ibspan.waw.pl`

Abstract

In the present paper an exact analysis of the error of sensitivity for a model of Timoshenko beam is considered. The analysis gives a deep insight into the nature of the inaccuracy problem and enables us to devise methods by which the severe error of the sensitivity can be substantially reduced or removed for the model problem.

Keywords

Sensitivity analysis, inaccuracy problem, semi-analytical analysis

1 INTRODUCTION

In the present paper the inaccuracy problem through an exact analysis of a model problem is studied. The consideration of the problem is based on the exact analytical solution of the global set of finite element equations for the semi-analytical design sensitivity analysis problem. It is shown that exact sensitivities are obtained for the model problem if the pseudo-loads are computed via exact, analytical differentiation of the stiffness elements with respect to the design variable. If the pseudoloads are determinated via numerical differentiation, the relative error of sensitivity increases with the fourth power of the number of finite elements.

2 MODEL PROBLEM

Finite element setting

The model problem pertains to Timoshenko beam of constant bending stiffness EI and variable length L that is loaded by a given, concentrated bending moment M at the free end. We have only one design variable a in the model since only the total beam length may vary. In this paper the design variable a will be considered, namely

$$a = \ell, \text{ where } \ell = L/n \, . \tag{1}$$

Here ℓ is the element length resulting from a uniform subdivision of the beam into n finite elements. It is the aim of our study to establish expressions for the exact sensitivity $\frac{\delta u_n}{\delta L}$ and the approximate sensitivity $\frac{\Delta u_n}{\Delta L}$ through finite element analysis using $a = \ell$. We note that

$$\frac{\delta u}{\delta L} = C_a \frac{\delta u}{\delta a} \qquad \text{where } C_a = \frac{1}{n}. \tag{2}$$

As Olhoff and Rasmussen (1991a, 1991b) and Cook, Malkus and Plesha (1989) we diskretize the beam into a total number of n finite elements of equal length, see Figure 1, $\ell = L/n$. With only one design variable ℓ, the global equilibrium equations may be written as

$$[S(\ell)]\,[D] = [F], \text{ where } S \text{ is the stiffness matrix.} \tag{3}$$

Here the nodal displacements $\{D\}$ and external loads $\{F\}$ are defined as

$$[D] = [u_1, Q_1, \ldots, u_n, Q_n]^T \tag{4}$$

$$[F] = [p_1, M_1, \ldots, p_n, M_n]^T \tag{5}$$

where the components u_i, Q_i in (4) and p_i, M_i in (5) refer to the right-hand nodal point of the i-th element in the Figure 1.

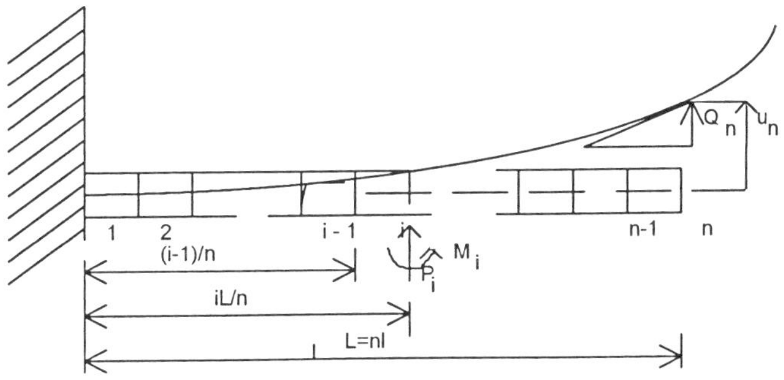

Figure 1 Global finite element model.

Figure 2 shows the i-th beam element used for the finite element structure in Figure 1, with definitions and sign conventions for the element nodal forces p_{1i} and p_{21}, moments m_{1i} and m_{2i}, translations u_{1i} and u_{21} and rotations Q_{1i} and Q_{21}.

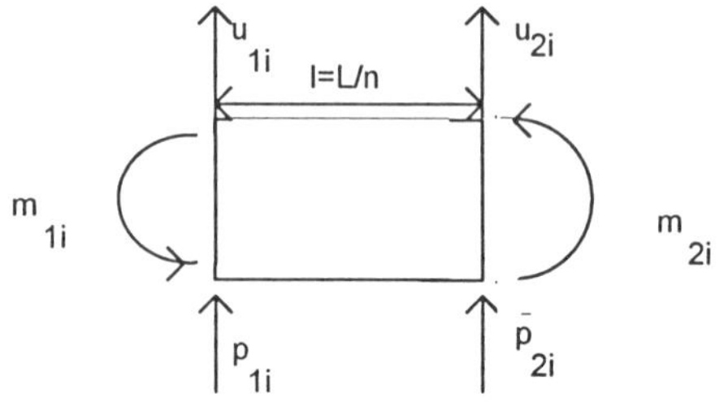

Figure 2 Timoshenko beam element.

The element has the length ℓ and bending stiffness EI. The element stiffness relations are:

$$\begin{Bmatrix} p_{1i} \\ m_{1i} \\ p_{2i} \\ m_{2i} \end{Bmatrix} = \begin{bmatrix} s_{11} & s_{12} & -s_{11} & s_{12} \\ s_{12} & s_{22} & -s_{12} & s_{24} \\ -s_{11} & -s_{12} & s_{11} & -s_{12} \\ s_{12} & s_{24} & -s_{12} & s_{22} \end{bmatrix} \begin{Bmatrix} u_{1i} \\ Q_{1i} \\ u_{2i} \\ Q_{2i} \end{Bmatrix} \tag{6}$$

$$= \begin{bmatrix} p_{11}+r_{11} & p_{12}+r_{12} & -p_{11}-r_{11} & p_{12}+r_{12} \\ p_{12}+r_{12} & p_{22}+r_{22} & -p_{12}-r_{12} & p_{22}+r_{24} \\ -p_{11}-r_{11} & -p_{12}-r_{12} & p_{11}+r_{11} & -p_{12}-r_{12} \\ p_{12}+r_{12} & p_{22}+r_{24} & -p_{12}-r_{12} & p_{22}+r_{22} \end{bmatrix} * \begin{Bmatrix} u_{1i} \\ Q_{1i} \\ u_{2i} \\ Q_{2i} \end{Bmatrix}$$

where the values of p_{11}, p_{12}, p_{22} and r_{11}, r_{12}, r_{22} and r_{24} are given by

$$p_{11} = KAG\tfrac{1}{\ell}(1-\gamma)^2, \quad p_{12} = \tfrac{1}{2}\,\mathrm{KAG}(1-\gamma)^2, \quad p_{22} = KAG\tfrac{\ell}{4}(1-\gamma)^2, \quad r_{11} = EI\frac{12\gamma^2}{\ell^3},$$

$$r_{12} = 6EI\frac{\gamma^2}{\ell^2}, \quad r_{22} = EI\frac{1+3\gamma^2}{\ell}, \quad r_{24} = EI\frac{3\gamma^2-1}{\ell}. \tag{7}$$

The coefficient γ can be computed as

$$\gamma = \frac{1}{1+12\dfrac{EI}{KAG\ell^2}} \tag{8}$$

KAG is the shear stiffness.

From the expression (6) and according to the relations
$u_{2i} = u_{1i+1}, \quad u_{1i} = u_{2i-1}, \quad Q_{1i} = Q_{2i-1}, \quad Q_{2i} = Q_{1i+1}$ we obtain:

$$\left.\begin{array}{l} -s_{11}u_{i-1} - s_{12}Q_{i-1} + 2s_{11}u_i - s_{11}u_{i+1} + s_{12}Q_{i+1} = P_i \\ s_{12}u_{i-1} + s_{24}Q_{i-1} + 2s_{22}Q_i - s_{12}u_{i+1} + s_{24}Q_{i+1} = M_i \end{array}\right\} \text{for i} = 1,\ldots,n-1$$
$$\left.\begin{array}{l} -s_{11}u_{n-1} + s_{12}Q_{n-1} + s_{11}u_n - s_{12}Q_n = P_n \\ s_{12}u_{n-1} + s_{24}Q_{n-1} - s_{12}u_n + s_{22}Q_n = M_n \end{array}\right\} \text{for n.} \tag{9}$$

By means of (8) we can easily identify the form of the global stiffness matrix [*S*] in (3) in terms of the element stiffness components s_{11}, s_{12}, s_{22} and s_{24} defined in (6). Now, assuming the components $u_{i,}\, Q_i, i = 1,\ldots,n$ of the model displacement vector [*D*] in (3) and (9) to be known subject to a given model load vector [*F*] we may write (9) in the form

$$\begin{aligned} P_i &= s_{11}f_{i1} + s_{12}f_{i2} \\ M_i &= s_{12}g_{i1} + s_{22}g_{i2} + s_{24}g_{i4} \end{aligned} \tag{10}$$

where, for example, the coefficient f_{il} is given by

$$f_{i1} = -u_{i-1} + 2u_i - u_{i+1} \tag{11}$$

and so on.

Analytical sensitivity analysis problem statement

Let us denote by $\delta\{D\}/\delta l$ the vector of model displacement sensitivities with respect to the design variable ℓ, where $\ell = \frac{L}{n}$. This vector of sensitivities constitutes the solution to the equation:

$$[S(\ell)]\frac{\partial [D]}{\partial \ell} = [F]_{\partial \ell} \tag{12}$$

which we obtain by analytical differentiation of the equilibrium equations (3) with respect to the design variable ℓ. Differentiating the expression (3) we obtain

$$\frac{\partial S}{\partial \ell}D + S\frac{\partial D}{\partial \ell} = \frac{\partial F}{\partial \ell}. \tag{13}$$

Because the external load is independent on the design, so $\frac{\partial F}{\partial \ell} = 0$ and

$[F]_{\partial \ell} = [S(\ell)]\frac{\partial [D]}{\partial \ell} = -\frac{\partial S(\ell)}{\partial \ell}[D]$. $\{F\}_{\partial \ell}$ in (12) is termed *pseudo load vector* and it has a form:

$$[F]_{\partial \ell} = \{[\, P_{1\partial \ell}, M_{1\partial \ell},\ldots,P_{n\partial \ell}, M_{n\partial \ell}\,]\}^T. \tag{14}$$

From the analogy between (12) and (3) and the equivalent form (10) and (3) it follows, that the components of the pseudo load vector $[F]_{\partial \ell}$ in (12) and (14) can be computed as

$$\begin{aligned} P_{i\,\partial\ell} &= -\frac{\partial s_{11}}{\partial \ell} f_{i\,1} - \frac{\partial s_{12}}{\partial \ell} f_{i\,2} \\ M_{i\,\partial\ell} &= -\frac{\partial s_{12}}{\partial \ell} g_{i\,1} - \frac{\partial s_{22}}{\partial \ell} g_{i\,2} - \frac{\partial s_{24}}{\partial \ell} g_{i\,4} \quad i = 1,\dots,n \end{aligned} \tag{15}$$

Semi-analytical problem statement

The only difference between the analytical and the semi-analytical sensitivity analysis problems is that while the pseudo loads are determined via analytical differentiation of the stiffness matrix in the former, in the latter they are obtained via numerical differentiation. The pseudo load vector

$$\{F\}_{\Delta\ell} = \left\{P_{1\Delta\ell}, M_{1\Delta\ell}, \dots, P_{n\Delta\ell}, M_{n\Delta\ell}\right\}^T \tag{16}$$

is based on approximate differentiation of the stiffness matrix. The components of $\{F\}_{\Delta\ell}$ can then be computed as:

$$\begin{aligned} P_{i\Delta\ell} &= -\frac{\Delta s_{11}}{\Delta \ell} f_{i1} - \frac{\Delta s_{12}}{\Delta \ell} f_{i\,2} \\ M_{i\Delta\ell} &= -\frac{\Delta s_{12}}{\Delta \ell} g_{i1} - \frac{\Delta s_{22}}{\Delta \ell} g_{i2} - \frac{\Delta s_{24}}{\Delta \ell} g_{i4} \qquad i = 1,\dots,n \end{aligned} \tag{17}$$

Exact solutions to the finite element equations for displacements and design sensitivities

The well-known Bernoulli-Euler transverse deflection and slope for this problem are given by

$$u(x) = \frac{Mx^2}{2EI}; \qquad Q(x) = u'(x) = \frac{Mx}{EI} \tag{18}$$

In the finite element setting (18) can be expressed as

$$\begin{aligned} u_i(M) &= \frac{ML^2}{2EI}\left(\frac{i}{n}\right)^2 \\ Q_i(M) &= \frac{ML}{EI}\frac{i}{n}, \qquad i = 0,\dots,n \end{aligned} \tag{19}$$

Now we can determine the expressions f_{i1}, f_{i2}, f_{n1}, f_{n2}, g_{i1}, g_{i2}, g_{i4}, g_{n1}, g_{n2}, g_{n4} in (10). For example

$$\begin{aligned} f_{i\,1} &= -u_{i-1} + 2u_i - u_{i+1} = -\frac{ML^2}{EI}\frac{1}{n^2} \\ f_{i\,2} &= -Q_{i-1} + Q_{i+1} = 2\frac{ML}{EI}\frac{1}{n} \end{aligned} \tag{20}$$

So let us solve analytically the beam equation for the deflection u_n at the beam end subject to a given concentrated force P_i and moment M_i. Then let us apply the superposition principle to obtain the displacement u_n subject to the application of the external loads P_i, M_i in all nodal points i = 1,...,n. We have

$$u_n = \sum_{i=1}^{n} u(M_i) + \sum_{i=1}^{n} u(P_i) = \sum_{i=1}^{n}\left[\frac{M_i L^2}{2EI}\left(\frac{i}{n}\right)\left(2-\frac{i}{n}\right)\right] + \sum_{i=1}^{n} P_i\left[KAGL + \frac{L^3}{6EI}\left(\frac{i}{n}\right)^2\left(3-\frac{i}{n}\right)\right] =$$
$$= L\left\{\sum_{i=1}^{n} M_i \frac{L}{2EI}\left(\frac{i}{n}\right)\left(2-\frac{i}{n}\right) + P_i\left[KAG + \frac{L^2}{6EI}\left(\frac{i}{n}\right)^2\left(3-\frac{i}{n}\right)\right]\right\} \tag{21}$$

From the analogy between (3) and (12) it follows, that

$$\frac{\partial u_n}{\partial \ell} = L\sum_{i=1}^{n}\left[p_{i\partial\ell}\left[KAG + \frac{L^2}{6EI}\left(\frac{i}{n}\right)^2\left(3-\frac{i}{n}\right)\right] + M_{i\partial\ell}\frac{L}{2EI}\left(\frac{i}{n}\right)\left(2-\frac{i}{n}\right)\right] \tag{22}$$

and similarly the component $\dfrac{\Delta u_n}{\Delta \ell}$ is determined by:

$$\frac{\partial u_n}{\Delta \ell} = L\sum_{i=1}^{n}\left[p_{i\Delta\ell}\left[KAG + \frac{L^2}{6EI}\left(\frac{i}{n}\right)^2\left(3-\frac{i}{n}\right)\right] + M_{i\Delta\ell}\frac{L}{2EI}\left(\frac{i}{n}\right)\left(2-\frac{i}{n}\right)\right] \tag{23}$$

3 PSEUDO LOADS AND DESIGN SENSITIVITY OBTAINED BY ANALYTICAL DIFFERENTIATION OF THE STIFFNESS MATRIX

We have calculated the values of expressions s_{11}, s_{12}, s_{22}, s_{24} and $\dfrac{\partial s_{11}}{\partial \ell}, \dfrac{\partial s_{12}}{\partial \ell}, \dfrac{\partial s_{22}}{\partial \ell}, \dfrac{\partial s_{24}}{\partial \ell}$.

Using these values we obtain:

$$\begin{aligned}
P_{i\partial\ell} &= -\frac{12AGKMn^2}{AGKL^2 + 12EIn^2} \\
M_{i\partial\ell} &= -\frac{12AGKLni}{AGKL^2 + 12EIn^2} \\
P_{n\partial\ell} &= \frac{6AGKMn^2(2n-1)}{AGKL^2 + 12EIn^2} \\
M_{n\partial\ell} &= -\frac{2Mn(AGKL^2(3n-2) - 6EIn^2)}{L(AGKL^2 + 12EIn^2)}
\end{aligned} \tag{24}$$

Now we can obtain the value of expression (22)

$$\frac{\partial u_n}{\partial \ell}=\frac{LM(AGKL^2+6EIn(n+1))}{EI(AGKL^2+12EIn^2)} \tag{25}$$

We can notice, that if $K\to\infty$ we have

$$\lim_{K\to\infty}\frac{\partial u_n}{\partial \ell}=\lim_{K\to\infty}\frac{LM}{EI}\frac{AGL^2+\frac{6EIn(n+1)}{K}}{AGL^2+\frac{12EIn^2}{K}}=\frac{LM}{EI} \tag{26}$$

The result (26) is the same as the exact result in the case of Bernoulli-Euler beam.

4 EXACT ERROR ANALYSIS OF THE SEMI -ANALYTICAL SENSITIVITY PROBLEM

We now focus on the semi-analytical sensitivity analysis problem stated in section 2, adopting first -order forward difference approximations to the derivatives of the stiffness components that constitute the basis for the computation of the approximate pseudo-load vector $[F]_{\Delta\ell}$. Thus we obtain the following expressions for the forward finite difference approximations to the derivatives of stiffness. We use the relation

$$-\frac{\Delta s_{ij}}{\Delta\ell}=\frac{s_{ij}(\ell)-s_{ij}(\ell+\Delta\ell)}{\Delta\ell} \quad \text{for i =1, 2 and j =1, 2, 4} \tag{27}$$

So we obtain the expressions $-\frac{\Delta s_{11}}{\Delta\ell},-\frac{\Delta s_{12}}{\Delta\ell},-\frac{\Delta s_{22}}{\Delta\ell},-\frac{\Delta s_{24}}{\Delta\ell}$. We may now express in a similar fashion the approximate pseudo load components $P_{i\Delta\ell}, P_{n\Delta\ell}$, $M_{i\Delta\ell}$ and $M_{n\Delta\ell}$.

$$P_{i\Delta\ell}=-\frac{12KAGLMn^2}{(KAG((\Delta\ell)n+L)^2+12EIn^2)((\Delta\ell)n+L)} \tag{28}$$

$$P_{n\Delta\ell}=\frac{6KAGLMn^2(2n-1)}{(KAG((\Delta\ell)+L)^2+12EIn^2)((\Delta\ell)n+L)} \tag{29}$$

$$M_{i\Delta\ell}=-\frac{12KAGLMni}{KAG((\Delta\ell)n+L)^2+12EIn^2} \tag{30}$$

$$M_{n\Delta\ell}=\frac{Mn(KAG((\Delta\ell)n-2L(3n-2))((\Delta\ell)n+L)+12EIn^2)}{(KAG((\Delta\ell)n+L)^2+12EIn^2)((\Delta\ell)n+L)} \tag{31}$$

Now we may calculate the semi-analytical displacement sensitivity $\frac{\Delta u_n}{\Delta L}$ that is the subject of our model problem:

$$\frac{\Delta u_n}{\Delta L} = \frac{L^2 M\left[KAG((\Delta\ell)^2 n^2 - (\Delta\ell)Ln(5n^2-4) + 2L^2) + 12EIn(n+1)\right]}{2EI\left[KAG((\Delta\ell)n + L)^2 + 12EIn^2\right]((\Delta\ell)n + L)} \tag{32}$$

Let us now establish the expression for the relative error

$$\varepsilon = \frac{\dfrac{\Delta u}{\Delta L} - \dfrac{\partial u}{\partial L}}{\dfrac{\partial u}{\partial L}} \tag{33}$$

We have obtained the result

$$\varepsilon = -\frac{\eta(K^2A^2G^2L^4(5n^2 + 2\eta^2 + 5\eta + 2) + 12KAGEIL^2n(5n^3 + n(\eta^2 + 2\eta + 1) + \eta^2 + 3\eta + 3)}{2(KAGL^2 + 6EIn(n+1))(KAGL^2(\eta+1)^2 + 12EIn^2(\eta+1))} \tag{34}$$

where $\eta = \dfrac{\Delta\ell}{\ell}$

As we may see ε is the function of n^4 and η^2, not n^2 and η, as it has been for Bernoulli-Euler case. But we may notice, that

$$\lim_{n\to\infty} \varepsilon = \frac{-\eta(5KAGL^2 + 12EI)}{12EI(\eta+1)} \tag{35}$$

This means that the error ε of the semi-analytical displacement sensitivity does not increase to infinity if n increases to infinity. We have tried to calculate also the boundary of $\dfrac{\Delta u}{\Delta L}$ for $K\to\infty$ and $\Delta\ell\to 0$. This result is $\dfrac{LM}{EI}$. As we can notice it is the same as for Bernoulli-Euler beam.

5 REFERENCES

Cook, R.D., Malkus, D.S. and Plesha, M.E. (1989) *Concepts and applications of finite element analysis*. Wiley & Sons, New York.

Olhoff, N. and Rasmussen, J. (1991a) Method of error elimination for a class of semi-analytical sensitive analysis problems, in *Engineering optimization in design processes*, (ed. Eschenauer, H.A., Mattheck, C., Olhoff, N.) pp.193-200, Springer, Berlin, Heidelberg, New York.

Olhoff, N. and Rasmussen, J. (1991b) Study of inaccuracy in semi-analytical sensitivity analysis a model problem. *Structural Optimization*, **3**, 203-213.

39

Memory effects and microscale

Małgorzata Peszyńska
Systems Research Institute, Polish Academy of Sciences
Newelska 6, 01-447 Warszawa, Poland
E-mail: mpesz@ibspan.waw.pl

Abstract

In many applications the models describing evolution phenomena with nonlocal effects have the form of PDEs with integral memory terms of Volterra convolution type. In this paper we present an overview of applications and indicate the related analytical and approximation issues. We show that, from computational point of view, in some cases it is advantageous to consider some auxiliary problem defined at microscale which is either imbedded in the definition of the problem or has to be introduced.

Keywords

Integro–partial differential equations, memory terms, microscale, convolution integrals

1 INTRODUCTION

In this paper we deal with two intersecting topics: memory effects and microscale. We want to exploit the mutual relation between these two phenomena which frequently are simultaneously present in a model. The purpose of the paper is expository: we present a collection of different results and applications from a new perspective. As the topic is very broad, we will restrict ourselves only to some representative contributions to the field.

We restrict our attention here to the memory effects which arise in evolution equations and have the form of convolution terms

$$\mathcal{Q}_\tau(u)(t) \stackrel{\text{def}}{=} \tau * \mathcal{D}u = \int_0^t \tau(t-s)\mathcal{D}u(s)ds.$$

Here u denotes the unknown solution to a differential problem (for convenience we omit spatial variables), $\mathcal{D}$ is a differential operator (in applications $\mathcal{D}$ can be the identity operator, the derivative with respect to the time variable, the Laplacian, or some nonlinear elliptic operator). The kernel $\tau : R \mapsto R$ is fixed for a given application and is typically a positive nonincreasing function i.e. the value of the integral depends more on the recent values of u than on the past ones; this property is called fading memory. The function τ can be bounded or unbounded at the origin; the presence of $\mathcal{Q}_\tau(u)$ in an equation affects its solutions in various ways (see section 2).

By *microscale* we mean any properties of the medium which are observable at a lower scale of observation than the *macroscopic* equation describing the quantity of interest to us. In numerous phenomena there are multiple scales present (micro, meso, macro, giga, ...). *Upscaling* techniques incorporate the information given on microscale into a law defined at a higher scale; we will denote this by by $\mathbf{m} \Rightarrow \mathbf{M}$. There exists a variety of upscaling techniques: e.g. the homogenization method, averaging, REV-based methods used in statistical mechanics, methods of asymptotic analysis. For example, the oscillating data of a PDE influence that equation at a microscopic scale while the average (in some sense) of this data enters a (different) PDE at a higher level. In some models we will deal with the coupling between the two scales i.e. $\mathbf{m} \Leftrightarrow \mathbf{M}$.

In some applications, originally of the form $\mathbf{m} \Leftrightarrow \mathbf{M}$, one can define a function τ and decouple the two scales. The macroscopic problem then is modified by the appearance of a memory term $\mathbf{m} \Rightarrow \tau; \tau \Rightarrow \mathbf{M}$. At first glance the decoupled system seems attractive from the point of view of analysis and approximation, because when dealing with it we do not have to resolve the complicated coupling on different scales. This approach, however, is only partially advantageous because of the computational issues arising in approximation of memory terms. See below for our model example and later sections 2 and 3 for details. It turns out then that the coupling with microscale can have some advantages over inclusion of memory terms.

Extrapolating this idea, suppose we are given a phenomenon governed by an evolution equation with a memory term $\tau \Leftrightarrow \mathbf{M}$ and that there is no direct relation to any microscale phenomena. For the reasons indicated above we propose to consider construction of some ("artificial") coupling to an auxiliary microscale $\mathbf{m} \Rightarrow \tau$ and then study the complexity of the original problem compared to the one with coupling to microscale, i.e., $\mathbf{m} \Leftrightarrow \mathbf{M}$.

The presentation below starts with a model example. Then in section 2 we discuss various issues related to the analysis and approximation of the memory terms arising in evolution equations. In section 3 we discuss why and how to exploit microscale present in the problem and give a brief review of some models with memory effects and their location in the framework of this paper.

Let us now present a model example. It comes from the study of (single phase, single component) fluid flow through a *fissured (fractured, double porosity)* medium (see Arbogast (1987), Hornung (1990)) and is derived by the homogenization method. The analysis as well as approximation of the two models as well as of their numerous multiphase and nonlinear extensions have been extensively studied, see references in Hornung (to appear).

$$\underline{\mathbf{m} \Leftrightarrow \mathbf{M}}$$

$$\begin{aligned} u_t - \nabla \cdot (\mathbf{D}\nabla u) &= q_{fm}(x,t), \quad x \in \Omega, t > 0, \\ v_t^x(y,t) - \nabla_y \cdot (\mathbf{d}\nabla_y v^x) &= 0, \quad y \in \Omega_x, t > 0, \\ v|_{\Gamma_x} &= u(x,t), \quad t > 0, \\ q_{fm}(x,t) &= \frac{1}{|\Omega_x|}\int_{\Omega_x} v_t^x(y,t)dy. \end{aligned}$$

The domain Ω is the macroscopic domain, e.g., a reservoir where the flow of a fluid of density u occurs. At each point $x \in \Omega$ there exists a *microscopic* domain Ω_x (a porous block) where the flow observable at a lower scale (of density v^x) occurs. All blocks Ω_x are isometric to a certain Ω_0. The pair $(u, \{v^x, x \in \Omega\})$ gives us the full information about the values of density in the fractures and in the blocks of the fissured medium Ω. The coefficients $\mathbf{D}, \mathbf{d}$ are the mobility coefficients. The equivalent model with explicit memory term has the form

$$\begin{array}{rcl} \underline{\tau \Leftrightarrow \mathbf{M}} & & \\ u_t - \nabla \cdot (\mathbf{D}\nabla u) & = & -u_t * \tau, x \in \Omega, \end{array}$$

where τ is obtained from a microscopic block problem

$$\begin{array}{rcl} \underline{\mathbf{m} \Rightarrow \tau} & & \\ r_t - \nabla_y \cdot (\mathbf{d}\nabla_y r) & = & 0, y \in \Omega_0, \\ r|_{\Gamma_0} & = & 1, r(y,0) = 0, \; y \in \Omega_0, \\ \tau(t) & = & -\frac{d}{dt}\frac{1}{|\Omega_0|}\int_{\Omega_0} r(y,t)dy. \end{array}$$

The values of the kernel τ in the latter system depend only on the shape of Ω_0 and on the coefficient $\mathbf{d}$. They can be computed analytically for some particular cases or, in a more general situation, approximated numerically. Once these values have been calculated, the coupling expressed in the first model by the term q_{fm} and boundary values is formally replaced by the memory term $u_t * \tau$. Note that by necessity the function τ is singular at the origin.

At first glance the latter model, as a single parabolic integro–differential equation, seems to be in a more convenient form for mathematical and numerical analysis. Practical evidence however (see Arbogast (1989,1990) and the following papers, references in Hornung (to appear)) suggests that the former (uncoupled) is more appropriate for applications. This is a typical instance of what we want to consider in this paper.

2 APPROXIMATION

In this section we want briefly to address the issues that one encounters when dealing with memory terms present in evolution equations: the qualitative and quantitative effect upon the solutions; the design of quadrature rules; the complexity of approximation.

We will consider the following typical cases of the convolution kernels, all of them nonnegative and nonincreasing: (1) trivial; $\tau(t) = 0$; (2) bounded; for ex. $\tau(t) = e^{-t}$; (3) unbounded at the origin, $L^2(0,T)$ integrable for every $T > 0$, for ex. $\tau(t) = t^{-\frac{1}{4}}$; (4) unbounded at the origin, non-L^2 but $L^1(0,T)$ integrable, for ex. $\tau(t) = \frac{1}{\sqrt{t}}$; (5) unbounded and not integrable near the origin, for ex. $\tau(t) = \frac{1}{t}$; (6) extreme; $\tau = \delta$ (Dirac) or $\tau = -\delta'$, "very" singular at the origin.

If the convolution kernel τ is $L^1(0,T)$, then the convolution operator $\mathcal{Q}_\tau$ sending a function $u \mapsto \tau * u$ is linear and continuous on $L^2(0,T)$. Additional assumptions on monotonicity of the kernels imply monotonicity of the operator. More precisely, if τ is an integrable nonconstant function with a continuous negative nondecreasing derivative, then one can prove the following property known as (strong) positivity of the kernel (see MacCamy (1972))

$$(u, u * \tau)_{L^2(0,T)} > 0, u \neq 0, u \in L^2(0,T).$$

This further implies that the operator $(I + \mathcal{Q}_\tau)^{-1}$ is a contraction (see Hornung (1990)). Similar assumptions yield another important property (see Peszynska (to appear))

$$(u, u_t * \tau)_{L^2(0,T)} \geq -\frac{1}{2}|u(0)|^2, u \in L^2(0,T).$$

Monotonicity and related properties are used in analysis of the problems with memory terms as well as in proofs of the convergence of the applicable numerical algorithms.

Let us go back to the model example from Introduction. Its well-posedness has been proved with the use of monotonicity techniques mentioned above. The smoothness of the solutions to it is not essentially affected by the presence of the memory terms because, without the coupling term (or memory term, respectively), the equations have purely parabolic character hence "infinite smoothing effects" can be observed. However, the quantative difference in solutions corresponding to the kernels of different degree of singularity is important (see Peszynska (1995)).

The situation changes when the type of the equation in which the convolution term appears is hyperbolic. The results reported in Dafermos (1989) for the viscoelasticity models show that the memory terms contribute to the smoothness of the solutions to these models. This impact becomes stronger with increasing degree of singularity of the convolution kernel, which has the meaning of the growing dissipative part of the equation.

The above phenomenon however, does not make the approximation of memory terms easier as a consequence of increasing degree of singularity of the kernel. The easiest case here is that of bounded kernels and for those most of the work has been done.

More specifically, we consider approximation of the term $\mathcal{Q}_\tau(u)(t)$ at $t = t_N$ so that N is the number of time steps (of variable or uniform length) that have elapsed. We seek a quadrature rule in the form

$$\mathcal{Q}_\tau(u)(t_N) \approx \sum_{k=1}^{N} \omega_{N,k}(\mathcal{D}u)_k.$$

In the right rectangular rule, for example, one sets $\omega_{N,k} = (t_k - t_{k-1})\tau(t_N - t_k), (\mathcal{D}u)_k = \mathcal{D}u(t_k)$. This rule as well as other typical numerical integration methods (rectangular, trapezoidal) consist in replacing the integrand by its polynomial interpolant. Such an approach is suitable for smooth kernels but fails to guarantee the stability in case of a singular τ. The methods proposed recently in McLean (1993) and Peszynska (to appear), applicable to unbounded kernels, are based on the *product integration method.* (see Linz (1985)). The idea here is to approximate only the well behaving part of the product and to integrate exactly the remaining part. In our case this would be, respectively, $\mathcal{D}u$ and τ. Additionally some adjustments must be made to make the quadrature rule consistent with the discretization of the original differential equation. One sets then $\omega_{N,k} = \frac{1}{t_k - t_{k-1}} \int_{t_{k-1}}^{t_k} \int_0^{t_N} \tau(t_N - s)ds$. For the details (the convergence proof, implementation and applications) see the respective papers.

All of the above mentioned approximation methods have one common characteristic: the weights $\omega_{N,k}$ have to be recomputed at each time step $t = t_N$. This implies further that we cannot calculate subtotals for the sum and reuse them at later steps. Rather, we need to store all the information about the "history" of the solution i.e. the values $(\mathcal{D}u)_1, (\mathcal{D}u)_2, \ldots (\mathcal{D}u)_N$ in the computer memory. That issue can be critical: note that the approximation methods necessary in applications must be combined with some discretization in space. Denote by N_h the number of nodes of spatial discretization (for compatibility it can be of order $O(N)$ or $O(N^2)$ in the finite element or finite difference approximation). Then at each time step we need to store vectors of length N_h. Remembering the whole "history" requires storing NN_h numbers. The order of magnitude of NN_h may be unacceptably large in a given implementation.

The direct approach to this issue by straightforward "cutting off" the "tail" of the kernel can lead to the loss of accuracy (see Peszynska (to appear)). On the other hand, the use of modern hardware (smart exploitation of different computer memory layers) can help in a particular implementation. The general and safe way to resolve the storage issue was proposed by Thomée and coworkers in Thomee (1992) (see also references in that paper) for bounded kernels and allows for storing only $\sqrt{N}N_h$ values of the solution; it is not clear though if that method would work for singular kernels.

In the following section we propose to exploit microscale in order to resolve that complexity problem.

3 MICROSCALE

In this section we first briefly show how the use of microscale helps in dealing with the complexity of approximation to the solutions of the model problem. Then we give an overview of applications and study the particular case of the convolution kernels in the form of Prony series. This serves as a motivation for a more general approach.

The solutions to the model problem in the form $\tau \Rightarrow \mathbf{M}$ can be approximated with the use of the algorithm suitable for unbounded kernels (see above) which has, however, the aforementioned drawback of the large storage complexity.

The alternative to the above is given by solving the equivalent problem $\mathbf{m} \Leftrightarrow \mathbf{M}$ where instead of a single memory term one deals with the coupling to microscale. If each node $1 \ldots N_h$ of the macro domain Ω is associated with a copy of microscopic domain Ω_x discretized with N_H nodes, then at each time step we need to store N_h values of the solution to the $\mathbf{M}$ problem and $N_H N_h$ values associated with the $\mathbf{m}$ problems. The key point is that N_H can be taken relatively small, for example of order 10. Hence, the storage totals to $(N_H + 1)N_h$. This is to be contrasted with the number NN_h for the model with memory term.

The price we pay for the computer storage savings by using the $\mathbf{m} \Leftrightarrow \mathbf{M}$ approach is a much bigger computational effort: in addition to the macroscopic problem one needs to solve N_h microscopic problems at each time step. This overhead can be reduced by solving the $\mathbf{m}$ problems (they are independent of each other) in parallel, The use of modern computer architectures is then a major advantage.

Let us now turn back to the general case. Table 1 contains a short overview of applications. The memory terms that arise from microscale are marked with an $\mathbf{m}$ in the first column. For these problems, in case of computer memory storage limitations, one might try to exploit the microscale and compare the efficiency of the two approaches, $\mathbf{m} \Leftrightarrow \mathbf{M}$ and $\tau \Rightarrow \mathbf{M}$. In other problems the memory terms come from constitutive laws and are identified from some empiric data. We shall propose a way to deal with it. As a motivation

	application how to find τ	$\mathcal{D}u$	type	refs
m	*single phase flow in fissured medium* the use of the heat kernel	u_t	parabolic	Hornung (1990)
	heat cond. in materials with memory constitutive equations	$u_t, \Delta u$	parabolic	Nunziato (1971)
	fading memory in viscoelasticity experimental data fitting	$(\phi(u_x))_x$	hyperbolic	Dafermos (1989)
m	*homogenization limits of conserv. laws* Young measures	u, u_{xx}	conserv. laws	Tartar (1990), Amirat (1989)
	nonlocal theory of dispersion/diffusion nonlocal effects in time/space Fourier and Laplace transform	Δu	convection– diffusion	Cushman (1994)
	control theory for phase transitions	general funct.	Stefan pbm	Hoffmann (1991)

Table 1 Applications.

let us consider a class of models of consolidation and creep of clay (see Thomas(1989)) where the kernel is sought in the form

$$\tau(t) = \sum_{k=1}^{K} \alpha_k e^{-\lambda_k t}, \quad \alpha_k, \lambda_k > 0,$$

(i.e. *Prony* series). This form is justified by the constitutive construction in which the clay medium behaves like a Hookean spring in series with K Kelvin (i.e. a spring in parallel to a dash–pot) bodies. In general many kernels of the fading memory type are expected to have the form of Prony series, with least-squares fit of the experimental data used to identify the coefficients.

The consequence of such a special form of the kernel is essential. Each term in $\mathcal{Q}_\tau(u)(t) = \sum_{k=1}^{K} \alpha_k(\mathcal{D}u(\bullet) * e^{-\lambda_k \bullet})(t)$ can be seen as α_k times the solution at time t of an ODE with the stiffness coefficient λ_k and the right hand side equal to $\mathcal{D}u(t)$. Hence, the value of the memory term can be computed through solutions of these ODEs; their approximation can be calculated by some discretization method appropriate for ODE. We want then to exploit the use of Prony series as some approximate representation of τ. In the framework of this paper that representation can be used to create the coupling with microscale. Microscale here should be understood as K separate ODEs. Note that these ODEs can be solved in parallel and that at each time step we need to store only K values of solutions of ODEs for each point corresponding to the spatial discretization of Ω i.e. we need to store KN_h numbers. This quota should be considerably less than NN_h or $\sqrt{N}N_h$ storage necessary in case of other methods, and so the microscale approach identified as Prony series looks attractive from the practical point of view. This idea requires careful analysis to be pursued in the forthcoming papers.

Let us now discuss the potential negative aspects of this approach. These concern mainly the difficulties with finding a proper Prony series and the stiffness of the system of ODEs.

In general, fitting the given experimental data requires solving a large system of nonlinear equations for $(\alpha_k, \lambda_k)_{k=1}^K$. One way to avoid this as well as to decrease the difficulties with the stiffness of the system of ODEs is by fixing the $(\lambda_k)_{k=1}^K$ to be integers from some interval. Suppose the kernel is $L^2(0,T)$ integrable. Then, by the change of variable $s = e^{-t}$ the problem of finding LSQ approximation of $\Upsilon(s) = \tau(-\ln s)$ by a polynomial of the form $P(s) = \sum_{k=1} \alpha_k s^{\lambda_k}$ is well posed. Setting $\lambda_k = k$ we need only to find the coefficients α_k what can be done by a standard LSQ algorithm. Scaling of the variable $s = e^{-\lambda_0 t}$ can change the range of sought exponents $\{\lambda_k\}_{k=1}^K$. The set of coefficients provides the best LSQ fit to the function $\Upsilon(s)$ and so there is a continuous dependence of the quality of approximation of $\tau(t)$ by $p(t) = P(e^{-t})$; the involved constants can be however potentially very large.

Another problem is that the Prony series takes a finite value at the origin equal to $p(0) = \sum_{k=1}^K \alpha_k$, while in many applications the kernel is unbounded there. However, in the absence of better methods, the approximation of τ as Prony series doesn't seem to be worse than the application of elementary quadrature schemes to unbounded kernels. The remaining issue is how to properly treat the case $\tau \in L^1, \tau \notin L^2$.

At the end we want to mention the exact representation of τ with the use of exponential terms (see Lubich (1988)). This is done with the Laplace transform inversion formula

$$\tau(t) = \frac{1}{2\pi i} \int_\Gamma \hat{\tau}(\lambda) e^{\lambda t} d\lambda,$$

where $\hat{\tau}$ is the Laplace transform of τ and Γ is some appropriately chosen contour in the complex plane. That approach leads however to a quadrature formula, hence, does not resolve the storage complexity problem and cannot be treated as an alternative to the above framework.

ACKNOWLEDGMENT

The significant part of the research presented in this paper was done while the author was visiting the Center for Applied Mathematics at Purdue University. The author would like to thank Professor Jim Douglas, Jr., the Director of the Center, for his invitation. The research is a result of many discussions at the Center, which the author would like gratefully to acknowledge, in particular with J. Douglas himself, and R.E. Showalter.

REFERENCES

Amirat, Y., Amdache, K. and Ziani, A. (1989) *Homogeneisation d'equations hyperboliques du premier ordre at applications aux ecoulements miscibles en milieux poreux*, Ann. Inst. Henri Poincare **6**, 397-417

Arbogast T. (1989) *Analysis of the Simulation of Single Phase Flow Through a Naturally Fractured Reservoir*, SIAM J. Numer. Anal. **26**, 12-29

Arbogast, T., Douglas (Jr.) J. and Hornung U. (1990) *Derivation of the Double Porosity Model of Single Phase Flow via Homogenization Theory*, SIAM J. Math. Anal. **21**, 823-836

Cushman, J.H., Hu, X. and Ginn T.R. (1994) *Nonequilibrium Statistical Mechanics of presasymptotic dispersion*, Journal of Statistical Physics **75** , 859-878

Dafermos, C., Ericksen, J,L. and Kinderlehrer, D. (1989) Amorphous Polymers and Non-Newtonian Fluids, IMA Volumes, Springer-Verlag

Hoffmann, K.H., Kenmochi, N. and Niezgodka, N. (1991) *Large Time Solutions of Two-Phase Stefan Problem with Delay*, Nonlinear Analysis: TMA

Hornung U. (to appear) Homogenization and Porous Media, *Interdisciplinary Applied Mathematics Series*, Springer, New York

Hornung, U. and Showalter R.E. (1990) *Diffusion Models for Fractured Media*, Jour. Math. Anal. Appl. **147**, 69-80

Linz P. (1985) Analytical and Numerical Methods for Volterra Equations, SIAM, Philadelphia

Lubich C. (1988) *Convolution Quadrature and Discretized Operational Calculus, Parts I & 2*, Numer. Math. **52**, 129-145 & 413-425

MacCamy, R.C. and Wong J.S. (1972) *Stability Theorems for Some Functional Differential Equations*, Trans. Amer. Math. Soc. **164**, 1-37

McLean, V., Thomée, V. and Wahlbin L.B. (1993) *Discretization with vaiable time steps of an evolution equation with a positive type memory term*, Applied Mathematics Report AMR93/18, December 1993, School of Math., The University of New South Wales

Nunziato J.W. (1971) *On heat conduction in materials with memory*, Quarterly Apppl. Math. **29**, 187–204

Peszyńska M. (1995) *On a model for nonisothermal flow in fissured media* Differential and Integral Equations **8**, 1497-1516

Peszyńska M. (to appear), *Finite element approximation of diffusion equations with integral convolution terms*, Math. Comp.

Tartar L. (1990) Memory effects and homogenization, Arch. Rat. Mech. Anal. **111**, 121–133

Thomas, H.R. and Bendani K. (1989) *Numerical solutions of one-dimensional rheological models of combined consolidation and creep*, Eng. Comput. **6**, 331-338

Thomée, V. and Wahlbin L.B. (1992) Long time numerical solution of a parabolic equation with memory, Dept. of Math, Chalmers University of Technology, The University of Göteborg, Preprint No 1992-14

40
The impact of elastic bodies upon a Timoshenko beam

*Yuriy A. Rossikhin and Marina V. Shitikova**
Voronezh State Academy of Construction and Architecture
Department of Theoretical Mechanics, ul.Kirova 3-75, Voronezh 394018, Russia. Phone: 0732-773992. E-mail: MVS@vgasa.voronezh.su

Abstract

The problem of the normal impact of an elastic sphere or bar upon a Timoshenko beam is considered. The impact process is accompanied by material local deformations and propagation of strong discontinuity wave surfaces in the beam. The dynamic deformations behind the wave fronts are determined by means of power series with variable coefficients (ray series). Within the contact region the stressed-strained state is taken into consideration variously according to the type of the striker: for the elastic sphere the contact force is governed by the Hertz's contact theory, but for the elastic bar the contact stress is found in terms of the theory of discontinuities. These two types of deformations are matched on the contact region boundary. The method of ray series and Hertz's contact theory allow one to determine analytically the main characteristics of the contact interaction.

Keywords

Impact, wave surfaces of strong discontinuity, ray method, Hertz's contact theory

1 INTRODUCTION

To construct a theory of the transverse impact upon beams, it is necessary to take into account not only local deformations but also wave phenomena occurring in contacting bodies during an impact. One of these theories, which takes into consideration both aspects and relates the Hertz's contact theory with the classical theory of beam vibrations, is the theory due to Timoshenko (1928). However, the classical theory of beams does not allow to construct the wave theory of impact, since the differential equations describing transverse vibrations of such bodies are not wave equations. As is known nonstationary flexural waves propagate in those bodies with an infinite large velocity. The transverse impact wave theory has become possible owing to the appearance of such equations for beams which take into account the rotary inertia and transverse shear deformations and, therefore, are wave equations. These wave equations were first introduced for a beam by Timoshenko (1928).

*The research described in this publication was made possible in part by Grant RJ5300 from the International Science Foundation and Russian Government and by Grant 94-01-00245a from the Russian Foundation for Fundamental Researches.

In the present paper, the wave theory of transverse impact upon Timoshenko beams is developed by the ray method, which was proposed by Achenbach and Reddy (1967) and has been extended by Rossikhin (1986) and Rossikhin and Shitikova (1992, 1993) to dynamic contact problems.

2 DETERMINING RELATIONS OF THE RAY THEORY FOR A TIMOSHENKO BEAM

The dynamic behaviour of an elastic homogeneous prismatic beam with due account for the rotary inertia and transverse shear deformations is described by the equations:

$$\partial M/\partial z - Q = -\rho I\dot{\beta}, \qquad \partial Q/\partial z = \rho F\dot{W}, \tag{1}$$

$$\dot{M} = -EI\,\partial\beta/\partial z, \qquad \dot{Q} = K\mu F(\partial W/\partial z - \beta), \tag{2}$$

where M is the bending moment, Q is the transverse force, W is the transverse displacement velocity of a beam central axis (velocity of deflection), β is the angular velocity of a cross section about the x-axis which is perpendicular to the plane of flexure $y-z$ (the axes z and y are directed along the beam axis and vertically down, respectively), E is Young's modulus, μ is the shear modulus, ρ is the density, I is the moment of inertia about the x-axis (vibrations occur in the y-direction), F is the cross-section area, K is the shear coefficient depending on the form of a cross section, and an over dot denotes the derivative with respect to time t.

Assume that as a result of the transverse impact upon the beam, a plane wave Σ of strong discontinuity propagates along the z-direction with the velocity G. Behind the wave surface Σ up to the boundary of the contact region, a certain desired function $Z(z,t)$ is represented by a series in terms of powers $t-(z-l)G^{-1} \geq 0$

$$Z(z,t) = \sum_{k=0}^{\infty} \frac{1}{k!}[Z_{,(k)}]\left(t - \frac{z-l}{G}\right)^k H\left(t - \frac{z-l}{G}\right), \tag{3}$$

where $[Z_{,(k)}] = Z^+_{,(k)} - Z^-_{,(k)} = [\partial^k Z/\partial t^k]$ are the jumps in kth derivatives of function Z with respect to time t on the wave surface Σ, i.e. at $t = z/G$, the upper indices $+$ and $-$ signify that the value is calculated ahead of and behind the wave front, respectively, $2l$ is the length of the contact area, and $H(t)$ is the Heaviside function.

To determine coefficients of the ray series (3) for the desired functions, it is necessary to differentiate the governing Eqs.(1) k times with respect to time, take their difference on the different sides of the wave surface Σ, and apply the condition of compatibility (Thomas, 1961) for discontinuities of $(k+1)$th order derivatives of a certain function Z with respect to time

$$G\left[\frac{\partial Z_{,(k)}}{\partial z}\right] = -[Z_{,(k+1)}] + \frac{d[Z_{,(k)}]}{dt}, \tag{4}$$

where d/dt is the total time derivative of a function defined on the moving surface Σ. As a result we obtain

$$\rho I\left(1 - \frac{E}{\rho G^2}\right)[\beta_{,(k+1)}] = -2EIG^{-2}\frac{d[\beta_{,(k)}]}{dt} - K\mu FG^{-1}[W_{,(k)}] + EIG^{-2}\frac{d^2[\beta_{,(k-1)}]}{dt^2}$$

$$+K\mu F G^{-1}\frac{d[W_{,(k-1)}]}{dt} - K\mu F[\beta_{,(k-1)}], \tag{5}$$

$$\rho F\left(1-\frac{K\mu}{\rho G^2}\right)[W_{,(k+1)}] =$$

$$-2K\mu F G^{-2}\frac{d[W_{,(k)}]}{dt} + G[\beta_{,(k)}] + \frac{d^2[W_{,(k-1)}]}{dt^2} - G\frac{d[\beta_{,(k-1)}]}{dt}. \tag{6}$$

From Eqs.(5) and (6), one can obtain the values $[\beta_{,(k)}]$ and $[W_{,(k)}]$ (k = -1,0,1,..) with an accuracy of arbitrary constants on the two waves: quasi-flexural wave and quasi-shear wave propagating with the velocities $G^{(1)} = (E/\rho)^{1/2}$ and $G^{(2)} = (K\mu/\rho)^{1/2}$, respectively. Arbitrary constants are determined from the condition of compatibility for deformations on the boundaries of the contact region.

3 IMPACT OF A THIN BAR UPON A TIMOSHENKO BEAM

Let a thin elastic bar of a rectangular cross section, whose axis coincides with the y-axis, move along this axis with the velocity V and bump by its end against the centre of a Timoshenko beam.

During impact two types of plane waves propagate along the beam, behind the wave fronts up to the contact area boundary the solution is determined by Eq.(3). At the same time, a longitudinal wave propagates along the thin bar with the velocity $G_b = (E_b/\rho_b)^{1/2}$ (E_b and ρ_b are the Young's modulus and density of the striking bar, respectively), representing itself a plane of strong discontinuity. In virtue of the fact that the jumps of derivatives of the displacement velocities of the bar's particles $[V_{,(k)}]$ = const, on this wave plane the dynamic condition of compatibility is satisfied what allows one to connect the stress σ^- and displacement velocity V behind the wave front with each other at every instant of time. Specifically, in the contact area, i.e. at $y = 0$, we have

$$\sigma' = \rho_b G_b(V - W), \tag{7}$$

where $\sigma' = \sigma|_{y=0}$ is the contact stress, $W = V|_{y=0}$ is the displacement velocity of the beam part being in contact.

In view of Eq.(7), the equation of motion of the beam part, which is in contact with the bar, may be written as

$$-2lF\rho\dot{W} + 2la\rho_b G_b(V - W) + 2Q = 0, \tag{8}$$

where a is the width of the beam.

The quantities W and Q entering into Eq.(8) are defined by Eqs.(3) where $z = l$. It is necessary to add the initial condition

$$W|_{t=o} = 0, \tag{9}$$

as well as the relation

$$\partial W/\partial z|_{z=l} = 0 \tag{10}$$

to Eq.(8). Substituting (3) into Eqs.(8) and (10) with due account of (9), and equating the coefficients associated with the same powers of t, we can find all necessary arbitrary constants. Thus the approximate solution of our problem can be obtained. As an example, we consider the transverse impact of a steel bar of the length s =120 mm and rectangular cross section 38.1 × 38.1 mm upon an aluminum beam with 38.1 × 38.1 mm cross section.

The initial velocity of impact is V=1 m/s. After calculation of ray series coefficients, the truncated ray series for the beam transverse displacement w, transverse force Q, and bending moment M can be written at $z = l$, respectively, as

$$w = 3.9\,10^5\tfrac{t^2}{2} - 3.07\,10^{11}\tfrac{t^3}{6} + 3.62\,10^{17}\tfrac{t^4}{24},$$
$$Q = -1.16\,10^{10}t + 9.11\,10^{15}\tfrac{t^2}{2} - 8.12\,10^{21}\tfrac{t^3}{6},$$
$$M = 1.12\,10^3 - 8.84\,10^8 t + 7.05\,10^{14}\tfrac{t^2}{2} - 6.55\,10^{20}\tfrac{t^3}{6}.$$

4 IMPACT OF A SPHERE UPON A TIMOSHENKO BEAM

Let an elastic sphere of the radius R and mass m move along the y-axis with the constant velocity V_0 towards an elastic beam. The impact occurs at $t = 0$.

When $t > 0$, the sphere displacement may be represented as

$$y = w + \alpha. \tag{11}$$

Then the equation of motion of the beam part being in contact without regard for an inertia term (due to infinitesimal of the contact region), and the equation of the sphere motion have the form

$$2Q + P(t) = 0, \tag{12}$$
$$m\ddot{y} = -P(t). \tag{13}$$

Equations (12) and (13) are solved with the following initial conditions to be taken into account:

$$y|_{t=0} = 0, \quad \dot{y}|_{t=0} = V_0, \quad w|_{t=0} = 0, \quad W|_{t=0} = 0. \tag{14}$$

The relation between the contact force $P(t)$ and penetration $\alpha(t)$ has the form

$$P = k\alpha^{3/2}, \tag{15}$$

where $k = 4R^{1/2}/3\pi(k_s + k_b)$, $k_s = (1 - \nu_s^2)/E_s$, $k_b = (1 - \nu^2)/E$, ν is the Poisson's ratio, and the indices s and b concern the sphere and beam, respectively.

The value Q entering into Eq.(12) is determined by the dynamic condition of compatibility as

$$Q = -\rho F G^{(2)} W. \tag{16}$$

This condition can be obtained if we interpret the discontinuity surface as the limiting layer of the width h at $h \to 0$, wherein the value Z to be found changes monotonically and infinitely from the magnitude Z^+ to the magnitude Z^-. Considering that on the wave surface (Thomas, 1961)

$$\frac{\partial}{\partial z} = \frac{d}{dn}, \qquad \frac{\partial}{\partial t} = -G\frac{d}{dn}, \tag{17}$$

where d/dn is the derivative with respect to the normal to the wave surface, and changing the partial derivatives in the second equation of (1) by its expressions (17), after the integration of the resulting relations with respect to n from $-h/2$ to $h/2$ and passage to the limit at $h \to 0$ we are led to the formula (16).

Substituting Eqs.(11), (15), and (16) into Eqs.(12)-(13), we arrive at the differential equation about the value $\alpha(t)$:

$$\ddot{\alpha} + \frac{k}{m}\alpha^{3/2} + \frac{3}{2}b\dot{\alpha}\alpha^{1/2} = 0, \quad b = \frac{k}{2\rho F G^{(2)}}. \tag{18}$$

Introducing $A = \dot{\alpha}$ and converting from the variable t to the new independent variable α, we are led to the equation

$$A\frac{dA}{d\alpha} + \frac{3}{2}bA\alpha^{1/2} = -\frac{k}{m}\alpha^{3/2} \tag{19}$$

with the initial conditions

$$\alpha|_{t=0} = 0, \quad \dot{\alpha}|_{t=0} = V_0. \tag{20}$$

Note that Eq.(19) is the Abel equation of the second kind (Kamke, 1959).

We seek the solution to Eq.(19) in the form

$$A = V_0 + \sum_{i=1}^{7} a_i\alpha^{(2i+1)/2} + \sum_{i=1}^{7} b_i\alpha^i + O(\alpha^8). \tag{21}$$

Substituting (21) into (19) and equating coefficients at equal powers of α, we determine the coefficients to be found

$$a_1 = -b, \quad a_2 = -\frac{2}{5}\frac{k}{V_0 m}, \quad a_3 = a_4 = b_1 = b_2 = b_3 = b_6 = 0,$$

$$b_4 = -\frac{5}{8}a_1a_2/V_0, \quad b_5 = -\frac{1}{2}a_2^2/V_0, \quad b_7 = -\frac{11}{14}a_1a_5/V_0,$$

$$a_5 = -\frac{8}{11}b_4a_1/V_0, \quad a_6 = -\left(\frac{10}{13}a_1b_5 + a_2b_4\right)/V_0, \quad a_7 = -\frac{2}{3}a_2b_5/V_0. \tag{22}$$

To obtain connection between the value α and the time, it is necessary to integrate Eq.(21).

The coefficients a_1 and a_2 are defined by the two processes being caused by the shock interaction: the coefficient a_1 is responsible for the dynamic processes arising in the beam during the propagation of the surfaces of discontinuity, but the coefficient a_2 answers for the quasi-static processes occurring at local bearing of the material due to the Hertz's contact theory. If $a_1 \to 0$, what realizes at an infinitely large velocity of shear wave propagation (Bernoulli-Euler beam), then the solution (21) for small α goes over into the quasi-static solution obtained by Timoshenko (1928).

REFERENCES

Achenbach, J.D. and Reddy, D.P. (1967) Note on wave propagation in linearly viscoelastic media. *ZAMP*, **18**, 141-4.

Kamke, E. (1959) *Differentialgleichungen Lösungsmethoden und Lösungen.* Leipzig.

Rossikhin, Yu.A. (1986) Impact of a rigid sphere onto an elastic half-space. *Soviet Applied Mechanics*, **22**, 403-9.

Rossikhin, Yu.A. and Shitikova, M.V. (1992) About Shock Interaction of Elastic Bodies with Pseudo Isotropic Uflyand-Mindlin Plates, in *Proceedings of the International Symposium on Impact Engineering* (ed. I. Maekawa), **2**, 623-8, Sendai, Japan.

Rossikhin, Yu.A. and Shitikova, M.V. (1994) A ray method of solving problems connected with a shock interaction. *Acta Mechanica*, **102**, 103-21.

Thomas, T.Y. (1961) *Plastic Flow and Fracture in Solids.* Academic Press, New York.

Timoshenko, S.P. (1928) *Vibrational Problems in Engineering.* Van Nostrand, New York (3d ed).

41

A shape optimization algorithm for the minimum drag problem in Stokes flow*

A. P. Suetov
Institute of Mathematics and Mechanics, Russian Academy of Science
620219 Ekaterinburg, S.Kovalevskoi Str. 16, Russia.
Phone: (3432)493143. Fax: (3432)442581.
E-mail: suetov@odu.imm.intec.ru

Abstract
This paper deals with the theory of shape (or domain) optimization. A model minimum drag problem in Stokes flow is considered. An algorithm for computation of a solution of the extended problem is given. The algorithm is based on reduction of the shape optimization problem to a family of coefficient optimization problems.

Keywords
Shape optimization, generalized solutions, computation algorithm

1 INTRODUCTION AND NOTATIONS

In the present paper a problem related to the theory of optimal shape design (Banichuk, 1983; Pironneau, 1984) is considered. From the mathematical point of view the shape is a subset in $\mathbf{R}^n$. The main feature of the problem is that the cost functional depends on the set by via the solution of the boundary value problem. We apply the approach used in (Suetov, 1994). According to this approach in Section 2 the known shape optimization problem is formulated and a definition of its generalized solutions is given. In Section 3 an extended shape optimization problem is formulated. In Section 4 a coefficient optimization problem is considered and some connections with the problems of Sections 2, 3 are established. In Section 5 an algorithm for approximate shape optimization is proposed. In this algorithm the problem of determination of an optimal shape is replaced by a family of coefficient optimization problems.

We will use the following notation:

- ∂M is the boundary of the set $M \subset \mathbf{R}^3$;
- meas(M) is the Lebesgue measure of a set $M \subset \mathbf{R}^3$;

*The research described in this publication was made possible in part by grant NMD300 from International Science Foundation and Russian Government

- $W_2^1(\Omega)$ is the Sobolev space ($\Omega \subset \mathbf{R}^3$ is open), $\mathbf{W}_2^1(\Omega) = (W_2^1(\Omega))^3$ (Adams, 1975);
- $\overset{\circ}{W}{}_2^1(\Omega)$ is the space of functions equal zero on $\partial\Omega$, $\overset{\circ}{\mathbf{W}}{}_2^1(\Omega) = \left(\overset{\circ}{W}{}_2^1(\Omega)\right)^3$;
- $V(\Omega) = \{\mathbf{u} \in \overset{\circ}{\mathbf{W}}{}_2^1(\Omega) : \operatorname{div} \mathbf{u} = 0\}$ (Temam, 1979) ;
- for $\mathbf{u} = (u_1, u_2, u_3) \in \mathbf{W}_2^1(\Omega)$ let $\nabla \mathbf{u} = \left(\frac{\partial u_i}{\partial x_j}\right)$, $\nabla \mathbf{u} \nabla \mathbf{v} = \sum_{i,j=1}^3 \frac{\partial u_i}{\partial x_j} \frac{\partial v_i}{\partial x_j}$;
- Δ is the Laplace operator, $\Delta \mathbf{u} = (\Delta u_1, \Delta u_2, \Delta u_3)$;
- $\|f|Y\|$ is the norm of a function f in a space Y.

2 A SHAPE OPTIMIZATION PROBLEM AND ITS GENERALIZED SOLUTIONS

We consider a problem of finding the shape of the body of unit volume which produces minimum drag when moving slowly through a viscous fluid at constant speed (Watson, 1971; Pironneau, 1973).

Let $\mathbf{T}$ be a bounded open set in $\mathbf{R}^3$, meas$(\mathbf{T}) > 1$, $\partial\mathbf{T} = \Gamma = \Gamma_1 \cup \Gamma_2$, Γ_1 is parallel to the plan Ox_2x_3 and Γ_2 is parallel to the axis Ox_1. Let $\mathbf{B}$ be a closed subset of $\mathbf{T}$, $\partial\mathbf{B} = S$ (see Figure 1).

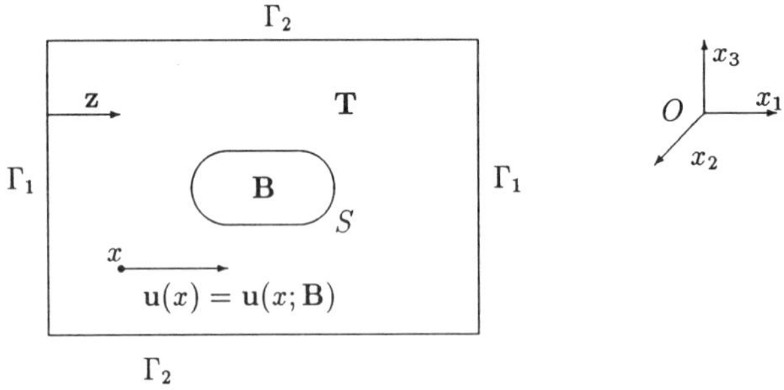

Figure 1 A body in the flow.

Let us consider the following boundary value problem: $\mathbf{u} = (u_1, u_2, u_3)$,

$$\Delta \mathbf{u} = \nabla p, \quad \operatorname{div} \mathbf{u} = 0 \quad \text{in} \quad \mathbf{T} \setminus \mathbf{B}; \qquad \mathbf{u}\,|_\Gamma = \mathbf{z} = (1, 0, 0), \quad \mathbf{u}\,|_S = 0 \tag{1}$$

We will denote by $\mathbf{u}(\mathbf{B})$ the weak solution of this problem. As the cost functional we accept the value which is proportional to the rate of energy dissipated by the fluid flow $\mathbf{u}(\mathbf{B})$:

$$J(\mathbf{B}) = F(\mathbf{u}(\mathbf{B})) = \int_{\mathbf{T}} \frac{1}{2} \sum_{i,j=1}^{3} \left(\frac{\partial u_i}{\partial x_j} + \frac{\partial u_j}{\partial x_i} \right)^2 \mathrm{dx}.$$

We define the class of admissible sets as

$\Phi = \{\mathbf{B} : \mathbf{B} \subset \mathbf{T},\ \mathbf{B}$ is closed, meas$(\mathbf{B}) \geq 1\}$.

Problem 1. It is required to find an admissible set $\mathbf{B}^* \in \Phi$, minimizing the functional $J(\mathbf{B})$ over the class of sets Φ:

$$\mathbf{B}^* = \inf_{\mathbf{B}\in\Phi} J(\mathbf{B}).$$

The question of existence of a solution of Problem 1 is not trivial (Osipov and Suetov, 1984; Pironneau, 1984), therefore the following definition is reasonable.

Definition. A generalized solution of the shape optimization problem is a weak limit point in $\mathbf{W}_2^1(\mathbf{T})$ of a sequence $\{\mathbf{u}(\mathbf{B}_i)\}$, where $\{\mathbf{B}_i\}$ is a minimizing sequence for Problem 1.

Statement 1 *There exists a generalized solution of Problem* 1.

Proof. A sequence $\{\mathbf{u}(\mathbf{B}_i)\}$, $\mathbf{B}_i \subset \mathbf{T}$, is bounded in space $\mathbf{W}_2^1(\mathbf{T})$ (Temam, 1979), therefore it has a weak limit point. □

Let $X = \{\mathbf{u} \in \mathbf{W}_2^1(\mathbf{T}) \;:\; \operatorname{div}\mathbf{u} = 0, \quad \mathbf{u}\,|_\Gamma = \mathbf{z}\}$ and $X(\mathbf{B}) = \{\mathbf{u} \in \mathbf{W}_2^1(\mathbf{T}) \;:\; \operatorname{div}\mathbf{u} = 0, \quad \mathbf{u}\,|_\mathbf{B} = 0, \quad \mathbf{u}\,|_\Gamma = \mathbf{z}\}$. It is clear that $X = X(\emptyset)$, $X(\mathbf{B}) \subset X \subset \mathbf{W}_2^1(\mathbf{T})$, $(X(\mathbf{B}) - \mathbf{z}) \subset V(\mathbf{T})$. Let us recall that $\mathbf{u}(\mathbf{B})$ is the solution of the variational identity

$$\mathbf{u} \in X(\mathbf{B}) \qquad \forall \mathbf{v} \in V(\mathbf{T}\setminus\mathbf{B}) \qquad \int_\mathbf{T} \nabla\mathbf{u}\cdot\nabla\mathbf{v}\mathrm{dx} = 0. \tag{2}$$

and, equivalently, is the solution of the minimization problem

$$E(\mathbf{u}(\mathbf{B})) = \min_{\mathbf{u}\in X(\mathbf{B})} E(\mathbf{u}), \qquad E(\mathbf{u}) = \int_\mathbf{T} (\nabla\mathbf{u})^2\mathrm{dx} \tag{3}$$

(Temam, 1979).

Lemma 1 *For every* $\mathbf{u} \in X$ *we have* $F(\mathbf{u}) = E(\mathbf{u})$.

Proof. Let us transform the energy dissipation formula

$$\frac{1}{2}\sum_{i,j=1}^{3}\left(\frac{\partial u_i}{\partial x_j} + \frac{\partial u_j}{\partial x_i}\right)^2 = \sum_{i,j=1}^{3}\left(\frac{\partial u_i}{\partial x_j}\right)^2 + \sum_{i,j=1}^{3}\frac{\partial u_i}{\partial x_j}\frac{\partial u_j}{\partial x_i} =$$

$$\sum_{i,j=1}^{3}\left(\frac{\partial u_i}{\partial x_j}\right)^2 + \sum_{j=1}^{3}\frac{\partial}{\partial x_j}\left(\sum_{i=1}^{3} u_i\frac{\partial u_j}{\partial x_i}\right) - \sum_{i=1}^{3} u_i\frac{\partial}{\partial x_i}\sum_{j=1}^{3}\frac{\partial u_j}{\partial x_j}$$

In virtue to $\operatorname{div}\mathbf{u} = 0$ the third term equals zero. Let us apply Green's formula to the integral of the second term

$$\int_{\mathbf{T}\setminus\mathbf{B}}\sum_{j=1}^{3}\frac{\partial}{\partial x_j}\left(\sum_{i=1}^{3} u_i\frac{\partial u_j}{\partial x_i}\right)\mathrm{dx} = \int_{\partial(\mathbf{T}\setminus\mathbf{B})}\sum_{j=1}^{3}\left(\sum_{i=1}^{3} u_i\frac{\partial u_j}{\partial x_i}\right) n_j\ \mathrm{ds}$$

Let us recall $\partial(\mathbf{T}\setminus\mathbf{B}) = S\cup\Gamma_1\cup\Gamma_2$, The condition $\mathbf{u}\,|_S = 0$ gives $\int_S \sum_{i,j=1}^{3} u_i\frac{\partial u_j}{\partial x_i} n_j\mathrm{ds} = 0$.

Since Γ_1 is parallel to Ox_2, Ox_3 and $\mathbf{u}\mid_{\Gamma_1}=\mathbf{z}$ we see that $\frac{\partial u_i}{\partial x_2}\mid_{\Gamma_1}=0, \frac{\partial u_i}{\partial x_3}\mid_{\Gamma_1}=0$ for all i,j. Equalities $\frac{\partial u_2}{\partial x_2}=0, \frac{\partial u_3}{\partial x_3}=0$ and $\operatorname{div}\mathbf{u}=0$ imply $\frac{\partial u_1}{\partial x_1}\mid_{\Gamma_1}=0$. From $\mathbf{u}\mid_{\Gamma_1}=(1,0,0)$ and $\mathbf{n}\mid_{\Gamma_1}=(n_1,0,0)$ we have $\sum_{i,j=1}^3 u_i\frac{\partial u_j}{\partial x_i}n_j=u_1\frac{\partial u_1}{\partial x_1}n_1=0$ at every point of Γ_1. Thus $\int_{\Gamma_1}\sum_{i,j=1}^3 u_i\frac{\partial u_j}{\partial x_i}n_j\mathrm{ds}=0$.

By the definition of $\mathbf{T}$ the surface Γ_2 is parallel to Ox_1 and $\mathbf{u}\mid_{\Gamma_2}=\mathbf{z}$, hence $\frac{\partial u_j}{\partial x_1}=0$ for all j. From $\mathbf{u}\mid_{\Gamma_2}=(1,0,0)$ and $\mathbf{n}\mid_{\Gamma_2}=(0,n_2,n_3)$ it follows that $\sum_{i,j=1}^3 u_i\frac{\partial u_j}{\partial x_i}n_j=u_1\left(\frac{\partial u_2}{\partial x_1}n_2+\frac{\partial u_3}{\partial x_1}n_3\right)=0$ at every point of Γ_2. Hence

$$\int_{\Gamma_2}\sum_{i,j=1}^{3}u_i\frac{\partial u_j}{\partial x_i}n_j\mathrm{ds}=0,\quad \int_{\mathbf{T}\setminus\mathbf{B}}\sum_{i,j=1}^{3}\frac{\partial u_i}{\partial x_j}\frac{\partial u_j}{\partial x_i}\mathrm{dx}=\int_{\partial(\mathbf{T}\setminus\mathbf{B})}\sum_{i,j=1}^{3}u_i\frac{\partial u_j}{\partial x_i}n_j\mathrm{ds}=0$$

and

$$\forall\mathbf{u}\in X\quad F(\mathbf{u})=\int_{\mathbf{T}\setminus\mathbf{B}}\frac{1}{2}\sum_{i,j=1}^{3}\left(\frac{\partial u_i}{\partial x_j}+\frac{\partial u_j}{\partial x_i}\right)^2\mathrm{dx}=\int_{\mathbf{T}\setminus\mathbf{B}}\sum_{i,j=1}^{3}\left(\frac{\partial u_i}{\partial x_j}\right)^2\mathrm{dx}=E(\mathbf{u}).\tag{4}$$

□

Remark 1 *It is clear from lemma 1 and (3) that* $J(\mathbf{B})=E(\mathbf{u}(\mathbf{B}))$ *and*

$$\inf_{\mathbf{B}\in\Phi}J(\mathbf{B})=\inf_{\mathbf{B}\in\Phi}\inf_{\mathbf{u}\in X(\mathbf{B})}E(\mathbf{u})=\inf_{\mathbf{u}\in U_1}E(\mathbf{u}),\qquad U_1=\bigcup_{\mathbf{B}\in\Phi}X(\mathbf{B})$$

This remark is similar to the variational principle from (Watson, 1971).

3 AN AUXILIARY OPTIMIZATION PROBLEM

Let us formulate an optimization problem similarly to (Pironneau, 1984, ch.3.2.2).

> **Problem 2.** Let $U=\{\mathbf{u}\in X:\operatorname{meas}(x\in\mathbf{T}:|\mathbf{u}(x)|=0)\geq 1\}$. It is required to find a function $\mathbf{u}^*\in U$ minimizing the functional $E(\mathbf{u})$ over the class U:
>
> $$E(\mathbf{u}^*)=\min_{\mathbf{u}\in U}E(\mathbf{u}).$$

Statement 2 *There exists a solution of Problem 2. A weak limit point of a minimizing sequence of Problem 2 is a strong limit point of the sequence.*

Proof. The existence follows from the coercivity and weakly lower semicontinuity of $E(\mathbf{u})$ on $\overset{\circ}{\mathbf{W}}{}^1_2(\mathbf{T})$, from compactness of the imbedding $\mathbf{W}^1_2(\mathbf{T})$ into $\mathbf{L}_2(\mathbf{T})$ and from strong closeness of the set U in $\mathbf{L}_2(\mathbf{T})$ (Pironneau, 1984). Let us prove the strong convergence. Let $\{\mathbf{u}_i\}$ be a sequence of functions from the set U, minimizing the functional $E(\mathbf{u})$ over U, $\mathbf{u}_i\rightharpoonup\mathbf{u}^*$ weakly in $\mathbf{W}^1_2(\mathbf{T})$ as $i\to\infty$ and $E(\mathbf{u}_i)\longrightarrow E(\mathbf{u}^*)$ as $i\to\infty$. If $\mathbf{u}\in X$ then $(\mathbf{u}-\mathbf{z})\in\overset{\circ}{\mathbf{W}}{}^1_2(\mathbf{T})$ and $E(\mathbf{u}-\mathbf{z})=E(\mathbf{u})$. The functional $(E(\mathbf{u}))^{1/2}$ is equivalent to the standard norm in $\overset{\circ}{\mathbf{W}}{}^1_2(\mathbf{T})$. Weak convergence and convergence of norms imply $(\mathbf{u}_i-\mathbf{z})\longrightarrow(\mathbf{u}^*-\mathbf{z})$ strongly in $\overset{\circ}{\mathbf{W}}{}^1_2(\mathbf{T})$ as $i\to\infty$. This implies strong convergence of the minimizing sequence in $\mathbf{W}^1_2(\mathbf{T})$. □

Remark 2 *Let U_2 be the close of U_1 in the strong topology of $\mathbf{W}_2^1(\mathbf{T})$. It is clear that $U_1 \subset U_2 \subseteq U$. By definitions of Φ and U from $\mathbf{B} \in \Phi$ it follows that $X(\mathbf{B}) \subset U$ and $\mathbf{u}(\mathbf{B}) \in U$, therefore*

$$\min_{\mathbf{u}\in U} E(\mathbf{u}) \leq \min_{\mathbf{u}\in U_2} E(\mathbf{u}) = \min_{\mathbf{u}\in U_1} E(\mathbf{u}) = \inf_{\mathbf{B}\in\Phi} J(\mathbf{B}).$$

Statement 3

(i) If $U_2 = U$ then $\inf_{\mathbf{B}\in\Phi} J(\mathbf{B}) = \min_{\mathbf{u}\in U} E(\mathbf{u})$.
(ii) If $U_2 = U$ then the set of generalized solutions of Problem 1 is equal to the set of solutions of Problem 2.
(iii) If $\inf_{\mathbf{B}\in\Phi} J(\mathbf{B}) = \min_{\mathbf{u}\in U} E(\mathbf{u})$ then a generalized solution of Problem 1 is a strong limit point of the corresponding minimizing sequences and is a solution of Problem 2.

Proof.

(i) Let $\mathbf{u}^* \in U$ and $E(\mathbf{u}^*) = \min_{\mathbf{u}\in U} E(\mathbf{u})$. Due to the assumption there exists $\mathbf{B}_i \in \Phi$ and $\mathbf{u}_i \in X(\mathbf{B}_i)$ such that $\mathbf{u}_i \longrightarrow \mathbf{u}^*$ strongly in $\mathbf{W}_2^1(\mathbf{T})$ $i \to \infty$. We have $\lim_{i\to\infty} E(\mathbf{u}_i) = \min_{\mathbf{u}\in U} E(\mathbf{u})$ and $E(\mathbf{u}_i) \geq E(\mathbf{u}(\mathbf{B}_i))$, thus $\inf_{\mathbf{B}\in\Phi} J(\mathbf{B}) \leq \min_{\mathbf{u}\in U} E(\mathbf{u})$. From this and the remark 2 we obtain the equality.

(ii) a) A solution of Problem 2 is a generalized solution of Problem 1. Let $\mathbf{u}^* \in U, E(\mathbf{u}^*) = \min_{\mathbf{u}\in U} E(\mathbf{u})$ and let $\mathbf{B}_i \in \Phi$ be a sequence of sets and $\mathbf{u}_i \in X(\mathbf{B}_i)$ be a sequence of functions such that $\mathbf{u}_i \longrightarrow \mathbf{u}^*$ strongly in $\mathbf{W}_2^1(\mathbf{T})$ as $i \to \infty$. The existence of the sequences follows from the assumption of the statement. It is obvious that $\lim_{i\to\infty} E(\mathbf{u}_i) = E(\mathbf{u}^*)$. In virtue to remark 1 and the property of solutions of (2) we have $J(\mathbf{B}_i) = E(\mathbf{u}(\mathbf{B}_i)) \leq E(\mathbf{u}_i)$, therefore $\lim_{i\to\infty} J(\mathbf{B}_i) \leq E(\mathbf{u}^*)$. But due to remark 2 we have $J(\mathbf{B}_i) \geq E(\mathbf{u}^*)$, therefore

$$\lim_{i\to\infty} J(\mathbf{B}_i) = \lim_{i\to\infty} E(\mathbf{u}(\mathbf{B}_i)) = \lim_{i\to\infty} E(\mathbf{u}_i) = \inf_{\mathbf{B}\in\Phi} J(\mathbf{B}) = \min_{\mathbf{u}\in U} E(\mathbf{u}).$$

The variational identity (2) implies equality $E(\mathbf{u}_i) - E(\mathbf{u}(\mathbf{B}_i)) = \|\nabla(\mathbf{u}(\mathbf{B}_i) - \mathbf{u}_i)|\mathbf{L}_2\|^2$. Because the seminorm $\|\nabla\mathbf{u}|\mathbf{L}_2\|$ is equivalent to the standard norm in $\overset{\circ}{\mathbf{W}}{}_2^1(\mathbf{T})$ and $(\mathbf{u}(\mathbf{B}_i) - \mathbf{u}_i) \in \overset{\circ}{\mathbf{W}}{}_2^1(\mathbf{T})$ we have that $\mathbf{u}(\mathbf{B}_i) \longrightarrow \mathbf{u}^*$ strongly in $\mathbf{W}_2^1(\mathbf{T})$ as $i \to \infty$ and $\mathbf{u}^*$ is a generalized solution of Problem 1.

(ii) b) A generalized solution of Problem 1 is a solution of Problem 2. Let $\{\mathbf{B}_i\}$ be a minimizing sequence for Problem 1 and $\mathbf{u}(\mathbf{B}_i) \to \mathbf{u}^\#$ weakly in $\mathbf{W}_2^1(\mathbf{T})$ as $i \to \infty$, so $\mathbf{u}^\#$ is a generalized solution of Problem 1. It follows from $\mathbf{B}_i \in \Phi$ that $\mathbf{u}(\mathbf{B}_i) \in U$ and $\mathbf{u}(\mathbf{B}_i) \longrightarrow \mathbf{u}^\#$ strongly in $\mathbf{L}_2(\mathbf{T})$ as $i \to \infty$, hence $\mathbf{u}^\# \in U$. According to (i) we have

$$\lim_{i\to\infty} E(\mathbf{u}(\mathbf{B}_i)) = \lim_{i\to\infty} J(\mathbf{B}_i) = \inf_{\mathbf{B}\in\Phi} J(\mathbf{B}) = \min_{\mathbf{u}\in U} E(\mathbf{u}) \leq E(\mathbf{u}^\#).$$

Weak lower semicontinuity of the functional E implies $\lim_{i\to\infty} E(\mathbf{u}(\mathbf{B}_i)) \geq E(\mathbf{u}^\#)$, therefore $E(\mathbf{u}^\#) = \min_{\mathbf{u}\in U} E(\mathbf{u})$ and $\mathbf{u}^\#$ is a solution of Problem 2.

(iii) Using the notation of part (ii) b) of the proof we can write

$$\inf_{\mathbf{B}\in\Phi} J(\mathbf{B}) = \lim_{i\to\infty} E(\mathbf{u}(\mathbf{B}_i)) \geq E(\mathbf{u}^\#) \geq \min_{\mathbf{u}\in U} E(\mathbf{u}),$$

hence $\|\nabla \mathbf{u}_i|\mathbf{L}_2\| \longrightarrow \|\nabla \mathbf{u}^{\#}|\mathbf{L}_2\|$ as $i \to \infty$. The strong convergence follows from the weak convergence and from the convergence of norms.

□

Remark 3 *It is known that the close of* $\bigcup_{\mathbf{B}\in\Phi} \overset{\circ}{\mathrm{W}}{}_2^1(\mathbf{T} \setminus \mathbf{B})$ *in the strong topology of* $\overset{\circ}{\mathrm{W}}{}_2^1(\mathbf{T})$ *equals* $\{u \in \overset{\circ}{\mathrm{W}}{}_2^1(\mathbf{T}) : \mathrm{meas}(x \in \mathbf{T} : u(x) = 0) \geq 1\}$ (Suetov, 1994).

4 A COEFFICIENT OPTIMIZATION PROBLEM

Let $A > 0$. We define the set of admissible coefficients

$$K_A = \{k \in L_\infty(\mathbf{T}) : \int_{\mathbf{T}} k(x)\mathrm{dx} \geq A, \quad \forall x \in \mathbf{T} \quad 0 \leq k(x) \leq A\}.$$

It is obvious that K_A is convex and weak compact in $\mathrm{L}_2(\mathbf{T})$. For every admissible coefficient $k(x)$ we can consider the boundary value problem

$$\triangle \mathbf{u} - k \cdot \mathbf{u} = \nabla p, \quad \operatorname{div} \mathbf{u} = 0 \quad \text{in} \quad \mathbf{T}; \qquad \mathbf{u}\,|_\Gamma = \mathbf{z} = (1, 0, 0). \tag{5}$$

We will denote by $\mathbf{u}(k)$ the weak solution of problem (5).

We define a cost functional on the set of admissible coefficients:

$$J_2(k) = E_2(\mathbf{u}(k), k), \qquad E_2(\mathbf{u}, k) = \int_{\mathbf{T}} \{(\nabla \mathbf{u})^2 + k \cdot \mathbf{u}^2\}\mathrm{dx}.$$

Problem 3. It is required to find an admissible coefficient $k_A \in K_A$ minimizing the functional $J_2(k)$:

$$J_2(k_A) = \min_{k \in K_A} J_2(k)$$

Statement 4 *Problem 3 has a solution.*

Proof. Statement 4 follows from weak compactness of K_A and weak continuity of the mapping $k \mapsto \mathbf{u}(k)$. □

It is known that $\mathbf{u}(k)$ minimizes the functional $E_2(\mathbf{u}, k)$ in the argument $\mathbf{u}$ over the set X (Temam, 1979) :

$$E_2(\mathbf{u}(k), k) = \min_{\mathbf{u} \in X} E_2(\mathbf{u}, k). \tag{6}$$

Remark 4 *It is clear from (6) that Problem 3 is equivalent to the minimization problem for the functional* $E_2(\mathbf{u}, k)$ *in both arguments:*

$$\min_{k \in K_A} J_2(k) = \min_{\mathbf{u} \in X, k \in K_A} E_2(\mathbf{u}, k).$$

Let $h(\mathbf{u}) = \sup\{\delta : \text{meas}(x \in \mathbf{T} : |\mathbf{u}(x)| \leq \delta) \leq 1\}$.

Theorem 1 *Let k_A be a solution of Problem 3, $\mathbf{u}_A = \mathbf{u}(k_A)$. Then*

$$k_A(x) = \begin{cases} A & \text{if } |\mathbf{u}_A(x)| \leq h(\mathbf{u}_A), \\ 0 & \text{otherwise,} \end{cases}$$

$\text{meas}(x \in \mathbf{T} : |\mathbf{u}_A(x)| \leq h(\mathbf{u}_A)) = 1$, $h(\mathbf{u}_A) \longrightarrow 0$ *as* $A \to \infty$.

Proof. The proof is based on Remark 4. □

Lemma 2 *Let k_A be a solution of Problem 3 for every $A > 0$ and $\mathbf{u}^*$ is a weak limit point of $\{\mathbf{u}(k_A)\}$ as $A \to \infty$. Then* $\text{meas}(x \in \mathbf{T} : \mathbf{u}^*(x) = 0) \geq 1$ □

Theorem 2 *Let k_A be a solution of Problem 3 for every $A > 0$ and $\mathbf{u}^*$ is a weak limit point of $\{\mathbf{u}(k_A)\}$ as $A \to \infty$. Then $\mathbf{u}^*$ is a solution of Problem 2, $\mathbf{u}^*$ is a strong limit point of $\{\mathbf{u}(k_A)\}$ and $\lim_{A\to\infty} J_2(k_A) = \min_{\mathbf{u}\in U} E(\mathbf{u})$.*

Proof. According to lemma 2 we have $\mathbf{u}^* \in U$, therefore it is sufficient to prove that $E(\mathbf{u}^*) = \min_{\mathbf{u}\in U} E(\mathbf{u})$. Let $\mathbf{v} \in U$ and

$$\chi_{\mathbf{v}}(x) = \begin{cases} 1 & \text{if } |\mathbf{v}(x)| = 0, \\ 0 & \text{otherwise.} \end{cases}$$

Then $A \cdot \chi_{\mathbf{v}} \in K_A$ and the following inequalities hold:

$$\forall \mathbf{v} \in U \qquad E(\mathbf{u}(k_A)) \leq E_2(\mathbf{u}(k_A), k_A) = \min_{\mathbf{u}\in X, k\in K_A} E_2(\mathbf{u}, k) \leq E_2(\mathbf{v}, A \cdot \chi_{\mathbf{v}}) = E(\mathbf{v}),$$

hence $E(\mathbf{u}(k_A)) \leq E_2(\mathbf{u}(k_A), k_A) \leq \min_{\mathbf{u}\in U} E(\mathbf{u})$. Weak lower semicontinuity of the functional $E(\mathbf{u})$ imply the inequalities

$$\min_{\mathbf{u}\in U} E(\mathbf{u}) \leq E(\mathbf{u}^*) \leq \liminf_{A\to\infty} E(\mathbf{u}(k_A)) \leq \limsup_{A\to\infty} E(\mathbf{u}(k_A)) \leq \min_{\mathbf{u}\in U} E(\mathbf{u}).$$

Thus $E(\mathbf{u}^*) = \lim_{A\to\infty} E(\mathbf{u}(k_A)) = \lim_{A\to\infty} E_2(\mathbf{u}(k_A), k_A) = \min_{\mathbf{u}\in U} E(\mathbf{u})$ and $\mathbf{u}^*$ is a solution of the Problem 2. The strong convergence follows from the weak convergence and from the convergence of norms (see the proof of statement 3). □

5 A SHAPE OPTIMIZATION ALGORITHM

Theorems 1 and 2 suggest the following method for computation of a suboptimal shape:
1) choosing a large number A;
2) solving Problem 3, let k_A is a solution, $\mathbf{u}_A = \mathbf{u}(k_A)$;
3) taking a set $\mathbf{B}_A = \{x \in \mathbf{T} : |\mathbf{u}_A(x)| \leq h(\mathbf{u}_A)\}$ as a suboptimal shape.

Let me recall connections between Problem 1 and Problem 3. If the hypothesis

the close of $\bigcup_{\mathbf{B}\in\Phi} X(\mathbf{B})$ in the strong topology of $\mathbf{W}_2^1(\mathbf{T})$ is equal to U

is true then from statement 3 and theorem 2 it follows that limit points of a family $\{\mathbf{u}(k_A)\}$ are generalized solutions of Problem 1.

Without any hypothesis we can use the following a posteriori argument. It is clear that $J_2(k_A) < \min_{\mathbf{u}\in U} E(\mathbf{u}) \leq \inf_{\mathbf{B}\in U} \leq J(\mathbf{B}_A)$ (see the proof of theorem 2), hence inequalities $0 < J(\mathbf{B}_A) - \inf_{\mathbf{B}\in U} J(\mathbf{B}) \leq J(\mathbf{B}_A) - \min_{\mathbf{u}\in U} E(\mathbf{u}) \leq |J(\mathbf{B}_A) - J_2(k_A)| + |J_2(k_A) - \min_{\mathbf{u}\in U} E(\mathbf{u})|$ are true. Due to theorem 2 $\lim_{A\to\infty} |J_2(k_A) - \min_{\mathbf{u}\in U} E(\mathbf{u})| = 0$. Therefore if the parameter A is big enough and the value $|J(\mathbf{B}_A) - J_2(k_A)|$ is small then $J(\mathbf{B}_A)$ is close to the optimal value $\inf_{\mathbf{B}\in U} J(\mathbf{B})$.

It is obvious that Problem 3 is non-convex. We describe an algorithm for computation of a function that satisfies to a necessary condition of optimality for Problem 3. Let

$$\kappa_-(x;\mathbf{u}) = \begin{cases} 1 & \text{if} \quad |\mathbf{u}(x)| < h(\mathbf{u}), \\ 0 & \text{otherwise,} \end{cases} \qquad \kappa(x;\mathbf{u}) = \begin{cases} 1 & \text{if} \quad |\mathbf{u}(x)| \leq h(\mathbf{u}), \\ 0 & \text{otherwise} \end{cases}$$

and for every $\mathbf{u} \in X$ let $\chi(x;\mathbf{u})$ be a function such that

$$\forall x \in \mathbf{T} \quad \kappa_-(x;\mathbf{u}) \leq \chi(x;\mathbf{u}) \leq \kappa(x;\mathbf{u}), \qquad \int_{\mathbf{T}} \chi(x;u)\mathrm{dx} = 1.$$

We note that $A \cdot \chi(\mathbf{u}) \in K_A$.

Theorem 3 *Let $k_0 \in K_A$, $\mathbf{u}_i = \mathbf{u}(k_i)$, $k_{i+1} = A\cdot\chi(\mathbf{u}_i)$ and a subsequence $\{k_{i_j}\}$ converges to k_* weakly in $\mathrm{L}_2(\mathbf{T})$. Then the sequence $\{\mathbf{u}_{i_j}\}$ converges to a function $\mathbf{u}_*$ strongly in $\mathbf{W}_2^1(\mathbf{T})$, $\mathbf{u}_* = \mathbf{u}(k_*)$ and*

$$k_*(x) = \begin{cases} A & \text{if} \quad |\mathbf{u}_*(x)| < h(\mathbf{u}_*), \\ 0 & \text{if} \quad |\mathbf{u}_*(x)| > h(\mathbf{u}_*), \end{cases}$$

therefore $(\mathbf{u}_*, k_*)$ is a critical point of $E_2(\mathbf{u}, k)$ on $X \times K_A$. □

6 CONCLUSION

In this work an algorithm for solving of a shape optimization problem is proposed. An important feature of the algorithm is that the corresponding boundary value problems can be solved on the same domain and grid, with only distinction in the lowest term coefficients of equations. The idea of the algorithm can be applied to other shape optimization problems for minimization of the energy functional of a system.

REFERENCES

Adams R. (1975) *Sobolev spaces.* Acad.Press, New York.

Banichuk, N.V. (1983) *Problems and methods of optimal structural design.* Plenum Press, New York.

YU.S.Osipov, A.P.Suetov (1984) On a problem of J.-L.Lions *Soviet Math. Dokl.*, Vol. **29**, No 3, p.487-91.

Suetov, A. (1994) A shape optimization algorithm for an elliptic system *Control and Cybernetics,* **23**, No 3, 565-74.
Pironneau, O. (1984) *Optimal shape Design for Elliptic Systems.* Springer-Verlag, New-York.
Pironneau O. (1973) On optimal profile in Stokes flow. *J. Fluid Mech.* **59**, part 1, 117-28.
Temam R. (1979) *Navier-Stokes equations. Theory and numerical analysis.* North-Holland Publishing Company, Amsterdam.
Watson S. R. (1971) Toward the Minimum Drag on a Body of Given Volume in Slow Viscous Flow. *J. Inst. Math. Applics* **7**, No 3, 367-376.

INDEX OF CONTRIBUTORS

KEYWORD INDEX